Integrated disease management of wheat and barley

It is widely recognised that agriculture is a significant contributor to global warming and climate change. Agriculture needs to reduce its environmental impact and adapt to current climate change whilst still feeding a growing population, i.e. become more 'climate-smart'. Burleigh Dodds Science Publishing is playing its part in achieving this by bringing together key research on making the production of the world's most important crops and livestock products more sustainable.

Based on extensive research, our publications specifically target the challenge of climate-smart agriculture. In this way we are using 'smart publishing' to help achieve climate-smart agriculture.

Burleigh Dodds Science Publishing is an independent and innovative publisher delivering high quality customer-focused agricultural science content in both print and online formats for the academic and research communities. Our aim is to build a foundation of knowledge on which researchers can build to meet the challenge of climate-smart agriculture.

For more information about Burleigh Dodds Science Publishing simply call us on +44 (0) 1223 839365, email info@bdspublishing.com or alternatively please visit our website at www.bdspublishing.com.

Related titles:

Achieving sustainable cultivation of wheat Volume 1: Breeding, quality traits, pests and diseases
Print (ISBN 978-1-78676-016-6); Online (ISBN 978-1-78676-018-0, 978-1-78676-019-7)

Achieving sustainable cultivation of wheat Volume 2: Cultivation techniques
Print (ISBN 978-1-78676-020-3); Online (ISBN 978-1-78676-022-7, 978-1-78676-023-4)

Chapters are available individually from our online bookshop: https://shop.bdspublishing.com

climate•SMART•publishing
IN AGRICULTURAL SCIENCE

BURLEIGH DODDS SERIES IN AGRICULTURAL SCIENCE

NUMBER 19

Integrated disease management of wheat and barley

Edited by Professor Richard Oliver, Curtin University, Australia

burleigh dodds
SCIENCE PUBLISHING

Published by Burleigh Dodds Science Publishing Limited
82 High Street, Sawston, Cambridge CB22 3HJ, UK
www.bdspublishing.com

Burleigh Dodds Science Publishing, 1518 Walnut Street, Suite 900, Philadelphia, PA 19102-3406, USA

First published 2019 by Burleigh Dodds Science Publishing Limited

Library of Congress Control Number: 2018957803

British Library Cataloguing in Publication Data
A catalogue record for this book is available from the British Library

ISBN 978-1-78676-216-0 (Print)
ISBN 978-1-78676-219-1 (PDF)
ISBN 978-1-78676-218-4 (ePub)
ISSN 2059-6936 (print)
ISSN 2059-6944 (online)

DOI 10.19103/AS.2018.0039

Typeset by Deanta Global Publishing Services, Chennai, India
Printed by Lightning Source

Contents

Part 2 Key challenges in integrated disease management of cereals

Series list

Title	Series number
Achieving sustainable cultivation of maize - Vol 1 From improved varieties to local applications Edited by: Dr Dave Watson, CGIAR Maize Research Program Manager, CIMMYT, Mexico	001
Achieving sustainable cultivation of maize - Vol 2 Cultivation techniques, pest and disease control Edited by: Dr Dave Watson, CGIAR Maize Research Program Manager, CIMMYT, Mexico	002
Achieving sustainable cultivation of rice - Vol 1 Breeding for higher yield and quality Edited by: Prof. Takuji Sasaki, Tokyo University of Agriculture, Japan	003
Achieving sustainable cultivation of rice - Vol 2 Cultivation, pest and disease management Edited by: Prof. Takuji Sasaki, Tokyo University of Agriculture, Japan	004
Achieving sustainable cultivation of wheat - Vol 1 Breeding, quality traits, pests and diseases Edited by: Prof. Peter Langridge, The University of Adelaide, Australia	005
Achieving sustainable cultivation of wheat - Vol 2 Cultivation techniques Edited by: Prof. Peter Langridge, The University of Adelaide, Australia	006
Achieving sustainable cultivation of tomatoes Edited by: Dr Autar Mattoo, USDA-ARS, USA & Prof. Avtar Handa, Purdue University, USA	007
Achieving sustainable production of milk - Vol 1 Milk composition, genetics and breeding Edited by: Dr Nico van Belzen, International Dairy Federation (IDF), Belgium	008
Achieving sustainable production of milk - Vol 2 Safety, quality and sustainability Edited by: Dr Nico van Belzen, International Dairy Federation (IDF), Belgium	009
Achieving sustainable production of milk - Vol 3 Dairy herd management and welfare Edited by: Prof. John Webster, University of Bristol, UK	010
Ensuring safety and quality in the production of beef - Vol 1 Safety Edited by: Prof. Gary Acuff, Texas A&M University, USA & Prof. James Dickson, Iowa State University, USA	011
Ensuring safety and quality in the production of beef - Vol 2 Quality Edited by: Prof. Michael Dikeman, Kansas State University, USA	012
Achieving sustainable production of poultry meat - Vol 1 Safety, quality and sustainability Edited by: Prof. Steven C. Ricke, University of Arkansas, USA	013
Achieving sustainable production of poultry meat - Vol 2 Breeding and nutrition Edited by: Prof. Todd Applegate, University of Georgia, USA	014
Achieving sustainable production of poultry meat - Vol 3 Health and welfare Edited by: Prof. Todd Applegate, University of Georgia, USA	015

Acknowledgements

We wish to acknowledge the following for their help in reviewing particular chapters:

- Chapter 12: Dr Frank Vandenbosch, Rothamsted Research, UK

Introduction

Diseases remain a serious problem in wheat and barley cultivation. It has been estimated that around 20% of global crop production is lost to diseases. Leading fungal diseases affecting wheat and barley include rusts, Septoria blotches, powdery mildew, tan spot, spot blotch, net blotch, scald and Fusarium species. Conventional control using fungicides faces a number of challenges such as increasing regulation and the spread of fungicide resistance. The chapters in this volume sum up the wealth of research addressing this challenge. The volume reviews key recent research on the main fungal diseases, their modes of infection and potential strategies for dealing with them. The volume also looks at key challenges in developing integrated approaches to disease management, from breeding more resistant varieties to improved identification of disease and the use of natural antifungal compounds.

Part 1 Fungal diseases of cereals

The first part of the volume reviews the latest research on understanding the main fungal diseases of cereals. Chapter 1 examines rusts affecting wheat and barley. The chapter explores the incidence and impact of wheat and barley rust diseases. The chapter then examines genes for rust resistance and the pathogen biology of rusts, before going on to consider integrated methods of rust control. Finally, the chapter addresses the challenge of achieving durable resistance to rusts in wheat.

Moving from rusts to another class of fungal diseases, Chapter 2 considers the biology and management of Fusarium diseases. The chapter reviews yield and quality losses caused by Fusarium. It then summarises recent findings on Fusarium-cereal interactions, mode of infection and host-plant responses. The chapter then discusses the way this understanding can be used to improve integrated disease management of Fusarium. As well as assessing what we know about understanding and improving genetic resistance, it also discusses work to identify new fungicide molecules that aim for specific fungal metabolic pathways during infection at an early stage. The chapter also discusses the effectiveness of integrated management techniques such as ways to prevent initial infection, changes to tillage and fertilization practices, the role of rotations, more targeted use of fungicides and the use of biological control agents. Finally, the chapter also reviews what we know about the mycotoxins produced by Fusarium fungi and what can be done to reduce them.

The subject of Chapter 3 is Septoria tritici blotch (STB), a fungal disease affecting wheat. STB is one of the most common and economically-important diseases of wheat worldwide. Fungicides have been the primary means of disease management since the 1980s, but resistance has rendered some compounds ineffective. Qualitative or quantitative resistance genes have now been found on all 21 wheat chromosomes, giving a rich but complex source of genes for plant improvement. Molecular markers are now available for many resistance genes making marker-assisted selection for increased resistance an achievable goal for the future. The first STB resistance gene has now been cloned but the mechanism of resistance remains a mystery. Despite the huge progress made during the past twenty years, STB remains a difficult disease that is likely to require substantial future efforts for its control. Building on the latest research on host-pathogen interactions, the chapter evaluates current methods of disease management, particularly in improving resistance, ongoing challenges and future trends in managing the disease.

Chapter 4 moves on to consider the characteristics and effects of powdery mildew. Powdery mildew pathogens are among the most important cereal disease-causing organisms, and can result in significant losses in yield and reductions in grain quality. They pose a significant challenge for barley and wheat agricultural production systems. The chapter provides an overview of current knowledge on wheat and barley resistance to powdery mildew. The chapter discusses current research on disease biology and epidemiology and how a better understanding of the basic molecular mechanisms of cereal-mildew interactions can be used in disease management. The chapter reviews the efficacy of management techniques. These include altering seeding rates and irrigation to create a more open crop canopy structure and less disease-conducive micro-climate, splitting nitrogen applications, removing volunteer plants, more targeted use of fungicides as well as the potential the use of biopesticides. A particular focus is improving resistance. The chapter reviews new approaches for fast gene isolation in wheat and barley, the identification of resistance genes and research into the molecular mechanisms underlying resistance. The chapter includes a case study on the identification, functional characterization and applications of the Pm3 wheat powdery mildew resistance genes in wheat breeding.

The next fungal disease of wheat to be considered, in Chapter 5, is tan spot. The chapter focuses on disease management, including cultural practices, chemical control, and the impact of the choice of wheat varieties. The chapter presents a detailed case study of the management of tan spot in Brazil, and considers the importance of host-pathogen interactions and the genetics of host resistance. Finally, the chapter includes a case study of ToxA-assisted breeding for tan spot resistance in Australia.

The focus of Chapter 6 is Septoria nodorum blotch. The chapter describes the causal agent of Septoria nodorum blotch, the fungus *Parastagonospora nodorum*, and outlines methods of disease management. The chapter examines genetic factors that produce resistance to Septoria nodorum blotch, and explains the process of the genetic mapping of resistance and the use of markers in breeding. Finally, the chapter presents a detailed case study on resistance to Septoria nodorum blotch in Norwegian wheat. Moving on to a lesser-known pathogen of wheat, Chapter 7 considers the challenge posed by *Bipolaris sorokiniana*. The chapter describes the geographic distribution of the disease and describes how *Bipolaris sorokiniana* causes disease in wheat, focusing on the fungus's pathogenic variability as well as methods of disease management.

Chapter 8 considers wheat blast, describing the characteristics, origin and survival of this fungus, based on the experience of managing the disease in Brazil (where losses are particularly high). The chapter describes the disease's origin, disease symptoms and conditions for wheat blast development, and the latest research on its causal agent. Finally, the chapter considers both genetic and chemical control options for wheat blast. The subject of Chapter 9 is net blotches. The chapter addresses the available control measures for net blotch, considering host-pathogen genetic interactions in both the spot and net forms of net blotch. The chapter includes a detailed case study on the subject of breeding for durable resistance to the net form of net blotch in South Australia.

The final chapter of the section, Chapter 10, deals with scald (leaf blotch), which is caused by the hemi¬biotrophic pathogen *Rhynchosporium commune*. Scald is one of the major diseases of barley worldwide. The chapter reviews the origins, epidemiology and other characteristic features of scald, and considers the agricultural consequences of the pathogen's biology. The chapter then considers resistance breeding programs, in which more than a dozen major resistance genes as well as quantitative trait loci (QTLs) have been identified, and discusses strategies to minimize the damage caused by the disease, including agricultural practices and combinations of different fungicides.

Part 2 Key challenges in integrated disease management of cereals

The second part of the volume discusses key challenges in integrated disease management of wheat and barley. These challenges include developing new fungicides, the problem of fungicide resistance, breeding disease-resistance varieties, improving disease identification and the use of natural antifungal compounds. The focus of Chapter 11 is the key challenges in developing new fungicides to treat cereal diseases. The chapter considers the current status of global wheat production, the impact of crop loss on food security, and the emergence of the current regulatory environment surrounding pesticides. The chapter then discusses the current status of the global fungicide market, and some of the major issues in its future longevity: the development of resistance to active ingredients, the impact of legislation that may restrict or remove active ingredients from the market, and the rising costs of developing new active ingredients.

Chapter 12 moves on to examine the occurrence of fungicide resistance in cereal diseases and ways to avoid it. The targeted use of fungicides can help minimise the yield losses caused to grain cereals by plant pathogenic fungi. However, the long-term use of fungicide classes based on a single mode of action has imposed consistent selection pressures on certain pathogens. Over time, this has resulted in many cases of fungicide resistance, leading to a reduction or the total loss of field performance of a range of active ingredients. There is a pressing need to develop better anti-resistance management strategies. The chapter briefly reviews the development of resistance in cereal pathogens, before presenting and evaluating a variety of anti-resistance strategies. The chapter includes detailed case studies of how data from field trials have helped to illustrate the effectiveness of different control strategies.

Complementing the previous chapter's focus on breeding for resistance, Chapter 13 concentrates on improving genetic resources in the breeding of disease-resistant varieties of wheat and other cereals. The chapter highlights recent advances in phenotyping, genotyping and other breeding technologies that have the potential to speed up the discovery and utilization of new sources of disease resistance for wheat and barley. Technologies covered in the chapter include Focused identification of germplasm strategy (FIGS), speed breeding, Genome wide association studies (GWAS) and Genomic selection (GS).

The focus of Chapter 14 is the improvement of diagnostic techniques for disease identification in wheat and other cereal crops. Cereal pests and diseases can significantly reduce potential crop yields. Correct and timely identification in-field is key to their control. The chapter summarises and evaluates some of the many available methods for detection of pests and diseases, ranging from classical methods of plant pathology, such as visual or microscopic recognition, to new DNA sequencing technologies. The chapter discusses how protein-based diagnostics that rely on specificity of antibodies, such as variations on ELISA detection and dipstick or lateral flow devices, have been extended to protein profiling by MALDI-TOF mass spectrometry. Nucleic acid-based amplification techniques using various forms of PCR are also now widely implemented, including isothermal amplification and in field pathogen detection. The chapter shows how the molecular diagnostic techniques are being rapidly supplemented with image-based diagnostics, which rely on hyperspectral imaging, applied via satellites or drones for regional or field scale monitoring of disease incidence and progression.

Moving from diagnostic techniques to disease control, Chapter 15 examines the role of natural antifungal compounds in controlling diseases in wheat and other cereals. Secondary metabolites are essential when considering the ecological context in which an organism operates. They have an array of functions, related to signalling and dealing with biotic and abiotic stress situations. For defence against pathogenic fungi, plants can synthesise various metabolites. Wheat, like several other cereals, produces various benzoxazinoids which are also associated with herbicide detoxification and the defence against herbivores and microorganisms. Apart from these benzoxazinoids, wheat seems to produce only a limited number of other metabolites with known antifungal activity when compared to other cereals. Nevertheless, wheat is still capable of successfully defending itself against pathogenic fungi by outsourcing its defences, i.e. through the mobilization of the microbial community. The chapter provides an overview of the way in which these natural antimicrobial compounds contribute to the defence of wheat against fungal pathogens, as well as considering the counter defensive actions undertaken by some of those fungal pathogens.

The volume's final chapter, Chapter 16, continues the theme of disease management by examining the role of tillage practices, rotations and intercropping in combatting diseases of wheat and other cereals. The great majority of modern agroecosystems comprise rotations of a small number of crop species where each year's crop is a genetically uniform monoculture. The narrow genetic basis for disease resistance in modern cultivars increases the potential for pathogens to overcome this resistance and attack a large number of plants at a similar state of development. The chapter presents a number of strategies for limiting foliar disease development in wheat and barley. The chapter discusses how crop rotations with non-cereal species can substantially reduce inoculum sources for residue-borne cereal leaf diseases. The chapter then examines other strategies, such as intercropping, gene deployment and conservation tillage. The chapter discusses the effectiveness of each strategy against particular foliar diseases. The chapter concludes that there is potential for residue-borne cereal leaf diseases to be managed under conservation tillage via crop rotation, fungicide application and careful choice of variety.

Part 1

Fungal diseases of cereals

Chapter 1

Diseases affecting wheat and barley: rusts

R. F. Park, University of Sydney, Australia

1 Introduction

Rust diseases are caused by pathogenic fungi of the order Pucciniales. Those that attack important agricultural and horticultural species are among the most destructive of all plant pathogens. Three rust diseases occur in common wheat (*Triticum aestivum*): stem or black rust (caused by *Puccinia graminis* f. sp. *tritici*; *Pgt*), stripe or yellow rust (caused by *P. striiformis* f. sp. *tritici*; Pst) and leaf or brown rust (caused by *P. triticina*) (Table 1). All three pathogens can also infect durum wheat (*Triticum durum*), which in some regions is also infected by a second leaf rust causing pathogen, *P. recondita*. Barley is affected by four rust diseases: stem or black rust (caused by *P. graminis*), stripe or yellow rust (caused by *P. striiformis* f. sp. *hordei*; Psh), leaf or brown rust (caused by *P. hordei*), and crown rust (caused by *P. coronata* f. sp. *hordei*; Pch) (Table 1).

Prior to the use of fungicides in broad-acre cereal farming, the rust diseases were controlled by integrating agronomic practices such as removing self-sown (volunteer) cereal plants during the non-cropping phase (the 'green bridge') with genetic resistance. Green bridge removal has proven especially important in reducing the survival of these obligate biotrophic pathogens from one cropping cycle to the next in areas such as Australia in which winters are mild. Combining agronomy and genetic resistance was highly effective in reducing the damage caused by these pathogens for many years. More recently, in some countries, the use of fungicides became an additional component of integrated control of cereal diseases,

http://dx.doi.org/10.19103/AS.2018.0039.01

Table 1 The rust diseases of wheat and barley, causal agents and host ranges

Disease	Causal agent	Main host	Other cereal hosts	Distribution
Wheat stem rust	*P. graminis* f. sp. *tritici*	*Triticum aestivum*	*T. turgidum* subsp. *durum, Hordeum vulgare, Triticosecale*	Global
Wheat stripe rust	*P. striiformis* f. sp. *tritici*	*Triticum aestivum*	*T. turgidum* subsp. *durum, Triticosecale*	Global
Wheat leaf rust	*P. triticina*	*Triticum aestivum*	*T. turgidum* subsp. *durum, Triticosecale*	Global
Wheat leaf rust	*P. recondita*	*T. turgidum* subsp. *durum*	Nil	Morocco, Portugal, Italy
Barley stem rust	*P. graminis**	*Hordeum vulgare*	See below*	Global
Barley stripe rust	*P. striiformis* f. sp. *hordei*	*Hordeum vulgare*	Nil	Western Europe, Middle East, South Asia, East Africa, South America, North America
Barley leaf rust	*P. hordei*	*Hordeum vulgare*	Nil	Global
Barley crown rust	*P. coronata* f. sp. *hordei*	*Hordeum vulgare*	Nil	North America

* Several forms of the stem rust pathogen *P. graminis* can infect barley: *P. graminis* f. sp. *tritici* (the wheat stem rust pathogen), *P. graminis* f. sp. *secalis* (the rye stem rust pathogen), and in Australia, a somatic hybrid between these two forms known locally as the 'scabrum' rust.

including the rusts. Part of the increased interest in using fungicides in this way has been due to the wide spectrum of pathogens they control. This is, however, shrinking as insensitivity to some fungicide classes has been observed to develop in an increasing number of pathogens. Fortunately, extensive knowledge of the biology of rust pathogens and classical genetics of rust resistance in cereals has enabled rapid progress in the development of new tools that continue to accelerate rust resistance breeding and improve the efficiency of developing varieties with durable resistance, the importance of which is expected to grow as chemical use becomes more regulated and insensitivity develops in even more cereal pathogens.

2 The incidence and impact of wheat rust diseases

2.1 *P. graminis* f. sp. *tritici*

Of the three rust diseases of wheat, stem rust (also known as black rust) is potentially the most damaging, having the ability to cause total crop failure. An overview of global crop losses caused by the three wheat rust pathogens indicated varying regional significance (Saari and Prescott 1985). Wheat stem rust was identified as historically important in Australasia, the United States, Canada, northern Africa, South Africa and, to a lesser extent, Europe. Apart from epidemics in Ethiopia in 1993 and 1994, the incidence of stem rust in wheat crops around the world was low from the early 1970s until the late 1990s, when it re-emerged as a serious problem in east Africa. Severe stem rust infection was observed in wheat nurseries in Uganda in 1998, and greenhouse assays of a single sample of stem rust collected from these nurseries (accession 'Ug99') in South Africa identified the presence of a new pathotype (Pretorius et al. 2000), which has since become known widely as 'Ug99' and has the North American race designation TTKSK (Jin et al. 2008). This pathotype (race) has the ability to overcome many of the known resistance genes, including *Sr31*, a gene for which virulence has not previously been detected (Singh et al. 2011). Analyses carried out on samples of stem rusted wheat collected from across a wide area have since shown that pathotype TTKSK is a member of a family of closely related pathotypes that is now known as the 'Ug99' lineage. In addition to Uganda, one or more of these pathotypes are present in Egypt, Eritrea, Ethiopia, Iran, Kenya, Mozambique, Rwanda, South Africa, Sudan, Tanzania, Yemen and Zimbabwe (http://www.RustTracker.cimmyt.org). The 'Ug99' lineage comprises at least seven pathotypes that differ in virulence on resistance genes *Sr21*, *Sr24*, *Sr31* and *Sr36* (Jin et al. 2008, 2009; Pretorius et al. 2010). Studies using microsatellite markers showed that many of these pathotypes have identical fingerprints, consistent with them being recently derived from a common ancestor via single-step mutation (Pretorius et al. 2010). Significantly, surveys in Turkey, Pakistan and India over recent years have failed to detect any of these pathotypes, and three *Pgt* isolates collected from Pakistan in 2009 were clearly shown to differ from pathotype TTKSK in their SSR fingerprint (Karaoglu, H. and Park, R. F. unpublished) and virulence profile (Fetch, T. unpublished).

Severe but localized stem rust epidemics in Ethiopia in 2013 and 2014 were due to a distinctive pathotype, TKTTF, which was virulent on a widely grown variety Digalu (Singh et al. 2015). While not related to the Ug99 family of pathotypes, TKTTF overcame the resistance gene *SrTmp* in Digalu and rendered it highly susceptible with losses of up to 100% on over 10000 ha of smallholder farmer's fields.

2.2 *P. striiformis* f. sp. *tritici*

Stripe rust (also known as yellow rust) is a very destructive and widespread disease of wheat. The causal pathogen Pst infects leaves and leaf sheaths, and can cause severe infection of glumes (hence the former name, *P. glumarum*). While the primary telial/uredinial host of Pst is common wheat, it has also been recorded on durum wheat, cultivated and wild emmer wheat, and on triticale (*Triticosecale*). It has also been reported from cultivated barley (*Hordeum vulgare*), rye (*Secale cereale*) and a range of grass species including wild barley grass (*H. glaucum*, *H. leporinum*), *Elymus*, *Leymus*, *Agropyron*, *Phalaris* and *Bromus*.

For many years, Pst was considered to lack a sexual phase. Studies of genetic diversity within Pst reported high levels of diversity in both Pakistan and China, which was later explained by the discovery of barberry as the alternate host (Jin et al. 2010), and of infection of barberry by Pst in nature (Zhao et al. 2013). Around 33 species of *Berberis* and *Mahonia* have since been confirmed as susceptible to Pst (Zhao et al. 2016).

While historically considered to be a disease of cooler climates, widespread stripe rust epidemics since 2000 have occurred in warmer regions and were attributed to the emergence and spread of two closely related pathotypes with increased aggressiveness and adaptation to warmer temperatures (Hovmoller et al. 2008; Milus et al. 2009). Of all the cereal rust diseases, stripe rust has been the most damaging over much of the past 40 years. Beddow et al. (2015) estimated that globally, stripe rust was responsible for annual losses of about 5.47 million tonnes of wheat grain (US$979 million), with a 90% chance of at least 4.70 million tonnes (US$842 million) being lost in any given year.

2.3 *P. triticina* and *P. recondita*

Leaf rust (also known as brown rust) of common wheat is caused by *P. triticina*. This is the most common and widely distributed wheat rust disease. The causal fungus is heteroecious, with common wheat being the telial/uredinial host and *Thalictrum speciosissimum* or *Isopyrum fumaroides* the alternate (pycnial/aecial) host. The telial/uredinial phase of *P. triticina* has also been reported on tetraploid durum wheat (*Triticum turgidum* ssp. *durum*), wild emmer (*Triticum dicoccoides*), domesticated emmer wheat (*Triticum dicoccum*) and triticale.

P. triticina principally attacks the leaf blades, although it can also infect the leaf sheath and glumes in highly susceptible varieties. If infection occurs early, it can cause yield losses of more than 30%. The damage due to leaf rust tends to be less than that caused by stem rust or stripe rust; however, due to its more frequent and widespread occurrence, globally it is believed to cause higher overall losses.

P. triticina can also cause leaf rust on durum wheat. Singh et al. (2004) reported the presence of a new race of *P. triticina*, BBG/BN, which could infect many durum wheat varieties including Altar C84 which had been grown in Mexico and had remained resistant for some 16 years. Epidemics caused by the new race were estimated to have caused a loss of at least US $32 million in damage. Field testing demonstrated a high degree of susceptibility among a majority of varieties from 31 countries including the United States and Canada, plus most CIMMYT durum wheat germplasm. *P. triticina* race BBG/BN carries virulence for *Lr10, Lr14b, Lr20, Lr23* and probably *Lr28* (Singh et al. 2004). A very similar race, virulent on *Lr10*, *Lr20* and *Lr28* (*Lr14b* was not tested) was identified by Park and Felsenstein (1998) from leaf rust isolates from southern France, where it comprised some 35% of isolates race typed from there. Comparative pathogenicity tests showed that this race was indistinguishable from several isolates obtained from Morocco, suggesting that it may have originated from northern Africa. RAPD fingerprinting of three isolates of the southern France race showed that it was genetically very distinct from 58 isolates chosen to represent the diversity of races identified in western Europe in 1995 (Park et al. 2000).

A second leaf rust disease of durum wheat has been reported from Morocco (Ezzahiri et al. 1992), Portugal (D'Oliveira and Samborski 1966) and southern Italy (Casulli and Pasquini 1992). The pathogen responsible was described by Anikster et al. (1997) as *P. recondita* Group II Type A. This rust pathogen alternates on *Anchusa italica*, a member of the Boraginaceae, and has been reported as a likely source of primary inoculum in Portugal, Morocco and Spain (Ezzahiri et al. 1992, 1994).

3 The incidence and impact of barley rust diseases

3.1 Stripe rust (caused by *P. striiformis* f. sp. *hordei*)

While Pst is capable of infecting barley, susceptible genotypes are rare and losses in commercial barley crops are not known to have been caused by this pathogen. Two other distinct forms of *P. striiformis* infect barley, viz. the barley stripe rust pathogen *P. striiformis* f. sp. *hordei* (Psh) and the barley grass stripe rust pathogen *P. striiformis* f. sp. *pseudo-hordei* (Psph). Of these, Psh poses the much greater threat.

Psh was prevalent for many years in parts of Western Europe, the Middle East, South Asia and East Africa (Stubbs 1985). The highly virulent race 24 of Psh was first detected in

South America near Bogota, Colombia, in 1975, and had spread to nearly all commercial barley-growing regions in South America, except Brazil, by 1982 (Dubin and Stubbs 1986). In 1987, it was detected for the first time in Mexico where it caused severe damage to barley, with yield losses of up to 50%. In 1991, Psh was also reported in the United States, where it reached epidemic levels in parts of California and the Pacific Northwest (Brown et al. 2001). While Psh has not been reported in Australia, it is known that a great deal of Australian barley germplasm is highly susceptible, suggesting that if it were introduced, significant losses would occur (Wellings et al. 2000b).

Samples of *P. striiformis* collected from wild *Hordeum* species in Australia in 1998, mostly from southern NSW and also from other regions in eastern Australia, included isolates that were pathogenically distinct from Pst in being avirulent for many of the stripe rust differential genotypes (Wellings et al. 2000a). Evidence from DNA-based studies demonstrated further the distinctiveness of this rust from Australian isolates of Pst, and also revealed genetic variability among isolates of the rust collected the following year, suggesting that more than one pathogen genotype was introduced into Australia (Keiper et al. 2003). The rust was described as a new *forma specialis* of *P. striiformis*, *P. striiformis* f. sp. *pseudo-hordei* (Psph) (Wellings 2007). Rust testing of Australian barley varieties established that about 90% were resistant to Psph despite a high frequency of susceptibility observed in adult plant testing of the same varieties at Toluca in Mexico with Psh race 24 (Wellings et al. 2000a). Despite some initial concern, Psph has not caused economic losses in commercial crops of barley or wheat in Australia. Given the possibility of increased pathogenicity on barley and/or wheat via mutation or even somatic hybridisation with Pst, the pathogen continues to be monitored closely in annual pathogenicity surveys.

3.2 Leaf rust (caused by *P. hordei*)

Although the economic importance of barley leaf rust depends on the region in the world, and differs from year to year (Niks et al. 2000), it appears that it has increased in recent years (Park et al. 2014). While its incidence is sporadic, it is probably the most common and widely distributed rust disease of barley, occurring throughout all the barley-growing areas of the eastern and mid-western United States, North Africa, Europe, New Zealand, Australia and some parts of Asia, where severe yield losses occur in susceptible varieties, particularly in areas where crops mature late (Park et al. 2014). Significant yield losses have been reported from New Zealand (Arnst et al. 1979), Australia (Cotterill et al. 1992), North America (Griffey et al. 1994), the Czech Republic (Dreiseitl and Steffenson 2000), the United Kingdom (Jenkins et al. 1972), Ethiopia (Semeane et al. 1996) and South Africa (van Niekerk et al. 2001). Although total crop loss does not occur, under epidemic conditions, yield reductions of up to 62% have been reported in susceptible varieties (Clifford 1985; Cotterill et al. 1992; Griffey et al. 1994; Tan 1978) (see the case study for further details).

3.3 Crown rust (caused by *P. coronata* f. sp. *hordei*)

Crown rust of barley was first reported in the United States, when an outbreak of the disease was found in Nebraska, although there is evidence that the pathogen was present in the Great Plains region as early as the 1950s (Jin et al. 1992). The causal agent was identified by Jin and Steffenson (1999) as a variant of the crown rust pathogen, viz. *P. coronata* var. *hordei*. It appears that this pathogen is restricted to North America. Agrama et al. (2004) reported that the disease was sporadic in occurrence in the Great Plains of the

United States and the Prairie Provinces of Canada, only occasionally reaching levels that cause reductions in yield (Niu et al. 2014).

3.4 Stem rust (caused by *Puccinia graminis*)

Stem rust tends to be a problem on barley only when inoculum levels are high due to epidemics in wheat and/or triticale, because epidemics on these crops exert substantial inoculum pressure on barley crops (Dill-Macky et al. 1990). Due to the widespread planting of resistant wheat varieties and conditions unfavourable for rust development, losses to stem rust in barley have been rare in many production areas (Park 2007). In Australia, barley crops may be affected by three physiological forms of *P. graminis*; the form virulent on wheat and/or triticale (*Pgt*); the form virulent on cereal rye (*P. graminis* f. sp. *secalis*; Pgs); and a putative somatic hybrid between these two *formae speciales*, referred to locally as the 'scabrum' rust (Park 2007). In Canada and the Northern Great Plains of the United States, the wheat stem rust fungus *Pgt* has caused a number of significant epidemics on both wheat and barley (Kilian et al. 1997). Although stem rust has not been a major problem in the United States for over 70 years, it is still considered the most destructive cereal rust disease (Roelfs 1985).

4 Rust resistance genes

Since Biffin (1905) first demonstrated that resistance to stripe rust in wheat was inherited as a single Mendelian trait, many other genes conferring rust resistance have been characterised and catalogued. For convenience, these genes are referred to as either seedling/major/all stage resistance (ASR) genes, or minor/adult plant resistance (APR) genes, based on the growth stage at which they become effective. While useful, this division is to some extent an artificial one because a number of resistance genes that are typically only expressed at adult plant growth stages can be detected at early growth stages under certain conditions (e.g. the stripe rust resistance of the Australian wheat variety Cook; Park and Rees 1989). Resistance genes have been used extensively in breeding rust-resistant wheat and barley and served as the foundation of integrated control of these diseases in cereals.

Some 62 genes/alleles at 55 loci that confer resistance to *Pgt* in wheat have been catalogued to date. These genes have been given the designations *Sr2* through *Sr58*; the designations *Sr1*, *Sr3*, *Sr4* and *Sr9c* were abandoned due to a lack of reference stocks or duplication with previously described loci. Only three of these 55 loci comprise series of alleles (*Sr7*, alleles *Sr7a* and *Sr7b*; *Sr8*, alleles *Sr8a* and *Sr8b*; *Sr9*, alleles *Sr9a*, *Sr9b*, *Sr9d*, *Sr9e*, *Sr9f*, *Sr9g* and *Sr9h*). Of the remaining 52 loci, 23 were found in *Triticum aestivum*, and 30 were introgressed to *Triticum aestivum* from related grass species such as *Aegilops umbellulata*, *Thinopyrum intermedium* and *Th. ponticum*.

4.1 Wheat leaf rust

To date, 70 genetic loci involving some 77 resistance genes/alleles conferring resistance to *P. triticina* have been designated in wheat (*Lr1–Lr74*; the symbols *Lr4*, *Lr5*, *Lr6*, *Lr7*, *Lr8*, *Lr40*, *Lr41* and *Lr43* were abandoned due to a lack of reference stocks or duplication with previously described loci). About half of the designated resistance genes were found in *Triticum aestivum*, with the remainder being introduced into common wheat from other grass species. Allelic variation has been recorded at five loci (*Lr2*, *Lr3*, *Lr14*, *Lr17* and *Lr22*); however, the two alleles reported at the *Lr14* locus (*Lr14a* and *Lr14b*) were combined in a later study, showing that they were very closely linked and not true alleles. The two genes *Lr27* and *Lr31* act in a complementary manner (Singh and McIntosh 1984). Wheats possessing *Lr27* only are susceptible to *P. triticina* but can usually be identified by the presence of the linked stem rust resistance gene *Sr2* (Singh and McIntosh 1984). Interestingly, Singh et al. (1999) demonstrated that *Lr31* is the either the same as, or is completely linked with, the APR gene *Lr12*. If these genes are in fact the same, then *Lr12*/*Lr31* and *Lr27* must function as typical complementary genes at seedling growth stages. Because *Lr12* is effective on its own in adult plants, the complementarity must only operate at seedling growth stages. If so, *Lr27* could be regarded as a modifier gene, required by *Lr12* for expression in seedlings. Whether or not *Lr12* and *Lr31* are the same or different genes could be resolved if an isolate of the leaf rust pathogen differing in pathogenicity for *Lr27* + *Lr31* and *Lr12* was found (all isolates characterised so far, with evidence from Australia, Mexico, South Africa and Argentina, have been virulent or avirulent for both), or by cloning the genes.

4.2 Wheat stripe rust

A total of 78 genes/alleles at 74 loci (*Yr1*–*Yr76*; *Yr26* is considered to be synonymous with *Yr24*) conferring resistance to Pst have been catalogued to date. Three alleles have been reported at the *Yr3* locus, two at the *Yr4* locus and recently *Yr5* and *Yr7* were shown to be allelic. Two genes (*Yr73* and *Yr74*) are complementary; that is, they only confer resistance when present together (Dracatos et al. 2015). A majority of these genes/alleles were reported from *Triticum aestivum*, with others from other species of *Triticum*, *Aegilops*, *Th. intermedium* (*Yr50*) and *Secale cereale* (*Yr9*).

4.3 Barley stem rust

Eight stem rust resistance genes have been described in barley. Those conferring resistance to *Pgt* are *Rpg1* (in the varieties Peatland, Chevron and Kindred (Powers and Hines 1933; Shands 1939; Steffenson 1992), *Rpg2* (Hietpas-5; Patterson et al. 1957), *Rpg3* (PI 382313; Jedel et al. 1989), *Rpg4* (Q21861; Jin et al. 1994), *RpgU* (Peatland; Fox and Harder 1995) and Rpg6 (line 212Y1, a *H. vulgare* line with introgressions from *H. bulbosum*; Fetch et al. 2009). Genes conferring resistance to the rye stem rust pathogen Pgs include the recessive APR gene rpgBH (known formerly as rpgS, Black Hulless; Steffenson et al. 1984) and *Rpg5* (known formerly as RpgQ, Q21861; Sun et al. 1996).

4.4 Barley leaf rust

To date, 24 loci conferring hypersensitive seedling resistance (*Rph1 Rph19*, *Rph21* and *Rph22*) and three APR genes (*Rph20*, *Rph23* and *Rph24*) (Hickey et al. 2011; Park 2003; Park and Karakousis 2002; Sandhu et al. 2012; Singh et al. 2015; Ziems et al. 2017) from *H. vulgare*, *H. vulgare* subsp. *spontaneum*, and *H. bulbosum* have been designated.

Parlevliet and Ommeren (1975) proposed the term 'partial resistance' to describe resistance to leaf rust in certain barley genotypes. They differentiated partial resistance from seedling resistance and APR, indicating that the host is susceptible at all growth stages but that the infection frequency, latent period, rate of spore production and period of spore production may vary. Selection for partial resistance is often difficult in field plots as all genotypes show a susceptible reaction (Parlevliet and Ommeren 1975). Whilst numerous QTLs have been identified for both partial resistance at the seedling stage and APR for resistance to *P. hordei*, only in a few instances have genetic stocks for such QTLs been catalogued and markers developed for MAS. In barley, only three genes conferring APR to *P. hordei* have been characterised and designated (*Rph20*, *Rph23* and *Rph24*, on chromosomes 5HS, 7HS and 6HS, respectively) based on consistent detection in diverse field nurseries over multiple seasons and QTL mapping analysis (Golegaonkar et al. 2010; Hickey et al. 2011; Singh et al. 2014; Ziems et al. 2017) (see the case study).

4.5 Barley stripe rust

Relatively few studies have been conducted on the genetic basis of resistance to Psh in barley. The barley varieties Bigo, Abyssinian 14, BBA 2890, Abed Binder 12 and I5 were all reported to carry single recessive genes to a European isolate of Psh race 24 (Nover and Scholz 1969). The genes in Bigo, Abyssinian 14 and BBA 2890 were shown to be allelic and designated 'yr'. The gene in Abed Binder 12 was designated *yr2*, and that in I5 *yr3*. Johnson (1968) reported a gene in Europa, Cambrinus and Deba Abed, which was designated *Yr4* and mapped to chromosome 1HS (Wettstein-Knowles 1992).

Field studies have detected genomic regions associated with stripe rust resistance on barley chromosomes 4H, 5H, 7H and 1H (Chen et al. 1994; Hayes et al. 1996; Thomas et al. 1995). Hayes et al. (1996) mapped a gene conferring APR to stripe rust to chromosome 7HL in the barley genotype CI10587; however, the identity of this gene relative to previously designated genes was not established. Loci conferring quantitative resistance to Psh have been discovered and some have been introgressed into new genetic backgrounds by using molecular marker-assisted selection (Chen et al. 1994; Toojinda et al. 1998). An investigation of 18 barley genotypes revealed the presence of 20 different resistance genes, three had single resistance genes and the remaining 15 genotypes had two or more genes for resistance. Interestingly, most of these genes were recessive (Chen and Line 1999).

5 Rust pathogen biology

The pathogens that cause rust diseases of wheat and barley have complex life cycles that include alternate dicot hosts. In most regions, the pathogen populations are largely asexual as the alternate hosts have been removed to reduce pathogen genetic variability and increase the longevity of host plant resistance. The existence of host genetic stocks carrying well-characterised resistance genes has allowed wheat and barley rust pathogen populations to be monitored for virulence, which has assisted resistance breeding efforts by allowing effective resistance genes to be targeted. The methods developed to identify rust pathotypes have not changed greatly over the past 90 years or so, and are based on greenhouse assays that take about 2–3 weeks. In addition to providing new tools to expedite wheat breeding, molecular genetics is beginning to provide insights into rust pathogens at the genomic level (e.g. Duplessis et al. 2011). To date, genome sequences have been generated for four of the rust pathogens that infect wheat and barley: *P. graminis*

f. sp. *tritici* (92 Mbp; Uphadyaya et al. 2015), *P. triticina* (135 Mbp; Cuomo et al. 2017), *P. striiformis* f. sp. *tritici* (131 Mbp; Zheng et al., 2013) and *P. hordei* (127 Mbp; Chen, 2017). While large for fungi, the genomes of these species are relatively small compared to some other rust species such as *P. vincae* (566.4 Mb; Talhinas et al. 2015). The genomes of all four rust species are highly heterozygous, comprise a large amount of repetitive elements, lack certain gene families (e.g. those involved in nitrate metabolism) and show expansion in others (e.g. carbohydrate degradation enzymes).

The complexity of these genomes has made accurate genome assembly and haplotype phasing difficult based on short-read NGS sequencing. It is to be expected that this will improve with newer long-read sequencing platforms.

6 Integrated methods of rust control

Improvements in understanding the genetic basis of disease resistance in wheat and barley and of host–pathogen interactions have greatly improved the efficiency of manipulating disease resistance genes in breeding programmes, allowing the development and use of high-yielding resistant varieties that in some cases have occupied vast areas. The integration of genetic resistance with agronomic practices has often been particularly important in improving the effectiveness of resistance in reducing yield losses, and in minimising the chance of rust pathogens changing to overcome resistance. In regions where rusts can persist all year round, resistant varieties also give rise to a genetically resistant green bridge, reducing the ability of rust pathogens to survive from one cropping cycle to another.

Fungicides have provided support in situations where resistance genes have been overcome by rust pathogen changes or where the level of resistance conferred by a gene or genes has been inadequate in preventing yield loss. The latter has occurred where minor gene resistance or APR has been used; these resistances do not provide protection from rust infection at early growth stages, and often provide incomplete protection at adult plant growth stages. For example, studies in which the protective value of the wheat stripe rust APR gene *Yr18* was assessed using near isogenic paired lines with and without *Yr18* demonstrated yield losses of 74–94% in a line lacking *Yr18*, but 31–52% in a line carrying *Yr18* (Ma and Singh 1996).

The value of genetics, fungicides and cultural practices to rust control in wheat and barley in Australia was estimated by Murray and Brennan (2009a,b). In barley, genetics contributed AUD$69 million, fungicides $9 million and cultural practices $41 million to the overall annual savings of $120 million (Murray and Brennan 2009b). In wheat, genetics contributed $1.021 billion, fungicides $384 million and cultural practices $118 million to the overall annual savings of AUD$1.523 billion (Murray and Brennan 2009b). These estimates reinforce the importance of genetics as the foundation of integrated enduring rust control in cereal production.

7 Durable resistance

The concept of durable resistance was introduced about 40 years ago, following the breakdown of slow rusting/APR to stripe rust in several English winter wheats (e.g. Joss Cambier) and continued effectiveness of resistance in several other varieties (e.g. Cappelle

Desprez, Hybrid de Bersee) (Johnson 1978). The resistance in the latter was referred to as durable, with durable resistance defined as 'resistance that remains effective when a variety is grown widely in environments favouring disease development'.

Advances in DNA marker technologies since then have allowed the genetic dissection of a range of sources of durable resistance, and increasingly, the cloning and characterisation of resistance genes contributing to both durable and non-durable resistance. To date, 12 rust resistance loci have been isolated from wheat, and three from barley. Of the 12 rust resistance genes that have been cloned from wheat, nine encode nucleotide-binding and leucine-rich repeat (NLR) receptor proteins [*Lr1*, Cloutier et al. (2007); *Lr10*, Feuillet et al. (2003); *Lr21*, Huang et al. (2003); *Sr22*, Steuernagel et al. (2016); *Sr33*, Periyannan et al. (2013); *Sr35*, Saintenac et al. (2013); *Sr45*, Steuernagel et al. (2016); *Sr50*, Mago et al. (2015); *Yr10*, Liu et al. (2014)]. With the exception of *Sr33*, which has not been tested widely (McIntosh et al. 1995), virulence has been detected for all of these genes, indicating a tendency for NLR-encoding resistance genes to be non-durable. The three remaining loci, *Yr36* (Fu et al. 2009), *Lr34/Yr18/Sr57* (Krattinger et al. 2009) and *Lr67/Yr46/Sr55* (Moore et al. 2015), confer APR and encode a chloroplast-localised protein with kinase and steroidogenic acute regulatory protein-related transfer (START) lipid-binding domains, an ATP-binding cassette transporter, and a protein that has lost hexose transport function, respectively. The loci *Lr34/Yr18/Sr57* and *Lr67/Yr46/Sr55*, along with *Lr46/Yr29/Sr58*, confer resistance against all three rust pathogens, and are associated with a linked morphological trait known as leaf tip necrosis. Experience to date suggests that these pleiotropic APR genes have some intrinsic durability, and that combinations of multiple effective resistance genes contribute to durability by lowering the chance of virulence matching gene combinations developing. In many instances, pleiotropic APR genes have been used to great effect in conjunction with ASR genes, providing backbone resistance that may also contribute to increasing the durability of ASR genes (Park 2015).

7.1 Enduring rust control in agriculture

The term durable resistance is sometimes mistakenly equated to enduring rust control in agriculture. Clearly, growing only varieties that carry high levels of durable resistance at a large scale would be expected to provide enduring rust control across agroecological zones, continents and possibly beyond. However, it is important to appreciate that resistance that has proven durable may not remain effective forever (Johnson 1984), stressing the importance of genetic diversity in the resistances deployed. The durability of resistance genes when deployed over large areas is complex, being determined not just by the ability of the pathogen to acquire matching virulence, but also other traits in the pathogen and host that can impact on overall disease epidemiology. For example, on the pathogen side, long-term surveys of pathogenicity of wheat rust pathogens in Australia have provided many examples where certain pathogen genotypes seem to have greater fitness, which is independent of virulence for resistance genes (e.g. a major shift in the composition of the wheat stem rust population that followed an exotic incursion in 1954; Watson 1970). On the host side, a change to growing early maturing wheat varieties developed by William Farrer in Australia had a huge impact in reducing losses to stem rust through 'disease escape' (Watson and Butler 1984). Both of these factors can influence the overall size of the pathogen population, and in so doing affect the timing of epidemic onset, disease pressure on varieties carrying incomplete levels of resistance and how frequently virulent mutant pathotypes emerge.

While a great deal of knowledge now exists from how individual host and pathogen genes interact and how resistance genes function in plants, much remains to be learned at the interface of the genetics of resistance and crop physiology. The cloning of resistance genes and corresponding avirulence genes have indicated considerable complexity not only in structure but also in the way in which gene products interact and trigger resistance. At the crop level and beyond, few large-scale studies have addressed in detail the effects of genetic diversity in crops in reducing disease incidence. A study in which the control of *Magnaporthe grisea* was examined in more than 3000 ha of rice crops in China, found very clear benefits from genetic diversity, with varietal mixtures resulting in an 89% yield increase in susceptible varieties (Zhu et al. 2000). Clearly, genetic diversity is important in sustained control of plant diseases, and it has been argued that even in cases where specific or major resistance genes are used, genetic diversity provides insurance against lack of durability and reduces genetic vulnerability (Watson 1970). The durability of some single gene resistances, despite extensive use, and the preferential survival of some pathogen genotypes in the absence of no obvious advantage in terms of virulence that has been well documented with wheat rust pathogens in Australia, indicates that the interactions between hosts and pathogens in nature are indeed complex.

7.2 Case study: durable resistance to leaf rust in barley

Barley leaf rust is a major disease of barley crops worldwide. Since the late 1990s, it has emerged as a significant limitation to production in most Australian barley-growing areas, and was estimated by Murray and Brennan (2009a) to cost the industry on average $21 million per year. Research initiated at the University of Sydney in 1989 has led to a sound understanding of the pathogenicity of the causal agent *P. hordei* and of major gene resistance in Australian barley cultivars. This research documented the loss of two ASR genes following their deployment in commercial barley varieties, viz. Rph12 in 1991, following its initial deployment in 1989 (variety Franklin) (Park 2003) and Rph3 in 2009, following its initial deployment in 2005 (variety Yarra) (Park et al. 2015).

Research was initiated in 1998 to identify new sources of resistance to leaf rust in barley, which led to the discovery that the Dutch cultivar Vada and a range of other European and Australian barley genotypes carried APR (Golegaonkar et al. 2009). Subsequent genetic studies showed that many of these varieties shared the same resistance gene, which was then mapped to chromosome 5HS and designated Rph20 (Golegaonkar et al. 2010; Hickey et al. 2011). Two additional loci controlling APR to leaf rust have since been mapped, viz. *Rph23* (Singh et al. 2015) and *Rph24* (Ziems et al. 2017).

The characterisation of three loci controlling APR to leaf rust in barley, along with linked markers, allowed critical appraisals to be made of the level of protection provided by each gene singly and in combination. These studies have shown clear evidence of additivity and a high level of protection across environments in barley genotypes carrying all three genes (Park, R. F. et al. unpublished).

Additional genomic regions associated with APR to leaf rust were identified by Singh et al. (2017). Future research will focus on resolving the identities of these resistances, assessing the potential of each in contributing additive resistance, and also looking for potential beneficial pleiotropic effects on resistance to other biotic stresses. Based on experience in wheat, where multi-genic resistance based on combinations of genes conferring both ASR and APR, it is hoped that combining effective resistance genes using marker-assisted selection will lead to the enduring control of leaf rust in barley.

8 Future trends in research

Many of the catalogued genes in wheat and barley that confer ASR, and some conferring APR, have been overcome by the development of new pathogen races. Resistance nonetheless remains the foundation of integrated rust control in both crops. A number of resistance genes conferring durable resistance have been identified, including three in wheat that confer slow rusting APR to all three rust pathogens (*Lr67/Yr46/Sr55*; *Lr34/Yr18/Sr57*; *Lr46/Yr29/Sr58*). Whether or not such loci exist in barley remains to be seen.

The number of resistance genes for which tightly linked high-throughput DNA markers are available continues to increase, allowing marker-assisted selection and greater efficiency in assembling gene combinations. The ability to incorporate resistance genes by transformation is also a reality, and combining multiple genes in a single construct will also become possible as more genes are cloned. An important consideration in this approach is, however, possible suppression of resistance genes. For example, recent studies of powdery mildew resistance genes in wheat showed that the susceptibility allele *Pm3C*, and three resistance alleles (*Pm3a*, *Pm3b* and *Pm3f*) all suppressed the rye derived resistance gene *Pm8*, which had been introgressed into wheat (Hurni et al. 2014). The strength of the suppression was shown to vary with the *Pm3* allele used, raising the possibility that gene editing could be used to modify alleles so that suppression did not occur. While this example may be an exception, clearly any strategy based on combining cloned resistance genes will need to take this possibility into consideration.

9 Where to look for further information

9.1 Key publications

Bushnell, W. R. and Roelfs, A. P. (1984). *The Cereal Rusts, Volume 1. Origins, Specificity, Structure and Physiology.* Academic Press Inc., Orlando, FL.

Knott, D. R. (1989). *The Wheat Rusts - Breeding for Resistance.* Monographs on Theoretical and Applied Genetics 12. Springer Verlag, Berlin.

McIntosh, R. A., Wellings, C. R. and Park, R. F. (1995). *Wheat Rusts. An Atlas of Resistance Genes*, p. 200. CSIRO, Melbourne.

McIntosh, R. A. and Pretorius, Z. A. (Guest editors). (2011). Special issue: Borlaug Global Rust Initiative provides momentum for wheat rust research. *Euphytica* 179: 1–207.

Park, R. F., Bariana, H. S. and Wellings, C. R. (Coordinating editors). (2007). Special issue: Global landscapes in cereal rust control. *Australian Journal of Agricultural Research* 58: 467–655.

Roelfs, A. P. and Bushnell, W. R. (1985). *The Cereal Rusts, Volume 2. Diseases, Distribution, Epidemiology and Control.* Academic Press Inc., Orlando, FL.

Simmonds, N. W. and Rajaram, S. (1988). *Breeding Strategies for Resistance to the Rusts of Wheat.* CIMMYT, Mexico.

9.2 Websites

Borlaug Global Rust Initiative: http://www.globalrust.org/.

Catalogue of wheat symbols for wheat: http://www.shigen.nig.ac.jp/wheat/komugi/genes/macgene/supplement2012.pdf.

Global Rust Reference Center: http://wheatrust.org/.
GrainGenes: A database for Triticeae and Avena: http://wheat.pw.usda.gov.
International Center for Agricultural Research in Dry Areas: http://www.icarda.org/.
International Center for Maize and Wheat Improvement (CIMMYT): http://www.cimmyt.org/.
Marker Assisted Selection in Wheat: http://maswheat.ucdavis.edu/.
RustTracker.org A global wheat rust monitoring system: http://rusttracker.cimmyt.org/.

10 Acknowledgements

The author would like to acknowledge financial support from the Coffey family and the Australian Grains Research and Development Corporation.

11 References

Agrama, H. A., Dahleen, L., Wentz, M., Jin, Y. and Steffenson, B. J. (2004). Molecular mapping of the crown rust resistance gene Rpc1 in barley. *Phytopathology* 94: 858–61.

Anikster, Y., Bushnell, W. R., Eilam, T., Manisterski, J. and Roelfs, A. P. (1997). *Puccinia recondita* causing leaf rust on cultivated wheats, wild wheats, and rye. *Canadian Journal of Botany* 75: 2082–96.

Arnst, B. J., Martens, J. W., Wright, G. M., Burnett, P. A. and Sanderson, F. R. (1979). Incidence, importance and virulence of *Puccinia hordei* on barley in New Zealand. *Annals of Applied Biology* 92: 185–90.

Beddow, J. M., Pardey, P. G., Chai, Y., Hurley, T. M., Kriticos, D. J., Braun, H. J., Park, R. F., Cuddy, W. S. and Yonow, T. (2015). Research investment implications of shifts in the global geography of wheat stripe rust. *Nature Plants* 1: 15132.

Biffin, R. H. (1905). Mendel's laws of inheritance and wheat breeding. *Journal of Agricultural Science* 1: 4–48.

Brown, W. M., Hill, J. P. and Velasco, V. R. (2001). Barley yellow rust in North America. *Phytopathology* 39: 367–84.

Casulli, F. and Pasquini, M. (1992). Virulence of *Puccinia recondita* f. sp. *tritici* in Italy. *Vortrage Pflanzenzuchtung* 24: 87–9.

Chen, J. (2017). Genomic studies of two cereal rust fungi with a focus on avirulence gene studies. PhD Thesis, The University of Sydney, p. 149.

Chen, X. and Line, R. F. (1999). Recessive genes for resistance to *Puccinia striiformis* f. sp. *hordei* in barley. *Phytopathology* 89: 226–32.

Chen, F. Q., Prehn, D., Hayes, P. M., Mulrooney, D., Corey, A. and Vivar, H. (1994). Mapping genes for resistance to barley stripe rust (*Puccinia striiformis* f. sp. *hordei*). *Theoretical and Applied Genetics* 88: 215–19.

Clifford, B. C. (1985). Barley leaf rust. In: *The Cereal Rusts, Volume II. Diseases, Distribution, Epidemiology and Control* (A. P. Roelfs and W. R. Bushnell (Eds)). Academic Press, New York, pp. 173–205.

Cloutier, S., McCallum, B. D., Loutre, C., Banks, T. W., Wicker, T., Feuillet, C., Keller, B. and Jordan, M. C. (2007). Leaf rust resistance gene Lr1, isolated from bread wheat (*Triticum aestivum* L.) is a member of the large psr567 gene family. *Plant Molecular Biology* 65: 93–106.

Cotterill, P. J., Rees, R. G., Platz, G. J. and Dill-Macky, R (1992). Effects of leaf rust on selected Australian barleys. *Australian Journal of Experimental Agriculture* 32: 747–51.

Cuomo, C. A., Bakkeren, G., Khali, H. B., Panwar, V., Joly, D., Linning, R., Sakthikumar, S., Song, X., Adiconis, X., Fan, L., Goldberg, J. M., Levin, J. Z., Young, S., Zeng, Q., Anikster, Y., Bruce, M., Wang, M., Yin, C., McCallum, B., Szabo, L. J., Hulbert, S., Chen, X. and Fellers, J. P. (2017). Comparative analyses highlights variable genome content of wheat rusts and divergence of the mating loci. *G3: Genes, Genomes, Genetics* 7: 361–76.

d'Oliveira, B. and Samborski, D. J. (1966). Aecial stage of *Puccinia recondita* on Ranunculaceae and Boraginaceae in Portugal. In: *Proceedings of the First Cereal Rust Conference, Cambridge, UK, 29 June–2 July, 1964* (R. C. F. Macer and M. S. Wolfe (Eds)). Plant Breeding Institute, Cambridge, UK, pp. 133–50.

Dill-Macky, R., Rees. R. G. and Platz, G. J. (1990). Stem rust epidemics and their effects on grain yield and quality in Australian barley cultivars. *Australian Journal of Agricultural Research* 41: 1057–63.

Dracatos, P. M., Zhang, P., Park, R. F., McIntosh, R. A. and Wellings, C. R. (2015). Complementary resistance genes in wheat selection 'Avocet R' confer resistance to stripe rust. *Theoretical and Applied Genetics* 129(1): 65–76. doi:10.1007/s00122-015-2609-7.

Dreiseitl, A. and Steffenson, B. J. (2000). Postulation of leaf-rust resistance genes in Czech and Slovak barley cultivars and breeding lines. *Plant Breeding* 119: 211–14.

Dubin, H. J. and Stubbs, R. W. (1986). Epidemic spread of barley stripe rust in South America. *Plant Disease* 70: 141–4.

Duplessis, S., Cuomo, C. A., Lin, Y. C., Aerts, A., Tisserant, E., Veneault-Fourrey, C., Joly, D., Hacquard, S., Amselem, J., Cantarel, B., Chiu, R., Coutinho, P. M., Feau, N., Field, M., Frey, P., Gelhaye, E., Goldberg, J., Grabherr, M. G., Kodira, C. D., Kohler, A., Kües, U., Lindquist, E. A., Lucas, S. M., Mago, R., Mauceli, E., Morin, E., Murat, C., Pangilinan, J. L., Park, R., Pearson, M., Quesneville, H., Rouhier, N., Sakthikumar, S., Salamov, A. A., Schmutz, J., Selles, B., Shapiro, H., Tanguay, P., Tuskan, G. A., Henrissat, B., Van de Peer, Y., Rouzé, P., Ellis, J. G., Dodds, P. N., Schein, J. E., Zhong, S., Hamelin, R. C., Grigoriev, I. V., Szabo, L. J. and Martin, F. (2011). Obligate biotrophy features unraveled by the genomic analysis of rust fungi. *Proceedings of the National Academy of the USA* 108: 9166–71.

Ezzahiri, B., Diouri, S. and Roelfs, A. P. (1992). *Anchusa italica* as an alternate host for wheat leaf rust in Morocco. *Plant Disease* 76: 1185.

Ezzahiri, B., Diouri, S. and Roelfs, A. P. (1994). Pathogenicity of *Puccinia recondita* f. sp. *tritici* in Morocco during 1985, 1988, 1990 and 1992. *Plant Disease* 78: 407–10.

Fetch, T., Johnston, P. A. and Pickering, R. (2009). Chromosomal location and inheritance of stem rust resistance transferred from *Hordeum bulbosum* into cultivated barley (*H. vulgare*). *Phytopathology* 99: 339–43.

Feuillet, C., Travella, S., Stein, N., Albar, L., Nublat, A. and Keller, B. (2003). Map-based isolation of the leaf rust disease resistance gene Lr10 from the hexaploid wheat (*Triticum aestivum* L.) genome. *Proceedings of the National Academy of Sciences USA* 100: 15253–8.

Fox, S. L. and Harder, D. E. (1995). Resistance to stem rust in barley and inheritance of resistance to race QCC. *Canadian Journal of Plant Science* 75: 781–8.

Fu, D., Uauy, C., Distelfeld, A., Blechl, A., Epstein, L., Chen, X., Sela, H., Fahima, T. and Dubcovsky, J. (2009). A kinase-START gene confers temperature-dependent resistance to wheat stripe rust. *Science* 323: 1357–60.

Golegaonkar, P. G., Park, R. F. and Singh, D. (2009). Evaluation of seedling and adult plant resistance to *Puccinia hordei* in barley. *Euphytica* 166: 183–97.

Golegaonkar, P. G., Park, R. F. and Singh, D. (2010). Genetic analysis of adult plant resistance to *Puccinia hordei* in barley. *Plant Breeding* 129: 162–6.

Griffey, C. A., Das, M. K., Baldwin, R. E. and Waldenmaier, C. M. (1994). Yield losses in winter barley resulting from a new race of *Puccinia hordei* in North America. *Plant Disease* 78: 256–60.

Hayes, P. M., Prehn, D., Vivar, H., Blake, T., Comeau, A., Henry, I., Johnston, M., Jones, B., Steffenson, B. and St Pierre, C. A. (1996). Multiple disease resistance loci and their relationship to agronomic and quality loci in a spring barley population. *Journal of Quantitative Trait Loci* (Online). http://probe.nalusda.gov:8000/otherdocs/jqtl/index.htm.

Hickey, L. T., Lawson, W., Platz, G. J., Dieters, M., German, S., Fletcher, S., Park, R. F., Singh, D., Pereya, S. and Franckowiak, J. (2011). Mapping Rph20: A gene conferring adult plant resistance to *Puccinia hordei* in barley. *Theoretical and Applied Genetics* 123: 44–68.

Hovmoller, M. S., Yahyaoui, A. H., Milus, E. A. and Justesen, A. F. (2008). Rapid global spread of two aggressive strains of a wheat rust fungus. *Molecular Ecology* 17: 3818–26.

Huang, L., Brooks, S. A., Li, W., Fellers, J. P., Trick, H. N. and Gill, B. S. (2003). Map-based cloning of leaf rust resistance gene Lr21 from the large and polyploid genome of bread wheat. *Genetics* 164: 655–64.

Hurni, S., Brunner, S., Stirnweis, D., Herren, G., Peditto, D., McIntosh, R. A. and Keller, B. (2014). The powdery mildew resistance gene Pm8 derived from rye is suppressed by its wheat orthologue Pm3. *The Plant Journal* 79: 904–13.

Jedel, P. E., Metcalfe, D. R. and Martens, J. W. (1989). Assessment of barley accessions PI 382313, PI 382474 and PI 382976 for stem rust resistance. *Crop Science* 29: 1473–7.

Jenkins, J. E., Melville, S. C. and Jemmett, J. L. (1972). The effect of fungicides on leaf diseases and on yield in spring barley in south-west England. *Plant Pathology* 21: 49–58.

Jin, Y. and Steffenson, B. J. (1999). *Puccinia coronata* var. *hordei* var. nov: Morphology and pathogenicity. *Mycologia* 91: 877–84.

Jin, Y., Steffenson, B. J. and Miller, J. D. (1994). Inheritance of resistance to pathotypes QCC and MCC of *Puccinia graminis* f. sp. *tritici* in barley line Q21861 and temperature effects on the expression of resistance. *Phytopathology* 84: 452–5.

Jin, Y., Steffenson, B. J., Oberthur, L. E. and Baenziger, P. S. (1992). *Puccinia coronata* on barley. *Plant Disease* 76: 1283.

Jin, Y., Szabo, L. J., Pretorius, Z. A., Singh, R. P., Ward, R. and Fetch, T. (2008). Detection of virulence to resistance gene Sr24 within race TTKS of *Puccinia graminis* f. sp. *tritici*. *Plant Disease* 92: 923–6.

Jin, Y., Szabo, L. J., Rouse, M., Fetch, T., Pretorius, Z. A., Singh, R. P., Wanyera, R. and Njau, P. (2009). Detection of virulence to resistance gene Sr36 within race TTKS lineage of *Puccinia graminis* f. sp. *tritici*. *Plant Disease* 93: 367–70.

Jin, Y., Szabo, L. J. and Carson, M. (2010). Century-old mystery of *Puccinia striiformis* life history solved with the identification of Berberis as an alternate host. *Phytopathology* 100: 432–5.

Johnson, R. (1968). Genetics of resistance of barley to stripe rust. In: *First International Congress of Plant Pathology*, London, p. 99 (Abstract).

Johnson, R. (1978). Practical breeding for durable resistance to rust diseases in self- pollinating cereals. *Euphytica* 27: 529–40.

Johnson, R. (1984). A critical analysis of durable resistance. *Annual Review of Phytopathology* 22: 309–30.

Keiper, F. J., Hayden, M. J., Park, R. F. and Wellings, C. R. (2003). Molecular genetic variability of Australian isolates of five cereal rust pathogens. *Mycological Research* 107: 545–56.

Kilian, A., Chen, J., Han, F., Steffenson, B. and Kleinhofs, A. (1997). Towards map-based cloning of the barley stem rust resistance genes Rpg1 and rpg4 using rice as an intergenomic cloning vehicle. *Plant Molecular Biology* 35: 187–95.

Krattinger, S. G., Lagudah, E. S., Spielmeyer, W., Singh, R. P., Huerta-Espino, J., McFadden, H., Bossolini, E., Selter, L. L. and Keller, B (2009). A putative ABC transporter confers durable resistance to multiple fungal pathogens in wheat. *Science* 323: 1360–3.

Liu, W., Frick, M., Huel, R., Nykforuk, C. L., Wang, X., Gaudet, D. A., Eudes, F., Conner, R. L., Kuzyk, A., Chen, Q., Kang, Z. and Laroche, A. (2014). The stripe rust resistance gene Yr10 encodes an evolutionary-conserved and unique CC–NBS–LRR sequence in wheat. *Molecular Plant* 7: 1740–55.

Ma, H. and Singh, R. P. (1996). Contribution of adult plant resistance gene Yr18 in protecting wheat from yellow rust. *Plant Disease* 80: 66–9.

Mago, R., Zhang, P., Vautrin, S., Šimková, H., Bansal, U., Luo, M. C., Rouse, M., Karaoglu, H., Periyannan, S., Kolmer, J., Jin, Y., Ayliffe, M. A., Bariana, H., Park, R. F., McIntosh, R., Doležel, J., Bergès, H., Spielmeyer, W., Lagudah, E. S., Ellis, J. G. and Dodds, P. N. (2015). The wheat Sr50 gene reveals rich diversity at a cereal disease resistance locus. *Nature Plants* 1: 15186. doi:10.1038/NPLANTS.2015.186

McIntosh, R. A., Wellings, C. R. and Park, R. F. (1995). *Wheat Rusts: An Atlas of Resistance Genes.* CSIRO Publications, Melbourne, 200pp.

Milus, E. A., Kristensen, K. and Hovmoller, M. S. (2009). Evidence for increased aggressiveness in a recent widespread strain of *Puccinia striiformis* f. sp. *tritici* causing stripe rust of wheat. *Phytopathology* 99: 89–94.

Moore, J. W., Herrera-Foessel, S., Lan, C., Schnippenkoetter, W., Ayliffe, M., Huerta-Espino, J., Lillemo, M., Viccars, L., Milne, R., Periyannan, S., Kong, X., Spielmeyer, W., Talbot, M., Bariana, H., Patrick, J. W., Dodds, P., Singh, R. and Lagudah, E. (2015). A recently evolved hexose transporter variant confers resistance to multiple pathogens in wheat. *Nature Genetics* 47: 1494–8.

Murray, G. M. and Brennan, J. P. (2009a). *The Current and Potential Costs of Diseases of Barley in Australia.* GRDC, Canberra, p. 60.

Murray, G. M. and Brennan, J. P. (2009b). *The Current and Potential Costs of Diseases of Wheat in Australia.* GRDC, Canberra, p. 70.

Niks, R. E., Walther, U., Jaiser, H., Martlnez, F., Rubiales, D., Andersen, O., Flath, K., Gymer, P., Heinrichs, F. and Jonsson, R. (2000). Resistance against barley leaf rust (*Puccinia hordei*) in West-European spring barley germplasm. *Agronomie* 20: 769–82.

Niu, Z., Puri, K. D., Chao, S., Jin, Y., Sun, Y., Steffenson, B. J., Maan, S. S., Xu, S. S. and Zhong, S. (2014). Genetic analysis and molecular mapping of crown rust resistance in common wheat. *Theoretical and Applied Genetics* 127: 609–19.

Nover, I. and Scholz, F. (1969). Genetische Untersuchungen zur Resistenz der Gerste gegen Gelbrost (*Puccinia striiformis* West.). *Theoretical and Applied Genetics* 39: 150–5.

Park, R. F. (2003). Pathogenic specialization and pathotype distribution of *Puccinia hordei* in Australia, 1992 to 2001. *Plant Disease* 87: 1311–16.

Park, R. F. (2007). Stem rust of wheat in Australia. *Australian Journal of Agricultural Research* 58: 558–66.

Park, R. F. (2015). Long term surveys of pathogen populations underpin sustained control of the rust diseases of wheat in Australia. *Journal of the Royal Society of New South Wales* 148: 15–27.

Park, R. F. and Felsenstein, F. G. (1998). Physiological specialization and pathotype distribution of *Puccinia recondita* in western Europe, 1995. *Plant Pathology* 47: 157–64.

Park, R. F. and Karakousis, A. (2002). Characterisation and mapping of gene Rph19 conferring resistance to *Puccinia hordei* in the cultivar Reka 1 and several Australian barleys. *Plant Breeding* 121: 232–6.

Park, R. F. and Rees, R. G. (1989). Expression of adult plant resistance and its effect on the development of *Puccinia striiformis* f. sp. *tritici* in some Australian wheat cultivars. *Plant Pathology* 38: 200–8.

Park, R. F., Jahoor, A. and Felsenstein, F. G. (2000). Genetic variation in European populations of *Puccinia recondita* f. sp. *tritici* using pathogenicity and molecular markers. *Journal of Phytopathology* 148: 169–80.

Park, R. F., Mohler, V. M., Nazari, K. and Singh, D. (2014). Characterisation and mapping of gene Lr73 conferring seedling resistance to *Puccinia triticina* in common wheat. *Theoretical and Applied Genetics* 127: 2042–9.

Park, R. F., Golegaonkar, P. G., Derevnina, L., Sandhu, K. S., Karaoglu, H., Elmansour, H. M., Dracatos, P. M. and Singh, D (2015). Leaf rust of cultivated barley: Pathology and control. *Annual Review of Phytopathology* 53: 565–89.

Parlevliet, J. E. and Ommeren, A. (1975). Partial resistance of barley to leaf rust, *Puccinia hordei*. II. Relationship between field trials, micro plot tests and latent period. *Euphytica* 24: 293–303.

Patterson, F. L., Shands, R. G. and Dickson, J. G. (1957). Temperature and seasonal effects on seedling reactions of barley varieties to three races of *Puccinia graminis* f. sp. *tritici*. *Phytopathology* 47: 395–402.

Periyannan, S., Moore, J., Ayliffe, M., Bansal, U., Wang, X., Huang, L., Deal, K., Luo, M., Kong, X., Bariana, H., Mago, R., McIntosh, R., Dodds, P., Dvorak, J. and Lagudah, E. (2013). The gene Sr33, an ortholog of barley Mla genes, encodes resistance to wheat stem rust race Ug99. *Science* 341: 786–8.

Powers, L. and Hines, L. (1933). Inheritance of reaction to stem rust and barbing awns in barley crosses. *Journal of Agricultural Research* 46: 1121–9.

Pretorius, Z. A., Singh, R. P., Wagoire, W. W. and Payne, T. S. (2000). Detection of virulence to wheat stem rust resistance gene Sr31 in *Puccinia graminis* f. sp. *tritici* in Uganda. *Plant Disease* 84: 203.

Pretorius, Z. A., Bender, C. M., Visser, B. and Terefe, T. (2010). First report of a *Puccinia graminis* f. sp. *tritici* race virulent to the Sr24 and Sr31 wheat stem rust resistance genes in South Africa. *Plant Disease* 94: 1163.

Roelfs, A. P. (1985). Wheat and rye stem rust. In: *The Cereal Rusts, Volume I. Diseases, Distribution, Epidemiology and Control* (A. P. Roelfs and W. R. Bushnell (Eds)). Academic Press Inc., Orlando, FL, pp. 3–37.

Saari, E. E. and Prescott, J. M. (1985). World distribution in relation to economic losses. In: *The Cereal Rusts, Volume II. Diseases, Distribution, Epidemiology and Control* (A. P. Roelfs and W. R. Bushnell (Eds)). Academic Press, Orlando, FL, pp. 259–98.

Saintenac, C., Zhang, W., Salcedo, A., Rouse, M. N., Trick, H. N., Akhunov, E. and Dubcovsky, J. (2013). Identification of wheat gene Sr35 that confers resistance to Ug99 stem rust race group. *Science* 341: 783–6.

Sandhu, K. S., Forrest, K. L., Kong, S., Bansal, U. K., Singh, D., Hayden, M. J. and Park, R. F. (2012). Inheritance and molecular mapping of a gene conferring seedling resistance against *Puccinia hordei* in the barley cultivar Ricardo. *Theoretical and Applied Genetics* 125: 1403–11.

Semeane, Y., Hundie, B., Woldeab, G. and Tadesse, D. (1996). Disease survey and loss assessment studies on barley. In: *Barley Research in Ethiopia: Past Work and Future Prospects* (H. Gebre and J. van Leur (Eds)). IAR, Addis Ababa, Ethiopia, pp. 105–15

Shands, R. G. (1939). Chevron, a barley variety resistant to stem rust and other diseases. *Phytopathology* 29: 209–11.

Singh, R. P. and McIntosh, R. A. (1984). Complementary genes for reaction to *Puccinia recondita* I. Genetic and linkage studies. *Canadian Journal Genetics and Cytology* 26: 723–35.

Singh, D., Park, R. F. and McIntosh, R. A. (1999). Genetic relationship between the adult plant resistance gene Lr12 and the complementary gene Lr31 for seedling resistance to leaf rust in common wheat. *Plant Pathology* 48: 567–73.

Singh, R. P., Huerta-Espino, J., Pfeiffer, W. and Figueroa-Lopez, P. (2004). Occurrence and impact of a new leaf rust race on durum wheat in northwestern Mexico from 2001 to 2003. *Plant Disease* 88: 703–8.

Singh, R. P., Hodson, D. P., Huerta-Espino, J., Jin, Y., Bhavani, S., Njau, P., Herrera-Foessel, S., Singh, P. K., Singh, S. and Govindan, V. (2011). The emergence of Ug99 races of the stem rust fungus is a threat to world wheat production. *Annual Review of Phytopathology* 49: 465–81.

Singh, D., Dracatos, P., Derevnina, L., Zhou, M. and Park, R. F. (2015a). Rph23: A new designated additive adult plant resistance gene to leaf rust in barley on chromosome 7H. *Plant Breeding* 134: 62–9.

Singh, R. P., Hodson, D. P., Jin, Y., Lagudah, E. S., Ayliffe, M. A., Bhavani, S., Rouse, M. N., Pretorius, Z. A., Szabo, L. J., Huerta-Espino, J., Basnet, B. R., Lan, C. and Hovmoller, M. S. (2015b). Emergence and spread of new races of wheat stem rust fungus: Continued threat to food security and prospects of genetic control. *Phytopathology* 105: 872–84.

Singh, D., Dracatos, P. M., Loughman, R. and Park, R. F. (2017). Genetic mapping of resistance to *Puccinia hordei* in three barley doubled-haploid populations. *Euphytica* 213: 16. doi:10.1007/s10681-016-1799-7.

Steffenson, B. J. (1992). Analysis of durable resistance to stem rust in barley. *Euphytica* 63: 153–67.

Steffenson, B. J., Wilcoxson, R. D. and Roelfs, A. P. (1984). Inheritance of resistance to *Puccinia graminis* f. sp. secalis in barley. *Plant Disease* 68: 762–3.

Steuernagel, B., Periyannan, S. K., Hernandez-Pinzon, I., Witek, K., Rouse, M. N., Yu, G., Hatta, A., Ayliffe, M., Bariana, H., Jones, J. D. G., Lagudah, E. S. and Wulff, B. H. (2016). Rapid cloning of disease-resistance genes in plants using mutagenesis and sequence capture. *Nature Biotechnology* 34: 652–5.

Stubbs, R. W. (1985). Stripe rust, pp. 61–101. In: *The Cereal Rusts, Volume II. Diseases, Distribution, Epidemiology, and Control* (A. P. Roelfs and W. R. Bushnell (Eds)). Academic Press, Orlando, FL, 606pp.

Sun, Y., Steffenson, B. J. and Jin, Y. (1996). Genetics of resistance to *Puccinia graminis* f. sp. *secalis* in barley line Q21861. *Phytopathology* 86: 1299–302.

Talhinas, P., Ramos, A. P., Tavaeres, D., Tavares, S. and Loureiro, J. (2015). Genome size estimates for six rust (Pucciniales) species. *Revista de Ciencias Agrarias* 38: 176–83

Tan, B. H. (1978). Verifying the genetic relationships between three leaf rust resistance genes in barley. *Euphytica* 27: 317–23.

Thomas, W. T. B., Powell, W., Waugh, R., Chalmers, K. J., Barua, U. M., Jack, P., Lea, V., Forster, B. P., Swanston, J. S. and Ellis, R. P. (1995). Detection of quantitative trait loci for agronomic, yield, grain and disease characters in spring barley (*Hordeum vulgare* L.). *Theoretical and Applied Genetics* 91: 1037–47.

Toojinda, T., Baird, E., Booth, A., Broers, L., Hayes, P., Powell, W., Thomas, W., Vivar, H. and Young, G. (1998). Introgression of quantitative trait loci (QTLs) determining stripe rust resistance in barley: An example of marker-assisted line development. *Theoretical and Applied Genetics* 96: 123–31.

Upadhyaya, N. M., Garnica, D. P., Karaoglu, H., Sperschneider, J., Nemri, A., Xu, B., Mago, R., Cuomo, C., Rathjen, J. P., Park, R. F., Ellis, J. G. and Dodds, P. N. (2015). Comparative genomics of Australian isolates of *Puccinia graminis* f. sp. *tritici* reveals extensive polymorphism in candidate effector genes. *Frontiers in Plant Science* 5: Article 759. doi:10.3389/fpls.2014.00759.

van Niekerk, B. D., Pretorius, Z. A. and Boshoff, W. H. P. (2001). Occurrence and pathogenicity of *Puccinia hordei* on barley in South Africa. *Plant Disease* 85: 713–17.

Watson, I. A. (1970). Changes in virulence and population shifts in plant pathogens. *Annual Review of Phytopathology* 8: 209–30.

Watson, I. A. and Butler, F. C. (1984). Wheat rust control in Australia. In: *National Conferences and Other Initiatives and Developments*. The University of Sydney, Sydney, NSW.

Wellings, C. R. (2007). *Puccinia striiformis* in Australia: A review of the incursion, evolution, and adaptation of stripe rust in the period 1979–2006. *Australian Journal of Agricultural Research* 58: 567–75.

Wellings, C. R., Read, B. J. and Moody, D. B. (2000a). Stripe rust affecting barley in Australia – current and potential threats. In: *Proceedings of the 8th International Barley Genetics Symposium* (S. Logue and L. Malcolm (Eds)). Grains Research and Development Corporation (Australia), Adelaide, Australia, pp. 197–9.

Wellings, C. R., Burdon, J. J., McIntosh, R. A., Wallwork, H., Raman, H. and Murray, G. M. (2000b). A new variant of *Puccinia striiformis* causing stripe rust on barley and wild Hordeum species in Australia. *Plant Pathology* 49: 803.

Wettstein-Knowles, P. V. (1992). Cloned and mapped genes: Current status. In: *Barley: Genetics, Biochemistry, Molecular Biology and Biotechnology* (P. R. Shewry (Ed.)). CAB International, Wallingford, UK, pp. 73–98.

Zhao, J., Wang, L., Wang, Z. Y., Chen, X. M., Zhang, H. C., Yao, J., Zhan, G., Chen, W., Huang, L. and Kang, Z. (2013). Identification of eighteen Berberis species as alternate hosts of *Puccinia striiformis* f. sp. *tritici* and virulence variation in the pathogen isolates from natural infection of barberry plants in China. *Phytopathology* 103: 935–40.

Zhao, J., Wang, M., Chen, X. and Kang, Z. (2016). Epidemiology and pathogen variation of cereal rusts. *Annual Review of Phytopathology* 54: 207–28.

Zheng, W., Huang, L., Huang, J., Wang, X., Chen, X., Zhao, J., Guo, J., Zhuang, H., Qiu, C., Liu, J., Liu, H., Huang, X., Pei, G., Zhan, G., Tang, C., Cheng, Y., Liu, M., Zhang, J., Zhao, Z., Zhang, S., Han, Q., Han, D., Zhang, H., Zhao, J., Gao, X., Wang, J., Ni, P., Dong, W., Yang, L., Yang, H., Xu, J.-R., Zhang, G. and Kang, Z. 2013. High genome heterozygosity and endemic genetic recombination in the wheat stripe rust fungus. *Nature Communications* 4: 2673.

Zhu, Y., Chen, H., Fan, J., Wang, Y., Li, Y., Chen, J., Fan, J. X., Yang, S., Hu, L., Leung, H., Mew, T. W., Teng, P. S., Wang, Z. and Mundt, C. C. (2000). Genetic diversity and disease control in rice. Nature 406: 718–22.

Ziems, L. A., Hickey, L. T., Platz, G. J., Franckowiak, J. D., Dracatos, P. M., Singh, D. and Park, R. F. (2017). Characterization of Rph24: A gene conferring adult plant resistance to *Puccinia hordei* in barley. *Phytopathology* 107: 834–41.

Chapter 2

Fusarium diseases: biology and management perspectives

Edward C. Rojas, Hans J. L. Jørgensen, Birgit Jensen and David B. Collinge, University of Copenhagen, Denmark

1 Introduction

Wheat and barley are crucial crops for world food supply and represent two of the most important agricultural commodities in temperate areas. World wheat production has averaged 632 million metric tonnes annually in the last 25 years, while barley production has remained constant at around 142 million tonnes (FAOSTAT). Fungal diseases directly affect crop productivity by reducing yield, but also affect the quality of cereals. A major problem is caused by a number of fungal species in the genus *Fusarium*. Fusarium head blight (FHB), also known as 'head scab' or 'wheat scab', and Fusarium crown rot (FCR), also known as 'seedling blight' or 'Fusarium root rot', are two diseases caused by *Fusarium* species in small-grain cereals. FHB is a late disease that occurs during flowering and kernel development, whereas FCR causes damping off following infection of the leaf sheaths, stem base and roots, mainly of seedlings and in early stage plants. Yield losses vary widely,

http://dx.doi.org/10.19103/AS.2018.0039.02

but can reach up to 70%. However, the most important aspect of *Fusarium* diseases in small-grain cereals is the production of toxic specialized (or secondary) metabolites like trichothecenes, fumonisins and oestrogenic metabolites by the causal fungi. These are harmful toxins for livestock and humans: consumption poses an important health risk, especially in the long term.

The main strategy for the control of *Fusarium*-caused diseases in cereals is currently based on the premise of reducing inoculum to prevent initial infection. Traditionally, these measures are accompanied with the use of fungicides that reduce the rate of infection and result in lower mycotoxin accumulation. However, these actions do not guarantee consistent control, particularly against such an unpredictable disease as FHB. Simultaneously, agricultural practices such as reduced tillage and continuous cereal rotations that favour the disease are under increasing focus (Mangalassery et al. 2014; Gan et al. 2015). Likewise, an increase in average temperatures and more stochastic rainfall pattern caused by climate change would affect pathogen populations and could change the geographical patterns of disease epidemiology (Parikka et al. 2012).

Taken together, these factors could lead to major FHB outbreaks like those experienced early in the twentieth century. Moreover, overuse of chemical control will inevitably drive to the appearance of fungicide-resistant *Fusarium* strains. All these challenges require plant scientists and especially agronomists to update their knowledge about the disease and to rethink their management strategies. Understanding the biological mechanisms that govern the interaction between pathogens and plants can provide clues of the areas where this can be improved. In this chapter, we attempt to compile the recent findings on the *Fusarium*–cereal interaction, mode of infection and plant responses and analyse their implications in the development of complementary management strategies in the frame of integrated disease management.

2 *Fusarium* epidemiology and distribution

Fungal geneticists place the origin of the genus *Fusarium* around 91.3 million years ago in the middle Cretaceous as mainly wood saprophytes (O'Donnell et al. 2013). The genus saw a later explosion in diversity during the Miocene, coinciding with an increase in angiosperm diversity, thus explaining their strong association with plants (Watanabe et al. 2011). While *Fusarium graminearum* (teleomorph *Gibberella zeae*) has been reported as the predominant species in wheat and barley, other species play important roles in specific regions. These include: *F. culmorum*, *F. avenaceum*, *F. pseudograminearum*, *F. triticum*, *F. poae* and *F. asiaticum*, among others (Pasquali et al. 2016). Some of the most related species are grouped under the name of '*Fusarium graminearum* species complex' (FGSC) or *Fusarium graminearum sensu lato*. This complex includes the taxa: *F. acaciae-mearnsii*, *F. aethiopicum*, *F. asiaticum*, *F. austroamericanum*, *F. boothii*, *F. brasilicum*, *F. cortaderiae*, *F. gerlachii*, *F. graminearum sensu stricto*, *F. louisianense*, *F. meridionale*, *F. mesoamericanum*, *F. nepalense*, *F. ussurianum* and *F. vorosii* (Aoki et al. 2012) although not all are universally recognized. The geographical distribution of FGSC has shown that several species can co-exist (Nielsen et al. 2011). FGSC populations can have separate evolutionary dynamics in relation to different hosts and should be considered as a group of relatively independent populations that built a meta-population globally (van der Lee et al. 2015).

These variations in population dynamics are mainly driven by niche-competition with other pathogens, disease management practices, weed control and canopy humidity (Xu and Nicholson 2009) and by climatic conditions such as temperature, rainfall patterns and crop rotations (van der Lee et al. 2015). For example, it is often reported that *F. graminearum* is predominant in warm areas, whereas *F. culmorum* and *F. avenaceum* are more abundant in colder areas. This is partially true, but some of these patterns have been shifting in the last years with the introduction of maize in crop rotations (especially in Europe), the great adaptability of the *Fusarium* spp. and rising temperatures due to climate change (Kim et al. 2016). Nonetheless, some studies have shown that long-term rotations reduce disease levels, but do not affect *Fusarium* population composition infecting stem bases (FCR), implying a more resilient population structure (Tillmann et al. 2017). Recently, several mycotoxin-producing *F. graminearum* and other species were isolated from symptomless grasses in North American prairies (Lofgren et al. 2017). This is evidence of lifestyle plasticity of *Fusarium* spp., which colonizes most grasses as endophytes, but can be virulent in cereal crops. In general, the large intrinsic genetic variation of *F. graminearum sensu lato* should be taken into account when assessing FHB epidemiology since *Fusarium* populations are shaped by environmental factors as well as host availability.

3 Disease cycle and infection

The FHB and FCR disease cycles (Fig. 1) commence when rising temperatures in early spring initiate the production of saprophytic hyphae in crop debris. Under warm and moist conditions, sexual ascospores or asexual conidia are produced. Although some *Fusarium* spp. have a known sexual stage, many pathogenic species, including *F. cerealis*, *F. culmorum*, *F. equiseti*, *F. poae* and *F. sporotrichioides*, have no known sexual stage (Kerényi et al. 2004). Indeed, asexual spores are the main cause of infections in most cultivated areas. Three kinds of asexual spores are produced depending on the species: macroconidia are produced in sporodochia, microconidia are produced on conidiophores and chlamydospores are produced inside hyphae (Dweba et al. 2017). Furthermore, sexual ascospores, produced in perithecia, are also able to produce infection and their role in generating genetic variation is key in the pathogenicity of *Fusarium* spp. (Vanheule et al. 2017).

Splashing is the main means of initial infection for FHB (Osborne and Stein 2007). After landing on floret tissue, spores can colonize the external surface of the glumes or infect directly through stomata, exposed anthers and lemma-palea openings (Champeil et al. 2004). The fungus forms specialized structures such as compound appressoria and infection cushions to penetrate the floral bracts and the ovary/grain tissues (Boenisch and Schäfer 2011). Infection is higher under conditions of high humidity and temperature. Initially, fungal hyphae grow in the apoplast without causing any visible symptoms. They reach the rachis and spread to neighbouring spikelets in both basipetal and acropetal directions, but not through the vascular tissue (Brown et al. 2010). As the disease advances, colonized tissue dies and symptoms appear (Brown et al. 2010). Recently, it was shown that barley trichomes play a role in trapping conidia and offering infection sites, suggesting a potential resistance mechanism (Imboden et al. 2017). In wheat, the fungus colonizes the spikelets, producing brown lesions on the glumes that quickly bleach the spikelets,

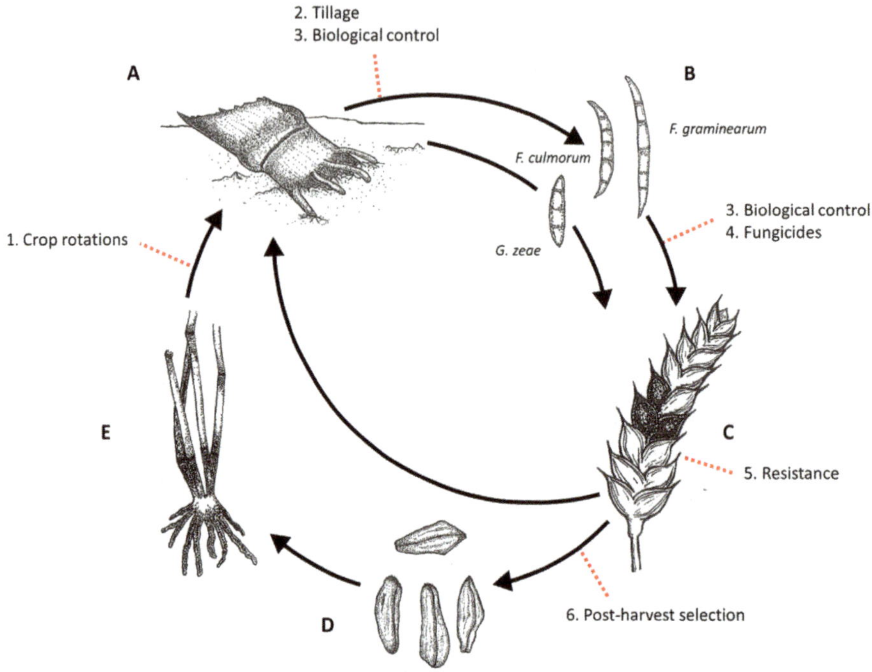

Figure 1 Disease cycle: A. Saprophytic phase on crop residues. B. Production of asexual macroconidia (*F. graminearum* and *F. culmorum*) or sexual ascospores (*G. zeae*). C. Infection of flowering spikelets and symptom appearance. D. Fusarium damaged kernels with mycotoxin contamination. E. Infected but visually acceptable grains can be used as seeds and may produce Fusarium crown rot on seedlings. Additional dotted lines represent points during the disease cycle where management strategies are able to disrupt the cycle: 1. Crop rotations are the best strategy for inoculum reduction. 2. Tillage can bury crop debris and reduce spore production. 3. Biocontrol agents on crop residues pre-sowing, applied to the head at anthesis or coated onto the seeds to prevent Fusarium crown rot are a potential control strategy. 4. Fungicide use at mid-flowering can reduce infection rates and severity, as well as mycotoxin accumulation. 5. Genetically resistant lines reduce initial infection, disease spread or mycotoxin accumulation. 6. Post-harvest grain selection can be performed adjusting combine harvester parameters or in storage facilities.

whereas in barley, these lesions can be dark or brownish which make real severity difficult to estimate (Janssen et al. 2018). In later stages of colonization, the spikelets turn pink or orange as the pathogen produces mycelia and synnema where conidia are produced over the necrotized tissue.

These spores and conidia can overwinter on crop residues, especially on tissues that do not degrade easily. Infected grains thus act as a disease vector when sown. These seeds germinate and may develop FCR. The infection in the crown area generates brown discoloured lesions on the coleoptiles and the roots which can turn black as they progress. In some cases, a pink/orange discoloration appears over the lower nodes due to production of spores (Scherm et al. 2013). This shows the great relevance of seed selection, as well as the use of agricultural practices in breaking the disease cycle. Likewise, it displays the crucial role of rainfall regimes as a driver of initial infection as well as a climatic variable.

4 Host–pathogen interaction

The interaction between *Fusarium* species and cereals is one of the most studied and important pathosystems in plant pathology (Dean et al. 2012). Full transcriptome analysis in the interactions between *F. graminearum* with cereals has been reviewed in depth recently by Kazan and Gardiner (2017). The fungal infection dynamics appear to be somewhat similar for both FHB and FCR. The fungus displays two phases: a biotrophic symptomless phase in which the fungus colonizes the intercellular tissue and a necrotrophic, symptomatic phase where the fungus degrades the infected tissue (Brown et al. 2017). However, its proven endophytic lifestyle and the latest findings at molecular level suggest a more complex interaction which raises questions about its ecology (Selosse et al. 2018).

Most studies have used *F. graminearum* as a model system for the interaction. It has been shown that during the first phase, fungal metabolism is entirely devoted to growth and plant response suppression. Several fungal proteins are secreted during this early phase (Yang et al. 2012). Fungal genes involved in transport, carbohydrate metabolism and detoxification are highly expressed during infection (Lysøe et al. 2011). Similarly, the production of extracellular siderophores such as TAFC (triacetylfusarinine C) and malonichrome has been shown to be activated during the first phase. These chelating molecules appear to play a role in coping with oxidative stress related to plant responses (Oide et al. 2014).

Likewise, the activation of genes involved in deoxynivalenol (DON) biosynthesis such as *TRI* enzymatic genes have been shown to increase during early stages. Interestingly, the production of DON has been shown to be activated by plant signals. Several polyamine compounds that are produced by the plant under infection activate DON biosynthesis in *F. graminearum* (Gardiner et al. 2009). Remarkably, the fungus appears to be able to hijack the plant metabolism to enhance DON production by activating the production of putrescine, a DON stimulator (Gardiner et al. 2010). Recently, it was shown that this mechanism is key for the sustenance of the biotrophic phase, as DON inhibits plant responses by binding to the small ribosomal units preventing protein translation (Brown et al. 2017). Many of the genes involved in DON production have been studied using reverse genetics, confirming the importance of genes like *Trichodiene synthase 5* (*Tri5*) in infection (Cuzick et al. 2008) and transcription factors such as *Tri6* and *Tri10* in pathogenicity (Jiang et al. 2016; Seong et al. 2009). Although important in enhancing infection, DON production is not vital for infection; either in heads (Jansen et al. 2005) or in seedlings (Powell et al. 2017).

During the necrotrophic phase, cell-degrading metabolism is primarily activated. Up to 50 Carbohydrate-Activated Enzymes (CAZymes) were found to be expressed during the interaction, but significantly increased during the necrotic phase (Zhang et al. 2012). These enzymes have been shown to be activated during starvation and saprophytic lifestyle. This suggests that the shift of infection mechanisms may be regulated by nutrient deprivation during the biotrophic phase (Brown et al. 2017). Indeed, autophagy mechanisms used to maintain homeostasis in fungal cells and recycle nutrients were shown to be necessary for growth and pathogenicity (Lv et al. 2017). Furthermore, Brown et al. (2017) found that the expression of several effector genes were upregulated during infection and interestingly, this secreted protein cocktail appears to be specific for each phase and their location in the genome aligns with high recombination regions.

To put it briefly, the pathogen can regulate its pathogenicity during the biotrophic to necrotrophic transition and this appears to be modulated by host signals and requires a tight transcriptional reprogramming in the fungus. The regulation involves plant response suppression mechanisms, stress-tolerance metabolism, effectors, polysaccharides degradation enzymes and mycotoxin production (Brown et al. 2017).

Like other plant–pathogen interactions, *F. graminearum* infection has been associated with enhanced expression of genes involved in primary metabolism and defence reactions in the host. Chitinases, reactive oxygen species and peroxidases as well as pathogenesis-related proteins (PR proteins) were activated rapidly in wheat after infection (Ding et al. 2011; Khaledi et al. 2016). Similarly, transcription factors associated with pathogen interactions such as *WRKY* and *bZIPs* and protein kinases involved in signal transduction were also expressed during FHB (Erayman et al. 2015). This indicates that deployment of a rapid and strong activation of defence responses is associated with increased resistance.

When comparing highly susceptible cultivars with moderately resistant, transport and detoxification of virulence factors such as effectors, toxins and enzymes contribute to increased resistance. ABC transporters, glucosyltransferases, proteinases and protein inhibitors make part of a second layer of defence mechanisms that are differentially expressed in plants with different levels of resistance (Gottwald et al. 2012). These defence mechanisms are energetically expensive (Martinez-Medina et al. 2016). Primary metabolism and carbon cycle genes are among the most abundantly expressed in wheat after *Fusarium* infection (Erayman et al. 2015). These genes play a key role in metabolic reprogramming during defence activation and their regulation represents potential targets for new breeding programmes (Hulsmans et al. 2016).

Plant hormones have received relatively little attention in the study of *Fusarium*–cereal interactions and the findings are not straightforward to interpret. Early activation of the jasmonate (JA) pathway appears to be crucial in response to *Fusarium* infection and together with the salicylic acid (SA) pathway; it has been associated with defence responses in tolerant cultivars (Ding et al. 2011). In contrast, Li and Yen (2008) suggested that SA did not play a role in defence. Furthermore, ethylene metabolism has been shown to enhance FHB resistance, while other studies associate ethylene with higher susceptibility (Xiao et al. 2013). Similarly, gibberellic acid and abscisic acid were shown to have antagonistic roles in FHB resistance, as exogenous application decreased and increased FHB symptoms, respectively (Buhrow et al. 2016). In brief, the plant response to infection is active and it includes several chemical mechanisms in two basic layers of defence. The activation of these mechanisms depends strongly on complex and subtle plant-hormone regulation and on activation of primary metabolism reconfiguration.

5 Genetic resistance

The use of *Fusarium*-resistant genetic material would be the most efficient and environmentally safe control practice. However, effective resistance has remained elusive for both wheat and barley, and no race-specific resistance genes have been described. Thus, resistance to both FHB and FCR appears to be scattered among many genes across chromosomes (He et al. 2016). Moreover, these genes appear to be subject to strong genotype-by-environment interactions (He et al. 2016). More than 250 quantitative trait loci (QTLs) have been described to confer some level of resistance, but the vast majority remain to be validated in reverse genetics experiments (Jia et al. 2017). Five types of resistance are recognized: Type I: resistance to initial infection; Type II: resistance to infection spreading; Type III: resistance to kernel infection; Type IV: tolerance to reduced yield and quality losses and Type V: resistance to mycotoxin accumulation (Mesterhazy

1995; Wegulo et al. 2015). Similarly, cleistogamous (closed flowering) lines show less susceptibility to FHB infection in both wheat and barley (Kubo et al. 2010).

Some of the most important QTLs have been characterized for their mode of action. Most of them appear to confer Type II resistance and in some cases by combining several mechanisms. For example, QTLs *QFhb1* and *QFhb2* have been associated with increased defence metabolites from the phenylpropanoid pathway, as well as activation of the JA pathway and detoxification of DON to less toxic compounds (Kazan and Gardiner 2017). QTL *Qfhs.ifa5a* conferred Type I resistance by enhanced lipid transfer protein, although the mechanism is unclear (Schweiger et al. 2013).

Using both native and exotic genetic diversity sources, several breeding programmes have achieved certain levels of success. The Chinese wheat line Sumai 3 shows consistent FHB resistance and has been used extensively in breeding. Several QTLs combining both Type I and Type II resistance confer its resistance. Similarly, several QTLs associated with the activation of SA, jasmonic acid and ethylene have been shown to confer resistance in the cultivar Wangshubai (Jia et al. 2017). Recently, a pore-forming toxin-like protein (*PFT*), which encodes a chimeric lectin, was shown to be responsible for the QTL *QFhb1*-mediated resistance (Rawat et al. 2016). The biochemical mechanisms behind this QTL were thought to be due to DON inactivation (Niwa et al. 2014). However, *PFT* is not involved in DON detoxification. Instead, it has been suggested that it might act as an antifungal protein that increases membrane permeability in the pathogen (Rawat et al. 2016).

Efforts in Europe have described other interesting QTLs such as *Qfhs.ndsu-3BS*, which improved Type II resistance. In the USA, native resistance genes have been successfully fixed onto resistant *QFhb1*-donor cultivars (Eckard et al. 2015). In the case of FCR, three QTLs (*Qcrs.cpi-3B*, *Qcrs.cpi-5D* and *Qcrs.cpi-2D*) have been shown to be consistent in reducing severity up to 60% when introduced in susceptible populations using gene pyramiding (Zheng et al. 2017). So, although effective resistance is as yet unavailable, the results suggest great potential for gene pyramiding strategies in the future. A list of partially resistant cultivars has been summarized recently by Shah et al. (2017).

Introducing foreign genes via transgenesis to reduce FHB is possible (Collinge et al. 2016). The antimicrobial protein bovine lactoferrin was introduced in the susceptible wheat cultivar Bobwhite. This transgenic wheat showed 75% less FHB incidence when compared to untransformed cultivars (Han et al. 2012). Similarly, the gene *NPR1* from rye (*Secale cereale*), which regulates the activation of systemic acquired resistance (SAR), was introduced into highly susceptible cultivar Ningmai. Transgenic plants showed earlier and higher levels of expression of PR-genes and a significantly higher resistance than the wild type (Yu et al. 2017). Likewise, overexpression of the *Arabidopsis thaliana NPR1* gene in wheat was shown to reduce severity by activating the SA pathway (Makandar et al. 2012).

Although considerable progress has been made in developing partial disease resistance by conventional plant breeding, total resistance probably cannot be achieved by this route. Promising results have been obtained by transgenic approaches and we can predict that these will implemented in some regions in the not-too-distant future.

6 Mycotoxins

Fusarium fungi associated with cereal diseases produce mycotoxins. These are diverse, specialized metabolites including trichotecenes, especially DON, nivalenol (NIV), 3-acetyldeoxynivalenol (3-ADON), T-2 toxin, HT-2 toxin, as well as fumonisins and oestrogenic metabolites such as zearalenone (ZEA) (Lee and Ryu 2017). Some of them are toxins in animals and represent an important health risk when consumed. Specific *Fusarium* mycotoxins are associated with nausea, vomiting, diarrhoea, abdominal pain and fever and under constant exposure they have been linked to neurological disorders, immunosuppression and cancer (Antonissen et al. 2014). ZEA has a rather different effect; being an oestrogen homologue, it causes reproductive disorders in farm animals and hyperoestrogenic responses in humans (Hueza et al. 2014). Mycotoxin-contaminated grain as animal fodder can cause anorexia, reduce weight gain rates and cause reproductive disorders in swine and poultry (Döll and Dänicke 2011).

Among all *Fusarium* mycotoxins, trichothecenes such as DON and its acetylated derivatives have been widely studied. The biosynthesis of these compounds commences with the cyclization of farnesyl pyrophosphate, catalyzed by the enzyme trichodiene synthase (*Tri5*), then it is modified through oxygenations (*Tri1*) and esterifications (*Tri16*) into DON (Boenisch et al. 2017). Individual isolates of *Fusarium* spp. can have different trichothecene profiles: specific gene expression patterns in the DON biosynthesis pathway (*TRI* genes) were observed during wheat infection with two *F. graminearum* strains with distinct chemotypes (Amarasinghe and Fernando 2016). It has also been shown that mutations in *Tri13* and *Tri7* genes can generate changes in chemotypes as these genes are involved in the conversion from DON to NIV and the acetylation of NIV to the more toxic 4NIV (Lee et al. 2002). However, DON remains the most common mycotoxin found in cereals. This confirms that DON production offers a fitness advantage when infecting cultivated cereals (Xu and Nicholson 2009) and can thus also be considered to be an (albeit weak) phytotoxin or virulence factor. Interestingly, a comparative genome analysis showed a richer genetic pool for the production of these secondary metabolites than previously estimated, as well as a capacity for horizontal gene transfer and other evolutionary mechanisms of diversification within the genus *Fusarium* (Ma et al. 2013). The occurrence of different mycotoxin chemotypes in harvested grains is directly related to the composition of *Fusarium* spp. in the population and it follows the same dynamics (Lee and Ryu 2017; van der Lee et al. 2015).

Many countries have legal restrictions on the permitted levels of mycotoxins, especially DON and ZEA in cereal-based products. The maximum permitted levels in the European Union (EU) are, for example, 0.75 ppm for DON and 0.075 ppm for ZEA, for flour, bran and germ for human consumption. Permitted levels for unprocessed wheat and barley are 1.7 and 1.5 ppm, respectively, for DON and 0.1 ppm for ZEA (European Commission 2006). There is no current legislation for NIV in the EU as NIV levels are closely related to DON (Nielsen et al. 2014). Mycotoxin detection methods are divided in two categories: immunochemical-based methods that allow a fast and simple screening of large number of samples. Several kits based on ELISA assays are commercially available and are included in most food safety protocols (Anfossi et al. 2016). Secondly, chromatography-based methods: these allow higher resolution of the mycotoxin profile, but require specialized personal to perform and interpret the results, and are consequently more expensive. These methods are often used as validation test to complement ELISA kits or to perform in-depth mycotoxin analysis (Capriotti et al. 2014). Mycotoxins are thermostable and remain in cereal products after processing (Antonissen et al. 2014), but contaminated grain can be removed to some degree during harvest by changing the fan speed and shutter opening in the combine harvester to reduce mycotoxin content significantly (Salgado et al. 2015). To sum up, mycotoxin production by *Fusarium* fungi is a key component of their survival and pathogenicity. They represent a present-day health threat in food chains that receives constant survey in the world.

7 Yield and quality losses

Mycotoxin contamination ultimately defines the price and purpose of the harvest since contaminated grains that surpass the maximum levels must be destroyed, while contaminated grain that does not exceed the permitted levels is at best penalized on its price in the market. Although growers assume most of the economic cost, higher prices and low quality costs are shared by grain traders, mills and bakers in the production chain. This makes total economic impact difficult to estimate. Annual average costs of FHB in wheat and barley in the USA has been estimated to approximately USD$27 million, representing 3.7% of the whole annual value of these crops (Nganje et al. 2004). Similarly for FCR, annual costs were estimated to be around USD$100 in Australia in both crops (Murray and Brennan 2010).

FHB yield losses have been assessed in multiple field studies. The values vary widely between 10% and 50%, depending on the weather conditions and *Fusarium* populations (Kikot et al. 2011; Xu and Nicholson 2009). Likewise, it has been shown that agricultural practices such as crop rotation, tillage, sowing date, fungicide treatment, cultivar, fertilizer source, irrigation regimes and overall production system (conventional vs organic) have an influence on the final DON concentration in both wheat and barley (Bernhoft et al. 2012; Wenda-Piesik et al. 2017). In the case of FCR, yield reductions can reach up to 26% in Australia and 50% in the USA (Liu and Ogbonnaya 2015). The grain yield reduction is mainly due to reduction in plant stand, and losses in kernel weight and other yield parameters are also important (Moya-Elizondo and Jacobsen 2016; Smiley et al. 2005). Additionally, it has been shown that Fusarium disease index and per cent of *Fusarium*-damaged kernels are two good indicators of DON concentration. However, since environmental variables largely affect DON levels, Fusarium disease index should not be used solely as a mycotoxin estimator (Paul et al. 2006). Recently, it has been shown that FHB total yield losses can be predicted using a linear mixed-model regression analyses to estimate the relationship between Fusarium disease index and yield parameters (Salgado et al. 2015).

Quality reductions in wheat are also associated to its diminished processing qualities. Mycotoxins were shown to reduce flour yield and brightness and most importantly, baking performance (Siuda et al. 2010). In the case of barley, effects of severe FHB are seen in the malting process. Reduced germination as well as deficient brewing parameters (gushing) and undesirable flavours have been associated with high levels of infected grains and DON concentrations (Oliveira et al. 2012). Recently, fungal biomass was correlated to low malt extract quality for both *F. poae*- and *F. langsethiae*-infected barley grains (Nielsen et al. 2014). This suggests that different mycotoxin levels, as well as final quality parameters, are affected by changes in *Fusarium* populations. Therefore, in addition to its potential health risks, FHB represents economic losses to farmers by reducing final yield. It also affects the industrial sector by reducing commercial quality parameters.

8 Disease management

8.1 Agricultural practices

Disease management in wheat and barley is based on agricultural practices that reduce the initial inoculum pressure. Crop rotations with non-cereal crops, especially avoiding rotations with maize, are the basis of integrated management strategies, particularly in areas with soil-humidity limitations (Qiu et al. 2016). Stubble burning (outlawed in the EU for cereal crops) or soil turnover through ploughing can also be used. These tillage operations can reduce disease incidence and severity as well as mycotoxin levels significantly (Scala et al. 2016). However, they may also lead to losses in soil quality parameters such as structure, porosity and moisture (Montgomery 2007) and therefore heavy-duty tillage operations are sometimes discouraged in intensive systems (Townsend et al. 2016) although they remain effective in reducing inoculum. In addition, it has been shown that avoidance of mineral fertilizer and herbicides can reduce the levels of DON in kernels, as these input factors can modify canopy microclimate by increasing humidity (Bernhoft et al. 2012). A decrease in nitrogen fertilization has also been shown to increase *Fusarium* infection in barley which suggests control of nitrogen fertilization as a possible way to minimize FHB in barley (Yang et al. 2010).

Remarkably, it has been shown that long-term rotations reduce disease levels, but do not affect *Fusarium* population composition, causing FCR (Tillmann et al. 2017). Likewise, other agronomical variables have been shown to be related to FHB. Delaying optimal sowing date by 20–25 days increased Fusarium disease index under central European conditions in both susceptible and tolerant winter wheat cultivars, whereas increasing sowing density from 400 to 600 germinated seeds/m^2 did not show differences (Gorczyca et al. 2018). However, in cultivars that do not need vernalization, a delay in sowing dates to late autumn or early spring did not have an effect on Fusarium disease index or DON contamination (Wenda-Piesik et al. 2017).

Organic grown barley showed lower severity than conventional barley and wheat systems in Norway. However, yield was reduced around 30% (Bernhoft et al. 2012). Similarly, in Switzerland, organic systems showed lower mycotoxin accumulation levels in barley, although the number of experimental units was low (Schöneberg et al. 2016). To conclude, agricultural practices, especially crop rotations, are the first and best tool to control FHB in cereals. They have an effect on FHB levels in two ways: by reducing pathogen inoculum pressure from debris and by modifying microclimate within the plot to decrease infection.

8.2 Fungicides

Fungicide use is common in intensive cereal cropping systems. In the case of *Fusarium* diseases, demethylation inhibitors (DMIs) such as propiconazole, prothioconazole, tebuconazole or combinations of these are often used at the early or mid-anthesis stage (Freije and Wise 2015). They can reduce infection and mycotoxin contamination after a single application, but their efficacy is usually lower than 50% (Lehoczki-Krsjak et al. 2015; Paul et al. 2008). Other fungicide groups such as quinone inhibitors, succinate dehydrogenase inhibitors and DMI molecules have very low efficacy for controlling Fusarium, and indeed their use can stimulate mycotoxin production in the infected plant (Marques et al. 2017; Wegulo et al. 2015; Zhang et al. 2009).

Many argue that fungicide efficacy is low due to technical gaps in the application that prevent fungicides to act in the right time and place. Indeed, *Fusarium* control efficacy depends significantly on the right timing, spray coverage and weather conditions during application (Lehoczki-Krsjak et al. 2015). Moreover, it has been shown that optimal fungicide spray timing for FHB reduction and low DON accumulation might not be the same (Yoshida et al. 2012). In barley, optimal time of application has been estimated to be 3 days after anthesis (Janssen et al. 2018) and double application of metconazole did not increase control rates (Tateishi et al. 2014). Interestingly, aerial irrigation simulating rain at different time points after fungicide application did not lead to statistically significant increases in Fusarium disease index or mycotoxins levels (Andersen et al. 2014). On the other hand, some triazole-based fungicides applied as seed treatments have given up to 50% reduction in FCR severity, but their action is limited in time (Moya-Elizondo and Jacobsen 2016).

Excessive use of fungicides will inevitably favour the development of fungicide-resistant strains of *Fusarium*. Fungicide resistance is especially important in FHB due to the limited number of effective active ingredients (Lucas et al. 2015). Natural resistance to fungicides in wild *Fusarium* populations has been documented in China (Chen and Zhou 2009), Europe and the USA (Dubos et al. 2013). Recently, a *F. graminearum* isolate was reported to be significantly less sensitive to tebuconazole, while being highly aggressive and toxigenic (Spolti et al. 2014). Moreover, multiple genes encoding for a 14 α-demethylase (*CYP51*) have been described in *Fusarium* spp. (Yin et al. 2009). This enzyme has been associated with reduced sensitivity to azole fungicides since they are involved in the synthesis of ergosterol (Fan et al. 2013). Remarkably, some genes that are required for pathogenicity such as ABC transporters (detoxification of plant response molecules) were also upregulated during azole treatments and have been proposed as a factor contributing to fungicide resistance (Ammar et al. 2013). So, although fungicides are widely used for *Fusarium* control, their efficiency remains low. Additionally, due to the low number of effective fungicides against *Fusarium* spp., monitoring resistance emergence and understanding its mechanisms are of primary importance in reducing FHB losses.

8.3 Biological control

The use of biological control has experienced increased interest in recent years. The latent risk of fungicide resistance and the environmental concerns over the use of fungicides have become priorities for farmers and consumers. Biological control can replace or complement current measures in integrated disease management plans (Jensen et al. 2016). This approach is also attractive since it has the potential for use during grain development when the use of fungicides is ineffective, undesirable and directly prohibited. Use of biocontrol in *Fusarium*–cereal interactions has been studied extensively in recent years and reviewed by Legrand et al. (2017).

Several biological control agents have been described to provide good results for both FHB and FCR. Among the most interesting agents is *Clonostachys rosea*, due to its validated results in the field. *C. rosea* strain ACM941, when sprayed to the spikes during anthesis, can reduce disease severity and the number of *Fusarium*-damaged kernels to the same levels of tebuconazole-treated plants (Xue et al. 2014). Similarly, *C. rosea* IK726 has been shown to reduce FCR caused by *F. culmorum* in field-grown wheat and barley (Jensen et al. 2000) and to act synergistically with other biocontrol organisms against insects (Keyser et al. 2016). Furthermore, *C. rosea* IK726 can degrade the *Fusarium* mycotoxin ZEA, which partly explains its biocontrol capacity against FCR (Kosawang et al. 2014).

Similarly, treatment with the root endophytic fungi *Serendipta* (formerly *Piriformospora*) *indica* has been shown to reduce FCR symptoms under greenhouse conditions and to decrease FHB incidence and severity up to 70% in wheat when inoculated in the soil at the time of sowing (Rabiey and Shaw 2016; Rabiey et al. 2015). *S. indica* has been shown to protect roots from necrotization by reducing antioxidant enzyme activity (Harrach et al. 2013). Likewise, bacterial strains such as *Bacillus amyloliquefaciens* FLN13 and *Lactobacillus plantarum* SLG17 have reduced Fusarium disease index up to 50% when inoculated simultaneously at flowering (Baffoni et al. 2015). *B. megaterium* BM1 showed similar results in South America (Pan et al. 2015). Also, *Pseudomonas fluorescens* LY1-8 showed reduced FCR and FHB severity by half in field trials when sprayed at flowering (Wang et al. 2015). Biocontrol agents can also be applied to crop debris to reduce inoculum pressure. Interestingly, *Streptomyces* sp. RC 87B reduced both FHB severity and DON concentrations when used in a dual application on crop debris before sowing and to the head during anthesis (Palazzini et al. 2017).

Nevertheless, only a few patents have been registered (*C. rosea* ACM941 and *Bacillus* sp. TrigoCor). Likewise, only two products are available on the market; Polyversum® based on *Pythium oligandrum* with exclusive use in France (Legrand et al. 2017) and Cerall® based on *Pseudomonas chlororaphis* as a seed treatment against FCR. Despite promising initial results, most potential biological control agents sprayed to the head or coated onto the seeds lack consistency in the field, or have limited industrial properties.

9 Future trends

New genomic tools have allowed faster genotyping and made genetic marker generation more efficient. QTL mapping has moved forward at the same speed and now marker-assisted selection using single-nucleotide polymorphisms is the norm in breeding programmes. However, gene validation for hundreds of QTLs remains one of the challenges for future breeders (Jia et al. 2017). Using high-throughput technologies, transcriptome and proteome analysis can help to close this gap in the strategy termed association mapping (Collinge 2018). Similarly, the use of new gene editing techniques such as CRISPR/Cas9 could bring new tools to the table (Collinge 2018). Theoretically, breeders could aim to inactivate or modify genes in the host that are targeted by fungal effectors and therefore reduce infection or generate genetic variation by specific mutations (Jung et al. 2017).

The transgenic technology Host-Induced Gene Silencing (HIGS) has given promising results for control of *F. graminearum* in barley by silencing key fungal genes such as *CYP51* using the host molecular machinery (Koch et al. 2013). Furthermore, its non-transgenic derivative Spray-Induced Gene Silencing (SIGS) has shown that external application of these long noncoding dsRNA reduced fungal growth in directly sprayed tissue as well as non-sprayed detached leaves (Koch et al. 2016). These pathogen-tailored methods are highly specific and could represent a new window in plant protection and food safety (Majumdar et al. 2017). Regardless of current limitations on the commercial use of transgenic crops, several genes have been validated in their capacity to confer resistance and their results should not be discarded (Collinge et al. 2016). New sources of resistance could come from macroscopic phenotypic features such as floret trichomes, thicker cuticular wax to more fundamental features, such as focus on core genes that regulate primary metabolism and the production of polyamide compounds during pathogen attacks.

The 'omics' era has helped us to identify specific genes and proteins that are involved in the molecular cross-talk between pathogen and host. This can be used to fine-tune the mechanisms of action of new fungicides, given the low number of active ingredients that can be used to prevent *Fusarium* diseases. New fungicide molecules should aim to interrupt initial infection by attacking primary metabolism in the pathogen and to prevent DON production in order to restrict its role in plant response suppression. They should also have translaminar movement and reach the apoplast, since that is where *Fusarium* hyphae colonize primarily. However, development of fungicide molecules with new modes of action is rare and enormous investments are required together with a thorough understanding of the biology of the pathogens in order to achieve success in this field. Additionally, spray technical parameters such as nozzle type, drop size, angle of application, coadjutants and so forth must be optimized for existing fungicides, since their delivery needs to be optimized to achieve high efficacy.

The full deployment of biological control alternatives is currently limited by a low reliability in the field. The classical method of isolation-confrontation (dual-culture assays) is insufficient to provide information for selection of agents that are expected to replace chemical solutions. *In planta* assays are more laborious, but refine the screening process by recreating real disease scenarios. Additionally, these screening processes must include industrial-scale production parameters such as mass production, shelf life, formulation, environmental risks and so forth in their selection (Collinge et al. 2018). In order to reduce problems in low establishment, potential organisms should be isolated directly from the conditions where they are expected to be used. In the same way, assessment of their compatibility with fungicides would open a window for integrated pest management strategies for combination of modes of action and would promote low-dose fungicide use.

We should conceive the occurrence of FHB and FCR as an exception in the ecological dynamics between cereal and its microbiome (plant-associated microbial communities). This balance can be modified by the pathogen causing disease, but can also be modified by humans through agronomic practices (i.e. crop rotation, tillage, cultivar, fungicides, etc.). It should be possible to modify this equilibrium in our favour by altering the composition and structure of the microbial community, to favour beneficial organisms to the detriment of pathogens. This concept has been recurrent in the past few years as a new framework in biological control strategies (Busby et al. 2017). New biocontrol alternatives must aim to modify the microbiome composition in order to achieve greater field consistency.

Finally, without any doubt, crop rotation is the single most effective strategy in reducing cereal diseases caused *by Fusarium* spp. This and other agricultural practices are mentioned as the basis of integrated pest management strategies to prevent disease and have been recommended largely by the EU (Barzman et al. 2015) (Fig. 1). These potential advances will not solve the *Fusarium* problem, but rather enrich our set of tools to complement agricultural practices in order to achieve an integrated disease management. The right combination of tools is the one that provides the disease control levels that each market requires while minimizing the environmental impacts of production. This will permit cereal production systems to adapt to new world tendencies of sustainable intensification, where environmentally friendly and consumer-based management approaches allow revenues to farmers.

10 Conclusion

Wheat and barley production is fundamental for the world economy. Constant improvement in production in a world with higher environmental standards is the challenge of contemporary agriculture and food professionals. FHB is a worldwide problem and its devastating effects have been recognized and studied for more than a century. Here, we have reviewed the current knowledge of the mechanisms involved in the interaction, its dynamics and its management methods. Currently, control of *Fusarium* diseases in cereals is far from flawless. However, recent scientific findings suggest that control of *Fusarium* diseases in cereal crops must be built upon good agricultural practices, crop rotations being the most important, and three main complementary pillars: increased genetic resistance, novel fungicide molecules and consistent biocontrol applications. These measures will require significant amounts of work and scientific creativity in the coming years. However, they will allow growers to adapt to the new realities of agricultural production.

11 Acknowledgements

This project has received funding from the European Union's Horizon 2020 research and innovation programme under the Marie Skłodowska-Curie grant agreement No. 674964.

12 Where to look for further information

Some of the most fundamental knowledge about *Fusarium* diseases and *Fusarium* mycotoxins can be found in:

- Leonard, K., & Bushnell, William R. (2003). *Fusarium head blight of wheat and barley*. St. Paul, Minn: APS Press.
- Desjardins, A. (2006). *Fusarium mycotoxins, chemistry, genetics, and biology*. St. Paul, Minn: APS Press.

A complete review of the problem in Latin American countries has been recently published in:

- Alconada Magliano, T., & Chulze, Sofia Noemi. (2013). *Fusarium head blight in Latin America*. Dordrecht: Springer.

The U.S. Wheat and Barley Scab Initiative (USWBSI) aims to provide effective control measures against Fusarium head blight. Their website is an excellent source for lab and field protocols, fundamental information as well as news from the field:

- www.scabusa.org

An updated database for regulations around the world can be found at:

- http://www.mycotoxins.info

The Fusarium database facilitates the identification of species using nucleotide BLAST queries online:

- http://isolate.fusariumdb.org/

13 References

Amarasinghe, C. C. and Fernando, W. G. D., 2016. Comparative analysis of deoxynivalenol biosynthesis related gene expression among different chemotypes of *Fusarium graminearum* in spring wheat. *Frontiers in Microbiology*, 7, pp. 1–10.

Ammar, G. A., , Tryono, R., Döll, K., Karlovsky, P., Deising, H. B. and Wirsel, S. G. R., 2013. Identification of ABC transporter genes of *Fusarium graminearum* with roles in azole tolerance and/or virulence. *PLoS ONE*, 8(11), pp. 1–13.

Andersen, K. F., Morris, L., Derksen, R. C., Madden, L. V. and Paul, P. A., 2014. Rainfastness of prothioconazole + tebuconazole for Fusarium head blight and deoxynivalenol management in soft red winter wheat. *Plant Disease*, 98(10), pp. 1398–406. Available at: http://apsjournals.apsnet.org/doi/10.1094/PDIS-01-14-0092-RE.

Anfossi, L., Giovannoli, C. and Baggiani, C., 2016. Mycotoxin detection. *Current Opinion in Biotechnology*, 37, pp. 120–6. Available at: http://dx.doi.org/10.1016/j.copbio.2015.11.005.

Antonissen, G., Martel, A., Pasmans, F., Ducatelle, R., Verbrugghe, E., Vandenbroucke, V., Li, S. J., Haesebrouck, F., Van Immerseel, F. and Croubels, S., 2014. The impact of *Fusarium* mycotoxins on human and animal host susceptibility to infectious diseases. *Toxins*, 6(2), pp. 430–52.

Aoki, T., Ward, T. J., Kistler, H. C. and O'Donnell, K., 2012. Systematics, phylogeny and trichothecene mycotoxin potential of *Fusarium* head blight cereal pathogens. *Mycotoxins*, 62(2), pp. 91–102.

Baffoni, L., Gaggia, F., Dalanaj, N., Prodi, A., Nipoti, P., Pisi, A., Biavati, B. and Di Gioia, D., 2015. Microbial inoculants for the biocontrol of *Fusarium* spp. in durum wheat. *BMC Microbiology*, 15(1), pp. 8–10. Available at: http://dx.doi.org/10.1186/s12866-015-0573-7.

Barzman, M., Bàrberi, P., Birch, A. N. E., Boonekamp, P., Dachbrodt-Saaydeh, S., Graf, B., Hommel, B., Jensen, J. E., Kiss, J., Kudsk, P., Lamichhane, J. R., Messéan, A., Moonen, A.-C., Ratnadass, A., Ricci, P., Sarah, J.-L. and Sattin, M., 2015. Eight principles of integrated pest management. *Agronomy for Sustainable Development*, 35(4), pp. 1199–215.

Bernhoft, A., Torp, M., Clasen, P. E., Løes, A. K. and Kristoffersen, A. B., 2012. Influence of agronomic and climatic factors on *Fusarium* infestation and mycotoxin contamination of cereals in Norway. *Food Additives and Contaminants – Part A Chemistry, Analysis, Control, Exposure and Risk Assessment*, 29(7), pp. 1129–40.

Boenisch, M. J. and Schäfer, W., 2011. *Fusarium graminearum* forms mycotoxin producing infection structures on wheat. *BMC Plant Biology*, 11(1), p. 110. Available at: http://www.biomedcentral.com/1471-2229/11/110.

Boenisch, M. J., Broz, K. L., Purvine, S. O., Chrisler, W. B., Nicora, C. D., Connolly, L. R., Freitag, M., Baker, S. E. and Kistler, H. C., 2017. Structural reorganization of the fungal endoplasmic reticulum upon induction of mycotoxin biosynthesis. *Scientific Reports*, 7, pp. 1–13. Available at: http://dx.doi.org/10.1038/srep44296.

Brown, N. A., Urban, M., van de Meene, A. M. L. and Hammond-Kosack, K. E., 2010. The infection biology of *Fusarium graminearum*: Defining the pathways of spikelet to spikelet colonisation in wheat ears. *Fungal Biology*, 114(7), pp. 555–71. Available at: http://dx.doi.org/10.1016/j.funbio.2010.04.006.

Brown, N. A., Evans, J., Mead, A. and Hammond-Kosack, K. E., 2017. A spatial temporal analysis of the *Fusarium graminearum* transcriptome during symptomless and symptomatic wheat infection. *Molecular Plant Pathology*, 18(9), pp. 1295–312.

Buhrow, L. M., Cram, D., Tulpan, D., Foroud, N. A. and Loewen, M. C. 2016. Exogenous abscisic acid and gibberellic acid elicit opposing effects on *Fusarium graminearum* infection in wheat. *Phytopathology*, 106, pp. 986–96.

Busby, P. E., Soman, C., Wagner, M. R., Friesen, M. L., Kremer, J., Bennett, A., Morsy, M., Eisen, J. A., Leach, J. E. and Dangl, J. L., 2017. Research priorities for harnessing plant microbiomes in sustainable agriculture Research priorities for harnessing plant microbiomes in sustainable agriculture. *PLoS Biology*, 15(3), pp. 1–14. Available at: http://dx.doi.org/10.1371/journal.pbio.2001793.

Capriotti, A. L., Cavaliere, C., Foglia, P., Samperi, R., Stampachiacchiere, S., Ventura, S. and Laganà, A., 2014. Multiclass analysis of mycotoxins in biscuits by high performance liquid chromatography-tandem mass spectrometry. Comparison of different extraction procedures. *Journal of Chromatography A*, 1343, pp. 69–78. Available at: http://dx.doi.org/10.1016/j.chroma.2014.04.009.

Champeil, A., Doré, T. and Fourbet, J. F., 2004. Fusarium head blight : Epidemiological origin of the effects of cultural practices on head blight attacks and the production of mycotoxins by Fusarium in wheat grains. *Plant Science*, 166, pp. 1389–415.

Chen, Y. and Zhou, M.-G., 2009. Characterization of *Fusarium graminearum* isolates resistant to both carbendazim and a new fungicide JS399-19. *Phytopathology*, 99(4), pp. 441–6.

Collinge, D. B., 2018. Transgenic crops and beyond : How can biotechnology contribute to the sustainable control of plant diseases? *European Journal of Plant Pathology*. Available at: https://doi.org/10.1007/s10658-018-1439-2.

Collinge, D. B., Mullins, E., Jensen, B. and Jørgensen, H. J. L. 2016. The status and prospects for biotechnological approaches for attaining sustainable disease resistance. In: D. B. Collinge (Ed.), *Plant Pathogen Resistance Biotechnology*. Hoboken: Wiley-Blackwell, pp. 1–20.

Collinge, D. B., Jørgensen, H. J. L., Latz, M. A. C., Manzotti, A., Ntana, F., Rojas, E. C. and Jensen, B. (2018). Searching for novel fungal biological control agents for plant disease control among endophytes. In: T. R. Hodkinson, F. M. Doohan, M. Saunders and B. R. Murphy (Eds), *Endophytes: For a Growing World*. Cambridge: Cambridge University Press (in press).

Cuzick, A., Urban, M. and Hammond-Kosack, K., 2008. *Fusarium graminearum* gene deletion mutants map1 and tri5 reveal similarities and differences in the pathogenicity requirements to cause disease on Arabidopsis and wheat floral tissue. *New Phytologist*, 177(4), pp. 990–1000.

Dean, R., van Kan, J. A. L., Pretorius, Z. A., Hammond-Kosack, K. E., Di Pietro, A., Spanu, P. D., Rudd, J. J., Dickman, M. B., Kahmann, R., Ellis, J. and Foster, G. D., 2012. The top 10 fungal pathogens in molecular plant pathology. *Molecular Plant Pathology*, 13(4), pp. 414–30.

Ding, L., Xu, H. B., Yi, H. Y., Yang, L. M., Kong, Z. X., Zhang, L. X., Xue, S. L., Jia, H. Y. and Ma, Z. Q., 2011. Resistance to hemi-biotrophic *F. graminearum* infection is associated with coordinated and ordered expression of diverse defense signaling pathways. *PLoS ONE*, 6(4), p. e19008.

Döll, S. and Dänicke, S., 2011. The *Fusarium* toxins deoxynivalenol (DON) and zearalenone (ZON) in animal feeding. *Preventive Veterinary Medicine*, 102(2), pp. 132–45. Available at: http://dx.doi.org/10.1016/j.prevetmed.2011.04.008.

Dubos, T., Pasquali, M., Pogoda, F., Casanova, A., Hoffmann, L. and Beyer, M., 2013. Differences between the succinate dehydrogenase sequences of isopyrazam sensitive Zymoseptoria tritici and insensitive *Fusarium graminearum* strains. *Pesticide Biochemistry and Physiology*, 105(1), pp. 28–35. Available at: http://dx.doi.org/10.1016/j.pestbp.2012.11.004.

Dweba, C. C., Figlan, S., Shimelis, H. A., Motaung, T. E., Sydenham, S., Mwadzingeni, L. and Tsilo, T. J., 2017. Fusarium head blight of wheat: Pathogenesis and control strategies. *Crop Protection*, 91, pp. 114–22. Available at: http://dx.doi.org/10.1016/j.cropro.2016.10.002.

Eckard, J. T., Gonzalez-Hernandez, J. L., Caffe, M., Berzonsky, W., Bockus, W. W., Marais, G. F. and Baenziger, P. S., 2015. Native Fusarium head blight resistance from winter wheat cultivars 'Lyman', 'Overland', 'Ernie' and 'Freedom' mapped and pyramided onto 'Wesley'-*Fhb1* backgrounds. *Molecular Breeding*, 35(1), p. 6.

Erayman, M., Turktas, M., Akdogan, G., Gurkok, T., Inal, B., Ishakoglu, E., Ilhan, E. and Unver, T., 2015. Transcriptome analysis of wheat inoculated with *Fusarium graminearum*. *Frontiers in Plant Science*, 6, pp. 1–17. Available at: http://journal.frontiersin.org/Article/10.3389/fpls.2015.00867/abstract.

European Commission., 2006. Commission Regulation (EC) No. 1881/2006 of 19 December 2006 setting maximum levels for certain contaminants in foodstuffs. *Official Journal of the European Communities*, L364(1881), pp. 5–24. Available at: http://eur-lex.europa.eu/legal-content/EN/ALL/?uri=CELEX:32006R1881 (accessed 14 March 2018).

Fan, J., Urban, M., Parker, J. E., Brewer, H. C., Kelly, S. L., Hammond-Kosack, K. E., Fraaije, B. A., Liu, X. L. and Cools, H. J., 2013. Characterization of the sterol 14α-demethylases of *Fusarium graminearum* identifies a novel genus-specific *CYP51* function. *New Phytologist*, 198(3), pp. 821–35.

Freije, A. N. and Wise, K. A., 2015. Impact of *Fusarium graminearum* inoculum availability and fungicide application timing on Fusarium head blight in wheat. *Crop Protection*, 77, pp. 139–47. Available at: http://dx.doi.org/10.1016/j.cropro.2015.07.016.

Gan, Y., Hamel, C., O'Donovan, J. T., Cutforth, H., Zentner, R. P., Campbell, C. A., Niu, Y. N. and Poppy, L., 2015. Diversifying crop rotations with pulses enhances system productivity. *Scientific Reports*, 5, pp. 1–14. Available at: http://dx.doi.org/10.1038/srep14625.

Gardiner, D. M., Kazan, K. and Manners, J. M., 2009. Novel genes of *Fusarium graminearum* that negatively regulate deoxynivalenol production and virulence. *Molecular Plant-Microbe Interactions*, 22(12), pp. 1588–600. Available at: http://apsjournals.apsnet.org/doi/10.1094/MPMI-22-12-1588.

Gardiner, D. M., Kazan, K., Praud, S., Torney, F. J., Rusu, A. and Manners, J. M., 2010. Early activation of wheat polyamine biosynthesis during Fusarium head blight implicates putrescine as an inducer of trichothecene mycotoxin production. *BMC Plant Biology*, 10(1), p. 289. Available at: http://www.biomedcentral.com/1471-2229/10/289.

Gorczyca, A., Oleksy, A., Gala-Czekaj, D., Urbaniak, M., Laskowska, M., Waśkiewicz, A. and Stępień, Ł., 2018. Fusarium head blight incidence and mycotoxin accumulation in three durum wheat cultivars in relation to sowing date and density. *Science of Nature*, 105(1–2), p. 2.

Gottwald, S., Samans, B., Lück, S. and Friedt, W., 2012. Jasmonate and ethylene dependent defence gene expression and suppression of fungal virulence factors: Two essential mechanisms of Fusarium head blight resistance in wheat? *BMC Genomics*, 13(1), 369.

Han, J., Lakshman, D. K., Galvez, L. C., Mitra, S., Baenziger, P. S. and Mitra, A., 2012. Transgenic expression of lactoferrin imparts enhanced resistance to head blight of wheat caused by *Fusarium graminearum*. *BMC Plant Biology*, 12(1), p. 33. Available at: http://www.biomedcentral.com/1471-2229/12/33.

Harrach, B. D., Baltruschat, H., Barna, B., Fodor, J. and Kogel, K.-H., 2013. The mutualistic fungus *Piriformospora indica* protects barley roots from a loss of antioxidant capacity caused by the necrotrophic pathogen Fusarium culmorum. *Molecular Plant-microbe Interactions : MPMI*, 26(5), pp. 599–605. Available at: https://apps.webofknowledge.com/full_record.do?product=UA&search_mode=GeneralSearch&qid=3&SID=U1jv6kHejfeRC3XMJXd&page=1&doc=3.

He, X., Lillemo, M., Shi, J., Wu, J., Bjørnstad, Å., Belova, T., Dreisigacker, S., Duveiller, E. and Singh, P., 2016. QTL characterization of Fusarium head blight resistance in CIMMYT bread wheat line Soru#1. *PLoS ONE*, 11(6), pp. 1–18.

Hueza, I. M., Raspantini, P. C. F., Raspantini, L. E. R., Latorre, A. O. and Górniak, S. L., 2014. Zearalenone, an estrogenic mycotoxin, is an immunotoxic compound. *Toxins*, 6(3), pp. 1080–95.

Hulsmans, S., Rodriguez, M., De Coninck, B. and Rolland, F., 2016. The *SnRK1* energy sensor in plant biotic interactions. *Trends in Plant Science*, 21(8), pp. 648–61. Available at: http://dx.doi.org/10.1016/j.tplants.2016.04.008.

Imboden, L., Afton, D. and Trail, F., 2018. Surface interactions of *Fusarium graminearum* on barley. *Molecular Plant Pathology*, 19(6), pp. 1332–42.

Jansen, C., Liu, C. and Van der Fels-Klerx, H. J., 2005. Infection patterns in barley and wheat spikes inoculated with wild-type and trichodiene synthase gene disrupted *Fusarium graminearum*. *Proceedings of the National Academy of Sciences*, 102(46), pp. 16892–7. Available at: http://www.pnas.org/cgi/doi/10.1073/pnas.0508467102.

Janssen, E. M., Liu, C. and Van der Fels-Klerx, H. J., 2018. *Fusarium* infection and trichothecenes in barley and its comparison with wheat. *World Mycotoxin Journal*, 11(1), pp. 33–46. Available at: http://www.wageningenacademic.com/doi/10.3920/WMJ2017.2255.

Jensen, B., Knudsen, I. M. B. and Jensen, D. F., 2000. Biological seed treatment of cereals with fresh and long-term stored formulations of *Clonostachys rosea*: Biocontrol efficacy against *Fusarium culmorum*. *European Journal of Plant Pathology*, 106(3), pp. 233–42.

Jensen, D. F., Karlsson, M., Sarrocco, S. and Vannacci, G., 2016. Biological control using microorganisms as an alternative to disease resistance. In: D. B. Collinge (Ed.), *Plant Pathogen Resistance Biotechnology*. Hoboken: Wiley-Blackwell, pp. 341–63.

Jia, H., Zhou, J., Xue, S., Li, G., Yan, H., Ran, C. and Ma, Z., 2017. A journey to understand wheat Fusarium head blight resistance in the Chinese wheat landrace Wangshuibai. *The Crop Journal*, 6, pp. 1–12. Available at: http://linkinghub.elsevier.com/retrieve/pii/S2214514117300971.

Jiang, C., Zhang, C., Wu, C., Sun, P., Hou, R., Liu, H., Wang, C. F. and Xu, J. R., 2016. *TRI6* and *TRI10* play different roles in the regulation of deoxynivalenol (DON) production by cAMP signalling in *Fusarium graminearum. Environmental Microbiology*, 18, pp. 3689–701.

Jung, C., Capistrano-Gossmann, G., Braatz, J., Sashidhar, N. and Melzer, S., 2017. Recent developments in genome editing and applications in plant breeding. *Plant Breeding*, 137, pp. 1–9. Available at: http://doi.wiley.com/10.1111/pbr.12526.

Kazan, K. and Gardiner, D. M., 2017. Transcriptomics of cereal–*Fusarium graminearum* interactions: What we have learned so far. *Molecular Plant Pathology*, 19(3), pp. 1–41. Available at: http://doi.wiley.com/10.1111/mpp.12561.

Kerényi, Z., Moretti, A., Waalwijk, C., and Ola, B., 2004. Mating type sequences in asexually reproducing *Fusarium* species mating type sequences in asexually reproducing *Fusarium* species. *Applied and Environmental Microbiology*, 70(8), pp. 4419–23.

Keyser, C. A., Jensen, B. and Meyling, N. V., 2016. Dual effects of *Metarhizium* spp. and *clonostachys rosea* against an insect and a seed-borne pathogen in wheat. *Pest Management Science*, 72(3), pp. 517–26.

Khaledi, N., Taheri, P. and Falahati-Rastegar, M., 2016. Reactive oxygen species and antioxidant system responses in wheat cultivars during interaction with *Fusarium* species. *Australasian Plant Pathology*, 45(6), pp. 653–70. Available at: http://dx.doi.org/10.1007/s13313-016-0455-y.

Kikot, G. E., Moschini, R., Consolo, V. F., Rojo, R., Salerno, G., Hours, R. A., Gasoni, L., Arambarri, A. M. and Alconada, T. M., 2011. Occurrence of different species of *Fusarium* from wheat in relation to disease levels predicted by a weather-based model in Argentina Pampas Region. *Mycopathologia*, 171(2), pp. 139–49.

Kim, D. W., Kim, G. Y., Kim, H. K., Kim, J., Jeon, S. J., Lee, C. W., Lee, H. B. and Yun, S. H., 2016. Characterization of nivalenol-producing *Fusarium culmorum* isolates obtained from the air at a rice paddy field in Korea. *Plant Pathology Journal*, 32(3), pp. 182–9.

Koch, A., Kumar, N., Weber, L., Keller, H., Imani, J. and Kogel, K.-H., 2013. Host-induced gene silencing of cytochrome P450 lanosterol C14-demethylase-encoding genes confers strong resistance to *Fusarium* species. *Proceedings of the National Academy of Sciences*, 110(48), pp. 19324–9. Available at: http://www.pnas.org/cgi/doi/10.1073/pnas.1306373110.

Koch, A., Biedenkopf, D., Furch, A., Weber, L., Rossbach, O., Abdellatef, E., Linicus, L., Johannsmeier, J., Jelonek, L., Goesmann, A., Cardoza, V., McMillan, J., Mentzel, T. and Kogel, K. H., 2016. An RNAi-based control of *Fusarium graminearum* infections through spraying of long dsRNAs involves a plant passage and is controlled by the fungal silencing machinery. *PLoS Pathogens*, 12(10), pp. 1–22.

Kosawang, C., Karlsson, M., Vélëz, H., Rasmussen, P. H., Collinge, D. B., Jensen, B. and Jensen, D. F., 2014. Zearalenone detoxification by zearalenone hydrolase is important for the antagonistic ability of *Clonostachys rosea* against mycotoxigenic *Fusarium graminearum. Fungal Biology*, 118(4), pp. 364–73.

Kubo, K., Kawada, N., Fujita, M., Hatta, K., Oda, S. and Nakajima, T., 2010. Effect of cleistogamy on Fusarium head blight resistance in wheat. *Breeding Science*, 60(4), pp. 405–11. Available at: http://joi.jlc.jst.go.jp/JST.JSTAGE/jsbbs/60.405?from=CrossRef.

Lee, H. J. and Ryu, D., 2017. Worldwide occurrence of mycotoxins in cereals and cereal-derived food products: Public health perspectives of their co-occurrence. *Journal of Agricultural and Food Chemistry*, 65(33), pp. 7034–51.

Lee, T., Han, Y. K., Kim, K. H., Yun, S. H. and Lee, Y. W., 2002. *Tri13* and *Tri7* determine deoxynivalenol- and nivalenol-producing chemotypes of *Gibberella zeae. Applied and Environmental Microbiology*, 68(5), pp. 2148–54.

Legrand, F., Picot, A., Cobo-Díaz, J. F., Chen, W. and Le Floch, G., 2017. Challenges facing the biological control strategies for the management of Fusarium head blight of cereals caused by *F. graminearum. Biological Control*, 113, pp. 26–38. Available at: http://dx.doi.org/10.1016/j.biocontrol.2017.06.011.

Lehoczki-Krsjak, S., Varga, M. and Mesterházy, Á., 2015. Distribution of prothioconazole and tebuconazole between wheat ears and flag leaves following fungicide spraying with different nozzle types at flowering. *Pest Management Science*, 71(1), pp. 105–13.

Li, G. and Yen, Y., 2008. Jasmonate and ethylene signaling pathway may mediate Fusarium head blight resistance in wheat. *Crop Science*, 48(5), pp. 1888–96.

Liu, C. and Ogbonnaya, F. C., 2015. Resistance to Fusarium crown rot in wheat and barley: A review. *Plant Breeding*, 134(4), pp. 365–72.

Lofgren, L. A., Leblanc, N. R., Certano, A. K., Nachtigall, J., Labine, K. M., Riddle, J., Broz, K., Dong, Y. H., Bethan, B., Kafer, C. W. and Kistler, H. C., 2017. *Fusarium graminearum*: Pathogen or endophyte of North American grasses? *New Phytologist*, pp. 1203–12.

Lucas, J. A., Hawkins, N. J. and Fraaije, B. A., 2015. The evolution of fungicide resistance. *Advances in Applied Microbiology*, 90, pp. 29-92. Available at: http://dx.doi.org/10.1016/bs.aambs.2014.09.001.

Lv, W., Wang, C., Yang, N., Que, Y., Talbot, N. J. and Wang, Z., 2017. Genome-wide functional analysis reveals that autophagy is necessary for growth, sporulation, deoxynivalenol production and virulence in *Fusarium graminearum*. *Scientific Reports*, 7(1), pp. 1–12. Available at: http://dx.doi.org/10.1038/s41598-017-11640-z.

Lysøe, E., Seong, K.-Y. and Kistler, H. C., 2011. The transcriptome of *Fusarium graminearum* during the infection of wheat. *Molecular Plant-Microbe Interactions*, 24(9), pp. 995–1000. Available at: http://apsjournals.apsnet.org/doi/10.1094/MPMI-02-11-0038.

Ma, L.-J., Geiser, D. M., Proctor, R. H., Rooney, A. P., O'Donnell, K., Trail, F., Gardiner, D. M. and Kazan, K., 2013. *Fusarium* pathogenomics. *Annual Review of Microbiology*, 67(1), pp. 399–416. Available at: http://www.annualreviews.org/doi/10.1146/annurev-micro-092412-155650.

Majumdar, R., Rajasekaran, K. and Cary, J. W., 2017. RNA Interference (RNAi) as a potential tool for control of mycotoxin contamination in crop plants: Concepts and considerations. *Frontiers in Plant Science*, 8, p. 200. Available at: http://journal.frontiersin.org/article/10.3389/fpls.2017.00200/full.

Makandar, R., Nalam, V. J., Lee, H., Trick, H. N., Dong, Y. and Shah, J., 2012. Salicylic acid regulates basal resistance to Fusarium head blight in wheat. *Molecular plant-microbe interactions : MPMI*, 25(3), pp. 431–9. Available at: http://apsjournals.apsnet.org/doi/abs/10.1094/MPMI-09-11-0232.

Mangalassery, S., Sjögersten, S., Sparkes, D. L., Sturrock, C. J., Craigon, J. and Mooney, S. J., 2014. To what extent can zero tillage lead to a reduction in greenhouse gas emissions from temperate soils? *Scientific Reports*, 4, pp. 1–8.

Marques, L. N., Pizzutti, I. R., Balardin, R. S., Dos Santos, I. D., Dias, J. V., Stefanello, M. T. and Serafini, P. T., 2017. Occurrence of mycotoxins in wheat grains exposed to fungicides on Fusarium head blight control in southern Brazil. *Journal of Environmental Science and Health – Part B Pesticides, Food Contaminants, and Agricultural Wastes*, 52(4), pp. 244–50. Available at: http://dx.doi.org/10.1080/03601234.2016.1270682.

Martinez-Medina, A., Flors, V., Heil, M., Mauch-Mani, B., Pieterse, C. M. J., Pozo, M. J., Ton, J., van Dam, N. M. and Conrath, U., 2016. Recognizing plant defense priming. *Trends in Plant Science*, 21(10), pp. 818–22. Available at: http://dx.doi.org/10.1016/j.tplants.2016.07.009.

Mesterhazy, A., 1995. Types and components of resistance to Fusarium head blight of wheat. *Plant Breeding*, 114(5), pp. 377–86.

Montgomery, D. R., 2007. Soil erosion and agricultural sustainability. *Proceedings of the National Academy of Sciences of the United States of America*, 104(33), pp. 13268–72. Available at: http://www.pnas.org/cgi/content/long/104/33/13268.

Moya-Elizondo, E. A. and Jacobsen, B. J., 2016. Integrated management of Fusarium crown rot of wheat using fungicide seed treatment, cultivar resistance, and induction of systemic acquired resistance (SAR). *Biological Control*, 92, pp. 153–63.

Murray, G. M. and Brennan, J. P., 2010. Estimating disease losses to the Australian barley industry. *Australasian Plant Pathology*, 39(1), pp. 85–96.

Nganje, W. E., Kaitibie, S., Wilson, W. W., Leistritz, F. L. and Bangsund, D. A., 2004. *Economic Impacts of Fusarium Head Blight in Wheat and Barley : 1993-2001*. Fargo, North Dakota: Agricultural Experiment Station North Dakota State University.

Nielsen, L. K., Jensen, J., Nielsen, G., Jensen, J., Spliid, N., Thomsen, I., Justesen, A. F., Collinge, D. B. and Jørgensen, H. J., 2011. Fusarium head blight of cereals in Denmark: Species complex and related mycotoxins. *Phytopathology*, 101(8), 960–9. doi:10.1094/PHYTO-07-10-0188..

Nielsen, L. K., Cook, D. J., Edwards, S. G. and Ray, R. V., 2014. The prevalence and impact of Fusarium head blight pathogens and mycotoxins on malting barley quality in UK. *International Journal of Food Microbiology*, 179, pp. 38–49. Available at: http://dx.doi.org/10.1016/j.ijfoodmicro.2014.03.023.

Niwa, S., Kubo, K., Lewis, J., Kikuchi, R., Alagu, M. and Ban, T., 2014. Variations for Fusarium head blight resistance associated with genomic diversity in different sources of the resistant wheat cultivar 'Sumai 3'. *Breeding Science*, 64(1), pp. 90–6. Available at: http://jlc.jst.go.jp/DN/JST.JSTAGE/jsbbs/64.90?lang=en&from=CrossRef&type=abstract.

O'Donnell, K., Rooney, A. P., Proctor, R. H., Brown, D. W., McCormick, S. P., Ward, T. J., Frandsen, R. J. N., Lysoe, E., Rehner, S. A., Aoki, T., Robert, V., Crous, P. W., Groenewald, J. Z., Kang, S. and Geiser, D. M., 2013. Phylogenetic analyses of *RPB1* and *RPB2* support a middle Cretaceous origin for a clade comprising all agriculturally and medically important fusaria. *Fungal Genetics and Biology*, 52, pp. 20–31. Available at: http://dx.doi.org/10.1016/j.fgb.2012.12.004.

Oide, S., Berthiller, F., Wiesenberger, G., Adam, G. and Turgeon, B. G., 2014. Individual and combined roles of malonichrome, ferricrocin, and TAFC siderophores in *Fusarium graminearum* pathogenic and sexual development. *Frontiers in Microbiology*, 5, pp. 1–15.

Oliveira, P. M., Mauch, A., Jacob, F., Waters, D. M. and Arendt, E. K., 2012. Fundamental study on the influence of Fusarium infection on quality and ultrastructure of barley malt. *International Journal of Food Microbiology*, 156(1), pp. 32–43. Available at: http://dx.doi.org/10.1016/j.ijfoodmicro.2012.02.019.

Osborne, L. E. and Stein, J. M., 2007. Epidemiology of Fusarium head blight on small-grain cereals. *International Journal of Food Microbiology*, 119(1–2), pp. 103–8.

Palazzini, J. M., Yerkovich, N., Alberione, E., Chiotta, M. and Chulze, S. N., 2017. An integrated dual strategy to control *Fusarium graminearum sensu stricto* by the biocontrol agent *Streptomyces* sp. RC 87B under field conditions. *Plant Gene*, 9(36), pp. 13–18. Available at: http://dx.doi.org/10.1016/j.plgene.2017.07.002.

Pan, D., Mionetto, A., Tiscornia, S. and Bettucci, L., 2015. Endophytic bacteria from wheat grain as biocontrol agents of *Fusarium graminearum* and deoxynivalenol production in wheat. *Mycotoxin Research*, 31, pp. 137–43.

Parikka, P., Hakala, K. and Tiilikkala, K., 2012. Expected shifts in Fusarium species' composition on cereal grain in Northern Europe due to climatic change. *Food Additives and Contaminants – Part A Chemistry, Analysis, Control, Exposure and Risk Assessment*, 29(10), pp. 1543–55.

Pasquali, M., Beyer, M., Logrieco, A., Audenaert, K., Balmas, V., Basler, R., Boutigny, A.-L., Chrpová, J., Czembor, E., Gagkaeva, T., González-Jaén, M. T., Hofgaard, I. S., Köycü, N. D., Hoffmann, L., Lević, J., Marin, P., Miedaner, T., Migheli, Q., Moretti, A., Müller, M. E. H., Munaut, F., Parikka, P., Pallez-Barthel, M., Piec, J., Scauflaire, J., Scherm, B., Stanković, S., Thrane, U., Uhlig, S., Vanheule, A., Yli-Mattila, T. and Vogelgsang, S., 2016. A European database of *Fusarium graminearum* and *F. culmorum* trichothecene genotypes. *Frontiers in Microbiology*, 7, 406.

Paul, P. A., Lipps, P. E. and Madden, L. V., 2006. Meta-analysis of regression coefficients for the relationship between Fusarium head blight and deoxynivalenol content of wheat. *Phytopathology*, 96(9), pp. 951–61. Available at: http://apsjournals.apsnet.org/doi/10.1094/PHYTO-96-0951.

Paul, P. A., Lipps, P. E., Hershman, D. E., McMullen, M. P., Draper, M. A. and Madden, L. V., 2008. Efficacy of triazole-based fungicides for fusarium head blight and deoxynivalenol control in wheat: A multivariate meta-analysis. *Phytopathology*, 98(9), pp. 999–1011. Available at: http://apsjournals.apsnet.org/doi/10.1094/PHYTO-98-9-0999.

Powell, J. J., Carere, J., Fitzgerald, T. L., Stiller, J., Covarelli, L., Xu, Q., Gubler, F., Colgrave, M. L., Gardiner, D. M., Manners, J. M., Henry, R. J. and Kazan, K., 2017. The Fusarium crown rot pathogen *Fusarium pseudograminearum* triggers a suite of transcriptional and metabolic changes in bread wheat (*Triticum aestivum* L.). *Annals of Botany*, 119(5), pp. 853–67.

Qiu, J., Dong, F., Yu, M., Xu, J. and Shi, J., 2016. Effect of preceding crop on Fusarium species and mycotoxin contamination of wheat grains. *Journal of the Science of Food and Agriculture*, 96(13), pp. 4536–41.

Rabiey, M. and Shaw, M. W., 2016. *Piriformospora indica* reduces fusarium head blight disease severity and mycotoxin DON contamination in wheat under UK weather conditions. *Plant Pathology*, 65(6), pp. 940–52.

Rabiey, M., Ullah, I. and Shaw, M. W., 2015. The endophytic fungus *Piriformospora indica* protects wheat from Fusarium crown rot disease in simulated UK autumn conditions. *Plant Pathology*, 64(5), pp. 1029–40.

Rawat, N., Pumphrey, M. O., Liu, S., Zhang, X., Tiwari, V. K., Ando, K., Trick, H. N., Bockus, W. W., Akhunov, E., Anderson, J. A. and Gill, B. S., 2016. Wheat *Fhb1* encodes a chimeric lectin with agglutinin domains and a pore-forming toxin-like domain conferring resistance to Fusarium head blight. *Nature Genetics*, 48(12), pp. 1576–80.

Salgado, J. D., Madden, L. V. and Paul, P. A., 2015. Quantifying the effects of fusarium head blight on grain yield and test weight in soft red winter wheat. *Phytopathology*, 105(3), pp. 295–306. Available at: http://apsjournals.apsnet.org/doi/10.1094/PHYTO-08-14-0215-R.

Scala, V., Aureli, G., Cesarano, G., Incerti, G., Fanelli, C., Scala, F., Reverberi, M. and Bonanomi, G., 2016. Climate, soil management, and cultivar affect Fusarium head blight incidence and deoxynivalenol accumulation in durum wheat of Southern Italy. *Frontiers in Microbiology*, 7, pp. 1–10.

Scherm, B., Balmas, V., Spanu, F., Pani, G., Delogu, G., Pasquali, M. and Migheli, Q., 2013. *Fusarium culmorum*: Causal agent of foot and root rot and head blight on wheat. *Molecular Plant Pathology*, 14(4), pp. 323–41.

Schöneberg, T., Martin, C., Wettstein, F. E., Bucheli, T. D., Mascher, F., Bertossa, M., Musa, T., Keller, B. and Vogelgsang, S., 2016. Fusarium and mycotoxin spectra in Swiss barley are affected by various cropping techniques. *Food Additives and Contaminants – Part A Chemistry, Analysis, Control, Exposure and Risk Assessment*, 33(10), pp. 1608–19. Available at: http://dx.doi.org/10.1080/19440049.2016.1219071.

Schweiger, W., Steiner, B., Ametz, C., Siegwart, G., Wiesenberger, G., Berthiller, F., Lemmens, M., Jia, H. Y., Adam, G., Muehlbauer, G. J., Kreil, D. P. and Buerstmayr, H., 2013. Transcriptomic characterization of two major *Fusarium* resistance quantitative trait loci (QTLs), *Fhb1* and *Qfhs.ifa-5A*, identifies novel candidate genes. *Molecular Plant Pathology*, 14(8), pp. 772–85.

Selosse, M.-A., Schneider-Maunoury, L. and Martos, F., 2018. Time to re-think fungal ecology? Fungal ecological niches are often prejudged. *New Phytologist*, 217(3), pp. 968–72. Available at: http://doi.wiley.com/10.1111/nph.14983.

Seong, K. Y., Pasquali, M., Zhou, X., Song, J., Hilburn, K., McCormick, S., Dong, Y. H., Xu, J.-R. and Kistler, H. C., 2009. Global gene regulation by *Fusarium* transcription factors *Tri6* and *Tri10* reveals adaptations for toxin biosynthesis. *Molecular Microbiology*, 72(2), pp. 354–67.

Shah, L., Ali, A., Zhu, Y., Wang, S., Si, H. and Ma, C., 2017. Wheat defense response to Fusarium head blight and possibilities of its improvement. *Physiological and Molecular Plant Pathology*, 98, pp. 9–17. Available at: http://dx.doi.org/10.1016/j.pmpp.2017.01.004.

Siuda, R., Grabowski, A., Lenc, L., Ralcewicz, M. and Spychaj-Fabisiak, E., 2010. Influence of the degree of fusariosis on technological traits of wheat grain. *International Journal of Food Science and Technology*, 45(12), pp. 2596–604.

Smiley, R. W., Gourlie, J. A., Easley, S. A., Patterson, L.-M. and Whittaker, R. G., 2005. Crop damage estimates for crown rot of wheat and barley in the Pacific Northwest. *Plant Disease*, 89(6), pp. 595–604. Available at: http://apsjournals.apsnet.org/doi/10.1094/PD-89-0595.

Spolti, P., Del Ponte, E. M., Dong, Y., Cummings, J. A. and Bergstrom, G. C., 2014. Triazole sensitivity in a contemporary population of *Fusarium graminearum* from New York wheat and competitiveness of a tebuconazole-resistant isolate. *Plant Disease*, 98(5), pp. 607–13. Available at: http://apsjournals.apsnet.org/doi/10.1094/PDIS-10-13-1051-RE.

Tateishi, H., Miyake, T., Mori, M., Sakuma, Y. and Saishoji, T., 2014. Effect of application timing of metconazole on Fusarium head blight development and mycotoxin contamination in wheat and barley. *Journal of Pesticide Science*, 39(1), pp. 1–6. Available at: http://jlc.jst.go.jp/DN/JST.JSTAGE/jpestics/D12-077?lang=en&from=CrossRef&type=abstract.

Tillmann, M., von Tiedemann, A. and Winter, M., 2017. Crop rotation effects on incidence and diversity of *Fusarium* species colonizing stem bases and grains of winter wheat. *Journal of Plant Diseases and Protection*, 124(2), pp. 121–30.

Townsend, T. J., Ramsden, S. J. and Wilson, P., 2016. How do we cultivate in England? Tillage practices in crop production systems. *Soil Use and Management*, 32(1), pp. 106–17.

van der Lee, T., Zhang, H., van Diepeningen, A. and Waalwijk, C., 2015. Biogeography of *Fusarium graminearum* species complex and chemotypes: A review. *Food Additives and Contaminants – Part A Chemistry, Analysis, Control, Exposure and Risk Assessment*, 32(4), pp. 453–60. Available at: http://dx.doi.org/10.1080/19440049.2014.984244.

Vanheule, A., De Boevre, M., Moretti, A., Scauflaire, J., Munaut, F., De Saeger, S., Bekaert, B., Haesaert, G., Waalwijk, C., van der Lee, T. and Audenaert, K., 2017. Genetic divergence and chemotype diversity in the Fusarium head blight pathogen *Fusarium poae*. *Toxins*, 9(9), pp. 1–19.

Wang, L. Y., Xie, Y. S., Cui, Y. Y., Xu, J., He, W., Chen, H. G. and Guo, J. H., 2015. Conjunctively screening of biocontrol agents (BCAs) against Fusarium root rot and Fusarium head blight caused by *Fusarium graminearum*. *Microbiological Research*, 177, pp. 34–42. Available at: http://dx.doi.org/10.1016/j.micres.2015.05.005.

Watanabe, M., Yonezawa, T., Lee, K., Kumagai, S., Sugita-Konishi, Y., Goto, K. and Hara-Kudo, Y., 2011. Molecular phylogeny of the higher and lower taxonomy of the *Fusarium* genus and differences in the evolutionary histories of multiple genes. *BMC Evolutionary Biology*, 11(1), pp. 1–16. Available at: http://link.springer.com/article/10.1186/1471-2148-11-322.

Wegulo, S. N., Baenziger, P. S., Hernandez Nopsa, J., Bockus, W. W. and Hallen-Adams, H., 2015. Management of Fusarium head blight of wheat and barley. *Crop Protection*, 73, pp. 100–7. Available at: http://dx.doi.org/10.1016/j.cropro.2015.02.025.

Wenda-Piesik, A., Lemańczyk, G., Twarużek, M., Błajet-Kosicka, A., Kazek, M. and Grajewski, J., 2017. Fusarium head blight incidence and detection of *Fusarium* toxins in wheat in relation to agronomic factors. *European Journal of Plant Pathology*, 149(3), pp. 515–31.

Xiao, J., Xiao, J., Jin, X., Jia, X., Wang, H., Cao, A., Zhao, W., Pei, H. Y., Xue, Z. K., He, L. Q., Chen, Q. G. and Wang, X., 2013. Transcriptome-based discovery of pathways and genes related to resistance against Fusarium head blight in wheat landrace Wangshuibai. *BMC Genomics*, 14(1), p. 1.

Xu, X. and Nicholson, P., 2009. Community ecology of fungal pathogens causing wheat head blight. *Annual Review of Phytopathology*, 47(1), pp. 83–103. Available at: http://www.annualreviews.org/doi/10.1146/annurev-phyto-080508-081737.

Xue, A. G., Chen, Y., Voldeng, H. D., Fedak, G., Savard, M. E., Längle, T., Zhang, J. X. and Harman, G. E., 2014. Concentration and cultivar effects on efficacy of CLO-1 biofungicide in controlling Fusarium head blight of wheat. *Biological Control*, 73, pp. 2–7. Available at: http://dx.doi.org/10.1016/j.biocontrol.2014.02.010.

Yang, F., Jensen, J. D., Spliid, N. H., Svensson, B., Jacobsen, S., Jørgensen, L. N., Jørgensen, H. J. L., Collinge, D. B. and Finnie, C., 2010. Investigation of the effect of nitrogen on severity of Fusarium head blight in barley. *Journal of Proteomics*, 73(4), pp. 743–52. Available at: http://dx.doi.org/10.1016/j.jprot.2009.10.010.

Yang, F., Yang, F., Jensen, J. D., Svensson, B., Jørgensen, H. J. L., Collinge, D. B. and Finnie, C., 2012. Secretomics identifies *Fusarium graminearum* proteins involved in the interaction with barley and wheat. *Molecular Plant Pathology*, 13(5), pp. 445–53.

Yin, Y., Liu, X., Li, B. and Ma, Z., 2009. Characterization of sterol demethylation inhibitor-resistant isolates of *Fusarium asiaticum* and *F. graminearum* collected from wheat in China. *Phytopathology*, 99(5), pp. 487–97.

Yoshida, M., Nakajima, T., Tomimura, K., Suzuki, F., Arai, M. and Miyasaka, A., 2012. Effect of the timing of fungicide application on fusarium head blight and mycotoxin contamination in wheat. *Plant Disease*, 96(6), pp. 845–51. Available at: http://apsjournals.apsnet.org/doi/10.1094/PDIS-10-11-0819.

Yu, G., Zhang, X., Yao, J., Zhou, M. and Ma, H., 2017. Resistance against Fusarium head blight in transgenic wheat plants expressing the *ScNPR1* gene. *Journal of Phytopathology*, 165(4), pp. 223–31. Available at: http://doi.wiley.com/10.1111/jph.12553.

Zhang, Y. J., Fan, P. S., Zhang, X., Chen, C. J. and Zhou, M. G., 2009. Quantification of *Fusarium graminearum* in harvested grain by real-time polymerase chain reaction to assess efficacies of fungicides on Fusarium head blight, deoxynivalenol contamination, and yield of winter wheat. *Phytopathology*, 99(1), pp. 95–100.

Zhang, X.-W., Jia, L.-J., Zhang, Y., Jiang, G., Li, X., Zhang, D. and Tang, W.-H., 2012. In planta stage-specific fungal gene profiling elucidates the molecular strategies of *Fusarium graminearum* growing inside wheat coleoptiles. *The Plant Cell*, 24(12), pp. 5159–76. Available at: http://www.plantcell.org/cgi/doi/10.1105/tpc.112.105957.

Zheng, Z., Gao, S., Zhou, M., Yan, G. and Liu, C., 2017. Enhancing Fusarium crown rot resistance by pyramiding large-effect QTL in common wheat (*Triticum aestivum* L.). *Molecular Breeding*, 37(9), p. 107.

Chapter 3

Diseases affecting wheat: *Septoria tritici* blotch

S. B. Goodwin, USDA-ARS, USA

1 Introduction

Septoria tritici blotch (STB), caused by the ascomycete fungus *Zymoseptoria tritici* (formerly known as *Mycosphaerella graminicola* for its sexual stage and *Septoria tritici* for its asexual stage), is one of the most economically important pathogens of wheat worldwide. This pathogen typically does not kill the wheat plants but, under favourable conditions of cool, wet weather, its tan-coloured necrotic lesions can cover entire leaves, which removes their photosynthetic benefit to the plant. This is especially severe when lesions extend to the flag leaves, when yield losses of 50–60% are often reported. STB infection reduces test weight and quality (McKendry et al., 1995); however, the pathogen does not secrete any compounds that are toxic to humans or livestock; so, grain from infected plants is at least safe to consume.

Although the incidence and severity of plant diseases vary from year to year, STB is typically the number one disease affecting wheat in many parts of the world (Fig. 1) including most countries of northern Europe (Jørgensen et al., 2014; Fones and Gurr, 2015), the Russian Federation (Sanin, 2012), the southern cone of South America (Madariaga and Scharen, 1985; Díaz de Ackernann et al., 1995; Kraan and Nisi, 1995), Ethiopia (Saari and Wilcoxson, 1974), and in bread and durum wheat-growing regions in North Africa (Saari and Wilcoxson, 1974; Berraies et al., 2014; Ramdani and Ibriz, 2000; Ramdani et al., 2011; Siah et al., 2014). In all of these areas losses can be 50% or more if the disease is not controlled with fungicides and/or resistant cultivars. In many other wheat-growing areas STB occurs every year but typically does not cause as much damage, with losses ranging from 1% to 25% depending on the year (Fig. 1) although it can be higher occasionally during rare epidemic years. These areas include most of North America (Shipton et al.,

http://dx.doi.org/10.19103/AS.2018.0039.03
Published by Burleigh Dodds Science Publishing Limited, 2019.

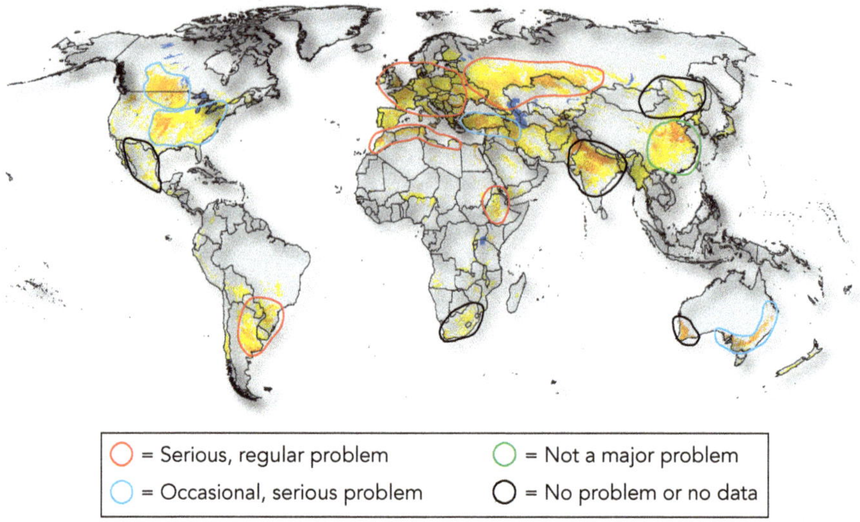

Figure 1 The importance of *Septoria tritici* blotch (STB) in wheat-producing areas worldwide. Coloured ovals indicate the importance of STB based on literature reports, where: red indicates that STB has been reported to be a serious problem during most growing seasons; blue indicates that it can cause severe epidemics occasionally; green indicates areas that should be favourable for the disease but where it is not reported as a problem; and black indicates areas where STB does not occur commonly, most likely due to unfavourable environmental conditions, or where there are no data. Shades of yellow to orange indicate increasing levels of wheat production. The grain production map by Kai Sonder (CIMMYT) was modified with permission.

1971; Shaner and Finney, 1982), the Middle East (Saari and Wilcoxson, 1974; Bastam et al., 2010), southwestern Eurasia (Saari and Wilcoxson, 1974) and southern Australia (Murray and Brennan, 2009). Areas in which STB incidence usually is low or it is not one of the major wheat diseases include China (Guo et al., 2014), India (Tyagi et al., 1969; Joshi et al., 1970) and northeastern and Western Australia (Murray and Brennan, 2009) (Fig. 1). In these areas STB epidemics occur rarely and the overall losses caused by this disease are very small relative to those precipitated by other diseases. STB incidence is thought to be rare in India, Pakistan and Afghanistan due to unfavourable temperatures and low rainfall (Saari and Wilcoxson, 1974). Why STB is not prevalent in the major wheat-growing country of China is not known.

Major problems in controlling STB result from difficulties in breeding for increased resistance (Brown et al., 2015), the rapid development of resistance to commonly applied fungicides in populations of the pathogen (Torriani et al., 2015) and a lack of understanding about the mechanisms of host–pathogen interactions.

Feeding a rapidly growing world population will become increasingly difficult in the future. During the past 60 years (for which good data are available), the rate of increase of food production has been linear (Fig. 2a), while human population sizes have increased exponentially (Fig. 2b). It simply will not be possible to continue to feed an exponentially growing human population with linear increases in food production. Controlling STB would go a long way towards achieving this short-term goal for production of the world's most widely grown grain crop.

Published by Burleigh Dodds Science Publishing Limited, 2019.

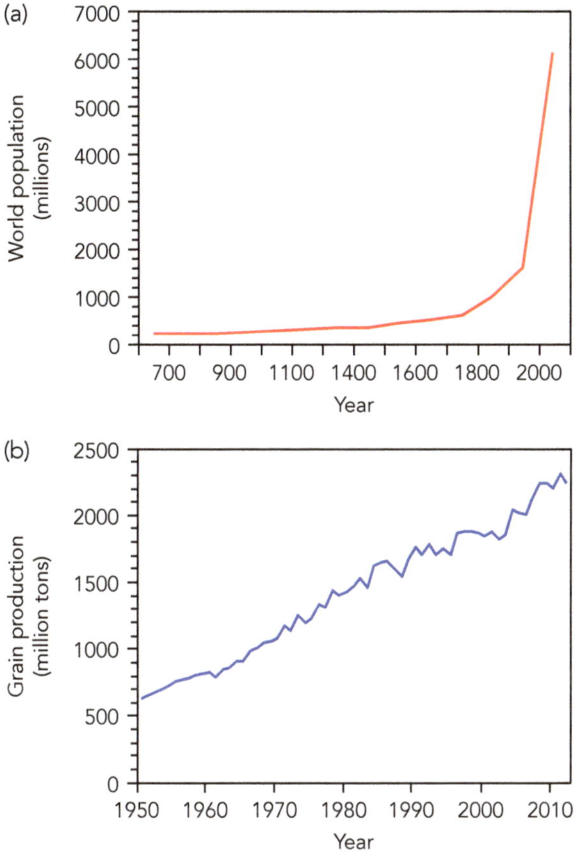

Figure 2 World population growth has been exponential, while the increase of food production is linear. Since about 1950 the world population has grown exponentially (a) while grain production during the same period has been essentially linear (b). It will not be possible to feed an exponentially growing population with linear increases in food production. Limiting the rate of human population growth is the only long-term solution to world hunger but can be alleviated during the short term by reducing food losses to plant diseases such as *Septoria tritici* blotch of wheat.

2 Use of fungicides for controlling STB

Fungicides have been the primary means for managing STB since its resurgence in Europe during the 1980s (Hardwick et al., 2001; Bearchell et al., 2005). Protectant fungicides such as mancozeb and chlorothalonil (Table 1) are effective against STB and typically affect multiple biological processes such as spore germination or growth on a leaf surface that are dependent on many gene products. For this reason, it is difficult for fungi to develop resistance to these compounds and they can be applied for many years without losing effectiveness. However, their utility is limited because they must make contact with the fungus to be effective and must be applied regularly as plants grow and the compounds

Table 1 Fungicides used to control *Septoria tritici* blotch on wheat

Fungicide group	Chemical group	Examples	Mode of action	First use	Resistance developed	Limitations
Dithiocarbamate	Mancozeb	Dithane, maneb, manzate	Multisite; affects spore germination and hyphal growth	1960s	–	Protectant only, has no effect once the fungus has penetrated its host
Chloronitrile	Chlorothalonil	Bravo, Echo, Daconil	Multisite; affects spore germination and hyphal growth	1960s	–	Protectant only, has no effect once the fungus has penetrated its host
MBC (Methyl benzimidazole carbamates)	Benzimidazoles, thiophanates	Benomyl, carbendazim, thiophanate-methyl	Single site; binds to β-tubulin	1970s	1984	Rapid development of resistance due to an amino acid substitution in the β-tubulin protein
DMI (Demethylation inhibitors)	Triazoles (azoles)	Cyproconazole, propiconazole, tebuconazole	Single site; affects the CYP51 protein involved in sterol biosynthesis	1979		Resistance developed due to multiple mutations affecting the structure or expression of the Cyp51 protein, or mutations in other genes that affect the rate of transport of DMIs out of cells
QoI (Quinone outside inhibitors)	Strobilurins	Azoxystrobin, pyraclostrobin, trifloxystrobin	Single site; binds to mitochondrial cytochrome b protein	1996	2000	Resistance developed rapidly due to mutations at amino acid position 143 and other locations in the cytochrome b protein
SDHI (Succinate dehydrogenase inhibitors)	Carboxamides	Bixafen, boscalid, isopyrazam	Single site; affects quinone-binding site of succinate dehydrogenase	2003	2015	Resistance develops due to multiple mutations, each of small effect, in the succinate dehydrogenase gene

are washed off by rain. These contact fungicides have little or no effect once the fungus has penetrated its host; so, they cannot cure infections once they have been initiated. Although these chemicals are effective against the STB pathogen, the need for numerous sprays and lack of curative ability make them impractical and uneconomic as the sole compound for wheat production where profit margins often are low. However, they still have a role when used in combination with other, site-specific fungicides, or where fungicide resistance is an issue.

Methyl benzimidazole carbamate (MBC) fungicides such as carbendazim were used against STB and other wheat diseases beginning in the 1970s. These fungicides attack the β-tubulin gene product and thereby interfere with mitotic cell division. Although they are systemic and have curative abilities, these compounds unfortunately lose effectiveness rapidly due to mutations in the β-tubulin gene, resulting in an amino acid substitution that renders the translated protein insensitive to the fungicide. Resistance was found within field populations of *Z. tritici* when it was first looked for in 1984 even in locations that had received four or fewer sprays of benzimidazoles during the previous nine years (Griffin and Fisher, 1985). Due to the high chance of developing resistance to MBC fungicides as seen in numerous other fungi (King and Griffin, 1985; Deising et al., 2008), they are used rarely for control of STB in commercial production except possibly in combination with other compounds that have different modes of action.

The demethylation inhibitor (DMI) fungicides (Table 1), also known as triazoles or azoles, were first applied to wheat beginning in 1979 (Estep et al., 2015). These compounds have both protectant and curative properties (Sanssené et al., 2011) and can provide effective protection with only 1–2 sprays per year (van den Berg et al., 2013; Estep et al., 2015; Wieczorek and Jørgensen, 2017). Their mode of action is to affect the cytochrome P450 enzyme sterol 14-demethylase (Cyp51), which performs a crucial step in the biosynthetic pathway for ergosterol, an essential component of cell membranes that is required for fungal growth and development. The widespread use of this fungicide class has resulted in a gradual erosion of its effectiveness over time, and isolates of *Z. tritici* showing resistance *in vitro* were detected beginning in 2002 (Mavroeidi and Shaw, 2005). At least 32 mutations in the Cyp51 gene leading to amino acid substitutions in its encoded protein have been identified, many of which have been shown to increase resistance to specific azole compounds (Cools and Fraaije, 2008; 2012). Genetic recombination within the Cyp51 gene during sexual reproduction can result in new combinations of mutations, and this mechanism is thought to be a major driver in the evolution of *Z. tritici* populations to greater resistance against DMI fungicides (Brunner et al., 2008). Mutations causing deletions of one or two amino acids from the 14α-demethylase protein also are known in populations of *Z. tritici* and can be associated with higher levels of resistance (Cools and Fraaije, 2012).

Some strains of *Z. tritici* that are resistant to DMI fungicides have no amino-acid-substituting mutations in their Cyp51 gene sequences. The mechanism of resistance in these isolates remained a mystery until it was discovered that they have mutations in the promoter sequence that result in overexpression of the Cyp51 protein compared to that seen in sensitive isolates – the extra copies of the 14α-demethylase enzyme allow ergosterol biosynthesis to continue even in the presence of DMIs (Cools et al., 2012). Strains that overexpress the Cyp51 gene also show complete cross resistance to multiple azole compounds (Cools et al., 2013), making them more problematic, especially if these mutations were to recombine with any of the numerous mutations affecting the amino acid composition of the 14α-demethylase.

A final mechanism of DMI resistance is through overexpression of transporters that remove the toxic compounds from cells. Strains of fungi that overexpress either ATP binding cassette (ABC) or major facilitator superfamily (MFS) transporters usually show resistance to multiple azole compounds (Paul and Moye-Rowley 2014). Even more troubling, overexpression of some transporters can provide protection against fungicides in different chemical classes in addition to azoles (Morschhäuser, 2010). Analysis of gene expression in strains of *Z. tritici* that showed a multidrug resistance (MDR) phenotype demonstrated that they were overexpressing the MFS transporter gene *MgMFS1* (Omrane et al., 2015). If mutations giving rise to overexpression of transporter genes become common in populations of *Z. tritici* and if they do not confer a fitness cost in the absence of fungicide, then chemical control of STB could become much more problematic.

A class of fungicide that was released with great expectations in 1996 was the quinone-outside inhibitors (QoIs), also known as strobilurins (Table 1). These fungicides work by binding to the cytochrome bc1 enzyme complex in mitochondria and thereby interfere with fungal respiration and concomitant energy production. The cytochrome b protein, encoded in the mitochondrial genome, is a major component of the bc1 complex (Trumpower, 1990). Unfortunately, it was soon discovered that a specific mutation in the cytochrome b gene, which resulted in a substitution of alanine for glycine at amino acid position 143 of the translated protein (denoted as G143A), conferred a high level of resistance to QoI fungicides (Gisi et al., 2002). This specific mutation appears to have arisen independently at least four times in European populations of *Z. tritici* and spread rapidly throughout the continent on the prevailing winds from west to east due to the high selection pressure exerted by frequent fungicide applications (Torriani et al., 2009). Other mutations to the cytochrome *b* gene also confer resistance but at a lower level than that conditioned by G143A (Gisi et al., 2002). The rapid spread of the G143A and other resistance-conferring mutations within populations of *Z. tritici* soon rendered QoI fungicides ineffective for control of STB (Torriani et al., 2015). Similar rapid replacements of sensitive by resistant populations have occurred or are occurring in North Africa (Siah et al., 2014) and, more recently, in western North America (Estep et al., 2015). To minimize resistance problems, it is recommended that QoIs be applied only in combination with other fungicides, particularly protectant fungicides with multisite modes of action.

The most recent class of chemicals being applied to manage STB is the succinate dehydrogenase inhibitors or SDHIs. Like the QoIs, these compounds also target fungal respiration, but instead of attacking the bc1 enzyme complex, the SDHIs interfere with the function of the nuclear-encoded, mitochondrial succinate dehydrogenase enzyme. This enzyme is heterotetrameric (Sierotzki and Scalliet, 2013), meaning it requires four subunits (designated A-D), each encoded by a different nuclear gene, to be assembled in the mitochondria to form the active enzyme. To date, only mutations of the B and C subunits have been discovered in *Z. tritici* (Kleemann and Mehl, 2017), although mutations have been identified in genes for the B, C and D subunits in other fungi (Sierotzki and Scalliet, 2013). These mutations had little or no effect on resistance and their frequencies within populations have been low (Dubos et al., 2013; Kirikyali et al., 2017). Recently, strains with much higher levels of resistance to SDHIs have been discovered in field populations in Ireland. These strains had independent mutations, leading to amino acid substitutions in both the C and D subunits of the succinate dehydrogenase enzyme; however, the effect of each mutation and the significance of their combination are not known (Dooley et al., 2016). Overexpression of MFS transporters also increases SDHI resistance, at least *in vitro* (Kirikyali et al., 2017). Although no disease-control failures due to SDHI resistance have been seen so far in the field (Dooley et al., 2016), it seems only a matter of time before this will occur. This process could be accelerated if recombination within or between the subunits during sexual reproduction joins different mutations to give new combinations with greater resistance. Due to these many problems with resistance to fungicides within populations of *Z. tritici*, it would be much better if the wheat host could be made more resistant to infection by the pathogen.

3 Breeding for increased resistance to STB

Breeding for increased resistance to STB has been hampered by a lack of mapped resistance genes and associated molecular markers plus a fundamental disagreement even about the nature of resistance. Major genes also tend to break down rapidly, and little to nothing is known about the mechanism of resistance. Even so, some successes have been noted and tremendous progress has been made in understanding the genetic basis of resistance during the past 20 years, which has set the stage for future gains.

From the 1950s through the 1990s, resistance was identified in numerous wheat cultivars and wild wheat relatives (reviewed in Goodwin, 2007; Goodwin, 2012). However, disagreement about whether the resistance was quantitative or qualitative could not be resolved until the nature of the interaction was identified more precisely.

The question of whether there were specific interactions between genes from wheat and *Z. tritici* was answered by Brading et al. (2002). This work demonstrated unequivocally that genes for avirulence in *Z. tritici* produced products that interact with those from a resistance gene in wheat. When the product of a specific wheat resistance gene recognizes the product of its corresponding avirulence gene, a resistance response will be initiated to prevent infection by the pathogen and the wheat plant will be resistant. Conversely, if the corresponding host resistance gene is not present, then no resistance response will be initiated and the pathogen can infect the susceptible host. This type of gene-for-gene relationship was first elucidated by Flor while working on flax rust in the 1940s (Flor, 1955; 1971) but is now known to occur in many host–pathogen systems, including *Z. tritici* and wheat. Genes showing a very strong interaction with pathogen avirulence gene products typically produce a high level of resistance and so are referred to as major genes and the resistance as vertical, because it produces a vertical bar on a histogram (van der Plank, 1966). Vertical resistance can be attractive to plant breeders because its strong phenotype makes it relatively easy to identify and manipulate; however, it tends to break down rapidly and so is not the best solution in the long term.

By the end of the last century, only four major genes for resistance to STB had been named (Wilson, 1985; Somasco et al., 1996), none had been mapped genetically and there were no linked markers to assist their selection in breeding programmes (reviewed in Goodwin, 2007). This situation has improved remarkably since then, and there are now 20 major STB resistance genes that have been named and mapped in bread wheat and one additional gene in the wild wheat relative *Triticum monococcum* (Brown et al., 2015). Although it took longer for these genes to be mapped than with many other plant pathogens, the pace since 2001 has been very rapid (Table 2), yielding a wealth of new information about resistance to STB in wheat.

The genetics of resistance to STB was discussed recently in an excellent review by Brown et al. (2015); therefore, only a brief overview will be provided below.

Table 2 Timeline for publication of 20 genes for resistance to *Septoria tritici* blotch in bread wheat and one in the wheat relative *Triticum monococcum* since 2001

Year	Gene	Chromosome[a]	Reference
2001	*Stb5*	7DS	Arraiano et al. (2001)
2002	*Stb6*	3AS	Brading et al. (2002)
2003	*Stb7*	4AL	McCartney et al. (2003)
	Stb8	7BL	Adhikari et al. (2003)
2004	*Stb1*	5BL	Adhikari et al. (2004a)
	Stb4	7DS	Adhikari et al. (2004b)
2005	*Stb10*	1DL	Chartrain et al. (2005b)
	Stb11	1BS	Chartrain et al. (2005a)
	Stb12	4AL	Chartrain et al. (2005b)
2006	*Stb13*	7BL	Cowling (2006)
	Stb14	3BS	Cowling (2006)
2007	*Stb15*	6AS	Arraiano et al. (2007)
2008	*TmStb1*	7AMS	Jing et al. (2008)
2009	*Stb9*	2BL	Chartrain et al. (2009)
	StbWW	1BS	Raman et al. (2009)
2011	*StbSm3*	3AS	Cuthbert (2011)
	Stb18	6DS	Tabib Ghaffary et al. (2011)
2012	*Stb16q*	3DL	Tabib Ghaffary et al. (2012)
	Stb17	5AL	Tabib Ghaffary et al. (2012)
2013	*Stb2*	1BS	Liu et al. (2013)
2015	*Stb3*	7AS	Goodwin et al. (2015)

[a] Wheat chromosomes are indicated by the homoeologous group (1–7), followed by the donor genome (A, B or D in bread wheat or M for *T. monococcum*) and the arm (short or long).

Resistance to STB via major resistance genes does not trigger a hypersensitivity response. Instead, STB resistance phenotypes are almost always leaky. All of the major genes listed in Table 2 condition a very strong resistance and many were scored qualitatively, but the resistant plants are almost never immune and there is always a range of susceptibility among segregating progeny that complicates disease phenotyping. This makes scoring of STB resistance for genetic analyses much more difficult than it is with many other diseases.

Most of the major STB resistance genes have not been used extensively in wheat breeding, in part because they can break down rapidly. For example, the cultivar gene in Oregon was highly resistant when first released in 1992 but became susceptible within three years (Cowger et al., 2000). Surveys of pathogenic variation to cultivar sets containing many of the named resistance genes identified isolates from many populations worldwide that could overcome each source of resistance (Eyal et al., 1985; Kema et al., 1996; Arraiano and Brown, 2006). Although the frequency of virulence could be low in many locations,

Published by Burleigh Dodds Science Publishing Limited, 2019.

this would change rapidly if resistant cultivars were grown over large areas. Even so, some resistances appear to have lasted for a long time. For example, the *Stb1* gene, originally from the cultivar Bulgaria 88, was bred into Indiana soft red winter wheat cultivars during the 1970s and still provides effective resistance (Goodwin, 2007). When scoring disease in the field, STB can be found on the lower leaves of modern Indiana wheat cultivars, but it rarely affects the flag leaves, and the overall damage is very low. However, when other, susceptible cultivars are grown nearby, they often become heavily infected, with extensive damage to the flag leaves (personal observation). Presumably, this difference is due to the long-term effects of *Stb1*, although other, unknown genes also could be important.

Leakiness of the STB resistance phenotype in the field could be due to heterogeneity within the environment, additional cryptic quantitative resistance genes or may be inherent to the type of resistance response. In field trials, variation for resistance also could be due to disease escape (Arraiano et al., 2009) or correlations with other traits. For example, wheat lines that are taller and later flowering often appear more resistant in the field (Eyal, 1981; Danon et al., 1982; Simón et al., 2005), particularly if disease scoring is performed at a single time point rather than a common developmental stage (Arama et al., 1999). This kind of variation greatly complicates genetic analysis of STB resistance.

There has been an interesting evolution in the terminology of host–pathogen interactions, over the past 70 years, that has impacted our progress with STB. Flor considered the pathogen to possess virulence genes, since some strains could infect the host (were virulent), while others could not (Flor, 1955, 1971). As more was learned about resistance responses, it became clear that they are triggered when the host recognizes a pathogen gene product, called a fungal avirulence gene product (Shaner et al., 1992). Therefore, resistance occurred only when a fungus possessing the avirulence gene interacted with a host that expressed the corresponding resistance gene; fungi without the avirulence gene would not be recognized by the host and could establish infections (were virulent).

As more was learned about the fungal side of the interaction, it has become apparent that most fungal 'avirulence' genes actually attack the resistance mechanisms of the host (Dodds and Thrall, 2009). Therefore, rather than being avirulence genes, they really are virulence genes (nowadays usually called effectors), just ones that can be recognized by specific host genotypes possessing matching resistance genes. Fungal populations typically contain a diversity of effector genes at different frequencies depending on selection exerted by resistance genes present in populations of the host. What this means for plant breeding is that most major, qualitative resistance genes will not remain effective in the long term. Once they are deployed in commercial cultivars and grown over large acreages, they exert selection pressure, which will increase the frequency of pathogen individuals that lack the effector, because in these hosts it functions as an avirulence gene. Once this occurs the resistance will cease to be useful and will have been effectively broken by the pathogen. Cycles of major resistance genes being deployed, only to be broken by the pathogen until a new resistance gene is introduced, have been referred to as 'boom and bust' (Priestley, 1978) and add great uncertainty to agricultural production from year to year.

Although the mechanisms of resistance and pathogenicity for *Z. tritici* and wheat are not known completely, all of the major genes for resistance to STB in Table 2 are predicted to break down if deployed over large areas. The one possible exception is gene *Stb16q*. This gene provided resistance against all isolates of *Z. tritici* against which it was tested. Therefore, it may not have specific interactions, which could indicate a broad spectrum of effectiveness (Tabib Ghaffary et al., 2012). However, the nature of this gene is not clear. It originated in an accession of *T. tauschii*, the D genome donor of hexaploid wheat (Tabib Ghaffary et al., 2012), and its location could not be identified more precisely than chromosome arm 3DL due to a suppression of recombination in the genomic region around the gene.

Unfortunately, lack of recombination complicates the analysis of this gene and its transfer into other wheat cultivars. It is possible, and maybe even likely, that *Stb16q* does not occur singly but with two or more genes on the same chromosome arm that are locked into acting as a single linkage block due to an inversion. If this is the case, then the apparent broad effectiveness of *Stb16q* may be an illusion, caused by the combined effects of multiple genes. Lack of recombination of the chromosome arm containing *Stb16q* also means that introgressing the gene into new wheat cultivars could be problematic if any genes with negative effects are present in the same linkage block. Clearly, much more research is needed to better understand the genetic basis of the *Stb16q* gene and of broad-spectrum resistance to STB in general.

Due to the high likelihood that major genes will break down, there has been an increased focus on quantitative genes for resistance to STB. A review by Brown et al. (2015) listed 187 quantitative trait loci (QTLs) that have been identified and mapped in wheat. These occur on all wheat chromosomes except for 5D (Brown et al., 2015). Unlike the major genes that are only effective against some strains of the pathogen, most of the QTLs are thought to be effective against all strains of the pathogen. The disadvantage is that they have only small effects; so, many QTLs are required to act together to have a similar reduction in disease as a single major gene. The need to simultaneously select for multiple genes on different chromosomes, each with a small effect, greatly complicates the transfer of QTLs for STB resistance into new wheat cultivars. However, the long-term benefits of stable resistance make this complication worth the price for plant breeders.

All of the previously mapped resistance genes, both qualitative and quantitative, were mapped by analyses of segregating progeny in biparental genetic crosses. This approach is slow because it can only analyse the genes in the parents and each progeny set can take many years to develop. To overcome these limitations, genome-wide association studies (GWASs) have been used recently. The goal of GWASs is to search for associations between molecular markers and traits of interest in a diverse population. This approach, first developed for humans (Hirschhorn and Daly, 2005), uses a large, diverse population and seeks associations between molecular markers and a trait; unlinked markers will only occur with the phenotype at random due to the large number of meioses that have occurred since the members of the GWAS population shared a common ancestor. Unlike classical genetic analyses, where only those genes segregating from two parents can be analysed at a time, in GWAS a much higher number of traits from a larger section of the species' diversity can be analysed simultaneously. Because it is based on association only and there is always a possibility of type I errors, important linkages should be verified by additional analyses. However, the power of GWAS to discover new associations between molecular markers and traits of interest has made it a high priority in genetics, including the identification and analysis of fungal resistance genes in plants.

So far, at least four GWAS analyses that included resistance against *Z. tritici* have been published. Miedaner et al. (2013) analysed 87 elite European breeding lines and over 1000 of their progeny for STB phenotype using the 90 K SNP array, a genotyping platform for single-nucleotide polymorphisms (SNPs) located within 9000 genes (Cavanagh et al., 2013). Phenotyping was done in unreplicated field experiments at two locations in one year, one of which was inoculated one time with a mixture of isolates, while the other relied on natural infection. The analysis identified eight SNPs that were associated with STB resistance, of which only four could be mapped (Miedaner et al., 2013), all to chromosomes already known to carry major STB resistance genes. In addition, this work confirmed the previously identified positive correlation between plant height and STB resistance.

A more thorough GWAS experiment by Kollers et al. (2013) identified more than 100 associations between microsatellite markers and STB phenotype among 372 European wheat cultivars tested in replicated field trials at one location in Germany over two years. The trials were inoculated twice, ten days apart, but the origin of the inoculum was not stated and could have been a single isolate. With this work, at least one major or minor gene for STB resistance has been located on all 21 wheat chromosomes representing all three progenitor genomes. As with previous research, STB resistance was positively correlated with plant height and heading date, i.e. taller and later plants were more resistant. However, this phenomenon may not indicate resistance, but could instead be due to escape (Arraiano et al., 2009), particularly in trials that were inoculated and/or scored at only one or two time points. Taller, later plants may head out after inoculum is applied and might be scored before the disease has a chance to develop, making them appear to be resistant when in fact they may be susceptible. A similar escape phenomenon could occur in the field if taller, later plants are filling their grain when weather has become less favourable for disease development (i.e. too dry and/or warm) or if they are too tall for effective splash dispersal of the asexual pycnidiospores (Arraiano et al., 2009). These two early analyses demonstrated the potential for GWAS to identify novel STB-resistant genes.

Recent analyses have extended these results using Diversity Arrays Technology (DArT) (Jaccoud et al., 2001) markers. An analysis of 525 DArT markers on a worldwide panel of 96 mostly winter wheat cultivars tested in two field experiments in Argentina identified 42 regions on 18 chromosomes that were associated with resistance to STB (Gerard et al., 2017). This analysis included tests at both seedling and adult plant stages and the inoculations of adult plants were scored after the flag leaves emerged to minimize any spurious associations that could arise from scoring at only one time point as seen in previous work. About 43% of these chromosomal regions were associated with resistance in seedlings, 38% were seen only in adult plants, while the remaining 19% manifested at both stages (Gerard et al., 2017). Many of the linked DArT markers were in chromosomal regions that had been associated with major genes for resistance to STB previously, but the majority represented potentially new sources of resistance, including, as with Kollers et al. (2013), one on chromosome 5D. Another highly replicated GWAS of 164 northern European wheat lines identified four significant QTLs for STB resistance, the most significant of which was on 5D (Vagndorf et al., 2017).

Another recent association analysis of STB resistance included 225 wheat cultivars and breeding lines from the United Kingdom tested over multiple years and locations in the field and assayed for DArT and simple-sequence repeat (SSR or microsatellite) markers (Arraiano and Brown, 2017). This analysis identified 11 QTLs associated with STB resistance. As in previous work, some of these probably represented named STB genes, while other resistances appeared to be new (Arraiano and Brown, 2017). An interesting side result of this research is that it identified the inadvertent introduction of genes for susceptibility to STB, while breeding for increased yield and rust resistance in the UK. Introgression of the Yr6 gene for yellow rust resistance during the 1950s appears to have unwittingly included a block of genes including at least one for susceptibility to STB; other genes for susceptibility to STB were added when the Pch1 gene for resistance to eyespot was bred into UK wheat cultivars during the 1980s (Arraiano and Brown, 2017). These genes were favoured and maintained in British wheat breeding programmes because they increased yield, as long as STB pressure was low. Based on these results, it seems likely that at least part of the resurgence of STB since the 1980s may be due to inadvertent introduction of genes for susceptibility while selecting for increased yield and resistance to other diseases. This case history highlights how a more complete understanding of all genes in a linkage block is needed when introgressing genes from unadapted wheat germplasm to minimize unexpected negative repercussions.

4 Understanding the molecular basis of the host–pathogen interaction

Much less is understood about the mechanisms of fungal pathogenicity, and host resistance in the interaction between wheat and *Z. tritici* than is known for many biotrophic pathogens such as those causing rusts and powdery mildews, or necrotrophic pathogens such as *Parastagonospora nodorum* and *Pyrenophora tritici-repentis* which produce well characterised necrotrophic effectors.

Necrotrophs typically kill the invaded parts of their hosts quickly, usually by producing host-specific toxins (or necrotrophic effectors), and then feed on the dead or dying host tissues. For these pathogens, resistance or susceptibility is determined by interactions between toxins produced by the fungus and sensitivity genes in their hosts; plants lacking the corresponding sensitivity gene to a particular pathogen toxin will be resistant.

Following penetration through the stomata, *Z. tritici* grows intercellularly without triggering an effective host defence response for usually 9–14 days, then the host tissue rapidly dies and the pathogen enters a phase of necrotrophic growth. However, the biotrophic phase is not like that for true biotrophs because of the intercellular growth without production of specialized feeding structures or cell penetration of any kind. During this phase it does not increase much in biomass and its source of nutrition is not understood. Nothing is known about what triggers the switch from biotrophy to necrotrophy, but clearly the result is that the mode of nutrition changes and the pathogen begins expressing genes for host cell catabolism. This phase is different from that of true necrotrophs because no clear genes for fungal-produced toxins or of host sensitivity to toxins have been identified. This murkiness in our understanding of the mechanisms of pathogenicity of *Z. tritici* to wheat has complicated efforts to control STB disease.

The pace of research has quickened since publication of the *Z. tritici* genomic sequence in 2011. The genome contains a reduced number of genes for cell-wall-degrading enzymes relative to those of other sequenced plant pathogens but a similar number of proteases as in endophytes, indicating that the stealth pathogenicity of *Z. tritici* could have evolved from an endophytic ancestor (Goodwin et al., 2011). The genome consists of 21 chromosomes, eight of which can be lost with no obvious effects on fitness and so are dispensable, and represents the first finished genome for a filamentous fungus with the highest number of dispensable (or accessory) chromosomes (Goodwin et al., 2011). The dispensable chromosomes have a higher content of repetitive DNA and far fewer genes relative to the core chromosomes, and appear to have originated by horizontal transfer from an unknown donor more than 10 000 years ago. Although they are not required for pathogen survival they must be of benefit at some stage of the life cycle or in specific environments, or they would not be retained within the species.

To better understand the pathogenicity of *Z. tritici*, several genes have been knocked out for functional analysis, and gene expression has been analysed by microarrays and RNA sequencing (RNA-seq) experiments. Knockout analyses have identified that one clear effector, *Mg3LysM*, which is involved in pathogenicity to wheat, blocks induction of chitin-triggered plant defence responses, and protects fungal hyphae from plant chitinases (Marshall et al., 2011). Other potential effectors have been identified through bioinformatics analyses, but either were not involved in pathogenicity or had a redundant function because their deletion strains were still pathogenic (reviewed in Kettles and Kanyuka, 2016). Many others have a possible role in pathogenicity but have not yet been tested by knockout analyses (McDonald et al., 2016).

Analysis of culture filtrates identified two heat-stable proteins that can induce necrosis in wheat leaves (Ben M'Barek et al., 2015). These proteins have some characteristics of the host-specific toxins seen in necrotrophic pathogens, but their role in pathogenicity of *Z. tritici* to wheat remains unknown (Kettles and Kanyuka, 2016). All of this work indicates that *Z. tritici* has a complicated interaction with its wheat host, but provides little information about the mode of nutrition during the early stages of infection. Keon et al. (2007) showed that the concentrations of sugars in apoplastic fluids did not change during the early stages of infection and concluded that the fungus was not catabolizing carbohydrates for growth and so may not be truly biotrophic. Lack of sugar catabolism is consistent with the relative paucity of genes for cell-wall-degrading enzymes in the genome sequence. However, plants regulate the concentration of sugars in their apoplastic spaces; so, this is not a good argument about initial nutrition.

Early RNA-seq analyses showed that there was much variability in gene expression over time (Yang et al., 2013; Kellner et al., 2014; Grandaubert et al., 2015); but they either had a very low sequencing depth or too few time points to draw firm conclusions. A very thorough analysis of gene expression confirmed the lack of carbohydrate catabolism during the biotrophic stage of infection but showed that many genes are induced during the necrotrophic phase (Rudd et al., 2015). Genes for lipid and fatty acid degradation were expressed during the early, biotrophic stage, suggesting that the stealth pathogenicity of *Z. tritici* could be facilitated by 'survival off' of stored lipids and fatty acids rather than acquisition of nutrients from the host during the initial phase of infection (Rudd et al., 2015). Peptidases and many other genes including those for cell-wall-degrading enzymes were expressed once the fungus switched to necrotrophic growth and no longer needed to hide its presence from the host. A more recent analysis of four isolates of *Z. tritici* at several time points on a single wheat cultivar showed strain-specific expression of many genes, some of which could be involved in pathogenicity (Palma-Guerrero et al., 2017).

Cultivar variability is another form of the environmental influence on gene expression noted by Rudd et al. (2015). Taken together, these analyses showed that gene expression in *Z. tritici* varies widely among individuals and environments (Yang et al., 2015), and relatively few generalities can be made about gene expression other than that gene expression is very low during the biotrophic phase and increases dramatically during necrotrophic growth, and that *Mg3LysM* and probably numerous other effectors are involved in the interactions with different wheat genotypes. Although these analyses have greatly increased our understanding of the molecular mechanisms of *Z. tritici* pathogenicity, the mode of nutrition during the asymptomatic, presumably biotrophic phase, the switch from this phase to necrotrophy, and whether toxins are involved during necrotrophic growth are still not well understood. Therefore, much remains to be understood.

Host resistance responses to *Z. tritici* are even less well understood than fungal pathogenicity. Early work with host gene expression showed that resistant wheat cultivars can respond to *Z. tritici* very quickly, within 3 hours after inoculation (Ray et al., 2003), and identified a protein disulphide isomerase that was associated with the resistance response. A lipoxygenase was downregulated during the time that pathogenesis-related proteins were at peak expression, indicating that repression of some genes could be part of the resistance response (Ray et al., 2003). Analysis of an extensive time course that included 12 samples from 0 to 27 days after inoculation on two resistant and two susceptible cultivars showed that many resistance-associated genes had two peaks of expression, one early (1–3 days after inoculation) at the time of penetration and a second, much more massive late peak during the necrotrophic stage (from 12 to 24 DAI) in at least one resistant cultivar (Adhikari et al., 2007). These genes included several pathogenesis-related (PR) proteins plus other genes identified in a large-scale cDNA-AFLP profiling experiment that had not been implicated in disease resistance previously. The late peak corresponds to the time when the fungus is growing exponentially in a susceptible interaction and indicates that late responses probably determine the outcome of the host–pathogen interaction (Adhikari et al., 2007). A more recent analysis of these and other genes confirmed some, but not all, of the previous results and showed a great variability in gene expression among cultivars and time points (Orton et al., 2017). Clearly, wheat can very rapidly sense and respond to contact with *Z. tritici*, but it appears that the late responses at the switch from biotrophic to necrotrophic growth probably determine whether the interaction results in resistance or susceptibility. However, the exact nature of the interaction and the mechanism of resistance are still not known.

Some insight into the molecular basis of resistance was obtained from the recent cloning of the *Stb6* gene (Saintenac et al., 2017), the first for any STB resistance gene. This gene, which does not confer complete resistance, is present in many European wheat cultivars. *Stb6* appears to be a wall-associated receptor kinase or similar gene. Susceptible lines have a point mutation in the kinase domain that presumably prevents signalling so that the host does not mount an effective resistance response (Saintenac et al., 2017). Although cloning of this gene does not tell much about the resistance mechanism itself, identifying genes that are activated by *Stb6* should help fill in the picture considerably in the future.

5 Conclusion and future trends

A substantial amount of progress has been made in understanding STB during the past two decades, yet our knowledge remains very incomplete and there is still much more to learn. One thing that is becoming obvious is that *Z. tritici* is a very formidable opponent and will not go down without a fight. This was demonstrated dramatically during and after the recent Millennium drought in Australia. From 2001 to 2009, southeastern Australia experienced its worst drought on record (van Dijk et al., 2013) and STB essentially disappeared. Once the rains returned, so did STB (McDonald et al., 2016). Apparently, survival was sufficient even under extremely trying conditions for the pathogen so that it was still poised to cause severe epidemics once conditions became favourable for infection. With climate change, even those areas where STB incidence currently is very low (Fig. 1) could suddenly start to have epidemics. It seems that it will never be possible to lower our guard against this disease.

Published by Burleigh Dodds Science Publishing Limited, 2019.

Fungicides have been the standard approach for management of STB during the past 40 years, but resistance has developed rapidly and two or more compounds must be used in combination for maximum effectiveness (Torriani et al., 2015). The population structure of *Z. tritici* makes it a high-risk pathogen for overcoming fungicides and new resistance genes (McDonald and Linde, 2002). Therefore, problems of fungicide resistance are likely to continue and there will always be a need for new strategies to maintain effectiveness of chemical control.

Breeding for increased resistance has been modestly effective, and some resistance genes have lasted longer than expected. However, often the most effective and durable resistance genes are those that have been introgressed from other species, and wheat breeders must be cognizant of all genes on a linkage block to avoid accidentally introducing genes for susceptibility as occurred in the UK (Arraiano and Brown, 2017). A similar phenomenon could have occurred in other areas but has not yet been discovered.

The best long-term approach for breeding wheat with increased resistance to STB will involve quantitative resistance. If qualitative genes are used, they should be pyramided with others and backed up with numerous QTLs to decrease spread and reproduction of the pathogen. Fortunately, a wealth of major genes and QTLs are available since at least one resistance gene has been identified on all 21 wheat chromosomes. Molecular markers are now available for many, if not most, of these genes. Development of methods for high-throughput genotyping for marker-assisted selection should help wheat breeders incorporate STB resistance into advanced cultivars, but these genes will have to be chosen carefully to minimize linkage drag of undesirable traits.

Breeding materials will need to be screened in each geographical area, as the local fungal populations will have different frequencies of pathogenicity genes. This will be particularly important if more quantitative resistance loci are host specific. Cultivars with good resistance will require fewer fungicide sprays, which should help reduce selection for resistance and extend the useful life of chemical compounds.

With an excellent reference genome of the pathogen and very good resources for wheat genome sequences, the tools are available for rapid advances in our understanding of their interaction. Recent cloning of the *Stb6* gene is a major step forward, but there is still much to be learned and our understanding of the mechanisms of resistance and pathogenicity remains primitive. If the pathogen survives off of stored reserves during biotrophic growth, then we would expect that ascospores and pycnidiospores should be packed with lipids, fatty acids or other energy stores. The trigger for the switch from apparently benign endophytic growth to aggressive pathogen is still not known, nor is whether toxins play a major role in disease aetiology. Answers to these questions should indicate how wheat can be engineered for better resistance and how chemicals can be better designed to exploit apparent weaknesses. With the extremely rapid progress over the past two decades, it seems likely that most of these questions will be answered, auguring for much better disease management in the future. However, this pathogen has demonstrated its resilience and so this is a war that will be fought for many years to come.

6 Where to look for further information

Several general review articles have been written (Shipton et al., 1971; King, 1983; Ponomarenko et al., 2011; Goodwin, 2007) and, more recently, an issue of the journal

Fungal Genetics and Biology, '*Septoria tritici* blotch disease of wheat: Tools and techniques to study the pathogen *Zymoseptoria tritici*' was devoted to articles about many aspects of STB and its causal organism, published as volume 79 in June 2015. Among these is the excellent review on the genetics of resistance by Brown et al. (2015).

7 References

Adhikari, T. B., Anderson, J. M. and Goodwin, S. B. (2003). Identification and molecular mapping of a gene in wheat conferring resistance to *Mycosphaerella graminicola*. *Phytopathology* 93, 1158–64.

Adhikari, T. B., Cavaletto, J. R., Dubcovsky, J., Gieco, J., Schlatter, A. R. and Goodwin, S. B. (2004a). Molecular mapping of the *Stb4* gene for resistance to *Septoria tritici* blotch in wheat. *Phytopathology* 94, 1198–206.

Adhikari, T. B., Yang, X., Cavaletto, J. R., Hu, X., Buechley, G., Ohm, H. W., Shaner, G. and Goodwin, S. B. (2004b). Molecular mapping of *Stb1*, a potentially durable gene for resistance to *Septoria tritici* blotch in wheat. *Theoretical and Applied Genetics* 109, 944–53.

Adhikari, T. B., Balaji, B., Breeden, J. and Goodwin, S. B. (2007). Resistance of wheat to *Mycosphaerella graminicola* involves early and late peaks of gene expression. *Physiological and Molecular Plant Pathology* 71(1), 55–68.

Arama, P. F., Parlevliet, J. E. and van Silfhout, C. H. (1999). Heading date and resistance to *Septoria tritici* blotch in wheat not genetically associated. *Euphytica* 106, 63–8.

Arraiano, L. S. and Brown, J. K. M. (2006). Identification of isolate-specific and partial resistance to *Septoria tritici* blotch in 238 European wheat cultivars and breeding lines. *Plant Pathology* 55, 726–38.

Arraiano, L. S. and Brown, J. K. M. (2017). Sources of resistance and susceptibility to Septoria tritici blotch of wheat. *Molecular Plant Pathology* 18, 276–92.

Arraiano, L. S., Worland, A. J., Ellerbrook, C. and Brown J. K. M. (2001). Chromosomal location of a gene for resistance to *Septoria tritici blotch* (*Mycosphaerella graminicola*) in the hexaploid wheat 'Synthetic 6x'. *Theoretical and Applied Genetics* 103, 758–64.

Arraiano, L. S., Chartrain, L., Bossolini, E., Slatter, H. N., Keller, B. and Brown J. K. M. (2007). A gene in European wheat cultivars for resistance to an African isolate of *Mycosphaerella graminicola*. *Plant Pathology* 56, 73–8.

Arraiano, L. S., Balaam, N., Fenwick, P. M., Chapman, C., Feuerhelm, D., Howell, P., Smith, S. J., Widdowson, J. P. and J. K. M. Brown. (2009). Contributions of disease resistance and escape to the control of *Septoria tritici* blotch of wheat. *Plant Pathology* 58, 910–22.

Bastam, S. V., Ramezanpour, S. S., Soltanloo, H., Kia, S., Kalate, M and Pahlevani, M. H. (2010). Inheritance of resistance to *Septoria tritici* blotch (STB) in some Iranian genotypes of wheat (Triticum aestivum L.). *International Journal of Genetics and Molecular Biology* 2, 34–42.

Bearchell, S. J., Fraaije, B. A., Shaw, M. W. and Fitt, B. D. L. (2005). Wheat archive links long-term fungal pathogen population dynamics to air pollution. *Proceedings of the National Academy of Sciences USA* 102, 5438–42.

Ben M'Barek, S., Cordewener, J. H. G., Tabib Ghaffary, S. M., van der Lee, T. A. J., Liu, Z., Mirzadi Gohari, A., Mehrabi, R., America, A. H. P., Robert, O., Friesen, T. L., Hamza, S., Stergiopoulos, I., de Wit, P. J. G. M. and Kema, G. H. J. (2015). FPLC and liquid- chromatography mass spectrometry identify candidate necrosis-inducing proteins from culture filtrates of the fungal wheat pathogen *Zymoseptoria tritici*. *Fungal Genetics and Biology* 79, 54–62. doi:10.1016/j.fgb.2015.03.015.

Berraies, S., Gharbi, M. S., Rezgui, S. and Yahyaoui, A. (2014). Estimating grain yield losses caused by septoria leaf blotch on durum wheat in Tunisia. *Chilean Journal of Agricultural Research* 74, 432–7.

Brading, P. A., Verstappen, E. C. P., Kema, G. H. J. and Brown, J. K. M. (2002). A gene-for-gene relationship between wheat and *Mycosphaerella graminicola*, the *Septoria tritici* blotch pathogen. *Phytopathology* 92, 439–45.

Brown, J. K. M., Chartrain, L., Lasserre-Zuber, P. and Saintenac, C. (2015). Genetics of resistance to *Zymoseptoria tritici* and applications to wheat breeding. *Fungal Genetics and Biology* 79, 33–41.

Brunner, P. C., Stefanato, F. L. and McDonald, B. A. 2008. Evolution of the CYP51 gene in *Mycosphaerella graminicola*: Evidence for intragenic recombination and selective replacement. *Molecular Plant Pathology* 9, 305–16.

Cavanagh, C. R., Chao, S., Wang, S., Huang, B. E., Stephen, S., Kiani, S., Forrest, K., Saintenac, C., Brown-Guedira, G. L., Akhunova, A., See, D., Bai, G., Pumphrey, M., Tomar, L., Wong, D., Kong, S., Reynolds, M., Lopez da Silva, M., Bockelman, H., Talbert, L., Anderson, J. A., Dreisigacker, S., Baenziger, S., Carter, A., Korzun, V., Morrell, P. L., Dubcovsky, J., Morell, M. K., Sorrells, M. E., Hayden, M. J. and Akhunov, E. (2013). Genome-wide comparative diversity uncovers multiple targets of selection for improvement in hexaploid wheat landraces and cultivars. *Proceedings of the National Academy of Sciences USA* 110, 8057–62.

Chartrain, L., Berry, S. T. and Brown, J. K. M. (2005a). Resistance of wheat line Kavkaz-K4500L.6.A.4 to *Septoria tritici* blotch controlled by isolate-specific resistance genes. *Phytopathology* 95, 664–71.

Chartrain, L., Joaquim, P., Berry, S. T., Arraiano, L. S., Azanza, F. and Brown, J. K. M. (2005b). Genetics of resistance to *Septoria tritici* blotch in the Portuguese wheat breeding line TE 9111. *Theoretical and Applied Genetics* 110, 1138–44.

Chartrain, L., Sourdille, P., Bernard, M. and Brown, J. K. M. (2009). Identification and location of Stb9, a gene for resistance to *Septoria tritici* blotch in wheat cultivars Courtot and Tonic. *Plant Pathology* 58, 547–55.

Cools, H. J. and Fraaije, B. A. (2008). Are azole fungicides losing ground against *Septoria* wheat disease? *Pest Management Science* 64, 681–4.

Cools, H. J. and Fraaije, B. A. (2012). Update on mechanisms of azole resistance in *Mycosphaerella graminicola* and implications for future control. *Pest Management Science* 69, 150–5.

Cools, H. J., Bayon, C., Atkins, S., Lucas, J. A. and Fraaije, B. A. (2012). Overexpression of the sterol 14α-demethylase gene (MgCYP51) in *Mycosphaerella graminicola* isolates confers a novel azole fungicide sensitivity phenotype. *Pest Management Science* 68, 1034–40.

Cools, H. J., Hawkins, N. J. and Fraaije, B. A. (2013). Constraints on the evolution of azole resistance in plant pathogenic fungi. *Plant Pathology* 62(Suppl. 1), 36–42.

Cowger, C., Hoffer, M. E. and Mundt, C. C. (2000). Specific adaptation by *Mycosphaerella graminicola* to a resistant wheat cultivar. *Plant Pathology* 49, 445–51.

Cowling, S. G. (2006). Identification and mapping of host resistance genes to *Septoria tritici* blotch of wheat. Ph.D. thesis, University of Manitoba, Winnipeg, Canada, 129pp.

Cuthbert, R. D. (2011). Molecular mapping of *Septoria tritici* blotch resistance in hexaploid wheat (*Triticum aestivum* L.). Ph.D. Thesis, University of Manitoba, Winnipeg, Canada, 139pp.

Danon, T., Sacks, J. M. and Eyal, Z. (1982). The relationship among stature, maturity class, and susceptibility to septoria leaf blotch of wheat. *Phytopathology* 72, 1037–42.

Deising, H. B., Reimann, S. and Pascholati, S. F. (2008). Mechanisms and significance of fungicide resistance. *Brazilian Journal of Microbiology* 39, 286–95.

Díaz de Ackernann, M., Stewart, S. and Ibañez, W. (1995). Pathogenic variability of *Septoria tritici* isolates from South America. In: S. Gilchrist, L. M. van Ginkel, A. McNab and G. H. J. Kema (Eds), *Proceedings of the Septoria tritici Workshop*. CIMMYT, Mexico, D.F., pp. 41–9.

Dodds, P. and Thrall, P. (2009). Recognition events and host–pathogen co-evolution in gene-for-gene resistance to flax rust. *Functional Plant Biology* 36, 395–408. http://doi.org/10.1071/FP08320

Dooley, H., Shaw, M. W., Mehenni-Ciz, J., Spink, J. and Kildea, S. (2016). Detection of *Zymoseptoria tritici* SDHI-insensitive field isolates carrying the SdhC-H152R and SdhD-R47W substitutions. *Pest Management Science* 72, 2203–7.

Dubos, T., Pasquali, M., Pogoda, F., Casanova, A., Hoffmann, L. and Beyer, M. (2013). Differences between the succinate dehydrogenase sequences of isopyrazam sensitive *Zymoseptoria tritici* and insensitive *Fusarium graminearum* strains. *Pesticide Biochemistry and Physiology* 105, 28–35.

Estep, L. K., Torriani, S. F. F., Zala, M., Anderson, N. P., Flowers, M. D., McDonald, B. A., Mundt, C. C. and Brunner, P. C. (2015). Emergence and early evolution of fungicide resistance in North American populations of *Zymoseptoria tritici*. *Plant Pathology* 64, 961–71.

Eyal, Z. (1981). Integrated control of *Septoria* diseases of wheat. *Plant Disease* 65, 763–8.

Eyal, Z., Scharen, A. L., Huffman, M. D. and Prescott, J. M. (1985). Global insights into virulence frequencies of *Mycosphaerella graminicola*. *Phytopathology* 75, 1456–62.

Flor, H. H. (1955). Host-parasite interaction in flax rust - its genetics and other implications. *Phytopathology* 45, 680–5.

Flor, H. H. (1971). Current status of the gene-for-gene concept. *Annual Review of Phytopathology* 9, 275–96.

Fones, H. and Gurr, S. (2015). The impact of *Septoria tritici* Blotch disease on wheat: An EU perspective. *Fungal Genetics and Biology* 79, 3–7.

Gerard, G. S., Börner, A., Lohwasser, U. and Simón, M. R. (2017). Genome-wide association mapping of genetic factors controlling *Septoria tritici* blotch resistance and their associations with plant height and heading date in wheat. *Euphytica* 213, 27.

Gisi, U., Sierotzki, H., Cook, A. and McCaffery, A. (2002). Mechanisms influencing the evolution of resistance to Qo inhibitor fungicides. *Pest Management Science* 58, 859–67.

Goodwin, S. B. (2007). Back to basics and beyond: Increasing the level of resistance to *Septoria tritici* blotch in wheat. *Australasian Plant Pathology* 36, 532–8.

Goodwin, S. B. (2012). Resistance in wheat to Septoria diseases caused by *Mycosphaerella graminicola* (*Septoria tritici*) and Phaeosphaeria (Stagonospora) nodorum, pp. 151–9. In: I. Sharma (Ed.), *Disease Resistance in Wheat*. CABI, Cambridge, MA.

Goodwin, S. B., Ben M'Barek, S., Dhillon, B., Wittenberg, A. H. J., Crane, C. F., Hane, J. K., Foster, A. J., Van der Lee, T. A. J., Grimwood, J., Aerts, A., Antoniw, J., Bailey, A., Bluhm, B., Bowler, J., Bristow, J., van der Burgt, A., Canto-Canché, B., Churchill, A. C. L., Conde-Ferràez, L., Cools, H. J., Coutinho, P. M., Csukai, M., Dehal, P., De Wit, P., Donzelli, B., van de Geest, H. C., van Ham, R. C. H. J., Hammond-Kosack, K. E., Henrissat, B., Kilian, A., Kobayashi, A. K., Koopmann, E., Kourmpetis, Y., Kuzniar, A., Lindquist, E., Lombard, V., Maliepaard, C., Martins, N., Mehrabi, R., Nap, J. P. H., Ponomarenko, A., Rudd, J. J., Salamov, A., Schmutz, J., Schouten, H. J., Shapiro, H., Stergiopoulos, I., Torriani, S. F. F., Tu, H., de Vries, R. P., Waalwijk, C., Ware, S. B., Wiebenga, A., Zwiers, L.-H., Oliver, R. P., Grigoriev, I. V. and Kema, G. H. J. (2011). Finished genome of the fungal wheat pathogen *Mycosphaerella graminicola* reveals dispensome structure, chromosome plasticity, and stealth pathogenesis. *PLoS Genetics* 7(6), e1002070.

Goodwin, S. B., Cavaletto, J. R., Hale, I. L., Thompson, I., Xu, S. S., Adhikari, T. B. and Dubcovsky, J. (2015). A new map location of gene Stb3 for resistance to *Septoria tritici* blotch in wheat. *Crop Science* 55, 35–43.

Grandaubert, J., Bhattacharyya, A. and Stukenbrock, E. H. (2015). RNA-seq-based gene annotation and comparative genomics of four fungal grass pathogens in the genus Zymoseptoria identify novel orphan genes and species-specific invasions of transposable elements. *G3: Genes, Genomes, Genetics* 5, 1323–33.

Griffin, M. J. and Fisher, N. (1985). Laboratory studies on benzimidazole resistance in *Septoria tritici*. *EPPO Bulletin* 15, 505–11.

Guo, J., Guo, J., Ma, X., Peng, J. and Zeng, F. (2014). Advances on infection and pathogenic mechanism of *Mycosphaerella graminicola*. *Chinese Agricultural Science Bulletin* 30, 20–5 (in Chinese with an English abstract).

Hardwick, N. V., Jones, D. R. and Slough, J. E. (2001). Factors affecting diseases of winter wheat in England and Wales, 1989–98. *Plant Pathology* 50, 453–62.

Hirschhorn, J. N. and Daly, M. J. (2005). Genome-wide association studies for common diseases and complex traits. *Nature Reviews Genetics* 6, 95–108.

Jaccoud, D., Peng, K., Feinstein. D. and Kilian, A. (2001). Diversity arrays: A solid state technology for sequence information independent genotyping. *Nucleic Acids Research* 29, E25.

Jing, H.-C., Lovell, D., Gutteridge, R., Jenk, D., Kornyukhin, D., Mitrofanova, O. P., Kema, G. H. J. and Hammond-Kosack, K. E. (2008). Phenotypic and genetic analysis of the Triticum monococcum–*Mycosphaerella graminicola* interaction. *New Phytologist* 179, 1121–32.

Jørgensen, L. N., Hovmøller, M. S., Hansen, J. G., Lassen, P., Clark, B., Bayles, R., Rodemann, B., Flath, K., Jahn, M., Goral, T., Czembor, J. J., Cheyron, P., Maumene, C., De Pope, C., Ban, R., Nielsen, G. C., and Berg, G. (2014). IPM Strategies and their dilemmas including an introduction to www.eurowheat.org. *Journal of Integrative Agriculture* 13, 265–81.

Joshi, L. M., Renfro, B. L., Saari, E. E., Wilcoxson, R. D. and Raychaudhuri, S. P. (1970). Diseases of wheat in India other than rusts and smuts. *Plant Disease Reporter* 54, 594–7.

Kellner, R., Bhattacharyya, A., Poppe, S., Hsu, T. Y., Brem, R. B. and Stukenbrock, E. H. (2014). Expression profiling of the wheat pathogen *Zymoseptoria tritici* reveals genomic patterns of transcription and host-specific regulatory programs. *Genome Biology and Evolution* 6(6), 1353–65.

Kema, G. H. J., Annone, J. G., Sayoud, R., van Silfhout, C. H., van Ginkel, M. and deBree, J. (1996). Genetic variation for virulence and resistance in the wheat–*Mycosphaerella graminicola* pathosystem I. Interactions between pathogen isolates and host cultivars. *Phytopathology* 86, 200–12.

Keon, J., Antoniw, J., Carzaniga, R., Deller, S., Ward, J. L., Baker, J. M., Beale, M. H., Hammond-Kosack, K. and Rudd, J. J. (2007). Transcriptional adaptation of *Mycosphaerella graminicola* to programmed cell death (PCD) of its susceptible wheat host. *Molecular Plant-Microbe Interactions* 20, 178–93. doi:10.1094/MPMI-20-2-0178.

Kettles, G. J. and Kanyuka, K. (2016). Dissecting the molecular interactions between wheat and the fungal pathogen *Zymoseptoria tritici*. *Frontiers in Plant Science* 7, 508. doi:10.3389/fpls.2016.00508.

King, J. E. and Griffin, M. J. (1985). Survey of benomyl resistance in *Pseudocercosporella herpotrichoides* on winter wheat and barley in England and Wales in 1983. *Plant Pathology* 34, 272–83.

King, J. E., Cook, R. J. and Melville, S. C. (1983). A review of *Septoria* diseases of wheat and barley. *Annals of Applied Biology* 103, 345–73.

Kirikyali, N., Diez, P., Luo, J., Hawkins, N. and Fraaije, B. A. (2017). Azole and SDHI sensitivity status of *Zymoseptoria tritici* field populations sampled in France, Germany and the UK during 2015. In: H. B. Deising, B. Fraaije, A. Mehl, E. C. Oerke, H. Sierotzki and G. Stammler (Eds), *Modern Fungicides and Antifungal Compounds*, Volume VIII. Deutsche Phytomedizinische Gesellschaft, Braunschweig, pp. 153–8.

Kleemann, J. and Mehl, A. (2017). SDHI sensitivity status of *Zymoseptoria tritici* and Botrytis cinerea. In: H. B. Deising, B. Fraaije, A. Mehl, E. C. Oerke, H. Sierotzki and G. Stammler (Eds), *Modern Fungicides and Antifungal Compounds*, Volume VIII. Deutsche Phytomedizinische Gesellschaft, Braunschweig, pp. 165–70.

Kollers, S., Rodemann, B., Ling, J., Korzun, V., Ebmeyer, E., Argillier, O., Hinze, M., Plieske, J., Kulosa, D., Ganal, M. W. and Röder, M. S. (2013). Genetic architecture of resistance to *Septoria tritici* blotch (*Mycosphaerella graminicola*) in European winter wheat. *Molecular Breeding* 32, 411–23.

Kraan, G. and Nisi, J. E. (1995). Septoriosis del trigo en la republica Argentina situacion del cultivo frente a la enfermedad. In: S. Gilchrist, L. M. van Ginkel, A. McNab and G. H. J. Kema (Eds), *Proceedings of the Septoria tritici Workshop*. CIMMYT, Mexico, D.F., pp. 1–8.

Liu, Y., Zhang, L., Thompson, I. A., Goodwin, S. B. and Ohm, H. W. (2013). Molecular mapping re-locates the Stb2 gene for resistance to *Septoria tritici* blotch derived from cultivar Veranopolis on wheat chromosome 1BS. *Euphytica* 190, 145–56.

Madariaga, R. and Scharen, A. L. (1985). *Septoria triciti* blotch in Chilean wild oat. *Plant Disease* 69, 126–7.

Marshall, R., Kombrink, A., Motteram, J., Loza-Reyes, E., Lucas, J., Hammond-Kosack, K. E., Thomma, B. P. H. J. and Rudd, J. J. (2011). Analysis of two in planta expressed LysM effector homologs from the fungus *Mycosphaerella graminicola* reveals novel functional properties and varying contributions to virulence on wheat. *Plant Physiology* 156 (2), 756–69.

Mavroeidi, V. I. and Shaw, M. W. (2005). Sensitivity distributions and cross-resistance patterns of *Mycosphaerella graminicola* to fluquinconazole, prochloraz and azoxystrobin over a period of 9 years. *Crop Protection* 24, 259–66.

McCartney, C. A., Brûlé-Babel, A. L., Lamari, L. and Somers, D. J. (2003). Chromosomal location of a race-specific resistance gene to *Mycosphaerella graminicola* in the spring wheat ST6. *Theoretical and Applied Genetics* 107, 1181–6.

McDonald, B. A. and Linde, C. (2002). Pathogen population genetics, evolutionary potential, and durable resistance. *Annual Review of Phytopathology* 40, 349–79.

McDonald, M. C., McGinness, L., Hane, J. K., Williams, A. H., Milgate, A. and Solomon, P. S. (2016). Utilizing gene tree variation to identify candidate effector genes in *Zymoseptoria tritici*. *G3: Genes, Genomes, Genetics* 6, 779–91.

McKendry, A. L., Henke, G. E. and Finney, P. L. (1995). Effects of Septoria leaf blotch on soft red winter wheat milling and baking quality. *Cereal Chemistry* 72, 142–6.

Miedaner, T., Zhao, Y., Gowda, M., Longin, C. F. H., Korzun, V., Ebmeyer, E., Kazman, E. and Reif, J. C. (2013). Genetic architecture of resistance to *Septoria tritici* blotch in European wheat. *BMC Genomics* 14, 858.

Morschhäuser, J. (2010). Regulation of multidrug resistance in pathogenic fungi. *Fungal Genetics Biology* 47, 94–106.

Murray, G. M. and Brennan, J. P. (2009). Estimating disease losses to the Australian wheat industry. *Australasian Plant Pathology* 38, 558–70.

Omrane, S., Sghyer, H., Audéon, C., Lanen, C., Duplaix, C., Walker, A.-S. and Fillinger, S. (2015). Fungicide efflux and the MgMFS1 transporter contribute to the multidrug resistance phenotype in *Zymoseptoria tritici* field isolates. *Environmental Microbiology* 17, 2805–23.

Orton, E. S., Rudd, J. J. and Brown, J. K. M. (2017). Early molecular signatures of responses of wheat to *Zymoseptoria tritici* in compatible and incompatible Interactions. *Plant Pathology* 66, 450–9.

Palma-Guerrero, J., Ma, X., Torriani, S. F. F., Zala, M., Francisco, C. S., Hartmann, F. E., Croll, D. and McDonald, B. A. (2017). Comparative transcriptome analyses in *Zymoseptoria tritici* reveal significant differences in gene expression among strains during plant infection. *Molecular Plant-Microbe Interactions* 30, 231–44.

Paul, S. and Moye-Rowley, W. S. (2014). Multidrug resistance in fungi: Regulation of transporter-encoding gene expression. *Frontiers in Physiology* 5, 143.

Ponomarenko A., Goodwin, S. B. and Kema, G. H. J. (2011). *Septoria tritici* blotch (STB) of wheat. *Plant Health Instructor*. doi:10.1094/PHI-I-2011-0407-01.

Priestley, R. H. (1978). Detection of increased virulence in populations of wheat yellow rust. In: P. R. Scott and A. Bainbridge (Eds), *Plant Disease Epidemiology*. Blackwell, Oxford, pp. 63–70.

Raman, R., Milgate, A. W., Imtiaz, M., Tan, M.-K., Raman, H., Lisle, C., Coombes, N. and Martin, P. (2009). Molecular mapping and physical location of major gene conferring seedling resistance to *Septoria tritici* blotch in wheat. *Molecular Breeding* 24, 153–64.

Ramdani, A. and Ibriz, H. (2000). Survey of wheat diseases in Morocco during three consecutive seasons, 1997, 98 and 99. *Option Méditerranéennes* 40, 407–11.

Ramdani, A., Halama, P., Elbekali, A. Y., Siah, A., Hafidi, M., Reignault, P., Tisserant, B. and Deweer, C. (2011). *Septoria tritici* blotch of wheat in Morocco: Current status and perspective. In: *Proceedings of the 8th International Symposium on Mycosphaerella and Stagonospora Diseases of Cereals*, 11–14 September 2011, México City. Mexico, p. 16.

Ray, S., Anderson, J. M., Urmeev, F. I. and Goodwin, S. B. (2003). Rapid induction of a protein disulfide isomerase and defense-related genes in wheat in response to the hemibiotrophic fungal pathogen *Mycosphaerella graminicola*. *Plant Molecular Biology* 53(5), 741–54.

Rudd, J. J., Kanyuka, K., Hassani-Pak, K., Derbyshire, M., Andongabo, A., Devonshire, J., Lysenko, A., Saqi, M., Desai, N. M., Powers, S. J., Hooper, J., Ambroso, L., Bharti, A., Farmer, A., Hammond-Kosack, K. E., Dietrich, R. A. and Courbot, M.(2015). Transcriptome and metabolite profiling of the infection cycle of *Zymoseptoria tritici* on wheat reveals a biphasic interaction with plant immunity involving differential pathogen chromosomal contributions and a variation on the hemibiotrophic lifestyle definition. *Plant Physiology* 167, 1158–85.

Saari, E. E. and Wilcoxson, R. D. (1974). Plant disease situation of high-yielding dwarf wheats in Asia and Africa. *Annual Review of Phytopathology* 12, 49–68.

Saintenac, C., Lee, W.-S., Cambon, F., Rudd, J. J., King, R., Marande, W., Bergès, H., Phillips, A. L., Uauy, C., Hammond-Kosack, K., Langin, T. and Kanyuka, K. (2017). Stb6 is a wall-associated kinase gene that provides gene-for-gene resistance against *Zymoseptoria tritici* in wheat. Abstract W795. In: *Plant & Animal Genome XX meetings*, San Diego, CA. https://pag.confex.com/pag/xxv/webprogram/Paper23713.html.

Sanin, S. S. (2012). Эпифитотии болезней зерновых культур: теория и практика. ГНУ Всероссийский научноисследовательскийинститут фитопатологии (ВНИИФ), Москва. [Epiphytotic diseases of cereal crops: Theory and practice. All-Russia Research Institute of Phytopathology (VNIIF), Moscow] (in Russian), 451pp.

Sanssené, J., Selim, S., Roisin-Fichter, C., Djerroud, L., Deweer, C. and Halama, P. (2011). Protective and curative efficacy of prothioconazole against isolates of *Mycosphaerella graminicola* differing in their in vitro sensitivity to DMI fungicides. *Pest Management Science* 67, 1134–40. doi:10.1002/ps.2163.

Shaner, G. and Finney, R. E. (1982). Resistance in soft red winter wheat to *Mycosphaerella graminicola*. *Phytopathology* 72, 154–8.

Shaner, G., Stromberg, E. L., Lacy, G. H., Barker, K. R. and Pirone, T. P. (1992). Nomenclature and concepts of pathogenicity and virulence. *Annual Review of Phytopathology* 30, 47–66.

Shipton, W. A., Boyd, W. R. J., Rosielle, A. A. and Shearer, B. I. (1971). The common *Septoria* diseases of wheat. *The Botanical Review* 37, 231–62.

Siah, A., Y. Elbekali, Y., Ramdani, A., Reignault, P., Torriani, F. F., Brunner, P. C. and Halama, P. (2014). QoI Resistance and mitochondrial genetic structure of *Zymoseptoria tritici* in Morocco. *Plant Disease* 98, 1138–44.

Sierotzki, H. and Scalliet, G. 2013. A review of current knowledge of resistance aspects for the next-generation succinate dehydrogenase inhibitor fungicides. *Phytopathology* 103, 880–7.

Simón, M. R., Perello, A. E., Cordo, C. A., Larran, S., van der Putten, P. E. L. and Struik, P. C. (2005). Association between *Septoria tritici* blotch, plant height, and heading date in wheat. *Agronomy Journal* 97, 1072–81.

Somasco, O. A., Qualset, C. O. and Gilchrist, D. G. (1996). Single-gene resistance to *Septoria tritici* blotch in the spring wheat cultivar 'Tadinia'. *Plant Breeding* 115, 261–7.

Tabib Ghaffary, S. M., Robert, O., Laurent, V., Lonnet, P., Margalé, E., van der Lee, T. A. J., Visser, R. G. F. and Kema, G. H. J. (2011). Genetic analysis of resistance to *Septoria tritici* blotch in the French winter wheat cultivars Balance and Apache. *Theoretical and Applied Genetics* 123, 741–54.

Tabib Ghaffary, S. M., Faris, J. D., Friesen, T. L., Visser, R. G. F., van der Lee, T. A. J., Robert, O. and Kema, G. H. J. (2012). New broad-spectrum resistance to *Septoria tritici* blotch derived from synthetic hexaploid wheat. *Theoretical and Applied Genetics* 124, 125–42.

Torriani, S. F. F., Brunner, P. C., McDonald, B. A. and Sierotzki, H. (2009). QoI resistance emerged independently at least 4 times in European populations of *Mycosphaerella graminicola*. *Pest Management Science* 65, 155–62.

Torriani, S. F. F., Melichar, J. P. E., Mills, C., Pain, N., Sierotzki, H. and Courbot, M. (2015). *Zymoseptoria tritici*: A major threat to wheat production, integrated approaches to control. *Fungal Genetics and Biology* 79, 8 12.

Trumpower, B. L. (1990). Cytochrome bc1 complexes of microorganisms. *Microbiological Reviews* 54, 101–29.

Tyagi, P. D., Joshi, L. M. and Renfro, B. L. (1969). Reaction of wheat varieties to *Septoria tritici* and report of an epidemic in north-western Punjab. *Indian Phytopathology* 22, 175–8.

Vagndorf, N., Nielsen, N. H., Edriss, V., Andersen, J. R., Orado, J., Jørgensen, L. N. and Jahoor, A. (2017). Genomewide association study reveals novel quantitative trait loci associated with resistance towards *Septoria tritici* blotch in north European winter wheat. *Plant Breeding* 136(4), 474–82. https://doi.org/10.1111/pbr.12490.

van den Berg, F., van den Bosch, F. and Paveley, N. D. (2013). Optimal fungicide application timings for disease control are also an effective anti-resistance strategy: A case study for *Zymoseptoria tritici* (*Mycosphaerella graminicola*) on wheat. *Phytopathology* 103, 1209–19.

van der Plank, J. E. (1966). Horizontal (polygenic) and vertical (oligogenic) resistance against blight. *American Potato Journal* 43, 43–52.

van Dijk, A. I. J. M., Beck, H. E., Crosbie, R. S., de Jeu, R. A. M., Liu, Y. Y., Podger, G. M., B. Timbal, B. and Viney, N. R. (2013). The Millennium Drought in southeast Australia (2001–2009): Natural and human causes and implications for water resources, ecosystems, economy, and society. *Water Resources Research* 49, 1040–57. doi:10.1002/wrcr.20123.

Wieczorek, T. M. and Jørgensen, L. N. (2017). Fungicide spray strategies avoiding resistance development in winter wheat pathogen *Zymoseptoria tritici*. In: H. B. Deising, B. Fraaije, A. Mehl, E. C. Oerke, H. Sierotzki and G. Stammler (Eds), *Modern Fungicides and Antifungal Compounds*, Volume VIII. Deutsche Phytomedizinische Gesellschaft, Braunschweig, pp. 301–6.

Wilson, R. E. (1985). Inheritance of resistance to *Septoria tritici* in wheat. In: A. L. Scharen (Ed.), *Septoria of Cereals – Proceedings of the Second International Septoria Workshop*, 2–4 August 1983, Bozeman, Montana, USA, pp. 33–5.

Yang, F., Li, W. and Jørgensen, H. J. L. (2013). Transcriptional reprogramming of wheat and the hemibiotrophic pathogen *Septoria tritici* during two phases of the compatible interaction. *PLoS ONE* 8(11), e81606. doi:10.1371/journal.pone.0081606.

Yang, F., Li, W., Derbyshire, M., Larsen, M. R., Rudd, J. J. and Palmisano, G. (2015). Unraveling incompatibility between wheat and the fungal pathogen *Zymoseptoria tritici* through apoplastic proteomics. *BMC Genomics* 16, 362.

Chapter 4

Diseases affecting wheat and barley: powdery mildew

Javier Sánchez-Martín, Salim Bourras and Beat Keller, University of Zürich, Switzerland

1 Introduction

Barley (*Hordeum vulgare* 2x = 14, HH) and bread wheat (*Triticum aestivum* L., 2x = 42, AABBDD), being a major source of animal feed and a staple human food, play a central role in agriculture (Oerke and Dehne, 2004). Under agricultural conditions, these crops are constantly challenged by abiotic and biotic stresses that reduce farm-level yield below the genetically determined potential. Among these constraints, fungal diseases reduce yield by 15% (Oerke and Dehne, 2004). Powdery mildew pathogens are among the most important cereal disease-causing organisms, and they can result in significant losses in yield and reduction of grain quality, posing a great challenge for barley and wheat agricultural production systems (Dean et al., 2012). In response to the need to counter this disease, breeding for quantitative and race-specific resistance and fungicide treatments have been the main means to control cereal mildews (Singh and Rajaram, 2002). In the past decade, powdery mildew resistance breeding has been increasingly supported by genomic and high-throughput genotyping technologies mostly on the host side, but there are emerging

http://dx.doi.org/10.19103/AS.2018.0039.04

applications on the pathogen side as well. These approaches have been implemented to accelerate the identification, selection and introduction of novel resistance sources in breeding programmes, as well as to gain more insight into basic pathogen evolution and population diversity.

This chapter aims at providing an overview of the current knowledge on wheat and barley resistance to powdery mildew. First, we will introduce basic biological concepts on disease epidemiology and management strategies specifically for mildew. Then, we will present emerging trends and technologies used in mildew resistance breeding and research, and how these can be combined into effective disease management strategies in the near future. Finally, we will propose additional directions for achieving durable resistance based on a deep understanding of the basic molecular mechanisms of cereal–mildew interactions.

2 The biology of the powdery mildew disease

2.1 The causal agent: *Blumeria graminis*

The powdery mildew diseases of cereals including barley and wheat are caused by the ascomycete *Blumeria graminis*. This fungus is the only species of the genus *Blumeria* and was previously classified as a member of the genus *Erysiphe*. Cereal powdery mildews are highly specific to their host and can be further divided into '*formae speciales*' (literally, special forms) corresponding to a subset of the pathogen species that can grow on one specific host species (Menardo et al., 2016). The powdery mildew diseases of barley and wheat are caused by *B. graminis* f. sp. *hordei* (*B.g. hordei*) and *B. graminis* f. sp. *tritici* (*B.g. tritici*) respectively, which have likely arisen and diverged shortly after the divergence of their hosts, wheat and barley, probably several millions of years ago (Wicker et al., 2013; Menardo et al., 2016). Hybridisation between cereal mildews has likely played an important role in the emergence and adaptation of new mildew forms. One prominent example is provided by *B.g. triticale*, a new powdery mildew pathogen that has emerged on triticale, an amphiploid (or 'artificial hybrid') between wheat and rye that remained immune to this disease until the 2000s. The *B.g. triticale* pathogen arose from a hybridisation event between wheat and rye powdery mildews, a process reminiscent of the emergence of its hybrid host (Menardo et al., 2016). Besides, *B.g. triticale* has retained infectivity towards wheat and rye, which is likely to favour the spread of this pathogen outside of the triticale cropping areas. Altogether, the evolutionary history of mildew as well as the rapid mechanisms leading to the emergence of new mildew forms substantiates the remarkable persistence of this disease in our cropping systems.

Powdery mildews are obligate biotrophic pathogens and they rely on living host cells to complete their life cycle (in contrast to necrotrophs which actively kill host cells). There are no morphological differences between wheat and barley mildews, and their life cycles are identical although they infect distinct cereal hosts (Fig. 1). Primary infections in the field can be initiated by the release of ascospores (sexual spores) from sexual fruiting bodies called chasmothecia, a highly resilient fungal structure. The sexual chasmothecia are produced towards the end of crop development upon mating of two mildew isolates with complementary mating types on the host. These dormant structures allow long-term survival of the primary mildew inoculum and produce new pathotypes resulting

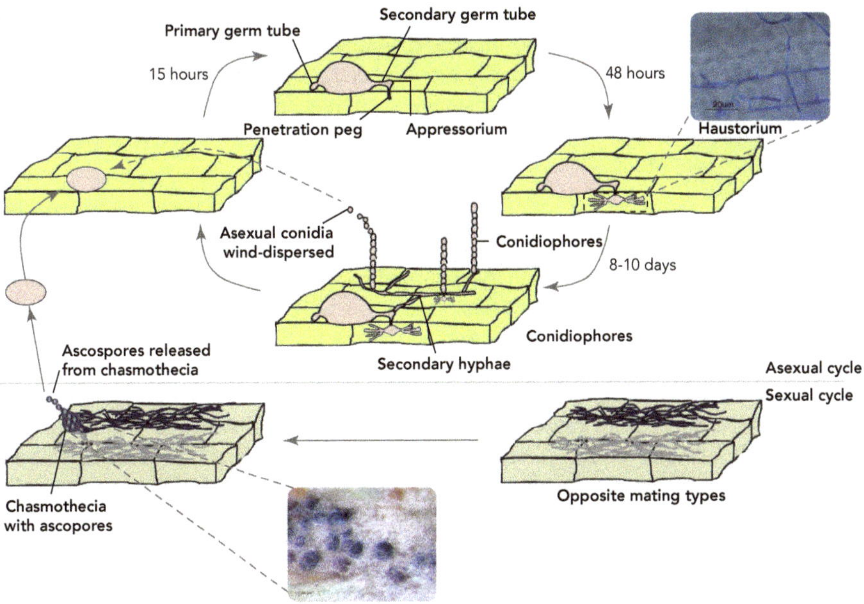

Figure 1 Life cycle of *Blumeria graminis*.

from sexual recombination, with new combinations of virulence. In addition, mildew can overwinter as chasmothecia on straw, or, in a Mediterranean context, oversummer as mycelium and conidia on volunteers, plants that serve as alternate host creating 'green bridges', providing inoculum for the following crop season.

The mildew life cycle typically starts by the landing of asexual conidia on a susceptible host leaf and then follows a highly coordinated sequence of events. Conidia best germinate at nearly saturated levels of relative humidity (~100%), and first form a primary germ tube, thought to be involved in sensing the host surface and probably gaining access to water (Green, Carver and Gurr, 2002). Subsequently, a secondary germ tube emerges, elongates and differentiates into a hooked, apical structure called the appressorium, which breaches the plant cuticle as a result of physical pressure and enzymatic degradation (Hückelhoven and Panstruga, 2011). Haustoria can only be found in epidermal cells, from which they produce exclusively aerial secondary hyphae that colonise the plant surface. Successful formation of this fungal structure is critical to mildew survival, in that it mediates highly regulated interactions with the host to ensure permanent suppression of host immunity, cell-to-cell exchanges and nutrient uptake. After six days, the aerial hyphae start differentiating to form branches of contiguous conidia called conidiophores, which reach maturity at 8–10 days. At this point, the first asexual cycle is completed, and the newly formed asexual conidiospores can initiate another cycle of asexual propagation.

Early symptoms of the powdery mildew disease are typically visible on the lower leaves and sheaths of wheat and barley seedlings, and symptoms can also include chlorotic flecks surrounding nascent mildew pustules. Under disease-conducive conditions (cyclic wet and dry periods), the rapid succession of asexual cycles, lasting only 7–10 days, yields large amounts of inoculum in a relatively short period of time. Patches of white mycelium, heavily loaded with conidiophores (hence the white powder), appear on the upper surface of the above-ground plant organs, including leaves, stems and ears, and sometimes even reach spikes. Conidia can be easily dispersed by wind, resulting in rapid spread of the disease to the whole field. Long-distance dispersal of the mildew inoculum is also common and assumed to play a role in the rapid adaptation of mildew population to the cereal landscape (Brown, 2002).

2.2 Geographical distribution of mildew disease and yield losses

Powdery mildew fungi are long-lasting, persistent diseases in cereal agricultural ecosystems. The global incidence of wheat mildew has remained stable over time on a worldwide basis since the 1970s and mildew is one of the most destructive diseases of barley (Jørgensen, 1988; Morgounov et al., 2012). While powdery mildews occur in all barley- and wheat-growing regions worldwide, the disease has been particularly damaging in areas of temperate, moist and cold climate like Europe, eastern North America and parts of Asia (Bennett, 1984; Cowger et al., 2012). In the last decades, the pathogen has spread to warmer and drier regions as a result of the adoption of modern agricultural practices (high sowing densities, intensive irrigation and nitrogen fertilisation) which facilitate mildew infection development (Thompson, Brown and Woodward, 1993; Cowger et al., 2012). For instance, mildew has become the most damaging disease of barley in Western Australia, while wheat mildew is an emerging problem in areas that have remained 'mildew-free', for example China or the Mid-Atlantic regions of the United States (Tucker et al., 2013; Chen et al., 2007; Cowger et al., 2016).

As biotrophic organisms depending on a living host, powdery mildew infections rarely prove fatal to the barley and wheat plants. Yet, cereal mildews are responsible for significant yield losses and significant depreciation of grain quality in these crops worldwide. Yield losses ranging from 5% to 34% have been reported in wheat and up to 40% in barley (Bennett, 1984; Chaure, Gurr and Spanu, 2000). This relates in particular to a reduced photosynthetic area due to fungal growth and dense sporulation. The impact on the yield is particularly damaging when disease severity is highest between the flag leaf emergence and the heading/flowering stages, resulting in the production of empty seed tillers and reduced malt quality grain (Cunfer, 2002; Tucker et al., 2013).

3 Agricultural management strategies to control mildews

For a disease to occur, three factors must be present simultaneously: a susceptible host, a virulent pathogen and a suitable environment. The interaction between these factors was conceptualised in the model of 'The disease triangle' by Stevens (1960). Therefore, an integrated disease management strategy must take into account all these factors.

In this section we will briefly describe cultural practices that may influence the environment in ways that reduce the ability of the pathogen to cause disease. In subsequent sections we first focus on the host, discussing recent technological advances that allow the identification of new sources of resistance and their genetic and molecular characterisation, and then on the pathogen, discussing recent progress made in the discovery of effectors. Our emphasis will be on the advances made in the understanding of the cereal-powdery mildew pathosystem based on recent advances in genomics and genetics and how this knowledge can be translated to increase the durability of host resistance in agriculture.

We differentiate two types of environments in agroecosystems. On the one hand, there is the atmospheric environment. It is expected that the undergoing climate change will lead to milder winters, drier summers and higher humidity. This will result in multiple effects on the occurrence and severity of plant diseases, which are difficult to predict. The rising CO_2 concentration levels could reduce cereal mildew incidence via an increase of pathogen latent periods and host resistance, as CO_2 has been reported to promote papilla formation that arrests powdery mildew at the penetration stage (Coakley, Scherm and Chakraborty, 1999). On the contrary, it is predicted that cereal mildews will become more severe via higher survival during winter and earlier epidemics during spring (West et al., 2015). Therefore, it is difficult to forecast the impact of climate change on cereal mildews.

On the other hand, there is the environment within the crop canopy. Today's agroecosystems primarily aim at crop yield based on the use of semi-dwarf varieties planted in dense systems supported with high nitrogen fertilisation rates and, when possible and needed, irrigation. This results in thicker canopies, which, in turn, leads to higher relative humidity levels, and decreased wind velocity. Together, these factors create a conducive microclimate for mildew development via the enhancement of spore germination and the formation of appressoria (Thompson, Brown and Woodward, 1993; Cowger et al., 2012). Therefore, it is advisable to implement practices that would create a more open crop canopy structure and less disease-conducive microclimate to counter the disease, such as moderate seeding rates or irrigation levels (Olesen et al., 2003). However, it seems unlikely that farmers would be open to such measures considering that the main driver in agrosystems is crop yield. Besides, since cereal mildews are biotrophic pathogens, any factor that promotes a proper nutritional and physiological status of the plant, may, in turn, foster powdery mildew development. Nitrogen is the most important macronutrient for cereal crops and a direct link between nitrogen levels in the plant and powdery mildew incidence has been reported, especially with single spring applications (Last, 1954; Broscious, Frank and Frederick, 1985). Therefore, it is recommended to split nitrogen applications in winter and spring (Olesen et al., 2003). Finally, crop rotation is ineffective as cereal mildews are spread by windborne spores (Brown, 2002). However, barley and wheat powdery mildew can overwinter on volunteers that may serve as a source of inoculum. Therefore, removing the volunteers by grazing or other means will assist in reducing inoculum loads (Cunfer, 2002).

The remarkable advances in powdery mildew disease control in wheat and barley are based on two practices: the use of fungicides and breeding for resistance. Chemical control of powdery mildews by broad-spectrum fungicides is possible. Indeed, several highly effective systemic fungicides are available to avoid powdery mildew epidemics. However, their use represents a direct cost, and it is not always economically feasible (Leath and Bowen, 1989). Besides, the use of systemic fungicides has become a less efficient strategy to control cereal mildews as several studies have reported an increase of fungicide resistance in mildews and difficulties with correct timing of applications. Furthermore, the use and introduction of new formulations has become highly constrained by environmental legislation, especially in Europe (Lucas et al., 2015; Hillocks, 2012). Alternatively, the biological control of mildews, especially in vegetables and ornamental plants, by employing biopesticides is receiving more and more attention. The market is much less developed for cereal crops, though there are already tried and tested biopesticides for wheat (e.g. Regalia®, Su et al., 2012). There is basic research showing very encouraging developments on the identification of antagonistically acting organisms (Gafni et al., 2015; Gao et al., 2015). Although nascent, the use of biopesticides to control cereal mildews should be kept in mind as a promising new strategy in the near future, overall within the context of an increasing organic agricultural production.

Therefore, strategies based on conventional breeding of mildew-resistant varieties have become a major disease control approach (Huang and Röder, 2004). Host resistance to counter powdery mildew fungi can be placed into two broad categories: race-specific and broad-spectrum resistance.

4 Race-specific resistance: new approaches for fast gene isolation in wheat and barley

Race-specific resistance is genetically controlled by major resistance (*R*) genes, which confer resistance to some but not all powdery mildew races. Race-specific resistance can be described by the gene-for-gene model (Flor, 1971), and it is triggered when host immune receptors encoded by *R* genes recognise their cognate AVR proteins from the pathogen (Jones and Dangl, 2006). *R* genes usually encode nucleotide-binding, leucine-rich repeat receptor (NLR) proteins (Qi and Innes, 2013) and are widely present in plant genomes. In the particular case of barley and wheat genomes, 191 and 1185 *R*-type genes have been annotated, respectively (Mayer et al., 2012; Clavijo et al., 2016). Race-specific resistance remains a cornerstone of breeding to combat powdery mildew as most *R* genes show a simple dominant Mendelian inheritance, which offers an easy and rapid way to introgress resistance. However, *R*-mediated resistance tends to be short-lived. According to the gene-for-gene model, an *R* gene recognises its matching *Avr* gene of the pathogen. Furthermore, the interactions between host *R* genes and pathogen *Avrs* are thought to be highly specific. Therefore, a slight change in the *Avr* gene would lead to an unrecognised effector, resulting in a susceptible host. Under agricultural conditions with high selection pressure, powdery mildews, which have very dynamic and rapidly evolving populations, often adapt to such single resistance genes by modifying effectors recognised by the NLRs. The final result is the rapid breakdown of *R*-mediated resistance in the field (McDonald and Linde, 2002).

In barley, more than 85 race-specific resistance genes have been genetically identified and designated at 11 loci (reviewed in Jørgensen and Wolfe (1994), mostly to the *mildew resistance locus A* (*Mla*), which shows an extreme functional diversification with more than 30 alleles described showing different specificities to barley mildew, with many of them cloned (Seeholzer et al., 2010). In the case of wheat, to date, 77 *Pm* genes/alleles have been identified at 53 loci (*Pm1–Pm53; Pm18* = *Pm1c*, *Pm22* = *Pm1e*, *Pm23* = *Pm4c*, *Pm31* = *Pm21*, *Pm48* = *Pm46*), and assigned with an official name in the wheat gene catalogue, and 37 additional genes not yet genetically mapped have been reported (https://shigen.nig.ac.jp/wheat/komugi/genes/macgene/supplement2017.pdf).

Despite the extensive diversity of *R* genes, race-specific resistance deployed in barley and wheat breeding programmes is based on only a few *R* genes in practical breeding. In barley, the genetic base of this type of resistance rests primarily on the *Mlra*, *Mla6*, *Mla14* and *Ml(lv)* alleles, present in 72 (69.23%) out of the 104 European winter barley varieties commercially available. In the case of wheat, a similar situation arises as the *Pm2*, *Pm3a*,

Pm3b, *Pm4b*, *Pm5*, *Pm6* and *Pm8* are the alleles/genes on which race-specific resistance is based (Bundessortenamt, 2016).

The widespread use of these few major resistance genes in agriculture has exerted a high selection pressure on pathogen populations, leading to the rapid evolution of virulent races. Consequently, there has been a loss of the efficiency of *R*-based resistance against powdery mildew in European barley production systems (Brown, 2015) and the *Pm*-based resistance in almost all wheat-growing areas (Table 1).

Two strategies have been proposed to improve the durability of race-specific resistance: multilines/mixtures and gene pyramiding. The deployment of major *R* genes in rotations in space and time, in multilines or as cultivar mixtures reduces the chance of the pathogen to adapt to host resistance due to barrier effects or limited target tissue suitable to the pathogen (Wolfe, 1985; Pink, 2002). The use of barley mixtures resulted in the reduction of mildew infection of up to 80% in European barley agricultural production systems during the 1980s and 1990s (Wolfe, 1992). On the contrary, the deployment of this approach in wheat has been very limited due to the lack of uniformity among varieties (Pink, 2002).

Alternatively, the combination (pyramiding) of *R* genes into a single cultivar should lengthen the durability of the resistance as the pathogen would need to accumulate several mutations to break the combined resistance provided for the combination of *R* genes (Hsam and Zeller, 2002; McDonald and Linde, 2002). This practice has resulted in many commercial cultivars containing combinations of major genes conferring resistance to powdery mildew. As an example, among the 125 European winter and spring wheat varieties currently available for commercial cultivation, 79 have a combination of *Pm* genes, combining two or several of them. Similarly, in the case of barley, out of 66 two- and six-row varieties reported to have known *R* genes, 33 are pyramids of *R* genes conferring resistance against barley mildew (Bundessortenamt, 2016).

One could think that a gene/allele-pyramiding approach would lead to an additive resistance because of the combined resistance spectrum from each *R* gene. However,

Table 1 Distribution of *Pm* genes defeated in the main wheat-growing areas

Gene/Allele	Resistance defeated in	References
Pm1	China, Europe	Hsam and Zeller (2002) and Wang et al. (2005)
Pm2	UK, Europe	Hsam and Zeller (2002) and UKCPVS (2017)
Pm3a	China, USA, Europe	Hsam and Zeller (2002), Parks et al. (2008) and Wang et al. (2005)
Pm4a	USA	Cowger et al. (2009)
Pm4b	UK, USA, Europe	Hsam and Zeller (2002), Parks et al. (2008) and UKCPVS (2017)
Pm5	UK, China, USA, Europe	Hsam and Zeller (2002), Parks et al. (2008), UKCPVS (2017) and Wang et al. (2005)
Pm6	UK, Europe	Hsam and Zeller (2002); UKCPVS (2017)
Pm7	China, USA	Parks et al. (2008) and Wang et al. (2005)
Pm8	UK, China, USA, Europe	Hsam and Zeller (2002), Parks et al. (2008), UKCPVS (2017) and Wang et al. (2005)
Pm17	China, USA	Cowger et al. (2009) and Li et al. (2011)

practical breeding experience shows us that this does not hold true in all cases. There are three main challenges in implementing the combination strategy of *R* genes. The first one is the possible suppression of introgressed genes by their homologues in certain genetic backgrounds: this is the case of the *Pm8* rye-derived gene suppressed by its homologue *Pm3* in wheat (McIntosh et al., 2011; Hurni et al., 2014). On the other hand, genetic interactions (dominance and epistatic effects) between different genes/alleles pyramided may further complicate this approach (Hurni et al., 2014). Recent studies have started to shed light on the molecular mechanisms behind the suppression, dominance and epistatic phenomena between *R* genes/alleles by using transient expression assays in *Nicotiana benthamiana* via *Agrobacterium tumefaciens* infiltrations (Hurni et al., 2014; Stirnweis et al., 2014b, see Section 7.2). Finally, pyramided *R* genes are challenged by the high genetic variability that powdery mildew populations exhibit. Sexual recombination is considered as a relevant factor contributing to that variability (Drenth and Goodwin, 1999). As a result, virulence genes can be resorted by recombination, yielding novel virulence phenotypes that would overcome the resistance provided by the pyramid of *R* genes (McDonald and Linde, 2002).

These aspects represent limiting factors for resistance pyramiding and should be identified before the lengthy breeding process starts to exploit race-specific resistance in a more efficient manner. To that end, isolation of *R* genes in a rapid and efficient manner is an ineluctable prerequisite to test for these interactions between genes/alleles and further select the 'most compatible' alleles/genes in a pyramid. Besides, the molecular isolation of genes can circumvent another limitation that breeding practice has always had to cope with: linkage drag. Wild grasses and ancestors have served as a reservoir of genes with high agronomic value. For example, 44 out of 77 of the *Pm* genes are derived from wild relatives of bread wheat. However, only few of them have been employed in breeding. Often, these genes are embedded in large chromosomal fragments with additional genes, which often result in undesired traits. Consequently, these *Pm* genes have been discarded for their incorporation into elite wheat varieties: this is the case of the *Pm12*, *Pm16* and *Pm20* genes among others (Hsam and Zeller, 2002).

Isolation of *R* genes in barley and wheat based on traditional map-based cloning strategies is hampered by their large genomes (~5000 Mb and ~17 000 Mb, respectively) and their highly repetitive nature (Kunzel, Korzun and Meister, 2000; Choulet et al., 2014). In the particular case of wheat, this is exacerbated by the presence of homologues and gene duplications as a consequence of its allopolyploid nature (Choulet et al., 2014; IWGSC, 2014). Consequently, gene isolation traditionally was a burdensome and lengthy job, not to say impossible in many occasions, in particular from the extremely recombination-poor pericentromeric regions. However, recently the pace of *R* gene discovery in barley and wheat has accelerated owing to impressive developments in DNA sequencing and genomic technologies.

Although technically possible, sequencing whole genomes of parents differing for a trait of interest is still not practical in barley or wheat. As an alternative, the International Wheat Genome Sequencing Consortium proposed the use of flow-sorted chromosomes as a pivotal point to genome complexity reduction (Giorgi et al., 2013; Doležel et al., 2014). This genome complexity reduction combined with high-quality sequence information via long-range assembly has been recently successfully used for gene cloning in wheat (Thind et al., 2017). This technique, however, partially depends on recombination, a particular feature to bear in mind when working with barley and wheat genomes as these genomes have large segments with very low or even suppressed recombination (Choulet et al., 2014). Besides, most of the *R* genes tend to reside in complex clusters which hinder their functional dissection by recombination and comparison to a reference sequence (Meyers et al., 2003).

To overcome these obstacles, other approaches have been based on 'mutational genomics' to identify the gene of interest in one go. This strategy is built around mutagenesis: phenotypic screening of mutants and sequencing the fraction of the genome containing the gene of interest in multiple independent mutants. By looking at the mutation overlap using the wild type as a reference, a single candidate can be identified, the one that is mutated in each independent mutant. Combining resistance (*R*) gene exome sequence capture (RenSeq) with mutagenesis, dubbed 'MutRenSeq', Steuernagel et al. (2016) successfully cloned two stripe rust-conferring resistance genes in wheat. However, the drawback of this technique is that it is contingent upon gene annotation in a reference genome, which is often absent from the reference cultivar (Mago et al., 2014).

To counter the bias imposed by exome capture, Sánchez-Martín and colleagues (2016) followed a chromosome approach, named 'Mutant Chromosome Sequencing' (MutChromSeq), for a rapid major resistance gene isolation. The sequences of flow-sorted chromosomes of a handful of independently derived EMS-mutants are compared to the wild type to identify the gene of interest. The approach was successfully applied for the isolation of a wax biosynthesis gene in barley, and a major gene conferring resistance to powdery mildew in wheat, the *Pm2* gene. MutChromSeq can be used to tackle genes in areas of suppressed recombination, does not require fine mapping or/and construction of a physical map across the interval and enables reference-free forward genetics in barley and wheat underpinning genes of agronomic interest in non-reference cultivars. Nevertheless, the technique is limited to genes of interest that can be associated with a clear phenotype defined by mutagenesis and a prior knowledge of which chromosome the gene is on.

These techniques, each with advantages and disadvantages, provide innovative ways to isolate genes in crops with complex genomes – such as barley and wheat in a fraction of the time it would typically take for map-based approaches. Once *R* genes can be isolated quickly and efficiently, transient expression assays would allow testing for 'incompatibilities' among *R* genes before pyramiding them into cultivars.

Being able to isolate *R* genes rapidly does not have, however, a direct and practical impact on breeding if one is not able to generate homozygous lines quickly. Indeed, breeders have always had to contend with the generation time of plants as a major operational bottleneck within breeding practices. Single seed descent, off-season nurseries and the doubled haploids technology have been the major breakthroughs to speed up practical breeding in the last decades. Still, a minimum of 8–10 yr is required to develop a cultivar from an initial cross. Hence, shorter plant-generation times have always been a priority in breeding. Recently, Zheng and colleagues (2013) developed a procedure combining embryo culture with growth under constant light, among other factors, which allow the production of up to eight generations in wheat and nine generations of barley per annum. Besides, this procedure is done under controlled conditions, which represent an additional benefit to select traits whose selection can be subjected to environmental factors. This kind of advance represents a considerable step forward in shortening times with direct implication in practical breeding. For example, this procedure, and variants thereof are now being put into practice to introgress and phenotype adult plant resistance to leaf rust in wheat (Riaz et al., 2016).

In conclusion, the scene is now set for the cloning, testing for eventual incompatibilities and efficient introgression of major genes into homozygous elite cultivars in a quick and efficient way. The molecular cloning of *R* genes will also allow developing highly specific markers for marker-assisted selection in breeding programmes.

5 Quantitative disease resistance: identification of partially acting resistance genes and first insights into their molecular mechanism

The second type of resistance to powdery mildew is often referred to as 'quantitative resistance' (QR) as it is the result of the joint action of several genes (Niks, Qi and Marcel, 2015). QR has also been defined as slow-mildewing (Roberts and Caldwell, 1970). QR does not prevent the infection but retards the development of disease symptoms at different stages of the infection, either previous to cell penetration or post penetration, such as reduction of successful infections, prolonged latent period, smaller size of sporulating structures or reduced sporulation (Agrios, 2005; Li et al., 2014). Alternatively, QR is considered as a kind of 'growth-stage-specific' resistance as it occurs at the adult plant stage. Therefore, it is also frequently referred to as 'adult plant resistance' (Gustafson and Shaner, 1982).

QR is more durable and effective against a broader range of mildew races than *R*-based immunity as it does not depend on the recognition of rapidly evolving effector molecules of the pathogen (Lagudah, 2011). Indeed, we find some examples of success stories of commercial varieties with stable and lasting QR to powdery mildew (Table 2).

Moreover, QR gene combinations have been seen to provide near immunity to powdery mildew, which makes QR play key role in sustainable management of plant diseases (Singh et al., 2011; Mundt, 2014).

Although the multi-genetic nature of QR is widely accepted, single genes with major effects may also naturally control this type of resistance. Some of these rare, but very relevant, cases have been found in barley and wheat. Loss-of-function mutations of the *Mlo* gene in barley confer broad-spectrum resistance against virtually all known *Bgh* isolates (Jørgensen, 1992). Barley lines carrying *mlo*-recessive mutations arrest the infection at the penetration stage, preventing the pathogen from developing the haustorium. Importantly, the resistance is not associated with the death of the attacked host cell. The *mlo* powdery mildew resistance gene was found as a spontaneously occurring mutation in Ethiopian accessions, named *mlo-11* (Jørgensen, 1992). Since then, more than 40 *mlo* mutants have been generated by different approaches: gene silencing, induced mutagenesis and so forth (Reinstädler et al., 2010). Of these, the natural *mlo-11* and the induced *mlo-9* alleles have been widely deployed in spring barley cultivars, keeping the disease at bay during decades (Lyngkjær et al., 2000). As a side effect, the incidence of the disease has declined in importance in crops of winter barley, as it is assumed that spring barley would otherwise act as a source of inoculum (Makepeace et al., 2007). However, *mlo*-based resistance is currently not yet employed in winter barley varieties so as to reduce the opportunities for the pathogen to adapt to this unique resistance (Dreiseitl, 2016; Henryk et al., 2016).

Since the *mlo*-based resistance in barley against powdery mildew is durable and broad-spectrum, wheat researchers have always longed to establish a *mlo*-based strategy to

Table 2 List of commercial wheat varieties with QR to powdery mildew

Cultivar name	Origin	Resistance duration	Marker/gene	Reference
Bezostaya 1	Russia	>30 yr	*Lr34/Yr18/Sr57/Pm38*	Morgounov et al. (2012)
Courtot	France	>30 yr	*Xgwm397*, Chr 4A	Bougot et al. (2006) and Chantret et al. (2001)
Forno	Switzerland	>10 yr[1]	*Xgwm397*, Chr 4A	Keller et al. (1999)
Fukuho-komugi	Japan	>25 yr[1]	*Lr34/Yr18/Sr57/Pm38*	Liang et al. (2006)
Knox	EEUU	>30 yr		Shaner (1973)
Folke	Sweden	>30 yr		Lillemo et al. (2012)
Massey	EEUU	>30 yr		Liu et al. (2001)
Lumai	China	>20 yr		Lan et al. (2010)
Bainong 64	China	>20 yr		Lan et al. (2009)
Saar	CIMMYT	~15 yr[1]	*Lr34/Yr18/Sr57/Pm38*	Lillemo et al. (2008)
RE714	France	>20 yr[1]		Chantret et al. (2001) and Mingeot et al. (2002)
RE9001	France	>20 yr[1]		Bougot et al. (2006)
Pingyuan 50	China	~60 yr		Asad et al. (2014)
USG3209	EEUU	>20 yr		Tucker et al. (2007)

[1] Estimation calculated based on the year of release.

counter wheat powdery mildew. However, contrary to barley, the natural loss-of-function of the three wheat orthologues of barley *Mlo*, *TaMLO-A1*, *TaMLO-B1* and *TaMLO-D1*, has never been reported. Nevertheless, the advent of new technologies allows now the generation of *mlo*-based powdery mildew-resistant wheat lines. By knocking out all three *TaMLO* homologues via CRISPR/Cas9, Wang et al. (2014) generated mutant lines highly resistant to powdery mildew at seedling stage, similar to the triple *mlo* mutant combinations achieved via TILLING (Targeting Induced Lesions IN Genomes) by Acevedo-Garcia and colleagues (2017). These works pave the way towards the generation of commercial *mlo*-based mildew-resistant wheat varieties.

Unlike the *mlo*-based resistance, the apparent polygenic nature of QR to powdery mildew in wheat has always complicated its management in breeding programmes, most probably due to underlying complex polygene-for-polygene interactions (Parlevliet and Zadocks, 1977). Correct assessments of QR imply extensive and precise measurements (Gustafson and Shaner, 1982), which are subjected to environmental influences, as well as host background and plant development stage. Moreover, the difficulty in identifying and selecting QR genes increases in the presence of race-specific resistance genes, which can mask the effects of QR genes (Lillemo, Bjørnstad and Skinnes, 2012). In the absence of reliable phenotypic markers associated to QR – only *Lr34* and *Lr46* were reported to be associated with a leaf tip necrosis phenotype (Singh, 1992; Rosewarne et al., 2006) – the study and use of QR in breeding have largely relied on genetic markers, which allow the dissection of QR into its genetic components by quantitative trait locus (QTL) analysis (Young, 1996). During the last few years, many QTL mapping studies aiming at finding diagnostic markers that could assist in QR selection in breeding programmes have been conducted. As a result, there are over a hundred of QTLs conferring QR to powdery mildew described (reviewed in Li et al., 2014), for some of which there are diagnostic markers available that have been regularly used in breeding. Undoubtedly, the recent incorporation of SNP-based arrays is assisting the efficient tagging of these QTLs, circumventing the low resolution of previously used markers (e.g. Yang et al., 2017).

In the future, the ideal situation for QR management would be the isolation of the genes underlying this type of resistance. In fact, major breakthroughs have been recently achieved in this regard with the isolation of *Lr34* and *Lr67* (Krattinger et al., 2009; Moore et al., 2015). Both are multi-pathogen broad-spectrum resistance genes, conferring partial resistance to rusts and powdery mildews. Therefore, and contrary to what was believed until now, single genes may control QR to powdery mildew in wheat as well, similar to the *mlo*-based resistance in barley. The molecular cloning of these genes enabled the development of gene-specific markers that have led to the identification of the genes underlying QR in some of those commercial cultivars widely employed (see Table 2).

Furthermore, the molecular cloning of these genes significantly adds to our understanding of what the molecular nature of QR is. Interestingly, *Lr34* and *Lr67* genes encode for membrane-localised transporter proteins, where spontaneous mutations after domestication led to the emergence of QR. This intriguing phenomenon is also observable in other pathosystems. For example, the *Pi21* gene, which confers partial, broad-spectrum resistance to rice blast, encodes for a protein with a heavy metal transport/detoxification domain (Fukuoka et al., 2009). Thanks to these types of discoveries, the identification of *Lr34*- or *Lr67*-like transporters among natural accessions and wild relatives and further modification to create genes conferring QR not only to powdery mildew but also to more diseases is getting increasingly easy (Krattinger and Keller, 2016a).

6 Recent insights into the molecular basis of race-specific resistance against *Blumeria graminis*

Mildew virulence patterns are usually described by defining the interaction of a particular race with host plants carrying specific *R* genes which are present in tester lines. This allows defining the virulence in natural populations based on gene-for-gene interactions (Flor, 1971). There, mildew resistance is solely activated in the presence of an active *Avr* allele in the pathogen, and resistance efficiency is correlated to the frequency of *Avr* alleles in mildew populations. It was shown in a worldwide collection of wheat and triticale mildew races that the resistance reaction conferred by the wheat *Pm2* resistance gene is controlled by a single gene in the pathogen: *AvrPm2*. There, allelic variation based on presence/absence of polymorphism of *AvrPm2* explains 100% of the variation in virulence patterns of *B.g. tritici*, and *B.g. triticale* on *Pm2* wheat (Praz et al., 2017). Mildew virulence can also be determined by non-conventional gene-for-gene interactions involving more than one gene in the pathogen. In this case, *R* gene activation in the presence of its cognate *Avr* can be defeated by a third pathogen-encoded suppressor gene (*Svr*). In wheat the *Pm3f* resistance is controlled by such *Avr-R-Svr* interaction, and both avirulence and suppressor genes have been identified (*AvrPm3*$^{a2/f2}$ and *SvrPm3*$^{a1/f1}$, respectively) (Bourras et al., 2015;

Parlange et al., 2015). It was also shown in a natural population that mildew virulence patterns of *Pm3f* can be explained by specific *Avr-Svr* allelic combinations. In particular, there was a general trend for increased virulence on *Pm3f* in wheat in the presence of an active *SvrPm3*$^{a1/f1}$ suppressor allele (McNally et al., 2017). To date, four avirulence genes have been identified in wheat and barley mildews (*AvrPm3*$^{a2/f2}$, *AvrPm2*, Avr_{a1} and Avr_{a13}) (Bourras et al., 2015; Lu et al., 2016; Praz et al., 2017), thus providing the basis for monitoring the pathogen population based on allelic frequencies of *Avrs*.

All these developments together are increasing our knowledge on pathogen effectors as well as their modifiers and suppressors, and their distinctive characteristics are being defined (expression level, type of polymorphism, etc.). This enables the identification of *Avr* genes and the detection of redundancy in recognition spectra of *Pm* genes to be done very efficiently. Such knowledge can be used in the pre-breeding stage to develop effector-assisted breeding practices to evaluate at earlier stages the potential beneficial impact of *Pm* genes on barley and wheat breeding programmes against powdery mildews.

7 Case study: identification, functional characterisation and applications of the *Pm3* wheat powdery mildew resistance genes in wheat breeding

7.1 Molecular identification of wheat *Pm* genes: marker development and characterisation of novel *Pm3* alleles from gene bank accessions

Only four of the 53 known *Pm* genes have been cloned and molecularly studied to date: *Pm2* (Sánchez-Martín et al., 2016), *Pm3* and its alleles (Bhullar et al., 2009, 2010a; Srichumpa et al., 2005; Yahiaoui et al., 2006, 2009, 2004), *Pm8* and its suppressor (Hurni et al., 2013, 2014) and *Pm21* (Cao et al., 2011). With the exception of *Pm21*, which encodes a serine/threonine kinase protein, the other cloned *Pm* genes encode NLR (nucleotide-binding domain, leucine-rich repeat receptor) proteins. The cloning of the *Pm3* genes *Pm3a-Pm3g* revealed that they form a true allelic series of resistance genes. They are all derived from the same homologue in a gene family of related but distinct genes. Analysis of a diverse set of tetraploid wheat lines revealed the presence of only one active *Pm3* gene, *Pm3k* (Yahiaoui, Kaur and Keller, 2009), suggesting that much of the evolution of the *Pm3* allelic series occurred after wheat domestication (Krattinger and Keller, 2016b). The molecular identification of the *Pm3* alleles was the basis for the development of highly specific molecular markers which are widely used in breeding programmes (Tommasini et al., 2006).

Gene banks provide a rich source of genetic diversity, but they are not well characterised at the molecular level and for breeding value in particular. The number of stored accessions that could be studied is overwhelming [857 940 accessions for wheat; (Kilian and Graner, 2012)] and these resources are underused due to technical difficulties to identify useful accessions in these large numbers. The conservation of the functional *Pm3* alleles allowed us to search using an allele-mining approach for additional, functional alleles in the wheat gene pool. Two different strategies were applied: first, a targeted approach based on focused identification of germplasm strategy (FIGS) was used. This strategy is based on the assumption that geographic regions and ecological environments with high mildew disease pressure would represent a particularly good source for additional powdery mildew resistance genes. Based on FIGS and 323 geographic sites, 1320 wheat accessions mostly from Turkey (419 accessions), Iran (391), Afghanistan (292) and Pakistan (133) were studied (Kaur et al., 2008; Bhullar et al., 2009). To complement this focus on eco-geographical criteria, which resulted in few accessions from some important wheat-growing regions, for example, China or India, a second set of 733 wheat accessions from 20 countries in Asia, America, Africa and Europe was selected. This included 78 genotypes from Nepal which was suspected to also be a good source of mildew resistance. These combined approaches resulted in the identification of nine novel, functional *Pm3* resistance alleles (*Pm3l* – *Pm3t*; Bhullar et al., 2009, 2010a). Together with a single functional *Pm3* allele isolated from tetraploid wheat (*Pm3k*; Yahiaoui, Kaur and Keller, 2009), a total of 17 different *Pm3* alleles have been identified, making it one of the largest allelic series of disease resistance genes in plants.

The studies on this genetically very diverse set of accessions, as well as of 272 landraces specifically analysed from Afghanistan, also allowed us to draw conclusions on the origin of some of the original *Pm3a*–*Pm3g* alleles (Bhullar et al., 2010b). The most frequently found alleles were *Pm3b* and *Pm3c*, with *Pm3b* highly present in lines from Afghanistan, indicating that the gene originated there. Interestingly, *Pm3c* which is very similar at the sequence level to *Pm3b* was found mostly in lines from India and Nepal, indicating a possible origin in this region. *Pm3f* and *Pm3g* alleles were very rare, suggesting that the few point mutations in these genes actually originated only very recently in modern breeding material (Bhullar et al., 2010b). Similarly, the closely related *Pm3d* and *Pm3e* alleles were rare, but a related susceptible allele that is likely their ancestor gene was found at high frequencies in lines from Nepal, Bhutan and India, again indicating an origin in this region. This suggests that further analysis of gene bank accessions specifically from these geographical regions in the Himalayan range might allow the identification of further mildew resistance genes.

7.2 Molecular improvement of *Pm3* function based on knowledge of functional diversity

The diversity of *Pm3* alleles and the comparative analysis of their distinct race-specific resistance spectra have led to hypotheses on the molecular determinants of specificity and narrow vs. broad recognition spectra (Brunner et al., 2010). One important finding of this study was the successful intragenic allele pyramiding of resistance alleles *Pm3d* and *Pm3e*: the combination of their characteristic polymorphisms in comparison to the susceptible reference allele *Pm3CS* in a single gene resulted in a combined resistance specificity and a broader resistance spectrum. This represents one of the first examples of immune receptor engineering for improved resistance. Furthermore, the study of the natural *Pm3* alleles revealed two amino acids in the ARC2 subdomain of the nucleotide-binding domain, thus widening the specificity of *Pm3a* and *Pm3b* alleles in comparison to *Pm3f* and *Pm3c* alleles, respectively. Replacement of these two amino acids in Pm3f resulted in a broader resistance spectrum very similar to the one conferred by *Pm3a* (Stirnweis et al., 2014a). This work demonstrated the potential of comparative approaches to design novel genes with improved resistance. Furthermore, it showed that transient expression of the wheat *Pm3* alleles in the heterologous host *Nicotiana* is a highly valuable experimental system for the study of the *Pm3* allele's function.

Based on the experimental work in *Nicotiana*, and also detailed genetic studies in wheat, it was found that some *Pm3* alleles can suppress others when combined in an F1 hybrid plant or as stable transgenes (Stirnweis et al., 2014b). Furthermore, it was found that *Pm3CS* suppresses the *Pm8* gene (an ortholog of *Pm3*; Hurni et al., 2013) which was introgressed into wheat from rye (Hurni et al., 2014). Such interactions can be tested easily and rapidly in *Nicotiana*, allowing the prediction of suppression effects before gene combinations are actually made in breeding. Finally, the study of *Pm3* alleles has recently allowed the identification of the first avirulence genes in wheat powdery mildew (Bourras et al., 2015; Praz et al., 2017, see Section 6). The molecular identification of pathogen molecules recognised by Pm3 immune receptors promises to further stimulate approaches towards resistance gene engineering.

7.3 From the lab to the field: studies on transgenic wheat genotypes overexpressing *Pm3* alleles

Several of the isolated *Pm3* alleles were transformed into the spring wheat cultivar Bobwhite under the control of the very strong maize ubiquitin promoter. Homozygous lines were tested over several years using a specific research infrastructure designed for field studies of transgenic plants (www.protectedsite.ch). A multitude of studies were performed (Brunner et al., 2011, 2012) and experiments with pyramids of *Pm3* alleles are ongoing. The results from the field show a strong mildew resistance conferred by the transgenes and, therefore, the transgenic use of overexpressed *Pm3* genes is a promising research direction with potential applications for breeding.

8 Future directions

8.1 Gene discovery through genome-wide association studies

Recently, genome-wide association studies assisted by high-density genetic maps have become valuable tools to map loci controlling agronomically relevant traits in plants. However, few of the studies have resulted in the direct identification of gene(s) responsible for a particular trait of interest, but rather of the genomic region where the gene is located. We believe that the recently published high-quality reference genome assembly for barley (Mascher et al., 2017) and the ongoing work to achieve the same end for bread wheat (International Wheat Genome Sequencing Consortium, personal communication) will represent a precious genomic resource to anchor in a more precise way the available markers and assist in identifying single genes.

8.2 Genome editing

CRISPR/Cas9-based genome editing has emerged as a promising tool to improve crops. In the case of wheat, the technology has been successfully applied to promote resistance to powdery mildew at seedling stage (Wang et al., 2014; Zhang et al., 2017). Although these studies represent a remarkable feat, it remains to be seen how these genome-edited varieties perform under agronomic conditions. Besides, these studies targeted host 'susceptibility' genes, which are not easily identified in complex genomes. Therefore, the next step in developing mildew-resistant barley and wheat lines will be to apply this technology in resistant genes/alleles to enlarge the resistance spectrum of those against powdery mildew. With this aim in mind, recent works are elucidating the role of critical amino acids in the intermolecular interactions between the R and AVR proteins (Stirnweis et al., 2014a). The findings of this type of research could be used via CRISPR/Cas9 for the engineering of resistance genes with broader recognition spectra.

8.3 Towards a pangenome for wheat and barley

Despite the achievement of the availability of the reference sequence genomes of barley and wheat, the whole diversity of those species is not yet completely known. To address this, the barley and wheat community have the vision of sequencing a larger number of accessions towards the establishment of the pangenome of these cereals (B. Keller, personal communication). This resource will provide unprecedented insights to tag biologically meaningful natural variations that will undoubtedly assist molecular breeding by harnessing new useful haplotypes, functional resistance genes or guide the development of allele- and gene-diagnostic molecular markers for MAS.

8.4 Pathogen-informed strategies

The isolation and study of *Avr* genes corresponding to *R* genes is essential to deepen our knowledge about the *Avr/R* gene biology (Schulze-Lefert and Panstruga, 2003). The recent identification of different cognate *AvrR-R* gene pairs is assisting in the understanding of how *R* genes recognise their corresponding *Avr* genes. This knowledge is essential towards the identification of highly conserved effectors/domains required for pathogenicity, which would enable the establishment of pathogen-informed strategies for mildew resistance in breeding.

9 Where to look for further information

9.1 Peer-reviewed articles, reviews and books

Three excellent reviews have been written, with particular focus on wheat powdery mildew incidence and host resistance deployed in breeding, by Bennett (1984), Cowger et al. (2012) and Hsam and Zeller (2002). Two reviews of QTL for QR to powdery mildew have recently been written by Li et al. (2014) and Marone et al. (2013).

An updated review about the *mlo*-based resistance has been written by Kusch and Panstruga (2016). The reader is referred to the review written by Jørgensen and Wolfe (1994) to get a comprehensive view on genetics of powdery mildew resistance in barley.

The Powdery Mildews: A Comprehensive Treatise (Bélanger et al., 2002) is a delightful book to increase readers' understanding of plant diseases caused by powdery mildew fungi in cereal crops.

9.2 WEB information

The Komugi database is an excellent resource to gain genomic and genetic knowledge of wheat and related species, *Aegilops*. Very importantly, it also includes the *Catalogue of Gene Symbols for Wheat*, which provides an updated overview about permanently and temporarily designated genes as well as stocks harbouring a particular gene of interest. The website is accessible at https://shigen.nig.ac.jp/wheat/komugi/.

The Genetic Resources Information System for Wheat and Triticale (www.wheatpedigree.net) provides information about genes of agronomic relevance present in wheat and relative accessions.

The UK Cereal Pathogen Virulence Survey (www.niab.com) reports annually incidence of cereal rust and mildews in the United Kingdom.

10 References

Acevedo-Garcia, J., Spencer, D., Thieron, H., Reinst, A., Hammond-kosack, K., Phillips, A. L. and Panstruga, R. (2017), 'mlo -based powdery mildew resistance in hexaploid bread wheat generated by a non-transgenic TILLING approach', *Plant Biotechnology Journal*, 15, pp. 367–78. doi:10.1111/pbi.12631.

Agrios, G. (2005), '*Plant Pathology*', 5th Edition, San Diego: Academic Press, p. 922. doi:10.1016/j.plantsci.2005.02.019.

Asad, M. A., Bai, B., Lan, C., Yan, J., Xia, X., Zhang, Y. and He, Z. (2014), 'Identification of QTL for adult-plant resistance to powdery mildew in Chinese wheat landrace Pingyuan 50', *Crop Journal*, 2, pp. 308–14. doi:https://doi.org/10.1016/j.cj.2014.04.009.

Bélanger, R. R., Bushnell, W. R., Dik, A. J. and Carver, T. L. (2002), '*The Powdery Mildews: A Comprehensive Treatise*', St. Paul: American Phytopathological Society (APS Press).

Bennett, F. G. A. (1984), 'Resistance to powdery mildew in wheat: A review of its use in agriculture and breeding programmes', *Plant Pathology*, 33(3), Blackwell Publishing Ltd, pp. 279–300. doi:10.1111/j.1365-3059.1984.tb01324.x.

Bhullar, N. K., Street, K., Mackay, M., Yahiaoui, N. and Keller, B. (2009), 'Unlocking wheat genetic resources for the molecular identification of previously undescribed functional alleles at the Pm3 resistance locus', *Proceedings of the National Academy of Sciences of the United States of America*, 106(23), pp. 9519–24. doi:10.1073/pnas.0904152106.

Bhullar, N. K., Mackay, M. and Keller, B. (2010a), 'Genetic diversity of the Pm3 powdery mildew resistance alleles in wheat gene bank accessions as assessed by molecular markers', *Diversity*, 2(5), pp. 768–86. doi:10.3390/d2050768.

Bhullar, N. K., Zhang, Z., Wicker, T. and Keller, B. (2010b), 'Wheat gene bank accessions as a source of new alleles of the powdery mildew resistance gene Pm3: A large scale allele mining project', *BMC Plant Biology*, 10, p. 88. doi:10.1186/1471-2229-10-88.

Bougot, Y., Lemoine, J., Pavoine, M. T., Guyomar'ch, H., Gautier, V., Muranty, H. and Barloy, D. (2006), 'A major QTL effect controlling resistance to powdery mildew in winter wheat at the adult plant stage', *Plant Breeding*, 125, pp. 550–6. doi:10.1111/j.1439-0523.2006.01308.x.

Bourras, S., McNally, K. E., Ben-David, R., Parlange, F., Roffler, S., Praz, C. R., Oberhaensli, S., Menardo, F., Stirnwels, D., Frenkel, Z., Schaefer, L. K., Flückiger, S., Treier, G., Herren, G., Korol, A. B., Wicker, T. and Keller, B. (2015), 'Multiple avirulence Loci and allele-specific effector recognition control the *Pm3* race-specific resistance of wheat to powdery mildew', *The Plant Cell*, 27(10), pp. 2991–3012. doi: 10.1105/tpc.15.00171.

Broscious, S. C., Frank, J. A. and Frederick, J. R. (1985), 'Influence of winter wheat management practices on the severity of powdery mildew and septoria blotch in Pennsylvania', *Phytopathology*, 75(5), pp. 538–42. doi:10.1094/Phyto-75-538.

Brown, J. K. M. (2002), 'Aerial dispersal of pathogens on the global and continental scales and its impact on plant disease', *Science*, 297(5581), pp. 537–41. doi:10.1126/science.1072678.

Brown, J. K. M. (2015), 'Durable resistance of crops to disease: A Darwinian perspective', *Annual Review of Phytopathology*, 53, pp. 513–39. doi:10.1146/annurev-phyto-102313-045914.

Brunner, S., Hurni, S., Streckeisen, P., Mayr, G., Albrecht, M., Yahiaoui, N. and Keller, B. (2010), 'Intragenic allele pyramiding combines different specificities of wheat *Pm3* resistance alleles', *Plant Journal*, 64(3), pp. 433–45. doi:10.1111/j.1365-313X.2010.04342.x.

Brunner, S., Hurni, S., Herren, G., Kalinina, O., von Burg, S., Zeller, S. L., Schmid, B., Winzeler, M. and Keller, B. (2011), 'Transgenic *Pm3b* wheat lines show resistance to powdery mildew in the field', *Plant Biotechnology Journal*, 9(8), pp. 897–910. doi:10.1111/j.1467-7652.2011.00603.x.

Brunner, S., Stirnweis, D., Diaz Quijano, C., Buesing, G., Herren, G., Parlange, F., Barret, P., Tassy, C., Sautter, C., Winzeler, M. and Keller, B. (2012), 'Transgenic *Pm3* multilines of wheat show increased powdery mildew resistance in the field', *Plant Biotechnology Journal*, 10(4), pp. 398–409. doi:10.1111/j.1467-7652.2011.00670.x.

Bundessortenamt. (2016), *Beschreibende sortenliste Getreide, Mais, Öl- und Faserpflanzen, Leguminosen, Rüben, Zwischenfrüchte*, Osterfelddamm 80, 30627 Hannover, ISSN 21 90-61 30.

Cao, A., Xing, L., Wang, X., Yang, X., Wang, W., Sun, Y., Qian, C., Ni, J., Chen, Y., Liu, D., Wang, X. and Chen, P. (2011), 'Serine/threonine kinase gene Stpk-V, a key member of powdery mildew resistance gene *Pm21*, confers powdery mildew resistance in wheat', *Proceedings of the National Academy of Sciences*, 108(19), pp. 7727–32. doi:10.1073/pnas.1016981108.

Chantret, N., Mingeot, D., Sourdille, P., Bernard, M., Jacquemin, J. M. and Doussinault, G. (2001), 'A major QTL for powdery mildew resistance is stable over time and at two development stages in winter wheat', *Theoretical and Applied Genetics*, 103, pp. 962–71. doi:https://doi.org/10.1007/s001220100645.

Chaure, P., Gurr, S. J. and Spanu, P. (2000), 'Stable transformation of *Erysiphe graminis* an obligate biotrophic pathogen of barley', *Nature Biotechnology*, 18(2), pp. 205–7. doi:10.1038/72666.

Chen, Y., Zhang, F., Tang, L., Zheng, Y., Li, Y., Christie, P. and Li, L. (2007), 'Wheat powdery mildew and foliar N concentrations as influenced by N fertilization and belowground interactions with intercropped faba bean', *Plant and Soil*, 291(1), pp. 1–13. doi:10.1007/s11104-006-9161-9.

Choulet, F., Alberti, A., Theil, S., Glover, N., Barbe, V., Daron, J., Pingault, L., Sourdille, P., Couloux, A., Paux, E., Leroy, P., Mangenot, S., Guilhot, N., Le Gouis, J., Balfourier, F., Alaux, M., Jamilloux, V., Poulain, J., Durand, C., Bellec, A., Gaspin, C., Safar, J., Dolezel, J., Rogers, J., Vandepoele, K., Aury, J.-M., Mayer, K., Berges, H., Quesneville, H., Wincker, P. and Feuillet, C. (2014), 'Structural and functional partitioning of bread wheat chromosome 3B', *Science*, 345(6194), pp. 1249721. doi:10.1126/science.1249721.

Clavijo, B. J., Venturini, L., Schudoma, C., Garcia Accinelli, G., Kaithakottil, G., Wright, J., Borrill, P., Kettleborough, G., Heavens, D., Chapman, H., Lipcombe, J., Barker, T., Lu, F.-H., McKenzie, N., Raats, D., Ramirez Gonzalez, R. H., Coince, A., Peel, N., Percival-Alwyn, L., Duncan, O., Trösch, J., Yu, G., Bolser, D., Naamati, G., Kerhornou, A., Spannagl, M., Gundlach, H., Haberer, G., Davey, R. P., Fosker, C., Di Palma, F., Phillips, A., Millar, A. H., Kersey, P. J., Uauy, C., Krasileva, K. V, Swarbreck, D., Bevan, M. W. and Clark, M. D. (2016), 'An improved assembly and annotation of the allohexaploid wheat genome identifies complete families of agronomic genes and provides genomic evidence for chromosomal translocations', *Genome Research*, 27(5), pp. 885–96. http://biorxiv.org/content/early/2016/10/13/080796.

Coakley, S. M., Scherm, H. and Chakraborty, S. (1999), 'Climate change and plant disease management', *Annual Review of Phytopathology*, 37, pp. 399–426. doi:10.1146/annurev.phyto.37.1.399.

Cowger, C., Miranda, L., Griffey, C., Hall, M., Murphy, J. P. and Maxwell, J. (2012), 'Wheat powdery mildew', In: Sharma, I. (Ed.), *Disease Resistance in Wheat*. Wallingford: CABI Publishing, pp. 84–119. doi:10.1079/9781845938185.0084.

Cowger, C., Parks, R. and Marshall, D. (2009), 'Appearance of powdery mildew of wheat caused by *Blumeria graminis* f. sp. *tritici* on *Pm17*-bearing cultivars in North Carolina', *Plant Disease*, 93, pp. 1219. doi:https://doi.org/10.1094/PDIS-93-11-1219B.

Cowger, C., Parks, R. and Kosman, E. (2016), 'Structure and migration in U.S. *Blumeria graminis* f. Sp. Tritici populations', *Phytopathology*, 106(3), pp. 295–304. doi:10.1094/PHYTO-03-15-0066-R.

Cunfer, B. M. (2002), 'Powdery mildew', In: Curtis, B., Rajaram, S. and Macpherson, H. (Eds), *Bread Wheat: Improvement and Production*. No. 30. Rome: FAO Plant Production and Protection Series.

Dean, R., Van Kan, J. A. L., Pretorius, Z. A., Hammond-Kosack, K. E., Di Pietro, A., Spanu, P. D., Rudd, J. J., Dickman, M., Kahmann, R., Ellis, J. and Foster, G. D. (2012), 'The Top 10 fungal pathogens in molecular plant pathology', *Molecular Plant Pathology*, 13(4), pp. 414–30. doi:10.1111/j.1364-3703.2011.00783.x.

Doležel, J., Vrána, J., Cápal, P., Kubaláková, M., Burešová, V. and Šimková, H. (2014), 'Advances in plant chromosome genomics', *Biotechnology Advances*, 32(1), pp. 122–36. doi:10.1016/j.biotechadv.2013.12.011.

Dreiseitl, A. (2016), 'Emerging *Blumeria graminis* f. sp. hordei pathotypes reveal 'Psaknon' resistance in European barley varieties', *The Journal of Agricultural Science*, 154(6), pp. 1082–9. doi:10.1017/S0021859615001069.

Drenth, A. and Goodwin, S. B. (1999), 'Population structure of oomycetes', In: Worrall, J. J. (Ed.), *Structure and Dynamics of Fungal Populations*. Dordrecht: Springer Netherlands, pp. 195–224. doi:10.1007/978-94-011-4423-0_9.

Flor, H. H. (1971), 'Current status of the gene-for-gene concept', *Annual Review of Phytopathology*, 9(1), pp. 275–96. doi:10.1146/annurev.py.09.090171.001423.

Fukuoka, S., Saka, N., Koga, H., Ono, K., Shimizu, T., Ebana, K., Hayashi, N., Takahashi, A., Hirochika, H., Okuno, K. and Yano, M. (2009), 'Loss of function of a proline-containing protein confers durable disease resistance in rice', *Science*, 325(5943), pp. 998–1001. doi:10.1126/science.1175550.

Gafni, A., Calderon, C. E., Harris, R., Buxdorf, K., Dafa-berger, A. and Levy, M. (2015), 'Biological control of the cucurbit powdery mildew pathogen *Podosphaera xanthii* by means of the epiphytic fungus *Pseudozyma aphidis* and parasitism as a mode of action', *Frontiers in Plant Science*, 6, p. 132. doi:10.3389/fpls.2015.00132.

Gao, X., Gong, Y., Huo, Y., Han, Q., Kang, Z. and Huang, L. (2015), 'Endophytic *Bacillus subtilis* strain E1R-J Is a promising biocontrol agent for wheat powdery mildew', *BioMed Research International*, 2015, p. 8.

Giorgi, D., Farina, A., Grosso, V., Gennaro, A., Ceoloni, C. and Lucretti, S. (2013), 'FISHIS: Fluorescence *in situ* hybridization in suspension and chromosome flow sorting made easy', *PLoS ONE*, 8(2). doi:10.1371/journal.pone.0057994.

Green, J. R., Carver, T. L. W. and Gurr, S. J. (2002), 'The formation and function of infection feeding structures', In: Bélanger, R., Bushnell, W., Dik, A., and Carver, T. L. W. (Eds), *The Powdery Mildews: A Comprehensive Treatise*, St. Paul: American Phytopathological Society (APS Press), pp. 66–82.

Gustafson, G. D. and Shaner, G. (1982), 'Influence of plant age on the expression of slow-mildewing resistance in wheat', *Phytopathology*, 72, pp. 746–9.

Henryk, J., Instytut, C., Roslin, A., Hodowli, C. I., Roslin, A., Pietrusinska, A., Hodowli, I., Roslin, A., Centre, N., Genetic, P., View, R. and Czembor, J. H. (2016), 'Transfer of the mlo resistance gene into to the genome of winter barley', *Progress in Plant Protection*, 56(3), 379–87. doi:10.14199/ppp-2016-061.

Hillocks, R. J. (2012), 'Farming with fewer pesticides: EU pesticide review and resulting challenges for UK agriculture', *Crop Protection*, 31(1), pp. 85–93. doi:10.1016/j.cropro.2011.08.008.

Hsam, S. L. K. and Zeller, F. J. (2002), 'Breeding for powdery mildew resistance in common wheat', In: Bélanger, R., Bushnell, W., Dik, A., and Carver, T. L. W. (Eds), *The Powdery Mildews: A Comprehensive Treatise*. St. Paul: American Phytopathological Society (APS Press), p. 292.

Huang, X.-Q. and Röder, M. S. (2004), 'Molecular mapping of powdery mildew resistance genes in wheat: A review', *Euphytica*, 137, pp. 203–23. doi:10.1023/B:EUPH.0000041576.74566.d7.

Hückelhoven, R. and Panstruga, R. (2011), 'Cell biology of the plant-powdery mildew interaction', *Current Opinion in Plant Biology*, 14(6), pp. 738–46. doi:10.1016/j.pbi.2011.08.002.

Hurni, S., Brunner, S., Buchmann, G., Herren, G., Jordan, T., Krukowski, P., Wicker, T., Yahiaoui, N., Mago, R. and Keller, B. (2013), 'Rye *Pm8* and wheat *Pm3* are orthologous genes and show evolutionary conservation of resistance function against powdery mildew', *Plant Journal*, 76(6), pp. 957–69. doi:10.1111/tpj.12345.

Hurni, S., Brunner, S., Stirnweis, D., Herren, G., Peditto, D., McIntosh, R. A. and Keller, B. (2014), 'The powdery mildew resistance gene *Pm8* derived from rye is suppressed by its wheat ortholog *Pm3*', *Plant Journal*, 79(6), pp. 904–13. doi:10.1111/tpj.12593.

IWGSC (2014), 'A chromosome-based draft sequence of the hexaploid bread wheat (*Triticum aestivum*) genome', *Science*, 345(6194), pp. 1251788. doi:10.1126/science.1251788.

Jones, J. D. G. and Dangl, J. L. (2006), 'The plant immune system', *Nature*, 444(7117), pp. 323–9. doi:10.1038/nature05286.

Jørgensen, J. H. (1992), 'Discovery, characterization and exploitation of Mlo powdery mildew resistance in barley', *Euphytica*, 63(1–2), pp. 141–52.

Jørgensen, J. H. (1994), 'Genetics of powdery mildew resistance in barley', *Critical Reviews in Plant Sciences*, 13(1), pp. 97–119.

Jørgensen, J. H. (1988), 'Erysiphe graminis, powdery mildew of cereal and grasses', *Advances in Plant Pathology*, (6), pp. 137–57.

Jørgensen, J. H. and Wolfe, P. M. (1994), 'Genetics of powdery mildew resistance in barley', *Critical Reviews in Plant Sciences*, 13(1), pp. 97–119. doi:10.1080/07352689409701910.

Kaur, N., Street, K., Mackay, M., Yahiaoui, N. and Keller, B. (2008), 'Molecular approaches for characterization and use of natural disease resistance in wheat', In: Collinge, D. B., Munk, L. and Cooke, B. M. (Eds), *Sustainable Disease Management in a European Context*. Dordrecht: Springer Netherlands, pp. 387–97. doi:10.1007/978-1-4020-8780-6_18.

Keller, M., Keller, B., Schachermayr, G., Winzeler, M., Schmid, J. E., Stamp, P. and Messmer, M. M. (1999), 'Quantitative trait loci for resistance against powdery mildew in a segregating wheat × spelt population', *Theoretical and Applied Genetics*, 98, pp. 903–12. doi:https://doi.org/10.1007/s001220051149.

Kilian, B. and Graner, A. (2012), 'NGS technologies for analyzing germplasm diversity in genebanks', *Briefings in Functional Genomics*, II(1), pp. 38–50. doi:10.1093/bfgp/elr046.

Krattinger, S. G. and Keller, B. (2016a), 'Molecular genetics and evolution of disease resistance in cereals', *New Phytologist*, 94, 765–73. doi:10.1111/nph.14097.

Krattinger, S. G. and Keller, B. (2016b), 'Molecular genetics and evolution of disease resistance in cereals', *New Phytologist*, 212(2), 320–332. doi:10.1111/nph.14097.

Krattinger, S. G., Lagudah, E. S., Spielmeyer, W., Singh, R. P., Huerta-Espino, J., McFadden, H., Bossolini, E., Selter, L. L. and Keller, B. (2009), 'A putative ABC transporter confers durable resistance to multiple fungal pathogens in wheat', *Science*, 323(5919), pp. 1360–3. doi:10.1126/science.1166453.

Kunzel, G., Korzun, L. and Meister, A. (2000), 'Cytologically integrated physical restriction fragment length polymorphism maps for the barley genome based on translocation breakpoints', *Genetics*, 154(1), pp. 397–412.

Kusch, S. and Panstruga, R. (2016), 'Mlo-based resistance: An apparently universal 'weapon' to defeat powdery mildew disease', *Mpmi*, 30(3), pp. 1–47. doi:10.1094/MPMI-12-16-0255-CR.

Lagudah, E. S. (2011), 'Molecular genetics of race non-specific rust resistance in wheat', *Euphytica*, 179(1), pp. 81–91. doi:10.1007/s10681-010-0336-3.

Lan, C., Liang, S., Wang, Z., Yan, J., Zhang, Y., Xia, X. and He, Z. (2009), 'Quantitative trait loci mapping for adult-plant resistance to powdery mildew in Chinese wheat cultivar Bainong 64', *Phytopathology*, 99, pp. 1121–1126. doi:10.1094/PHYTO-99-10-1121.

Lan, C., Ni, X., Yan, J., Zhang, Y., Xia, X., Chen, X. and He, Z. (2010), 'Quantitative trait loci mapping of adult-plant resistance to powdery mildew in Chinese wheat cultivar Lumai 21', *Molecular Breeding*, 25, pp. 615–22. doi:10.1007/s11032-009-9358-8.

Last, F. T. (1954), 'The effect of time of application of nitrogenous fertilizer on powdery mildew of winter wheat', *Annals of Applied Biology*, 41, pp. 381–92. doi:10.1111/j.1744-7348.1954.tb01139.x.

Leath, S. and Bowen, K. L. (1989), 'Effects of powdery mildew, triadimenol seed treatment, and triadimefon foliar sprays on yield of winter wheat in North Carolina', *Phytopathology*, 79(2), pp. 152–5. doi:10.1094/Phyto-79-152.

Li, H.-J., Wang, X.-M., Song, F.-J., Wu, C.-P., Wu, X.-F., Zhang, N., Zhou, Y. and Zhang, X.-Y. (2011), 'Response to powdery mildew and detection of resistance genes in wheat cultivars from China', *Acta Agronomica Sinica*, 37, pp. 943–54.

Li, Z., Lan, C., He, Z., Singh, R. P., Rosewarne, G. M., Chen, X. and Xia, X. (2014), 'Overview and application of QTL for adult plant resistance to leaf rust and powdery mildew in wheat', *Crop Science*, 54(5), pp. 1907–25. doi:10.2135/cropsci2014.02.0162.

Liang, S. S., Suenaga, K., He, Z. H., Wang, Z. L., Liu, H. Y., Wang, D. S., Singh, R. P., Sourdille, P. and Xia, X. C. (2006), 'Quantitative trait loci mapping for adult-plant resistance to powdery mildew in bread wheat', *Phytopathology*, 96, pp. 784–9. doi: 10.1094/PHYTO-96-0784.

Lillemo, M., Asalf, B., Singh, R. P., Huerta-Espino, J., Chen, X. M., He, Z. H. and Bjørnstad, Å. (2008), 'The adult plant rust resistance loci Lr34/Yr18 and Lr46/Yr29 are important determinants of partial resistance to powdery mildew in bread wheat line Saar', *Theoretical and Applied Genetics*, 116, pp. 1155–66. doi:10.1007/s00122-008-0743-1.

Lillemo, M., Bjørnstad, Å. and Skinnes, H. (2012), 'Molecular mapping of partial resistance to powdery mildew in winter wheat cultivar Folke', *Euphytica*, 185(1), pp. 47–59. doi:10.1007/s10681-011-0620-x.

Liu, S., Griffey, C. A. and Saghai Maroof, M. A. (2001), 'Identification of molecular markers associated with adult plant resistance to powdery mildew in common wheat cultivar Massey', *Crop Science*, 41, pp. 1268–75.

Lu, X., Kracher, B., Saur, I. M. L., Bauer, S., Ellwood, S. R., Wise, R. and Yaeno, T. (2016), 'Allelic barley MLA immune receptors recognize sequence-unrelated avirulence effectors of the powdery mildew pathogen', *Proceedings of the National Academy of Sciences*, 113(42), pp. E6486–E6495. doi:10.1073/pnas.1612947113.

Lucas, J. A., Hawkins, N. J. and Fraaije, B. A. (2015), 'The evolution of fungicide resistance', *Advances in Applied Microbiology*, 90, pp. 29–92. doi:10.1016/bs.aambs.2014.09.001.

Lyngkjær, M. F., Newton, A. C., Atzema, J. L. and Baker, S. J. (2000), 'The Barley mlo-gene: An important powdery mildew resistance source', *Agronomie*, 20(7), pp. 745–56. doi:10.1051/agro:2000173.

Mago, R., Tabe, L., Vautrin, S., Šimková, H., Kubaláková, M., Upadhyaya, N., Berges, H., Kong, X., Breen, J., Doležel, J., Appels, R., Ellis, J. G. and Spielmeyer, W. (2014), 'Major haplotype divergence including multiple germin-like protein genes, at the wheat Sr2 adult plant stem rust resistance locus', *BMC Plant Biology*, 14(1), p. 379. doi:10.1186/s12870-014-0379-z.

Makepeace, J. C., Oxley, S. J. P., Havis, N. D., Hackett, R., Burke, J. I. and Brown, J. K. M. (2007), 'Associations between fungal and abiotic leaf spotting and the presence of mlo alleles in barley', *Plant Pathology*, 56(6), Blackwell Publishing Ltd, pp. 934–42. doi:10.1111/j.1365-3059.2007.01680.x.

Marone, D., Russo, M. A., Laidò, G., De Vita, P., Papa, R., Blanco, A., Gadaleta, A., Rubiales, D. and Mastrangelo, A. M. (2013), 'Genetic basis of qualitative and quantitative resistance to powdery mildew in wheat: From consensus regions to candidate genes', *BMC Genomics*, 14(1), p. 562. doi:10.1186/1471-2164-14-562.

Mascher, M., Gundlach, H., Himmelbach, A., Beier, S., Twardziok, S. O., Wicker, T., Radchuk, V., Dockter, C., Hedley, P. E., Russell, J., Bayer, M., Ramsay, L., Liu, H., Haberer, G., Zhang, X.-Q., Zhang, Q., Barrero, R. A., Li, L., Taudien, S., Groth, M., Felder, M., Hastie, A., Šimková, H., Staňková, H., Vrána, J., Chan, S., Muñoz-Amatriaín, M., Ounit, R., Wanamaker, S., Bolser, D., Colmsee, C., Schmutzer, T., Aliyeva-Schnorr, L., Grasso, S., Tanskanen, J., Chailyan, A., Sampath, D., Heavens, D., Clissold, L., Cao, S., Chapman, B., Dai, F., Han, Y., Li, H., Li, X., Lin, C., McCooke, J. K., Tan, C., Wang, P., Wang, S., Yin, S., Zhou, G., Poland, J. A., Bellgard, M. I., Borisjuk, L., Houben, A., Doležel, J., Ayling, S., Lonardi, S., Kersey, P., Langridge, P., Muehlbauer, G. J., Clark, M. D., Caccamo, M., Schulman, A. H., Mayer, K. F. X., Platzer, M., Close, T. J., Scholz, U., Hansson, M.,

Zhang, G., Braumann, I., Spannagl, M., Li, C., Waugh, R. and Stein, N. (2017), 'A chromosome conformation capture ordered sequence of the barley genome', *Nature*, 544(7651), pp. 427–33. doi: 10.1038/nature22043.

Mayer, K. F. X., Waugh, R., Langridge, P., Close, T. J., Wise, R. P., Graner, A., Matsumoto, T., Sato, K., Schulman, A., Muehlbauer, G. J., Stein, N., Ariyadasa, R., Schulte, D., Poursarebani, N., Zhou, R., Steuernagel, B., Mascher, M., Scholz, U., Shi, B., Langridge, P., Madishetty, K., Svensson, J. T., Bhat, P., Moscou, M., Resnik, J., Close, T. J., Muehlbauer, G. J., Hedley, P., Liu, H., Morris, J., Waugh, R., Frenkel, Z., Korol, A., Bergès, H., Graner, A., Stein, N., Steuernagel, B., Scholz, U., Taudien, S., Felder, M., Groth, M., Platzer, M., Stein, N., Steuernagel, B., Scholz, U., Himmelbach, A., Taudien, S., Felder, M., Platzer, M., Lonardi, S., Duma, D., Alpert, M., Cordero, F., Beccuti, M., Ciardo, G., Ma, Y., Wanamaker, S., Close, T. J., Stein, N., Cattonaro, F., Vendramin, V., Scalabrin, S., Radovic, S., Wing, R., Schulte, D., Steuernagel, B., Morgante, M., Stein, N., Waugh, R., Nussbaumer, T., Gundlach, H., Martis, M., Ariyadasa, R., Poursarebani, N., Steuernagel, B., Scholz, U., Wise, R. P., Poland, J., Stein, N., Mayer, K. F. X., Spannagl, M., Pfeifer, M., Gundlach, H., Mayer, K. F. X., Gundlach, H., Moisy, C., Tanskanen, J., Scalabrin, S., Zuccolo, A., Vendramin, V., Morgante, M., Mayer, K. F. X., Schulman, A., Pfeifer, M., Spannagl, M., Hedley, P., Morris, J., Russell, J., Druka, A., Marshall, D., Bayer, M., Swarbreck, D., Sampath, D., Ayling, S., Febrer, M., Caccamo, M., Matsumoto, T., Tanaka, T., Sato, K., Wise, R. P., Close, T. J., Wannamaker, S., Muehlbauer, G. J., Stein, N., Mayer, K. F. X., Waugh, R., Steuernagel, B., Schmutzer, T., Mascher, M., Scholz, U., Taudien, S., Platzer, M., Sato, K., Marshall, D., Bayer, M., Waugh, R., Stein, N., Mayer, K. F. X., Waugh, R., Brown, J. W. S., Schulman, A., Langridge, P., Platzer, M., Fincher, G. B., Muehlbauer, G. J., Sato, K., Close, T. J., Wise, R. P. and Stein, N. (2012), 'A physical, genetic and functional sequence assembly of the barley genome', *Nature*, 491(7426), 711–6. doi: 10.1038/nature11543.

McDonald, B. A. and Linde, C. (2002), 'The population genetics of plant pathogens and breeding strategies for durable resistance', *Euphytica*, 124, pp. 163–80. doi:10.1023/A:1015678432355.

McIntosh, R. A., Zhang, P., Cowger, C., Parks, R., Lagudah, E. S. and Hoxha, S. (2011), 'Rye-derived powdery mildew resistance gene *Pm8* in wheat is suppressed by the *Pm3* locus', *Theoretical and Applied Genetics*, 123(3), pp. 359–67. doi:10.1007/s00122-011-1589-5.

McNally, K. E., Menardo, F., Lüthi, L., Praz, C. R., Müller, M. C., Kunz, L., Ben-David, R., Chandrasekhar, K., Dinoor, A., Cowger, C., Meyers, E., Xue, M., Zeng, F., Gong, S., Yu, D., Bourras, S. and Keller, B. (2017), 'Distinct domains of the AVRPM3A2/F2 avirulence protein from wheat powdery mildew are involved in immune receptor recognition and putative effector function', *New Phytologist*.

Menardo, F., Praz, C. R., Wyder, S., Ben-David, R., Bourras, S., Matsumae, H., McNally, K. E., Parlange, F., Riba, A., Roffler, S., Schaefer, L. K., Shimizu, K. K., Valenti, L., Zbinden, H., Wicker, T. and Keller, B. (2016), 'Hybridization of powdery mildew strains gives rise to pathogens on novel agricultural crop species', *Nature Genetics*, 48(2), pp. 201–5. doi:10.1038/ng.3485.

Meyers, B. C., Kozik, A., Griego, A., Kuang, H. and Michelmore, R. W. (2003), 'Genome-wide analysis of NBS-LRR-encoding genes in Arabidopsis', *The Plant cell*, 15(4), pp. 809–34. doi:10.1105/tpc.009308.During.

Mingeot, D., Chantret, N., Baret, P. V., Dekeyser, A., Boukhatem, N., Sourdille, P., Doussinault, G. and Jacquemin, J. M. (2002), 'Mapping QTL involved in adult plant resistance to powdery mildew in the winter wheat line RE714 in two susceptible genetic backgrounds', *Plant Breeding*, 121, pp. 133–40. doi:10.1046/j.1439-0523.2002.00679.x.

Moore, J. W., Herrera-Foessel, S., Lan, C., Schnippenkoetter, W., Ayliffe, M., Huerta-Espino, J., Lillemo, M., Viccars, L., Milne, R., Periyannan, S., Kong, X., Spielmeyer, W., Talbot, M., Bariana, H., Patrick, J. W., Dodds, P., Singh, R. and Lagudah, E. (2015), 'A recently evolved hexose transporter variant confers resistance to multiple pathogens in wheat', *Nature genetics*, Nature Publishing Group, 47(12), pp. 1494–8. doi:10.1038/ng.3439.

Morgounov, A., Tufan, H. A., Sharma, R., Akin, B., Bagci, A., Braun, H. J., Kaya, Y., Keser, M., Payne, T. S., Sonder, K. and McIntosh, R. (2012), 'Global incidence of wheat rusts and powdery mildew during 1969–2010 and durability of resistance of winter wheat variety Bezostaya 1', *European Journal of Plant Pathology*, 132(3), pp. 323–40. doi:10.1007/s10658-011-9879-y.

Mundt, C. C. (2014), 'Durable resistance: A key to sustainable management of pathogens and pests', *Infection, Genetics and Evolution*, Elsevier B.V., 27, pp. 446–55. doi:10.1016/j.meegid.2014.01.011.

Niks, R. E., Qi, X. and Marcel, T. C. (2015), 'Quantitative resistance to biotrophic filamentous plant pathogens: Concepts, misconceptions, and mechanisms', *Annual Review of Phytopathology*, 53(1), pp. 445–70. doi:10.1146/annurev-phyto-080614-115928.

Oerke, E. C. and Dehne, H. W. (2004), 'Safeguarding production - loss in major crops and the role of crop protection', *Crop Protection*, 23, pp. 275–85.

Olesen, J. E., Jørgensen, L. N., Petersen, J. and Mortensen, J. V. (2003), 'Effects of rates and timing of nitrogen fertilizer on disease control by fungicides in winter wheat. 2. Crop growth and disease development', *Journal of Agricultural Science*, 140(1), pp. 15–29. doi:10.1017/S0021859602002897.

Parks, R., Carbone, I., Murphy, J. P., Marshall, D. and Cowger, C. (2008), 'Virulence structure of the Eastern U.S. wheat powdery mildew population', *Plant Disease*, 92, pp. 1074–82. doi:https://doi.org/10.1094/PDIS-92-7-1074.

Parlange, F., Roffler, S., Menardo, F., Ben-David, R., Bourras, S., McNally, K. E., Oberhaensli, S., Stirnweis, D., Buchmann, G., Wicker, T. and Keller, B. (2015), 'Genetic and molecular characterization of a locus involved in avirulence of *Blumeria graminis* f. sp. tritici on wheat *Pm3* resistance alleles', *Fungal Genetics and Biology*, 82, pp. 181–92. doi:http://dx.doi.org/10.1016/j.fgb.2015.06.009.

Parlevliet, J. E. and Zadocks, J. C. (1977), 'The integrated concept of disease resistance: A new view including horizontal and vertical resistance in plants', *Euphytica*, 26(1), pp. 5–21.

Pink, D. A. C. (2002), 'Strategies using genes for non-durable disease resistance', *Euphytica*, 124(2), pp. 227–36. doi:10.1023/A:1015638718242.

Praz, C. R., Bourras, S., Zeng, F., Sánchez-Martín, J., Menardo, F., Xue, M., Yang, L., Roffler, S., Böni, R., Herren, G., McNally, K. E., Ben-David, R., Parlange, F., Oberhaensli, S., Flückiger, S., Schäfer, L. K., Wicker, T., Yu, D. and Keller, B. (2017), 'AvrPm2 encodes an RNase-like avirulence effector which is conserved in the two different specialized forms of wheat and rye powdery mildew fungus', *New Phytologist*, 213(3), pp. 1301–14. doi:10.1111/nph.14372.

Qi, D. and Innes, R. W. (2013), 'Recent advances in plant NLR structure, function, localization, and signaling', *Frontiers in Immunology*, 4, 348. doi:10.3389/fimmu.2013.00348.

Reinstädler, A., Müller, J., Czembor, J. H., Piffanelli, P. and Panstruga, R. (2010), 'Novel induced mlo mutant alleles in combination with site-directed mutagenesis reveal functionally important domains in the heptahelical barley Mlo protein.', *BMC Plant Biology*, 10, p. 31. doi:10.1186/1471-2229-10-31.

Riaz, A., Periyannan, S., Aitken, E. and Hickey, L. (2016), 'A rapid phenotyping method for adult plant resistance to leaf rust in wheat', *Plant Methods*, 12(1), p. 17. doi:10.1186/s13007-016-0117-7.

Roberts, J. J. and Caldwell, R. M. (1970), 'General resistance slow mildewing to *Erysiphe graminis* f. sp. tritici in Knox wheat', *Phytopathology*, 60, p. 1310. doi:10.1094/Phyto-60-1831.

Rosewarne, G. M., Singh, R. P., Huerta-Espino, J., William, H. M., Bouchet, S., Cloutier, S., McFadden, H. and Lagudah, E. S. (2006), 'Leaf tip necrosis, molecular markers and β1-proteasome subunits associated with the slow rusting resistance genes Lr46/Yr29', *Theoretical and Applied Genetics*, 112(3), pp. 500–8. doi:10.1007/s00122-005-0153-6.

Sánchez-Martín, J., Steuernagel, B., Ghosh, S., Herren, G., Hurni, S., Adamski, N., Vrána, J., Kubaláková, M., Krattinger, S. G., Wicker, T., Doležel, J., Keller, B. and Wulff, B. B. H. (2016), 'Rapid gene isolation in barley and wheat by mutant chromosome sequencing', *Genome Biology*, 17(1), 221. doi:10.1186/s13059-016-1082-1

Schulze-Lefert, P. and Panstruga, R. (2003), 'Establishment of biotrophy by parasitic fungi and reprogramming of host cells for disease resistance', *Annual Review of Phytopathology*, 41(1), 641–67. doi:10.1146/annurev.phyto.41.061002.083300.

Seeholzer, S., Tsuchimatsu, T., Jordan, T., Bieri, S., Pajonk, S., Yang, W., Jahoor, A., Shimizu, K. K., Keller, B. and Schulze-Lefert, P. (2010), 'Diversity at the Mla powdery mildew resistance locus from cultivated barley reveals sites of positive selection', *Molecular Plant-Microbe Interactions*, Scientific Societies, 23(4), pp. 497–509. doi:10.1094/MPMI-23-4-0497.

Shaner, G. (1973), 'Evaluation of slow-mildewing resistance of Knox wheat in the field', *Phytopathology*, 63, pp. 867–872.

Singh, R. P. (1992), 'Association between gene Lr34 for leaf rust resistance and leaf tip necrosis in wheat', *Crop Science*, 32(4), p. 874. doi:10.2135/cropsci1992.0011183X003200040008x.

Singh, R. P. and Rajaram, S. (2002), 'Breeding for disease resistance in wheat', In B. C. Curtis, S. Rajaram and H. Gómez Macpherson: *Bread Wheat: Improvement and Production*. Rome: Food and Agriculture Organization of the United Nations, http://www.fao.org/docrep/006/Y4011E/y4011e00.htm#Contents.

Singh, R. P., Huerta-Espino, J., Bhavani, S., Herrera-Foessel, S. A., Singh, D., Singh, P. K., Velu, G., Mason, R. E., Jin, Y., Njau, P. and Crossa, J. (2011), 'Race non-specific resistance to rust diseases in CIMMYT spring wheats', *Euphytica*, 179(1), pp. 175–86. doi:10.1007/s10681-010-0322-9.

Srichumpa, P., Brunner, S., Keller, B. and Yahiaoui, N. (2005), 'Allelic series of four powdery mildew resistance genes at the *Pm3* locus in hexaploid bread wheat', *Plant physiology*, 139(2), pp. 885–95. doi:10.1104/pp.105.062406.

Steuernagel, B., Periyannan, S. K., Hernández-Pinzón, I., Witek, K., Rouse, M. N., Yu, G., Hatta, A., Ayliffe, M., Bariana, H., Jones, J. D. G., Lagudah, E. S. and Wulff, B. B. H. (2016), 'Rapid cloning of disease-resistance genes in plants using mutagenesis and sequence capture', *Nature Biotechnology*, 34(6), pp. 652–5. doi:10.1038/nbt.3543.

Stevens, R. (1960), 'Cultural practices in disease control', In: Horsfall, J. G. and Dimond, A. E. (Eds), *Plant Pathology*, Vol. 3. Academic Press, New York, Switzerland, International START Secretariat, Washington, DC, USA., pp. 357–430. doi:https://doi.org/10.1016/B978-0-12-395678-1.50016-3.

Stirnweis, D., Milani, S. D., Jordan, T., Keller, B. and Brunner, S. (2014a), 'Substitutions of two amino acids in the nucleotide-binding site domain of a resistance protein enhance the hypersensitive response and enlarge the PM3F resistance spectrum in wheat', *Molecular Plant-microbe Interactions : MPMI*, 27(3), pp. 265–76. doi:10.1094/MPMI-10-13-0297-FI.

Stirnweis, D., Milani, S. D., Brunner, S., Herren, G., Buchmann, G., Peditto, D., Jordan, T. and Keller, B. (2014b), 'Suppression among alleles encoding nucleotide-binding-leucine-rich repeat resistance proteins interferes with resistance in F1 hybrid and allele-pyramided wheat plants', *Plant Journal*, 79(6), pp. 893–903. doi:10.1111/tpj.12592.

Su, H., Blair, R. and Johnson, T. (2012), 'Regalia® bioprotectant in plant disease management', *Outlooks on Pest Management*, 23(5), pp. 30–4. doi:https://doi.org/10.1564/23feb09.

Thind, A. K., Wicker, T., Simkova, H., Fossati, D., Moullet, O., Brabant, C., Vrana, J., Dolezel, J. and Krattinger, S. G. (2017), 'Rapid cloning of genes in hexaploid wheat using cultivar-specific long-range chromosome assembly', *Nature Biotechnology*, 35(8), 793–6. Nature Publishing Group, a division of Macmillan Publishers Limited. All Rights Reserved., advance on. Available at: http://dx.doi.org/10.1038/nbt.3877.

Thompson, G. B., Brown, J. K. M. and Woodward, F. I. (1993), 'The effects of host carbon dioxide, nitrogen and water supply on the infection of wheat by powdery mildew and aphids', *Plant, Cell and Environment*, Blackwell Publishing Ltd,16(6), pp. 687–94. doi:10.1111/j.1365-3040.1993.tb00487.x.

Tommasini, L., Yahiaoui, N., Srichumpa, P. and Keller, B. (2006), 'Development of functional markers specific for seven *Pm3* resistance alleles and their validation in the bread wheat gene pool', *Theoretical and Applied Genetics*, 114(1), pp. 165–75. doi:10.1007/s00122-006-0420-1.

Tucker, D. M., Griffey, C. A., Liu, S., Brown-Guedira, G., Marshall, D. S. and Maroof, M. A. S. (2007), 'Confirmation of three quantitative trait loci conferring adult plant resistance to powdery mildew in two winter wheat populations', *Euphytica*, 155, pp. 1–13. doi:10.1007/s10681-006-9295-0.

Tucker, M. A., Jayasena, K., Ellwood, S. R. and Oliver, R. P. (2013), 'Pathotype variation of barley powdery mildew in Western Australia', *Australasian Plant Pathology*, 42(5), pp. 617–23. doi:10.1007/s13313-013-0226-y.

UKCPVS. (2017), 'UKCPVS Annual Report'. http://www.niab.com/pages/id/316/UKCPVS (last accessed 25 June 2017).

Wang, Y., Cheng, X., Shan, Q., Zhang, Y., Liu, J., Gao, C. and Qiu, J.-L. (2014), 'Simultaneous editing of three homoeoalleles in hexaploid bread wheat confers heritable resistance to powdery mildew', *Nature Biotechnology*, 32(9), pp. 947–51. doi:10.1038/nbt.2969.

Wang, Z. L., Li, L. H., He, Z. H., Duan, X. Y., Zhou, Y. L., Chen, X. M., Lillemo, M., Singh, R. P., Wang, H. and Xia, X. C. (2005), 'Seedling and adult plant resistance to powdery mildew in Chinese bread wheat cultivars and lines', *Plant Disease*, 89, pp. 457–63. doi:10.1094/PD-89-0457.

West, J. S., Fitt, B. D. L., Townsend, J. A., Stevens, M. and Edwards, S. G. (2015), *Impact of Climate Change on Diseases in Sustainable Arable Crop Systems: CLIMDIS*. Project Report No. 539 HGCA, https://cereals.ahdb.org.uk/media/561365/rd-2007-3399-final-project-summary.pdf

Wicker, T., Oberhaensli, S., Parlange, F., Buchmann, J. P., Shatalina, M., Roffler, S., Ben-David, R., Doležel, J., Šimková, H., Schulze-Lefert, P., Spanu, P. D., Bruggmann, R., Amselem, J., Quesneville, H., Ver Loren van Themaat, E., Paape, T., Shimizu, K. K. and Keller, B. (2013), 'The wheat powdery mildew genome shows the unique evolution of an obligate biotroph', *Nature Genetics*, 45(9), pp. 1092–6. doi:10.1038/ng.2704.

Wolfe, M. S. (1985), 'The current status and prospects of multiline cultivars and variety mixtures for disease resistance', *Annual Review of Phytopathology*, 23, pp. 251–73. doi:10.1146/annurev.py.23.090185.001343.

Wolfe, M. S. (1992), 'Barley diseases: maintaining the value of our varieties', In: L. Munk (Ed.), *Barley Genetics VI*. Vol 2. Copenhagen, Denmark: Munksgaard International Publishers, pp. 1055–67.

Yahiaoui, N., Brunner, S. and Keller, B. (2006), 'Rapid generation of new powdery mildew resistance genes after wheat domestication', *Plant Journal*, 47(1), pp. 85–98. doi: Doi 10.1111/J.1365-313x.2006.02772.X.

Yahiaoui, N., Kaur, N. and Keller, B. (2009), 'Independent evolution of functional *Pm3* resistance genes in wild tetraploid wheat and domesticated bread wheat', *Plant Journal*, 57(5), pp. 846–56. doi:10.1111/j.1365-313X.2008.03731.x.

Yahiaoui, N., Srichumpa, P., Dudler, R. and Keller, B. (2004), 'Genome analysis at different ploidy levels allows cloning of the powdery mildew resistance gene *Pm3b* from hexaploid wheat', *Plant Journal*, 37(4), pp. 528–38. doi:10.1046/j.1365-313X.2003.01977.x.

Yang, L., Zhang, X., Zhang, X., Wang, J., Luo, M., Yang, M., Wang, H., Xiang, L., Zeng, F., Yu, D., Fu, D. and Rosewarne, G. M. (2017), 'Identification and evaluation of resistance to powdery mildew and yellow rust in a wheat mapping population', C. Li. (Ed)., San Francisco, CA USA: Public Library of Science. *PLoS ONE*, 12(5), p. e0177905. doi:10.1371/journal.pone.0177905.

Young, N. D. (1996), 'QTL mapping and quantitative disease resistance in plants', *Annual Review of Phytopathology*, 34, pp. 479–501. doi:https://doi.org/10.1146/annurev.phyto.34.1.479.

Zhang, Y., Bai, Y., Wu, G., Zou, S., Chen, Y., Gao, C. and Tang, D. (2017), 'Simultaneous modification of three homoeologs of *TaEDR1* by genome editing enhances powdery mildew resistance in wheat', *The Plant Journal*, 91(4), 714–24. doi:10.1111/tpj.13599.

Zheng, Z., Wang, H. B., Chen, G. D., Yan, G. J. and Liu, C. J. (2013), 'A procedure allowing up to eight generations of wheat and nine generations of barley per annum', *Euphytica*, 191(2), pp. 311–16. doi:10.1007/s10681-013-0909-z.

Chapter 5

Diseases affecting wheat: tan spot

C. S. Moffat, Curtin University, Australia; and F. M. Santana, Embrapa Trigo, Brazil

1 Introduction

Tan spot is an economically important disease of wheat worldwide and is found throughout the major wheat-growing areas of the world. This foliar fungal disease is a challenge to wheat production as it reduces green leaf area, restricting photosynthesis and subsequently the ability to fill grain and set yield. Initially, leaf lesions appear as small tan-coloured brown flecks, which expand into oval or diamond-shaped spots, often with a yellow chlorotic margin (Fig. 1a). As the disease progresses, these spots increase in size, forming large blotches and irregular areas of dead tissue. A small dark brown eyespot near the centre of the lesion is common (Fig. 1b). The disease is particularly severe on older leaves, and entire leaves, spikes and whole plants may also be affected.

The causal agent is the necrotrophic fungal pathogen, *Pyrenophora tritici-repentis* (Died.) Drechs. [anamorph: *Drechslera tritici-repentis* (Died.) Shoem. (sym. *Helminthosporium tritici-repentis* (Died)]. *P. tritici-repentis* is commonly abbreviated to Ptr (or Dtr in some areas) and is a filamentous ascomycete fungus belonging to the class Dothideomycetes and order Pleosporales. Though initially described as an occasional wheat pathogen, the incidence of tan disease has increased markedly in recent times, and the pathogen exhibits a wide geographic range. Countries affected include those in North America (Canada, the USA and Mexico) (Ali and Francl, 2003; Singh and Hughes, 2006; Lamari et al., 1998; Friesen et al., 2005); South America (Colombia, Ecuador, Peru, Brazil, Uruguay and Argentina) (Gurung et al., 2013; Dubin, 1983; Gamba et al., 2012; dos Santos et al., 2002; Moreno et al., 2008); North Africa (Morocco, Algeria and Tunisia) (Lepoint et al., 2010; Cherif et al., 1994; Lamari et al., 1995); Europe (UK, Belgium, Luxembourg, Germany, Denmark, Czechia, Slovakia, Hungary, Poland, Romania, Bulgaria Lithuania, Latvia, Sweden and Finland) (Todorova, 2006; Sidrat et al., 2016; Abdullah et al., 2017; Gurung et al., 2013; Hudcovicova et al., 2015;

http://dx.doi.org/10.19103/AS.2018.0039.05
Published by Burleigh Dodds Science Publishing Limited, 2019.

Figure 1 (a) Typical early tan spot symptoms showing necrotic leaf lesions surrounded by chlorosis. (b) Dark eyespots visible within lesions.

Sierotzki et al., 2007; Cook and Yarham, 1989; Jalli et al., 2011; Leisova et al., 2008; Lepoint et al., 2010); Asia (Turkey, Syria, Georgia, Armenia, Azerbaijan, Russia, Iran, Turkmenistan, Uzbekistan, Tajikistan, Kyrgyzstan, Kazakhstan, Pakistan, India, Nepal, Bangladesh and Japan) (Singh, 2007; Ali and Francl, 2001; Gurung et al., 2013; Aboukhaddour et al., 2011; Momeni et al., 2014; Lamari et al., 2003, 2005; Mironenko et al., 2016; Lepoint et al., 2010) and Australasia (Australia and New Zealand) (Murray and Brennan, 2009; Hampton and Matthews, 1978). Both bread wheat (*Triticum aestivum*) and durum wheat (*T. turgidum*) are susceptible to tan spot. Furthermore, Ptr has been isolated from a number of other species including barley (*Hordeum vulgare*), oat (*Avena sativa*) and rye (*Secale cereale*), as well as other grasses which may serve as sources of inoculum. Initial infections early in the season originate from affected grass hosts or diseased crop residue in the soil, with symptoms usually visible five days after infection.

The tan spot disease cycle consists of both sexual and asexual stages. Ptr is able to survive saprophytically on wheat stubble and straw, upon which small, black, raised fruiting bodies (pseudothecia) containing sexual ascospores develop and mature (Fig. 2). These are commonly observed around the stem nodes. Ascospores are present throughout the growing season and serve as the primary inoculum infecting the lower leaves, particularly during tillering and early jointing. The lower, more mature leaves are usually infected first (Fig. 3), with disease radiating to the upper leaves and leaf sheaths when conditions are favourable. Infection occurs over a wide temperature range, with a longer wet period required during cold conditions. Free moisture on the leaf surface is crucial for infection. Under these damp conditions, asexual spores (conidia), with their distinct snakehead-like appearance, form within mature leaf lesions and act as a secondary inoculum, sustaining disease throughout the growing season. Rain and wind disperse the spores and spread the disease within fields and beyond. Although leaf infections are most common, seed infections can also occur usually following prolonged wet periods that promote disease during grain development. Infected seeds exhibit a red discoloured seed coat (red smudge) that will usually downgrade market grain.

Figure 2 Pseudothecia on wheat stubble.

Tan spot is a major constraint to maintaining wheat yield and quality, resulting in shrivelled grain and lower test weights. Yield losses of up to 50% can occur under exceptionally wet seasons, although losses of around 15% are more common, with the effect on yield greatest during grain fill, particularly if the upper leaves are affected.

2 Disease management considerations

A number of strategies are available for the management of tan spot, consisting of cultural modifications, chemical control and wheat variety genetics. However, a single strategy rarely provides total protection, and growers need to consider a combination of strategies to provide effective and consistent long-term management.

2.1 Cultural practices

Prior to seeding, disease information from previous seasons should be assessed and consideration be given to growing an alternative crop or variety, or using a different field with a lower disease risk. Field location, topography and prevailing wind direction should also be taken into consideration, as sowing downwind from a field with a history of tan spot may present a high disease risk since condia can travel up to 200 km (De Wolf et al., 1998). Maintaining quality pathogen-free seed with good germination and vigour is also an important part of a holistic disease management programme.

Tan spot has gained more importance with the increased adoption of minimum and no till agronomic practices which has coincided with disease outbreaks (Ciuffetti et al., 2014). Therefore, the management of crop debris is an effective means to limit tan spot by eliminating potential sources of primary inoculum. Practices that reduce stubble density, such as tillage, burning or grazing, will reduce the level of inoculum, but must be balanced against the increased risk of soil erosion and reduction in soil moisture.

Published by Burleigh Dodds Science Publishing Limited, 2019.

Figure 3 Characteristic tan spot infection, with the lower leaves most severely affected.

Stubble management will not, however, reduce disease caused by spores blown in from surrounding fields.

Another effective strategy is to diversify crops through rotation, which helps to break disease cycles and enable an overall rise in crop productivity. Monoculture of continuous wheat or wheat–fallow–wheat put the crop at an increased risk of infection, yet economic considerations tempt growers to restrict rotations and perpetuate the disease.

2.2 Chemical control

Fungicide application can be used to manage tan spot when environmental conditions favour disease development, particularly on susceptible varieties. However, prior to any fungicide use, it is essential to first ensure that observed symptoms are indeed due to tan spot and not caused by other factors such as nutritional deficiencies, environmental stress or physiological yellowing (Moffat et al., 2015). Seed treatment with a registered fungicide can address potential early season disease problems, particularly under continuous wheat production or if a susceptible variety is grown. Early crop scouting combined with readily available weather information can be indicative of whether an in-crop application of fungicide is warranted. The overall incidence of disease is strongly influenced by rainfall. If forecasts suggest that prolonged wet conditions will occur, a spray may provide up to three weeks of protection. However, little benefit will be gained from spraying when conditions are warm and dry.

Proper timing of fungicide application is essential to maximise yield and grain weight. A high disease pressure early in the season may require spraying between early tillering and stem elongation. This may be followed up by a second spray after flag leaf emergence, particularly when conditions are wet. For mid- to late-season disease, a single spray between flag leaf emergence and head emergence is generally the most profitable. Where there is minimal rainfall, fungicide applied after flag-leaf emergence is unlikely to provide economic returns.

2.3 Wheat variety choice

Another strategy for managing tan spot is to utilise variety resistance. Variety choice has a major influence on disease development with tan spot resistance ratings varying widely. Wheat breeding entities continue to release improved varieties that combine enhanced agronomic, quality and disease traits. However, highly resistant commercial varieties are not readily available and continue to be a breeding priority. Thus, varietal improvement offers the best approach for a long-term solution to yield stability. Increased germplasm resistance also reduces the use of fungicides which are costly, time-consuming and potentially hazardous. When choosing a variety, growers should aim to sow those with the best available disease resistance package, while considering adaptation to local growing conditions as well as agronomic and marketing requirements.

2.4 Case study: management of tan spot in Brazil

In Brazil approximately 6 million tonnes of wheat is grown over 2 million hectares, with around 90% of the production from the southern region (the states of Paraná, Santa Catarina and Rio Grande do Sul) (CONAB, 2017). Yields typically are between 2500 kg/ha and 3000 kg/ha. Losses due to tan spot have been reported as high as 48% (Reis and Casa, 2007), and it is estimated that for every 1% increase in tan spot incidence, yield can be reduced by 17 kg/ha (Reis et al., 2016).

For over 20 years, a no-till management system has been in place in Brazil to maximise soil conservation. Stubble is left in place and allowed to decompose, thus promoting the supply of organic material to the soil. Growers are urged to apply crop rotation, mainly soybean and maize during the summer, and wheat, oat, barley, triticale and canola during the winter. However, not all growers follow this, and consequently severe outbreaks of tan spot do occur, more frequently where wheat is grown in monoculture.

The application of chemicals is widely practised for tan spot control in Brazil, with growers applying fungicides, most typically triazoles and strobilurins, at least once or twice a season from tillering onwards. In Paraná, local practice is to spray fungicide up to five times a season when considering all possible diseases (Santana, unpublished). The threshold for application at the elongation stage is around 6%, meaning that fungicide should be applied before a 6% disease incidence in order to be most effective. Despite this, growers have started to report lower fungicide efficacy and tan spot appears to be becoming more problematic.

In terms of germplasm, currently there are over 100 commercially available wheat varieties in Brazil, which exhibit resistance to a wide range of diseases. Breeders based at Embrapa Trigo have developed and released over 30 wheat varieties, of which 15 are moderately resistant to tan spot. However, no molecular markers for tan spot resistance have yet been incorporated, and Brazilian variety selection follows a more traditional method via natural infection through field trials.

3 The host–pathogen interaction

Necrosis and chlorosis are two distinct symptoms of tan spot and result from highly specialised interactions between Ptr and the wheat host. Ptr is able to secrete necrotrophic effectors (also called host-selective toxins) that play a primary role in the development of disease, by inducing plant cell death and enabling fungal colonisation (Tan et al., 2010). So far, three Ptr effectors have been described: ToxA, ToxB and ToxC. ToxA has been particularly well studied and is responsible for the tan-coloured necrosis on sensitive wheat varieties, while ToxB and ToxC cause the yellowing chlorosis.

Just as wheat varieties differ in their susceptibility to tan spot, Ptr isolates (or strains) differ in their pathogenicity. Individual isolates of Ptr are currently classified into one of eight races based on their ability to produce the three effectors, either alone or in combination (Table 1). Races 1 and 2 appear to be the most predominant races, while races 3 and 4 are found at lower frequency. Race 5 has a relatively wide distribution, while races 6, 7 and 8 are comparatively rare.

ToxA appears to be the most prevalent Ptr effector worldwide and is rapidly internalised into mesophyll cells of sensitive wheat genotypes, where it localises to the chloroplast. Reduced photosynthesis follows, via the light-dependent accumulation of reactive oxygen species (ROS) and photosystem disruption (Pandelova et al., 2012). Cell death is rapid and has been detected as early as 9 h following ToxA treatment. Like ToxA, the chlorosis induced by ToxB is also light-dependent and involves chloroplast damage, ROS accumulation and inhibition of photosynthesis. However, ToxB-induced cell death is slower than that of ToxA, occurring after 24 h and does not eventuate in tissue collapse (Pandelova et al., 2012).

Table 1 The current race classification system of Ptr isolates is based on presence or absence of ToxA, ToxB or ToxC

Race	Effectors
1	AC
2	A
3	C
4	–
5	B
6	BC
7	AB
8	ABC

Published by Burleigh Dodds Science Publishing Limited, 2019.

ToxC remains to be characterised with little information currently known and is an obvious target for researchers. It is thought to be a metabolite and has been described as a low-molecular weight, polar and non-ionic molecule (Effertz et al., 2002). Since ToxA and ToxB are both proteins and products of identified genes, Ptr isolates can be tested for their presence/absence through simple molecular detection (PCR). However, currently the lack of an identified biosynthetic pathway for ToxC excludes PCR-based screening. Instead, the presence of ToxC in Ptr isolates of interest relies on spore inoculation of a particular differential wheat line (6B365). The elucidation of ToxC through metabolome-based studies will undoubtedly facilitate significant advances in tan spot research.

Currently, Ptr isolate races are determined by infection of a set of differential wheat lines that are sensitive to specific effectors. These are Glenlea (ToxA sensitive), 6B662 (ToxB sensitive), Katepwa (ToxA and ToxB sensitive), 6B365 (ToxC sensitive) and Auburn or Salamouni (insensitive to all three effectors). However, the ability to detect new races is limited by the effectiveness of the differential lines, and expansion of the differential varieties will facilitate the identification of new races. Indeed, there have been a number of reports of additional unidentified effectors that extend beyond the eight-race model and are the focus of ongoing research efforts (Moffat et al., 2014; Manning and Ciuffetti, 2015).

A number of studies have also examined the genetic diversity in Ptr populations. These have employed various molecular markers to assess regional or global collections of isolates and have revealed high population differentiation and genetic variability. This is further supported by comparative genomics which is suggestive of high levels of genome plasticity among isolates (Moolhuijzen et al., In Press).

The Ptr genome of a North American race 1 isolate was sequenced several years ago, and its release marked a significant evolution in tan spot research (Manning et al., 2013). However, extraordinary progress has been made in genome sequencing technologies in the years since, and the focus of current research will be to release new high-quality reference genomes that can resolve whole chromosomes. Capturing genomes of different races from diverse geographic regions will be important in the identification and isolation of novel effectors.

4 The genetics of host resistance

Dominant host sensitivity genes confer sensitivity to Ptr effectors and are considered as susceptibility loci as they are highly correlated with disease. The single dominant gene *Tsn1* (*Tan spot necrosis 1*) located on the long arm of chromosome 5B confers sensitivity to ToxA (Faris et al., 2010). Similarly, a single dominant gene *Tsc2* (*Tan spot chlorosis 2*) confers sensitivity to ToxB, on the short arm of chromosome 2B but is yet to be cloned. *Tsc1* confers sensitivity to ToxC; however, the *Tsc1*–ToxC interaction appears to be more complex. A quantitative trait locus (QTL) on the short arm of chromosome 1A has been associated with race 3-induced chlorosis (ToxC) and partially purified ToxC (Liu et al., 2017; Effertz et al., 2002); yet the mode of inheritance of sensitivity/insensitivity to ToxC remains to be determined. Identification of the sensitivity genes for both ToxB and ToxC with the release of perfect markers remains a high priority for researchers.

Numerous reports of additional QTLs with effects ranging from minor to significant, as well as QTLs for race-nonspecific resistance/susceptibility (Faris et al., 2013) exist. Many of these have been detected via plant infection assays and are not yet associated with reaction to specific effectors. The recent development of genetic manipulation tools in Ptr is a game changer, allowing researchers to manipulate and delete genes in Ptr in the hunt for novel effectors. For example, removal of the ToxA gene means researchers can tap into ToxA-sensitive wheat mapping populations to uncover QTLs, where previously ToxA-induced necrosis masked any additional symptoms (Moffat et al., 2014; Manning and Ciuffetti, 2015). As more evidence emerges it is becoming apparent that the genetics of tan spot resistance/susceptibility is more complicated than first thought, with both major and minor genes likely to be involved.

5 Case study: ToxA-assisted breeding for tan spot resistance in Australia

In Australia, annual wheat production is around 24 million tonnes, with approximately 12 million hectares under cultivation for wheat. Tan spot (also called yellow spot in Australia) is arguably the most costly wheat disease to growers, with yield losses estimated at $212 million (AUD) per annum, and annual control costs of $463 million (AUD) (Murray and Brennan, 2009). Wheat breeding programmes here have enthusiastically adopted an effector-based germplasm screening approach, in addition to the conventional use of molecular markers through marker-assisted selection. This exploitation of fungal effectors as functional markers has proven to be beneficial in selecting for disease resistance, particularly in the case of ToxA.

Insensitivity to ToxA generally correlates with higher levels of tan spot resistance in Australian wheat varieties (See et al., 2018). ToxA-sensitive varieties are consistently rated as very susceptible (VS) to moderately susceptible/susceptible (MSS), while insensitive varieties are found in the moderately susceptible (MS) to moderately resistant (MR) range, with only a handful of exceptions. Thus, a priority for breeders has been the elimination of the *Tsn1* sensitivity gene, since sensitivity to ToxA is sufficient to determine significant disease susceptibility.

ToxA has been synthetically produced via a microbial expression system and delivered to Australian wheat breeders since 2009, with many thousands of doses dispatched annually. By simply infiltrating wheat leaves with ToxA, breeders can quickly determine which lines are sensitive (i.e. necrotic) (Fig. 4). These leaf infiltration effector assays are flexible and can be carried out in glasshouses and field plots, on seedlings and adult plants. One breeder is able to test several thousand plants in a single day and obtain clear results within a few days (Oliver et al., 2016). Those lines identified as sensitive are likely to be susceptible to tan spot (and septoria nodorum blotch which also produces ToxA) and can be removed early to streamline the variety selection process. Effector-assisted screening can be undertaken at any stage in the breeding process, enabling screening for disease susceptibility in varieties almost ready for release, as well as in new crosses still early in the development pipeline. It offers substantial advantages over field testing approaches, saving time, space and resources. Furthermore, a wheat variety trial over two years and four locations in the Western Australian wheat belt found that no yield penalties were observed in ToxA-insensitive varieties, suggesting that the absence of the *Tsn1* gene has no obvious downside in Australian conditions (Oliver et al., 2014).

The uptake of effector-assisted selection by breeders in Australia has been rapid and the impact can already be seen. Between the 2009/10 and 2015/16 seasons, the area in Western Australia sown with ToxA-sensitive varieties was reduced by approximately

Figure 4 ToxA leaf infiltration of a sensitive wheat variety showing necrosis one week after infiltration.

1.4 million hectares from 37.5% to 8.3%. Based on a typical yield reduction of 0.3 t/ha and average wheat prices of $250 (AUD)/t, this equates to 438 000 t, translating to saving of approximately $109 million (AUD). Further savings are expected as effector assays become more widespread, and new effectors are discovered and made available.

The use of effectors to screen seedlings early in cereal breeding programmes provides a direct and optimised approach to identify insensitive varieties for advancement, while molecular marker assays require DNA to be prepared and costly assays performed. Markers for tan spot are often derived from weak disease QTLs and require substantial validation and optimisation prior to adoption by the breeders. In contrast, all of the major Australian wheat breeders now include ToxA screening in their selection strategies (Vleeshouwers and Oliver, 2014). This enthusiastic adoption has substantially reduced the cost of screening for disease resistance, allowing breeders to concentrate their efforts and resources on other traits, such as drought or frost resistance, or improved yield. Australia has welcomed a new era of effector-assisted selective breeding for improving wheat resistance to necrotrophic pathogens, whilst maintaining or improving other desirable traits with no observable disadvantages. The challenge now lies in extending this approach to other major wheat-growing countries.

6 Conclusion and future trends

Implementing an effective, economic and lasting tan spot management strategy is not a simple task, with many factors and risks to consider. Growers must make informed decisions based on a combination of management practices encompassing cultural practices, fungicides and varietal resistance. While chemicals may prove effective in the short term, a viable and enduring approach to tan spot control lies in the improvement of varietal disease resistance. Research is continually advancing our understanding of the mechanisms of pathogenicity, with a particular emphasis on effectors. As the success of ToxA-assisted breeding in Australia has proven, this knowledge can truly be translated out of the laboratory and into the field to provide quantifiable and significant progress. Genomics will no doubt aid in the search for novel effectors, and Ptr genetic manipulation will enable functional testing of these in the laboratory. The validation of QTLs and the incorporation of diagnostic molecular markers into breeding programmes, combined with effector-assisted selection, are likely to prove very effective towards long-term control of tan spot.

7 Where to look for further information

An abundance of disease fact sheets and management information is available from local agricultural offices, state departments and industry professionals. Proprietary decision guides are usually available from crop protection companies.

The UK-based Agriculture and Horticulture Development Board website (AHDB) is well worth exploring and publishes a wheat disease management guide which has useful information for control of a comprehensive number of wheat diseases (https://cereals.ahdb.org.uk/media/176167/g63-wheat-disease-management-guide-february-2016.pdf).

The CABI (Centre for Agriculture and Biosciences International) Plantwise Knowledge Bank (http://www.plantwise.org/KnowledgeBank/home.aspx) is also worth a visit as a source of tan spot factsheets, including diagnostic resources and best-practice control advice.

The Grains and Research Development Corporation (GRDC) is a great resource for tan (yellow) spot research in Australia and includes variety trials and disease research reports (https://grdc.com.au/).

For a comprehensive and thorough review of the genetics underlying tan spot resistance, see Faris et al. (2013).

8 References

Abdullah, S., Sehgal, S. K., Ali, S., Liatukas, Z., Ittu, M. and Kaur, N., 2017. Characterization of *Pyrenophora tritici-repentis* (tan spot of wheat) races in Baltic States and Romania. *Plant Pathology Journal* 33, 133–9.

Aboukhaddour, R., Cloutier, S., Lamari, L. and Strelkov, S. E., 2011. Simple sequence repeats and diversity of globally distributed populations of *Pyrenophora tritici-repentis*. *Canadian Journal of Plant Pathology-Revue Canadienne De Phytopathologie* 33, 389–99.

Ali, S. and Francl, L. J., 2001. First report of tan spot on wheat in Pakistan. *Plant Disease* 85, 1031.

Ali, S. and Francl, L. J., 2003. Population race structure of *Pyrenophora tritici-repentis* prevalent on wheat and noncereal grasses in the Great Plains. *Plant Disease* 87, 418–22.

Cherif, M., Harrabi, M. and Morjane, H. I., 1994. Distribution and importance of wheat and barley diseases in Tunisia, 1989 to 1991. *Rachis* 13(1/2), 25–34.

Ciuffetti, L. M., Manning, V. A., Pandelova, I., Faris, J. D., Friesen, T. L., Strelkov, S. E., Weber, G. L., Goodwin, S. B., Wolpert, T. J. and Figueroa, M., 2014. *Pyrenophora tritici-repentis*: A plant pathogenic fungus with global impact. In: Dean, R., Lichens-Park, A. and Kole, C. (Eds), *Genomics of Plant-Associated Fungi: Monocot Pathogens*. Berlin, Heidelberg: Springer.

Conab (Companhia Nacional de Abastecimento), 2017. Acompanhamento da safra brasileira: Grãos. http://www.conab.gov.br/OlalaCMS/uploads/arquivos/17_08_10_11_27_12_boletim_graos_agosto_2017.pdf.

Cook, R. J. and Yarham, D. J., 1989. Occurrence of tan spot of wheat caused by *Pyrenophora tritici-repentis* on wheat in England and Wales in 1987. *Plant Pathology* 38, 101–2.

De Wolf, E. D., Effertz, R. J., Ali, S. and Francl, L. J., 1998. Vistas of tan spot research. *Canadian Journal of Plant Pathology-Revue Canadienne De Phytopathologie* 20, 349–70.

Dos Santos, A. M. P. V., Matsumura, A. T. S. and Van Der Sand, S. T., 2002. Intraspecific genetic diversity of Drechslera tritici-repentis as detected by random amplified polymorphic DNA analysis. *Genetics and Molecular Biology* 25, 243–50.

Dubin, H. J., 1983. Occurrence of *Pyrenophora tritici-repentis* in the Andean Countries of South-America. *Plant Disease* 67, 1040.

Effertz, R. J., Meinhardt, S. W., Anderson, J. A., Jordahl, J. G. and Francl, L. J., 2002. Identification of a chlorosis-inducing toxin from *Pyrenophora tritici-repentis* and the chromosomal location of an insensitivity locus in wheat. *Phytopathology* 92, 527–33.

Faris, J. D., Liu, Z. H. and Xu, S. S., 2013. Genetics of tan spot resistance in wheat. *Theoretical and Applied Genetics* 126, 2197–217.

Faris, J. D., Zhang, Z. C., Lu, H. J., Lu, S., Reddy, L., Cloutier, S., Fellers, J. P., Meinhardt, S. W., Rasmussen, J. B., Xu, S. S., Oliver, R. P., Simons, K. J. and Friesen, T. L., 2010. A unique wheat disease resistance-like gene governs effector-triggered susceptibility to necrotrophic pathogens. *Proceedings of the National Academy of Sciences of the United States of America* 107, 13544–9.

Friesen, T. L., Ali, S., Klein, K. K. and Rasmussen, J. B., 2005. Population genetic analysis of a global collection of *Pyrenophora tritici-repentis*, causal agent of tan spot of wheat. *Phytopathology* 95, 1144–50.

Gamba, F. M., Strelkov, S. E. and Lamari, L., 2012. Virulence of *Pyrenophora tritici-repentis* in the Southern Cone Region of South America. *Canadian Journal of Plant Pathology-Revue Canadienne De Phytopathologie* 34, 545–50.

Gurung, S., Short, D. P. G. and Adhikari, T. B., 2013. Global population structure and migration patterns suggest significant population differentiation among isolates of *Pyrenophora tritici-repentis*. *Fungal Genetics and Biology* 52, 32–41.

Hampton, J. G. and Matthews D., 1978. Drechslera spp on New-Zealand certified cereal seed. *New Zealand Journal of Agricultural Research* 21, 539–42.

Hudcovicova, M., Matusinsky, P., Gubis, J., Leisova-Svobodova, L., Heinonen, U., Ondreickova, K., Mihalik, D., Gubisova, M., Majeska, M. and Jalli, M., 2015. DNA Markers for identification of *Pyrenophora tritici-repentis* and detection of genetic diversity among its isolates. *Romanian Agricultural Research* 32, 263–72.

Jalli, M., Laitinen, P. and Latvala, S., 2011. The emergence of cereal fungal diseases and the incidence of leaf spot diseases in Finland. *Agricultural and Food Science* 20, 62–73.

Lamari, L., Sayoud, R., Boulif, M. and Bernier, C. C., 1995. Identification of a new race in *Pyrenophora tritici-repentis*: Implications for the current pathotype classification system. *Canadian Journal of Plant Pathology-Revue Canadienne De Phytopathologie* 17, 312–8.

Lamari, L., Gilbert, J. and Tekauz, A., 1998. Race differentiation in *Pyrenophora tritici-repentis* and survey of physiologic variation in western Canada. *Canadian Journal of Plant Pathology-Revue Canadienne De Phytopathologie* 20, 396–400.

Lamari, L., Strelkov, S. E., Yahyaoui, A., Amedov, M., Saidov, M., Djunusova, M. and Koichibayev, M., 2005. Virulence of *Pyrenophora tritici-repentis* in the countries of the Silk Road. *Canadian Journal of Plant Pathology-Revue Canadienne De Phytopathologie* 27, 383–8.

Lamari, L., Strelkov, S. E., Yahyaoui, A., Orabi, J. and Smith, R. B., 2003. The identification of two new races of *Pyrenophora tritici-repentis* from the host center of diversity confirms a one-to-one relationship in tan spot of wheat. *Phytopathology* 93, 391–6.

Leisova, L., Hanzalova, A. and Kucera, L., 2008. Genetic diversity of *Pyrenophora tritici-repentis* isolates as revealed by AFLP analysis. *Journal of Plant Pathology* 90, 233–45.

Lepoint, P., Renard, M. E., Legreve, A., Duveiller, E. and Maraite, H., 2010. Genetic diversity of the mating type and toxin production genes in *Pyrenophora tritici-repentis*. *Phytopathology* 100, 474–83.

Liu, Z. H., Zurn, J. D., Kariyawasam, G., Faris, J. D., Shi, G., Hansen, J., Rasmussen, J. B. and Acevedo, M., 2017. Inverse gene-for-gene interactions contribute additively to tan spot susceptibility in wheat. *Theoretical and Applied Genetics* 130, 1267–76.

Manning, V. A. and Ciuffetti, L. M., 2015. Necrotrophic effector epistasis in the *Pyrenophora tritici-repentis*-Wheat interaction. *PLoS One* 10(4), e0123548.

Manning, V. A., Pandelova, I., Dhillon, B., Wilhelm, L. J., Goodwin, S. B., Berlin, A. M., Figueroa, M., Freitag, M., Hane, J. K., Henrissat, B., Holman, W. H., Kodira, C. D., Martin, J., Oliver, R. P., Robbertse, B., Schackwitz, W., Schwartz, D. C., Spatafora, J. W., Turgeon, B. G., Yandava, C., Young, S., Zhou, S., Zeng, Q., Grigoriev, I. V., Ma, L. J. and Ciuffetti, L. M., 2013. Comparative genomics of a plant-pathogenic fungus, *Pyrenophora tritici-repentis*, reveals transduplication and the impact of repeat elements on pathogenicity and population divergence. *G3-Genes Genomes Genetics* 3, 41–63.

Mironenko, N. V., Baranova, O. A., Kovalenko, N. M., Mikhailova, L. A. and Rosseva, L. P., 2016. Genetic structure of the Russian populations of *Pyrenophora tritici-repentis*, determined by using microsatellite markers. *Russian Journal of Genetics* 52, 771–9.

Moffat, C. S., See, P. T. and Oliver, R. P., 2014. Generation of a ToxA knockout strain of the wheat tan spot pathogen *Pyrenophora tritici-repentis*. *Molecular Plant Pathology* 15, 918–26.

Moffat, C. S., See, P. T. and Oliver, R. P., 2015. Leaf yellowing of the wheat cultivar Mace in the absence of yellow spot disease. *Australasian Plant Pathology* 44, 161–6.

Momeni, H., Aboukhaddour, R., Javan-Nikkhah, M., Razavi, M., Naghavi, M. R., Akhavan, A. and Strelkov, S. E., 2014. Race Identification of *Pyrenophora tritici-repentis* in Iran. *Journal of Plant Pathology* 96, 287–94.

Moolhuijzen, P., See, P. T., Hane, J. K., Liu, Z., Oliver, R. P. and Moffat, C. S., In Press. Comparative genetics of the wheat fungal pathogen *Pyrenophora tritici-repentis* reveals chromosomal variations and genome plasticity. *BMC Genomics*.

Moreno, M. V., Stenglein, S. A., Balatti, P. A. and Perello, A. E., 2008. Pathogenic and molecular variability among isolates of *Pyrenophora tritici-repentis*, causal agent of tan spot of wheat in Argentina. *European Journal of Plant Pathology* 122, 239–52.

Murray, G. M. and Brennan, J. P., 2009. Estimating disease losses to the Australian wheat industry. *Australasian Plant Pathology* 38, 558–70.

Oliver, R., Lichtenzveig, J., Tan, K. C., Waters, O., Rybak, K., Lawrence, J., Friesen, T. and Burgess, P., 2014. Absence of detectable yield penalty associated with insensitivity to Pleosporales necrotrophic effectors in wheat grown in the West Australian wheat belt. *Plant Pathology* 63, 1027–32.

Oliver, R. P., Tan, K. C. and Moffat, C. S., 2016. Necrotrophic pathogens of wheat. In: Wrigley, C. W., Corke, H., Seetharaman, K. and Faubion, J. (Eds), *Encyclopedia of Food Grains*. Academic Press, Oxford, pp. 273–8.

Pandelova I, Figueroa M, Wilhelm L. J., Manning, V. A., Mankaney, A. N., Mockler, T. C. and Ciuffetti, L. M., 2012. Host-selective toxins of *Pyrenophora tritici-repentis* induce common responses associated with host susceptibility. *PLoS One* 7(7), e40240.

Reis, E. M. and Casa, R. T., 2007. *Doenças dos cereais de inverno: diagnose, epidemiologia e controle*, 2nd ed. Lages: Graphel, p. 176.

Reis, E. M., Zoldan, S. M. and Germano, B. C., 2016. Controle de doenças do trigo e triticale. In: *Informações técnicas para trigo e triticale – safra 2016/9ª Reunião da Comissão Brasileira de Pesquisa de Trigo e Triticale*, p. 228.

See, P. T., Marathamuthu, K. A., Iagallo, E. I., Oliver, R. P. and Moffat, C. S., 2018. Evaluating the importance of the tan spot ToxA-*Tsn1* interaction in Australian wheat varieties. *Plant Pathology*. DOI: 10.1111/ppa.12835.

Sidrat, A., Ali, S., Liatukas, Z., Ittu, M., Sehgal, S. and Kaur, N., 2016. Race characterization of *Pyrenophora tritici-repentis* population from Lithuania and Romania. *Phytopathology* 106, 101.

Sierotzki, H., Frey, R., Wullschleger, J., Palermo, S., Karlin, S., Godwin, J. and Gisi, U., 2007. Cytochrome b gene sequence and structure of Pyrenophora teres and P-tritici-repentis and implications for QoI resistance. *Pest Management Science* 63, 225–33.

Singh, D. P., 2007. First report of tan spot of wheat caused by *Pyrenophora tritici-repentis* in the northern hills and northwestern plains zones of India. *Plant Disease* 91, 460.

Singh, P. K. and Hughes, G. R., 2006. Genetic similarity among isolates of *Pyrenophora tritici-repentis*, causal agent of tan spot of wheat. *Journal of Phytopathology* 154, 178–84.

Tan, K. C., Oliver, R. P., Solomon, P. S. and Moffat, C. S., 2010. Proteinaceous necrotrophic effectors in fungal virulence. *Functional Plant Biology* 37, 907–12.

Todorova, M., 2006. First report of tan spot caused by *Pyrenophora tritici-repentis* (anamorph Drechslera tritici-repentis) in Bulgaria. *Plant Pathology* 55, 305.

Vleeshouwers, V. G. and Oliver, R. P., 2014. Effectors as tools in disease resistance breeding against biotrophic, hemibiotrophic, and necrotrophic plant pathogens. *Molecular Plant-Microbe Interactions* 27, 196–206.

Published by Burleigh Dodds Science Publishing Limited, 2019.

Chapter 6

Diseases affecting wheat: Septoria nodorum blotch

A. K. Ruud and M. Lillemo, Norwegian University of Life Sciences, Norway

1 Introduction

1.1 Wheat

The global production of wheat was approximately 729 million tons in 2014, and around 70% of the production is used for human consumption (FAO, 2017), making it one of the largest food crops in the world. Due to its adaptability, it is grown in a wide range of climates. Bread wheat (*Triticum aestivum*, L.) accounts for roughly 95% of wheat production, durum (*T. durum*, L.) for the remaining 5%.

Plant diseases constitute a major threat to productivity, and the Green Revolution breeders were early advocates of incorporating general resistance against diseases (Borlaug, 1966; Caldwell, 1968), as a more durable strategy than race-specific resistance. Race non-specific and durable resistance has, to some degree, been deployed for the biotrophic diseases stripe, leaf and stem rusts and powdery mildew (Krattinger et al., 2009; Lillemo et al., 2008; Moore et al., 2015; Singh et al., 2011).

1.2 The Septoria leaf blotch disease complex

The '*Septoria* leaf blotch disease complex' includes Septoria nodorum leaf and glume blotch (SNB) caused by *Parastagonospora nodorum*, Septoria tritici leaf blotch (STB) caused by *Zymoseptoria tritici*, tan spot caused by *Pyrenophora tritici-repentis* and, of less importance,

http://dx.doi.org/10.19103/AS.2018.0039.06

Septoria avenae blotch caused by *Parastagonospora avenae*. Due to symptom similarity in the field, and the fact that these pathogens often occur together, even on the same leaf (Blixt et al., 2010), they are often referred to as 'Septoria leaf blotch' or simply 'leaf blotch'. One should, however, be aware that different pathogens are involved, and that resistance to one pathogen of the leaf blotch disease complex does not necessarily imply anything about resistance to any of the other pathogens. Microscopic evaluation of spores from leaf samples, or DNA-based identification, is necessary to reliably identify the causal agents.

1.3 Nomenclature

The classification and nomenclature of the leaf blotch pathogens have changed many times – an overview of distinctive features and synonyms is provided in Table 1.

1.4 The impacts of leaf blotch diseases

The Septoria leaf blotch diseases have only been recognized as major diseases since the introduction of high-yielding, semi-dwarf wheat cultivars (King et al., 1983; Scharen, 1999). The relative importance of the individual pathogen species in the complex varies between geographic regions and has changed over time.

In Europe, particularly in Mediterranean climate and the 'Maritime Zone' including Northern France, Germany and the UK, *Z. tritici* is the most important wheat leaf blotch pathogen (Eyal, 1999; Fones and Gurr, 2015), while *P. nodorum* is associated with the northern regions (Eyal, 1999). Approximately 70% of the share of fungicides used on wheat in Europe ($1.2bn, ~€ 1bn) is primarily targeted towards the management of *Z. tritici* (Torriani et al., 2015).

From the 1970s to the 1980s, the prevalence and importance shifted from *P. nodorum* to *Z. tritici* in the UK and Germany (Jones, 1985; Polley and Thomas, 1991; Royle et al., 1986; Shaw, 1999). Possible reasons for the shift are increased nitrogen fertilization, susceptible cultivars, early sowing, climate changes (higher summer rainfalls), altered atmospheric SO_2 concentration and differential response and resistance development towards fungicides (Bayles, 1991; Bearchell et al., 2005; Cools et al., 2011; Lovell et al., 1997; Meien-Vogeler et al., 1994; Pereira et al., 2017; Polley and Thomas, 1991; Shaw, 1999).

Tan spot caused by *P. tritici-repentis* first emerged as a disease on wheat in 1934–41 in Canada, USA and India (Barrus, 1942; Hosford, 1982; Johnson, 1942; Mitra, 1934; Oliver et al., 2008). The emergence of *P. tritici-repentis* as a wheat pathogen has been attributed to the horizontal gene transfer of the necrotrophic effector (NE)-gene *ToxA* from *P. nodorum* (Friesen et al., 2006). Tan spot has since become an important leaf blotch disease in major wheat-growing countries, especially in the US northern Great Plains and Australia, and can be an important component of the leaf blotch disease complex in Europe, Canada and USA (Bhathal et al., 2003; Carmona et al., 2006; Hosford, 1982; Lamari et al., 2005; Murray and Brennan, 2009).

P. nodorum, the causal agent of SNB, occurs in all regions where wheat is grown and can cause significant losses (Eyal, 1981; King et al., 1983). Although STB has become more important in some European countries, its incidence has increased in other regions (DePauw, 1995), and SNB is still a major disease in many areas (Solomon et al., 2006). In Western Australia, SNB can be responsible for yield losses up to 31% (Bhathal et al., 2003). In Norwegian trials, the estimated yield loss due to SNB in the susceptible cultivar Bjarne

Table 1 Nomenclature, causal agents, synonyms and main features of leaf blotch diseases in wheat

Disease	Pathogen	Main distinctive features (symptoms, spores)
Septoria nodorum (leaf and glume) blotch Stagonospora nodorum blotch SNB	***Parastagonospora nodorum*** *Phaeosphaeria* (Hedjar.) syn. *Leptosphaeria nodorum* (Müll.), syn. *Septoria nodorum* (Berk.), syn. *Stagonospora nodorum* (Berk.)	*Lesions:* Lens-shaped lesions, with or without pycnidia. Pycnidia also on stem, leaf sheaths, glumes and nodes *Conidiospores:* cylindrical, 0-3 septa, 15–32 × 2–4 μm  Photo: A. K. Ruud.
Septoria tritici blotch STB	***Zymoseptoria tritici*** syn. *Mycosphaerella graminicola*, syn *Septoria tritici*	*Lesions:* Rectangular lesions with or without pycnidia *Conidiospores:* (two forms) – macropycnidiospores 35–98 × 1–3 μm, 3–5 septa *Micropycnidiospores*: 8–10.5 × 0.8–1 μm, without septa 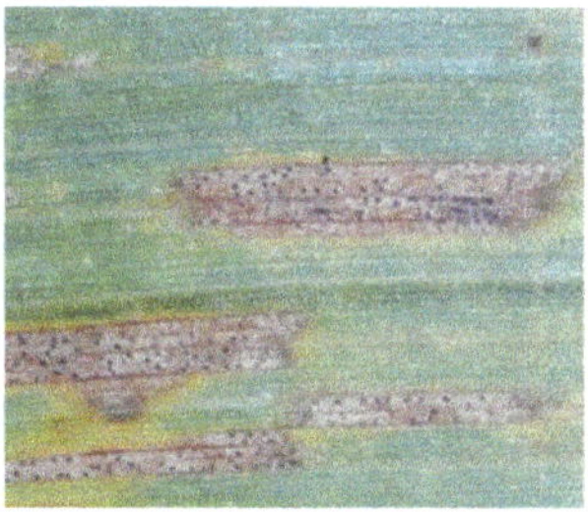 Photo: M. Lillemo.
Tan spot DTR, yellow (leaf) spot	***Pyrenophora tritici-repentis*** (Died) Shoem, syn. *Drechslera tritici-repentis* (Died.) Shoem. ***Helminthosporium tritici-vulgaris, H. tritici-repentis***	*Lesions:* Oval- to diamond-shaped lesions, with tan necrosis and extensive chlorosis. 'Eyespot' – chlorotic lesion with small dark brown spot. Reddish discoloration of kernels (red smudge) *Conidiospores:* cylindrical, 12–21 × 45–200 μm with 4–7 septa  Photo: North Dakota State University.

was calculated to be on average almost 25% based on data from 2009 to 2012 and a mean SNB severity of 20% (Abrahamsen, 2013). Quality measures like thousand kernel weight, hectolitre weight and grain filling were also well correlated with fungicide treatment.

2 *Parastagonospora nodorum* – the causal agent of Septoria nodorum blotch

Parastagonospora nodorum is a filamentous Ascomycete and member of the Dothideomycetes class, which includes several phytopathogens (Murray and Brennan, 2009; Quaedvlieg et al., 2013).

2.1 Life cycle and epidemiology

P. nodorum has a mixed reproduction system. The fungus is heterothallic with two mating types. Both mating types have to be present for sexual recombination to occur (Halama and Lacoste, 1991). The sexual fruiting bodies, pseudothecia (Fig. 1), contain numerous asci which release ascospores. These ascospores are wind borne over short and long distances (Bathgate and Loughman, 2001). The sexual stage is known from most areas where SNB is significant, including Norway (Bathgate and Loughman, 2001; Blixt et al., 2008; Cowger and Silva-Rojas, 2006; Ficke et al., 2011b). Asexual fruiting bodies, pycnidia, produce pycnidiospores which are splash dispersed within the canopy during rain events (Fig. 1) (Eyal et al., 1987; Solomon et al., 2006; Sommerhalder et al., 2011).

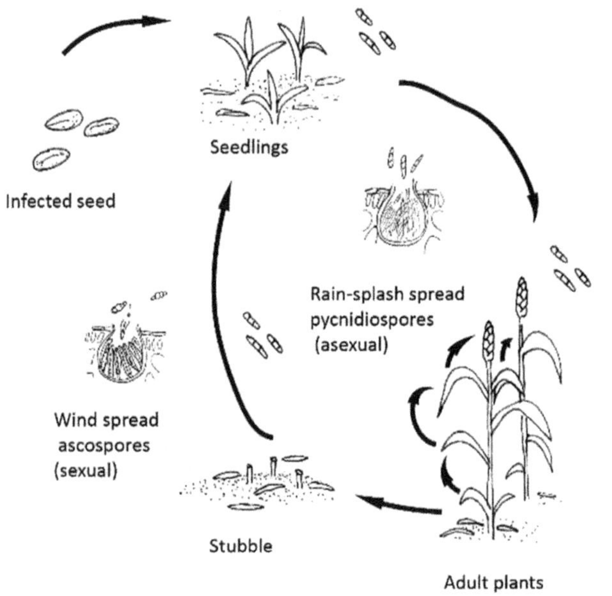

Figure 1 The life cycle of *P. nodorum*. Source: Adapted from Sommerhalder et al. (2011).

The mixed reproduction system provides both increased diversity due to the potential for variation through genetic recombination and fast clonal reproduction of favourable genotypes. Selection in different environments has likely given rise to high levels of variation in aggressiveness, and no single *P. nodorum* genotype is found to dominate in any environment (Ali and Adhikari, 2008; Blixt et al., 2008; Engle et al., 2006; Francki, 2013; Stukenbrock et al., 2006).

The pathogen survives on infected seeds and wheat stubble, which serve as primary inoculum sources (Fig. 1). Wind-borne ascospores are released from stubble, and sexual reproduction and formation of pseudothecia occur the whole growth season (Blixt et al., 2008; Sommerhalder et al., 2010). However, ascospore release also seems to be seasonal and is often assumed to be most important during fall and spring, coinciding with the emergence of wheat seedlings (Bathgate and Loughman, 2001; Bennett et al., 2007; Mittelstädt and Fehrmann, 1987). Rain-splash dispersed asexual pycnidiospores produced on the infected plants serve as primary and secondary inocula (Eyal et al., 1987). In order for efficient splash dispersal to occur, an intense rain shower of at least 5 mm and temperature >10°C, followed by at least 10 mm rainfall within the next 48 h, is necessary (Eyal et al., 1987) although dew and mist are sufficient to promote spore release (Bathgate and Loughman, 2001).

The disease progresses exponentially after Zadoks (Zadoks et al., 1974) growth stage 70–75 when the plant approaches maturity (Fig. 2), but with different slopes depending on factors like pre-crop, inoculum pressure, cultivar resistance and environmental conditions.

2.2 Symptoms

The *P. nodorum* germ tubes penetrate the leaf either directly through the cuticle or through open or closed stomata. Chlorosis at the infection site expands into oval lesions, often accompanied by necrosis. Pycnidia (Fig. 3, middle, top) can form in the infected tissue within a week under optimal conditions (Solomon et al., 2006).

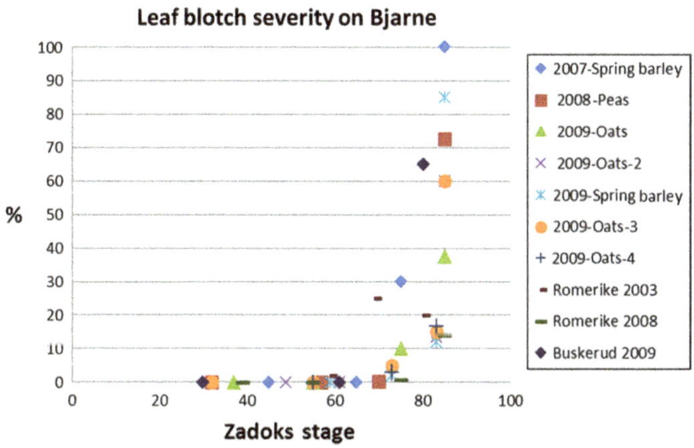

Figure 2 Percentage of SNB on the spring wheat cultivar Bjarne after various pre-crops. The disease develops exponentially from growth stage 70 to 75. Source: Adapted from Ficke et al. (2011b).

Figure 3 Left: Septoria nodorum leaf blotch symptoms in the field. Middle, top: pycnidia developing in the necrotic lesion. Middle, bottom: necrotic lesions and chlorosis on a flag leaf. Right: Septoria nodorum glume blotch. Photos: A. K. Ruud (left, middle), M. Lillemo (right).

In the field, symptoms of SNB first develop on the lower leaves and progress to the upper leaves through rain-splash dispersal. Under sufficiently long growth season and favourable weather conditions the pathogen will eventually reach the glumes and cause glume blotch (Fig. 3, right) (Eyal et al., 1987; McMullen and Adhikari, 2009; Solomon et al., 2006).

3 Disease management

SNB can be controlled through appropriate agricultural practices like crop rotation and tillage and by using fungicides and resistant cultivars. In later years, the recommended agricultural practice of reduced tillage to prevent soil erosion leads to increased disease pressure. The plant residues serve as primary inoculum in the subsequent growth season (Lillemo and Dieseth, 2011; McMullen and Adhikari, 2009).

3.1 Agricultural practice

Cultural practices have always been used to control disease pressure and optimize growth conditions for the crop. Rotation with crops that are non-hosts to *P. nodorum* is advised, since the pathogen survives on plant residues from the previous wheat crop. Rotations with resistant cultivars may also reduce the disease pressure since less inoculum is carried over from these (Krupinsky, 1999). Crop rotation is most effective to control diseases disseminated over short distances, like *P. nodorum* (Cunfer, 1998). However, political and economic incentives affect whether crop rotation is feasible for the farmer. In Norway, for example, farmers often grow wheat after wheat due to the larger interest in and price for this crop than for the alternative crops (Lillemo and Dieseth, 2011).

P. nodorum is further promoted if direct seeding or minimum tillage practices are applied (Krupinsky, 1999; Sutton and Vyn, 1990). This has become a common practice in modern agriculture in many parts of the world, in order to save costs, reduce soil erosion and limit water evaporation.

3.2 Fungicide control

Leaf blotch diseases are mainly controlled by application of fungicides at the heading stage. The main fungicide groups are strobilurins and triazoles.

Strobilurins inhibit fungal respiration by binding to cytochrome *b* complex III at the Q_0 site in the mitochondrial electron transport chain (Bartlett et al., 2002). Loss of strobilurin sensitivity is associated with a mutation in the pathogen's mitochondrial genome, leading to an amino acid change in cytochrome *b* (Gisi et al., 2002). In 2002, this mutation was found in *Z. tritici* isolates in several European countries (Leroux et al., 2007) and the same amino acid substitution was carried by the majority of Swedish *P. nodorum* isolates collected between 2003 and 2005 (Blixt et al., 2009).

Triazoles inhibit the 14-α-demethylation of lanosterol in the ergosterol biosynthesis pathway (Siegel, 1981; Vanden Bossche et al., 1995) and target *CYP51*, a cytochrome 450 enzyme responsible for the 14-α-demethylation. Reduced sensitivity to azoles can be caused by three mechanisms: Point mutation in the target gene *CYP51*, overexpression of *CYP51* and up-regulation of efflux proteins, leading to reduced accumulation of the fungicide inside the cells (Leroux et al., 2007).

Two non-synonymous amino acid substitutions in *CYP51* associated with reduced azole sensitivity were recently reported in European and Chinese *P. nodorum* isolates (Pereira et al., 2017), including 25% of Swedish isolates. A higher number of mutations has been found in *Z. tritici CYP51*, indicating that reduced sensitivity developed earlier in *Z. tritici* than in *P. nodorum*, and the effectiveness of all classes of this fungicide group is threatened (Cools et al., 2011) while still mostly effective against *P. nodorum* (Pereira et al., 2017).

Perhaps the shift in abundance from P. nodorum to Z. tritici in many European countries in recent years may be due to the greater speed that the latter pathogen adapted to triazole fungicides (McDonald, pers. comm.)? In Norway, *P. nodorum* remains the major leaf blotch pathogen in spring wheat. As in other countries, *P. nodorum* isolates collected in Norway have been reported with resistance to strobilurins and with reduced sensitivity to triazoles (Abrahamsen, 2013; Ficke et al., 2011a). The potential loss of effective fungicides underlines the need to control the disease by other and more sustainable methods.

3.3 Use of resistant cultivars

Growing cultivars with sufficient genetic resistance towards diseases is the most sustainable and economic strategy to prevent losses. Experiments conducted to measure the losses caused by leaf blotch disease in Norwegian wheat showed that the gain from fungicide treatment was lower in the most resistant cultivars like Zebra and, in particular, Mirakel (data for 2012). For these cultivars the yield was high in the untreated plots as well, illustrating the potential of cultivars with genetic resistance to reduce the need for fungicide spraying (Abrahamsen, 2013). A quantitative difference in genetic resistance to the disease allows a quantitative reduction in need for fungicide application.

4 The genetics of resistance

Breeding for resistance to SNB is challenged by the lack of major and dominant resistance genes. The inheritance of resistance is complex (Mullaney et al., 1982; Scharen and

Krupinsky, 1978) and strong genotype × environment (G × E) interactions can mask the relatively small contributions of the individual genes. Plant height and maturity are also associated with the development of the disease (Rosielle and Brown, 1980; Scott et al., 1982). However, significant residual resistance that is not associated with the confounding traits is also observed (Scott et al., 1982). This residual, 'true' genetic resistance and the discovery that host-specific gene-for-gene interactions are important determinants of susceptibility in the *P. nodorum*–wheat pathosystem (Liu et al., 2004a) provide the potential for genetic gain in SNB resistance breeding.

4.1 The gene-for-gene models

Flor's studies of the biotrophic pathosystem flax rust (*Melampsora lini*) and flax (*Linum marginale*) led to the classic gene-for-gene model of resistance (Flor, 1956, 1971) (Fig. 4a). The resistance is conferred when the product of an avirulence gene (*Avr*) from the pathogen is recognized by a resistance (*R*) gene in the host and hypersensitive response (HR) and programmed cell death are initiated (Fig. 4a).

4.2 Host-specific toxins and necrotrophic effectors

The first host-specific toxins (HSTs) were also discovered in the 1930–40s, for instance AK toxin produced by *Alternaria alternata* (Tanaka, 1933) and victorin produced by *Cochliobolus victoriae* (Meehan and Murphy, 1947). While resistance genes in the classical model are dominant (Fig. 4a), susceptibility is usually caused by a dominant susceptibility gene and is referred to as an inverse or mirror gene-for-gene model (Fig. 4b) (Friesen et al., 2007; Wolpert et al., 2002).

HSTs produced by fungi, like *Avr*-gene products in biotrophic systems, are also diverse in structure and biosynthetic mechanisms (Wolpert et al., 2002). Some peptide HSTs act as effectors by inducing HR (Faris et al., 2010; Oliver and Solomon, 2010) and are called NEs.

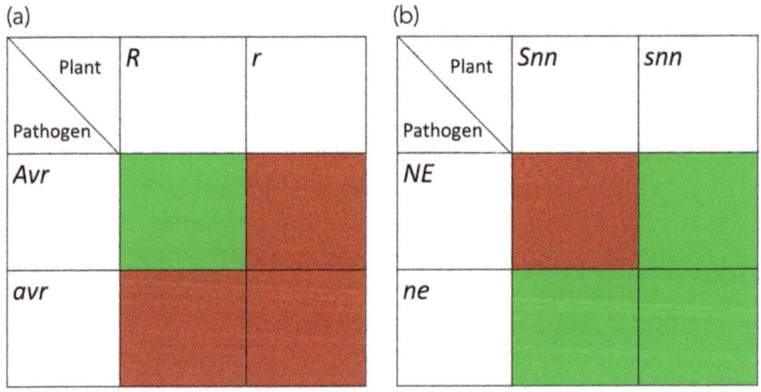

Figure 4 (a) The classical gene-for-gene model. Source: Adapted from: Flor (1971). In a biotrophic system, resistance is conferred when the product of a resistance (*R*) gene in the host recognizes an avirulence (*Avr*) gene product secreted by the pathogen. (b) The 'inverse' model. Source: Adapted from: Friesen et al. (2007). In a necrotrophic system, HR is induced upon recognition of a necrotrophic effector (NE) by the product of a sensitivity (*Snn*) gene and leads to increased susceptibility.

4.3 Necrotrophic effectors and sensitivity gene interactions in the *P. nodorum*–wheat pathosystem

It has been known for almost 40 years that *P. nodorum* produces phytotoxic compounds inducing plant cell death prior to hyphal growth (Bird and Ride, 1981). However, the role of these phytotoxins in relation to resistance was not understood by the end of the last century (Cunfer, 1999).

Liu et al. (2004a) characterized the first host-specific protein secreted by *P. nodorum*, and named it 'SnTox1'. Earlier, Tomas and Bockus (1987) had described that the causal agent of tan spot, *P. tritici-repentis*, secretes an HST, Ptr toxin, and later renamed it 'ToxA' (Tomas et al., 1990). The corresponding sensitivity locus *Tsn1/tsn1* was mapped to chromosome 5BL and was reported as a 'dominant susceptibility' locus (Faris et al., 1996). Friesen et al. (2006) discovered a *P. nodorum* gene that shared 99.7% sequence similarity to the *ToxA* in *P. tritici-repentis* described above, and the sensitivity also mapped to *Tsn1* (Liu et al., 2006). It was estimated that the *ToxA* gene was introduced from *P. nodorum* into *P. tritici-repentis* through horizontal gene transfer before 1941 when *P. tritici-repentis* emerged as a pathogen on wheat (Friesen et al., 2006).

The *ToxA* gene is also present in *Parastagonospora avenae* f.sp. *tritici*, closely related to *P. nodorum* (McDonald et al., 2012, 2013). Recently, the *ToxA* gene was also discovered in *Bipolaris sorokiniana*, the causal agent of spot blotch in wheat (McDonald et al., 2017). The *ToxA* region in *B. sorokiniana* showed more similarities with *P. tritici-repentis* than with *P. nodorum*.

The characterized *P. nodorum* NEs are small, secreted proteins, and virulence factors rather than true pathogenicity factors (Friesen et al., 2007), that is, they affect the degree of disease in the host. So far, at least eight NEs and nine corresponding *Snn* loci have been characterized (Table 2).

Table 2 Overview of the mapped and characterized necrotrophic effectors (NEs) from *P. nodorum* and corresponding host sensitivity loci in wheat

Necrotrophic effector (NE)	Protein size (kD)	Host sensitivity gene	Wheat chromosome arm	Type of gene/ mapping status	Reference
SnToxA	13.2	*Tsn1*	5BL	NBS-LRR-PK	Faris et al. (2010) and Liu et al. (2006)
SnTox1	9	*Snn1*	1BS	WAK	Liu et al. (2004a,b, 2012) and Shi et al. (2016b)
SnTox2	7–10	*Snn2*	2DS	Mapped	Friesen et al. (2007)
SnTox3	19	*Snn3-B1*	5BS	Fine-mapped	Friesen et al. (2008), Liu et al. (2009), Ruud et al. (2017) and Shi et al. (2016a)
		Snn3 D1	5DS	Fine-mapped	Zhang et al. (2011)
SnTox4	10–30	*Snn4*	1AS	Mapped	Abeysekara et al. (2009)
SnTox5	10–30	*Snn5*	4BL	Mapped	Friesen et al. (2012)
SnTox6	12	*Snn6*	6AL	Mapped	Gao et al. (2015)
SnTox7	10–30	*Snn7*	2DL	Mapped	Shi et al. (2015)

P. nodorum is a good model organism for genomic studies due to its willingness to grow on artificial media, its importance as a crop pathogen and its relatively small sized genome (≈37 Mb). The first *P. nodorum* reference genome was published by Hane et al. (2007), and it has later been re-sequenced (Syme et al., 2013) using isolates with different effector profiles and annotated (Syme et al., 2016). Naturally, an important objective of the genomic studies has been to investigate the genetic and evolutionary basis of the fungus' pathogenicity. Bioinformatics tools have been used to search for novel effector candidate genes and were successful in identifying the *SnTox1* gene (Liu et al., 2012). The criteria used to predict candidate effector genes are: (1) small secreted protein (<30 kDa), (2) being cysteine rich; (3) located near repetitive DNA regions or scaffold ends and (4) no blast matches. In addition, criteria like presence/absence of genes in virulent versus non-virulent isolates and evidence of positive selection can be applied when isolates with known differences in pathogenicity are compared (Syme et al., 2013). By applying these criteria on genomic data from three isolates [SN15, Sn4 and Sn79 (non-virulent)], the authors obtained a candidate list of 159 potential effector genes (Syme et al., 2013). Although two of the known effectors ranked top of the list, the number of genes that would need further investigation and validation is somewhat discouraging. Also, since few similarities have been identified between different effectors, as discussed above, the prediction criteria may not capture all the real candidate genes. In addition, although acknowledging that effector genes are often located in repetitive regions (Criterion 3), such regions have presented technical challenges and been filtered out in many next-generation-sequencing studies (Alkan et al., 2011), which means that there is a chance that many true candidate genes are missed (Treangen and Salzberg, 2011). More recently, technologies that also capture these regions, that is, sequence longer continuous pieces of DNA, like PacBio (Pacific Biosciences), have been developed (Goodwin et al., 2016).

The cloning of NEs and their corresponding host sensitivity genes has revealed quite distinct and diverse molecular functions. SnToxA was shown to be internalized into the chloroplasts of sensitive (*Tsn1*) host cells where it induces alterations in photosynthetic electron transport, leading to a burst in reactive oxygen species (Manning et al., 2009). However, the *Tsn1* protein does not interact directly with SnToxA (Faris et al., 2010). On the other hand, SnTox1 was found to be localized in the fungal cell walls where it likely serves to protect the pathogen against wheat chitinases, and it does not enter the plant cell (Liu et al., 2016). The effector is recognized directly by the Snn1 protein, which spans the plasma membrane and contains extra-cellular binding domains (Shi et al., 2016b). Although the outcome is the same – induction of programmed cell death – it appears that Snn1 activates PAMP-triggered immunity (PTI) upon direct recognition by SnTox1 at the host cell surface, while *Tsn1* activates effector-triggered immunity (ETI) upon indirect recognition of SnToxA inside the plant cell.

The relative contributions of NE-*Snn* interactions to disease under field conditions are still discussed (Francki, 2013), although more and more evidence supports that at least some of the interactions are important: SnToxA-*Tsn1* and SnTox2-*Snn2* were identified after spray inoculation of the flag leaf with a single *P. nodorum* isolate in the field (Friesen et al., 2009). The effect of SnToxA-*Tsn1* was also likely to underlie a significant quantitative trait locus (QTL) in the 05Y001 doubled-haploid mapping population one year, but not in the subsequent trial (Francki et al., 2011). Waters et al. (2011) found lesser differences in resistance rating between ToxA-insensitive and -sensitive cultivars than Oliver et al. (2009) and suggested that a reduction in the frequency of SnToxA-sensitive cultivars could have triggered a shift in the NE frequencies in the pathogen population. Waters et al. (2011)

also found a low, but significant correlation between sensitivity to SnTox3 and disease resistance ratings in Australian wheat cultivars. The identification of *Snn3* as a major susceptibility factor in the SHA3/CBRD × Naxos population in the field was the first to validate the importance of this locus in adult plant resistance (Ruud et al., 2017).

In Australia, SnToxA has been delivered to the breeders since 2009 (Vleeshouwers and Oliver, 2014). By 2012, 30000 doses of SnToxA and 6000 doses each of SnTox1 and SnTox3 were provided annually (Vleeshouwers and Oliver, 2014). The area of SnToxA-sensitive wheat in Australia fell from 30.4% in 2009–10 to 16.9% within three years. The estimated economic gain was approximately 50 million AUD, assuming a yield loss of 0.3 tons/ha in susceptible cultivars (Vleeshouwers and Oliver, 2014).

4.4 The nature of resistance and sensitivity genes

Most of the *R*-genes encode proteins with a nucleotide binding site (NBS) and leucine-rich repeats (LRRs). Upon direct or indirect recognition of a pathogen effector (i.e. the product of an *Avr*-gene), the NBS-LRR initiates signalling pathways, in most cases leading to HR and cell death (Jones and Jones, 1997; van't Slot et al., 2003).

Less is known about the genes conferring susceptibility to NEs. However, the molecular cloning of a number of sensitivity genes including *Tsn1* involved in ToxA sensitivity have shown that they often have NBS and LRR domains associated with ETI (Faris et al., 2010; Lorang et al., 2007; Nagy and Bennetzen, 2008). Recently, also *Snn1* conferring sensitivity to SnTox1 was cloned and shown to encode a wall-associated kinase (WAK) (Shi et al., 2016b). Receptor kinases are usually pattern recognition receptors involved in pathogen-associated molecular patterns (PAMP) triggered immunity. Responses to ETI and PAMP overlap, including the HR response (Dodds and Rathjen, 2010).

These gene-for-gene interactions, inducing HR upon recognition, form the framework for identifying resistance and sensitivity to SNB at the seedling stage. Single *P. nodorum* isolates can be grown in liquid medium and are expected to secrete NEs into the medium. When seedling leaves are infiltrated with filter-sterilized culture filtrate (CF) (Fig. 5), sensitive lines develop chlorosis or necrosis, while resistant lines remain healthy. The reaction types are usually scored on a 0–3 scale (Friesen and Faris, 2012) and the sensitivity locus can be genetically mapped if a segregating mapping population is used. Typically, symptoms develop after 3–5 days in the greenhouse.

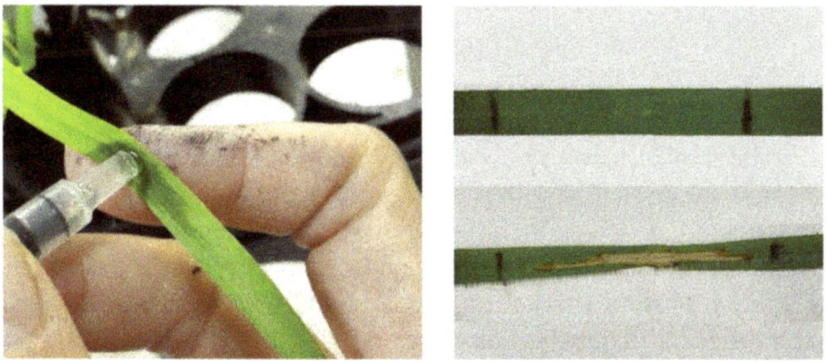

Figure 5 Left: infiltration with culture filtrate with a needleless syringe. Right, top: SnToxA-insensitive leaf 5 dpi (days past inoculation). Right, bottom: necrotic tissue developed in the infiltrated area in a SnToxA-sensitive leaf 5 dpi. Photos: A. K. Ruud (left), Min Lin (right).

Since the NE is a virulence factor, it should also have an effect on disease development after inoculation with a conidiospore suspension. Historically, different methods have been used to evaluate the role of different components of resistance at the seedling stage. These assessment methods include latency period, lesion expansion and development and number of pycnidia in the lesions (Czembor et al., 2003; Eyal and Scharen, 1977; Eyal et al., 1987). Quantitative measurements have also been used, for instance in Eyal and Scharen (1977), Karjalainen (1985) and Jönsson (1985). However, a reaction type scale from 0 to 5 (Liu et al., 2004b) emphasizing the extent of chlorosis and necrosis associated with the lesions is now commonly applied. This method is believed to be accurate in capturing the effect of potential NE–*Snn* interactions (Friesen and Faris, 2012).

Adult plant resistance to SNB is mainly quantitative and additive (Bostwick et al., 1993; Fried and Meister, 1987; Wicki et al., 1999). Dominant SNB resistance is also observed, and the segregation patterns of intermediate reactions can indicate the presence of modifier genes (Kleijer et al., 1977; Ma, 1993; Ma and Hughes, 1993, 1995). General mechanisms like pathogen production of cell wall degrading enzymes (Lehtinen, 1993; Magro, 1984), host lignification and papilla formation to reduce hyphae penetration (Bird and Ride, 1981) also explain variation in resistance.

Epistatic interactions among NEs seem to be common. For instance, it was shown that the SnToxA-*Tsn1* interaction is epistatic to SnTox3-*Snn3* and that SnTox3-*Snn3* is significant only in the presence of an incompatible SnTox2-*Snn2* interaction (Friesen et al., 2008). This was further corroborated by generation of knockout-strains of *P. nodorum* isolate SN15 whereby the *SnToxA*, *SnTox1* and *SnTox3* genes were sequentially deleted through genetic transformations (Tan et al., 2015). In the Calingiri × Wyalkatchem mapping population, the SnTox1-*Snn1* interaction was paramount when inoculated with SN15, and no effect was observed for the SnTox3-*Snn3* interaction. However, this interaction became significant when plants were inoculated with *tox1-6,* a knockout strain for *SnTox1* (Phan et al., 2016). Apparently, this was due to up-regulated *SnTox3* expression in *tox1-6* (Phan et al., 2016). The triple knockout strain (*toxa13*) lacking all three effector genes was still able to cause disease (Tan et al., 2015). When used in QTL mapping it unveiled a significant QTL on 2DS where *Snn2* is located, which was not detected with SN15 or *tox1-6* (Phan et al., 2016), which indicated that the 2DS QTL was epistatic to SnToxA-*Tsn1* and/or SnTox3-*Snn3*. Interestingly, the SnTox5-*Snn5* and SnTox6-*Snn6* interactions are also epistatic to SnTox3-*Snn3* (Friesen et al., 2012; Gao et al., 2015).

4.5 Escape mechanisms and environmental impacts on resistance

Morphological traits that reduce the contact between the pathogen and the plant can be identified as disease escapes (Parlevliet, 1977), and often lead to misinterpretation of true association with resistance. These include variation in plant height and timing of heading (earliness) (Scott et al., 1982). Taller plants may escape from rain-driven spread of disease in the canopy. Early maturing plants may escape the highest disease pressure (Francki, 2013), but on the other hand, later plants may appear more resistant at the time of disease scoring since the disease develops faster in more mature plants.

The development of disease is affected by weather conditions like temperature, rainfall and humidity. Variation in these factors within and between growth seasons can have a strong effect on the relative resistance rankings (Kim and Bockus, 2003).

5 Genetic mapping of resistance and use of markers in breeding

5.1 Molecular markers

Breeding for improvement of complex traits is not straightforward. The contribution of each individual locus is moderate and can be masked by other, dominant loci or epistatic effects. However, the development of molecular markers could provide a help in overcoming some of these difficulties. Co-dominant markers can distinguish between all genotypes (Tanksley, 1983). Markers can be either hybridization or polymerase chain reaction (PCR) based. In the first case, a probe is hybridized to the DNA. In PCR-based systems, small fragments of DNA are amplified with polymerase enzymes.

Simple sequence repeat (SSR) markers have many advantages. They can be non-anonymous, abundant and reproducible and show a high degree of inter- and intraspecific polymorphism (Mammadov et al., 2012; Semagn et al., 2014).

Single nucleotide polymorphisms (SNPs) are the most abundant of molecular markers, estimated to occur for every 100–300 bp in any genome (Gupta et al., 2001), and for wheat every 96–483 bp, depending on the stringency criteria and germplasm used (Manickavelu et al., 2012). Chip-based SNP platforms are oligonucleotide-based DNA microarrays that cover thousands of SNP markers that can be genotyped in one go. Recent genome sequencing efforts have enabled the development of low-cost high-density arrays in wheat, such as the Illumina 90K wheat chip (Wang et al., 2014) and the Affymetrix 35K breeders array (Allen et al., 2017).

Genotyping-by-sequencing (GBS, also called next-generation sequencing, NGS) can be an alternative to chip-based arrays. GBS is particularly suitable for projects where the genomes of several specimens are sequenced to discover large numbers of SNPs. No prior knowledge of the genome is necessary and its cost is lower than that of chip-based arrays (Elshire et al., 2011).

SNPs are quite easily transferred from one platform to another, and for genotyping of only a few markers on many samples, KASP (Kompetitive Allele Specific PCR) is a highly flexible and cost-effective alternative commonly used in wheat breeding (Semagn et al., 2014).

5.2 Linkage QTL mapping

Linkage mapping is the most widely used method to dissect complex traits and identify markers linked to them. The genomic regions associated with such a trait are called QTL. The development of molecular marker technologies in the 1980s facilitated the construction of genetic linkage maps (Collard et al., 2005), and complex traits could be separated into discrete QTL (Paterson et al., 1988).

Maps can be constructed for specific, segregating populations. Preferably, recombinant inbred lines (RILs) or double haploids (DHs) are used since they are homozygous and can be maintained and reproduced forever (Collard and Mackill, 2008). To identify QTL, the population is phenotyped for the traits of interest, and the phenotypic and genotypic data are analysed to uncover linkage between a certain phenotype and genetic regions.

After the initial detection, additional steps to confirm a QTL and to validate associated markers are usually required (Langridge et al., 2001). The effect and position of a QTL can be inaccurate due to sampling bias (Melchinger et al., 1998), and flanking markers may not be polymorphic in other genotypes.

Table 3 Overview of quantitative trait loci (QTL) for SNB resistance, including necrotrophic effector-sensitivity (NE-*Snn*) gene interactions

Plant tissue	Population	NE-*Snn*	QTL, chromosome	Markers	Reference
Seedling leaf	Liwilla × Begra, DH		*QSnl.ihar-2B*	*gwm501 – gwm410*	Czembor et al. (2003)
			QSnl.ihar-5B	*barc32 – gwm499*	
			QSnl.ihar-5D	*gwm205 – gwm212*	
	W7984 × Opata85	SnTox1-*Snn1*	1B	*mwg938 – snn1* *fcp618, psp3000*	Liu et al. (2004b)
			4B	*cdo1312*	
	Alba × Begra		*QSnl.ihar-6A*	*gwm570 – mwg934*	Arseniuk et al. (2004)
	BR34 × Grandin	SnToxA-*Tsn1*	5BL	*fcp1, fcp2, fcp394, fcp620*	Friesen et al. (2006)
	BR34 × Grandin	SnTox2-*Snn2*	2DS	*TC253803, cfd51*	Friesen et al. (2007)
		SnTox3-*Snn3-B1*	5BS	*gwm234, cfd20*	
	LDN × LDN (DIC-1B) (*T. turgidum*)		5BL	*bcd9 – fbb237*	Gonzalez-Hernandez et al. (2009)
	Arina × Forno	SnTox4-*Snn4*	1AS	*BG262267, BG26975, cfd58*	Abeysekara et al. (2009)
	Aegilops tauschii	SnTox3-*Snn3-D1*	*5D*	*cfd18 – hbg337*	Zhang et al. (2011)
	Lebsock × PI 94749 (*T. turgidum* subsp. *durum* × *T. turgidum* subsp. *carthlicum)*	SnTox5-*Snn5*	4BL	*wmc349 – cfd22, barc163*	Friesen et al. (2012)
	W7984 × Opata85	SnTox6-*Snn6*	6AL	*BE424987 – BE403326*	Gao et al. (2015)
	Chinese Spring × Timstein	SnTox7-*Snn7*	2D	*cfd44 – gwm311*	Shi et al. (2015)
	Calingiri × Wyalkatchem	SnTox1-*Snn1*	*Q.snb.fcu-1BS*	*gpw7059a – wPt-2654*	Phan et al. (2016)
			Qsnb.cur-2AS1	*gwm339 – gwm312*	
			Qsnb.cur-2DS	*cfd36 – wPt-669517*	
			Qsnb.cur-3AL	*tPt-1143 – wPt-4859*	
			Qsnb.fcu-4BL	*barc163 – wPt-4243*	
		SnTox3-*Snn3-B1*	*Qsnb.fcu-5BS*	*TC282809a – wPt-666323*	
	Calingiri × Wyalkatchem		*Qsnb.cur-2AS2*	*wmc382a – barc124a*	Rybak et al. (2017)
			Qsnb.cur-2DS	*cfd36 – wPt-669517*	
			Qsnb.cur-3AL	*tPt-1143 – wPt-4859*	
			Qsnb.cur-6AS	*gpw4329 – wPt-4270*	

SHA3/CBRD × Naxos		1A	*IWB52902*	Ruud et al. (2017)
		1B	*psp3000*	
		2D	*IWA8544*	
	SnTox3-*Snn3-B1*	5B	*IWB11709*	
		7B	*Wsnp_BE498662B_Ta_2_5*	
Altar 84 × Langdon (*T. turgidum* ssp. *durum*)	SnToxA-*Tsn1*	*Qsnb-fcu.5B*	*Tsn1*	Virdi et al. (2016)
GWAS 567 spring wheat landraces		2D	*wPt-665317*	Adhikari et al. (2011)
		3B	*wPt-6047*	
		5B	*wPt-1149*	
		6A	*wPt-7330*	
		7A	*wPt-4515*	
GWAS 528 spring wheat landraces		2D	*IWA7348*	Gurung et al. (2014)
		3A	*IWA4075*	
		5B	*IWA7024*	
GWAS 120 winter wheat cultivars and breeding lines		5A	*IWB67424*	Liu et al. (2015)
		5B	*IWB36366*	
		5B	*IWB38178*	
		5D	*IWB45668*	
GWAS 320 synthetic hexaploid wheat lines		7B	*1208964*	Jighly et al. (2016)
GWAS 121 spring wheat lines		1A	*IWB56511*	Ruud et al. (2018a)
		1B	*IWB6550*	
		3A	*IWB9350*	
		4B	*IWB6422*	
	SnTox3-*Snn3-B1*	5B	*IWB26869*	
	SnToxA-*Tsn1*	5B	*IWB67424, IWB67425, IWB3660*	
		6B	*IWB73405*	
		7A	*IWB21459*	
		7B	*IWB64015*	
		7D	*IWA732*	

(Continued)

Table 3 (*Continued*)

Plant tissue	Population	NE-*Snn*	QTL, chromosome	Markers	Reference
Seedling and adult plant leaf	BR34 × Grandin	SnTox2-*Snn2*	*QSnb.fcu-2DS*	*gwm614 – cfd53*	Friesen et al. (2009)
			QSnb.fcu-5AL	*barc151 – fcp13*	
		SnToxA-*Tsn1*	*QSnb.fcu-5BL*	*barc1116 – barc43*	
Adult plant leaf	Forno × Oberkulmer		*QSnl.eth-2D*	*psr932 – psr331a*	Aguilar et al. (2005)
			QSnl.eth-4B	*glk348 – psr921*	
			QSnl.eth-7B	*mwg710a – glk576*	
	WAWHT2074 × 6HRWSN125		*QSnl.daw-2D*	*cfd11 – gwm30*	Shankar et al. (2008)
	BR34 × Grandin		*QSnb.fcu-1BS*	*fcp267 – barc240*	Friesen et al. (2009)
	P92201D5 × P91193D1		*QSnl.daw-2A*	*gwm614a – wPt-7056*	Francki et al. (2011)
	EGA Blanco × Millewa		*QSnl.daw-1B*	*wPt-8949 – wPt-2575*	
			QSnl.daw-5B	*wPt-3457 – wPt-0935*	
	SHA3/CBRD × Naxos		*1B*	*wmc619*	Lu and Lillemo (2014)
			3AS	*gwm2*	
			3B	*wPt-4127*	
			3BL	*wPt-4933*	
			5BS	*wPt-5346*	
			5BL	*fcp1*	
			7A	*wmc603*	
			7B	*wPt-0963*	
	SHA3/CBRD × Naxos		*1A*	*IWA2995*	Ruud et al. (2017)
			1B	*SCM9*	
			3AS.1	*gwm2, IWB35234*	
			3AS.2	*IWB39383, IWB27319*	
			3BL	*wPt-4933*	
			5BS	*IWB11709*	
			5B.2	*wPt-5914*	
			7A	*IWB53887*	

	Calingiri × Wyalkatchem	SnTox1-*Snn1*	*Qsnb.fcu-1BS*	*gpw7059a – wPt-2654*	Phan et al. (2016)
			Qsnb.cur-6BS	*wPt-3168 – barc146a*	
	GWAS 121 spring wheat lines		2B	*IWB2427*	Ruud et al. (2018a)
			2D	*gwm301, IWB35134, IWB30879, IWB30880, IWB34359*	
			4B	*IWB6422*	
			7B	*IWB57207, IWB73685, IWB70085*	
Glume blotch	Arina × Forno		*QSng.sfr-3B*	*gwm389 – cfd79c*	Schnurbusch et al. (2003)
			QSng.sfr-4B	*gwm165 – glk335*	
	Arina × Forno and GWAS 44 winter wheat cultivars		*Qsng.sfr-3B*	*Sun2-3B*	Tommasini et al. (2007)
	Arina/3*Forno NILs		3BS1	*wmm756*	Shatalina et al. (2014)
			3BS2	*swm01210*	
	P92201D5 × P91193D1		*QSng.pur-2DL.1*	*gwm526a – cfd50b*	Uphaus et al. (2007)
			QSng.pur-2DL.2	*cfd50c – wPt9848*	
	WAWHT2074 × 6HRWSN125		*QSng.daw-4B*	*Rht1 – gwm495*	Shankar et al. (2008)
	Forno × Oberkulmer		*QSng.eth-5AL*	*psr1194 – psr918*	Aguilar et al. (2005)
	GWAS 320 synthetic hexaploid wheat lines		2AL	*wPt-1657*	Jighly et al. (2016)
			2BL	*1107710*	
			2DL	*wPt-7825*	
			3BS	*wPt2757, wPt-8079, wPt-3921*	
			4BL	*1094836*	
			5AL	*wPt-8262*	
			6B	*1019982*	
			7DS	*1263913*	
			7DL	*1216888, 1233921, 1227840*	

Source: Adapted from Francki (2013) and updated with more recent studies. *T. aestivum* unless otherwise noted. Adult plant resistance QTL from field trials are only listed if they have been reported significant in at least two environments (years, locations).

Table 3 shows an overview of SNB resistance QTL detected in QTL mapping studies. For field-based studies, only QTL significant (LOD ≥ 3 or threshold calculated by permutation test) in at least two environments are included.

5.3 Association mapping

Association mapping (AM, also called genome-wide association study, GWAS) emerged in the early 2000s as an alternative to biparental linkage mapping (Gupta et al., 2014). Only a handful of GWAS studies have investigated SNB resistance (Table 3). AM was used to fine-map a region on 3BS associated with Septoria nodorum glume blotch in 44 European winter wheat varieties (Tommasini et al., 2007). Adhikari et al. (2011) detected unique SNB seedling resistance QTL on 6A and 7A in a GWAS panel consisting of 576 land races from the USDA Small Grains Collection. A novel QTL on 3A was identified in a set of 528 spring wheat landraces from the same USDA Small Grains Collection in a study by Gurung et al. (2014), while two other QTL on 2D and 5B were described previously by Adhikari et al. (2011). Also, Liu et al. (2015) identified seedling QTL on 5A, 5B and 5D in a panel of 120 hard red winter wheats. To our knowledge, the GWAS on Nordic wheat germplasm (Ruud et al., 2018a) is the only study so far focusing on adult plant leaf resistance. It showed that field resistance is based on many small-effect QTL and only a few loci showed effect across field seasons (Table 3).

5.4 Marker-assisted selection and resistance breeding

Figure 6 shows the typical workflow for a genetic mapping study. A suitable population is genotyped with genetic markers (usually SNPs or SSRs) and phenotyped for the traits of interest. QTL mapping or GWAS is performed to identify significant marker-trait associations (MTAs). The most promising markers can then be validated in other

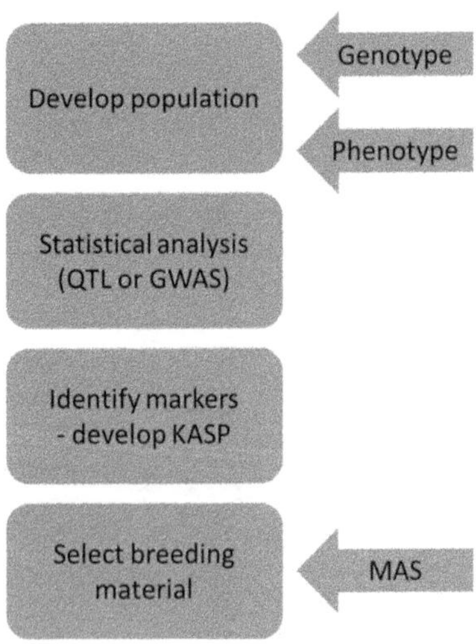

Figure 6 Typical workflow of a genetic mapping study with marker-assisted selection (MAS) as the goal.

populations and converted to economical KASP markers and used to screen and select breeding material (marker-assisted selection, MAS).

Economic and practical constraints decide whether MAS is feasible in a resistance breeding programme. Compared to phenotypic selection at the adult plant stage, MAS can provide higher accuracy and save time in the breeding cycle, if markers are closely linked or diagnostic for QTL that explain a substantial amount of the phenotypic variation (typically at least 10–20%). For quantitative disease resistance, the best strategy is suggested to be MAS followed by phenotypic selection. The subsequent step of phenotypic selection allows for the inclusion of minor QTL (Miedaner and Korzun, 2012).

5.5 Genomic prediction and selection

Genomic prediction and selection (Meuwissen et al., 2001) is an approach that predicts the best individuals based on genetic values. Compared to traditional models where only markers significantly linked to the trait are considered, genomic selection considers all markers across the genome to predict breeding values (BVs). Known QTL with large effects can be included in the genomic prediction models as fixed effects and further increase the prediction accuracy (Bernardo, 2014). The genomic selection is performed on a different population than the reference (training) set on which the genetic marker effects were calculated. Although initially used to predict BVs of animals, genomic selection also has the potential to improve genetic gain in crops like wheat (Crossa et al., 2010; Ornella et al., 2012; Storlie and Charmet, 2013).

For a complex trait like SNB seedling resistance, genomic prediction models performed better than a MAS approach using only the markers significantly associated with the trait (Juliana et al., 2017). The accuracy for prediction of adult plant resistance has not yet been reported for SNB. However, prediction values for adult plant resistance to comparable diseases like tan spot and Septoria tritici blotch showed that although the prediction accuracy was generally lower than for seedling resistance, genome-wide prediction models performed significantly better than the MAS approach which only considered the significant markers (Juliana et al., 2017).

6 A case study on SNB resistance in Norwegian wheat

SNB is one of the most important diseases in spring wheat in Norway (Ficke et al., 2011b; Lillemo and Dieseth, 2011). The disease is mainly controlled by fungicides since the use of other measures like crop rotation and autumn ploughing is limited. The potential loss of sensitivity towards fungicides (Abrahamsen, 2013; Ficke et al., 2011a) and concern about health and environmental risks underline the need for more sustainable control.

As outlined in Sections 4 and 5, the recent discoveries of the NEs and their role in the wheat-SNB pathosystem opened up new possibilities for resistance breeding by elimination of sensitivity loci in the germplasm (Vleeshouwers and Oliver, 2014; Zhang et al., 2009). On this background, research efforts were initiated in Norway in 2009 to characterize the local pathogen population, investigate the role of NEs in causing susceptibility under field conditions and map major resistance/susceptibly loci in relevant germplasm for wheat breeding. In the following section, we give a brief outline of the research approach and main results so far, with focus on implications for disease management and resistance breeding.

6.1 Survey of the pathogen population

During the 2010 and 2011 field seasons, a small survey was conducted by collecting leaf samples from unsprayed wheat fields with leaf blotch symptoms across the wheat-growing regions in Norway. These samples were analysed by PCR with species-specific primers, and the analysis showed that *P. nodorum* was present in almost all the samples collected from both spring wheat and winter wheat fields. *P. tritici-repentis* was detected in about half of the samples of both spring and winter wheat, while *Z. tritici* was less common and mostly found in winter wheat. These unpublished results proved a very similar situation as in Finland, where *P. nodorum* was also the dominating pathogen on spring wheat and was present in almost every sample that was analysed in 2009 (Jalli et al., 2011). The frequent occurrence of all three leaf blotch pathogens on the same samples was well in line with similar results obtained from winter wheat fields in Sweden (Blixt et al., 2010).

P. nodorum isolates were also collected from unsprayed fields in the major wheat-growing areas in 2012–14. The majority (69%) of the Norwegian *P. nodorum* isolates investigated (n = 62) harboured the *SnToxA* gene, which may be an adaptation to the sensitive host populations (Ruud et al., 2018b). The frequency was much higher in these isolates than in Swiss isolates (McDonald et al., 2013). The high frequency of the *SnToxA* gene in the pathogen population can also explain why the sensitivity was detected as significant under natural infection in our field trials. Also, the frequencies of isolates producing SnTox3 (76%) and SnTox1 (53%) were high in the Norwegian *P. nodorum* population (Ruud et al., 2018b). This demonstrated that the known effectors are common in the pathogen population, and that elimination of corresponding sensitivity alleles in the wheat germplasm might be a promising avenue to follow in resistance breeding.

6.2 Establishment of reliable field testing methodology

To identify genetic SNB resistance, the first steps are to collect and screen diverse germplasm in the field and in the greenhouse. In the field, one can either rely on natural infection by the ambient pathogen population, or inoculate with single isolates or a mixture of isolates. The reproducibility across locations or seasons may be higher when the same isolates are used. On the other hand, one or a few isolates may not be representative of the situation in the farmers' fields. Breeders usually rely on natural infection in the field for evaluation of leaf blotch resistance (Cowger and Murphy, 2007). Fraser et al. (2003) suggested that promotion of infection by natural inoculum gives a better estimate of host resistance under natural epidemics than inoculation of the nurseries with selected isolates. In the field trials that will be referred to, we relied on natural infection promoted by mist irrigation and naturally infected straw.

6.2.1 Correction for confounding traits

Development of SNB at the adult plant stage in the field is influenced by variation in plant height and earliness (timing of heading). It is important to account for the effects of these traits. This can be done in several ways:

1. Score the plants and/or spray inoculate the flag leaves at the same developmental stage, in the greenhouse or in a tunnel to avoid effects of plant height and earliness on disease ratings. These are good measures, but very labour intensive and time-consuming when large populations are screened.

2 Choose or develop mapping populations with little variation in earliness and height.
3 Score all traits of interest separately and only consider QTL for SNB resistance that does not co-locate with QTL for the confounding traits. However, true resistance QTL under the threshold can go undetected by this method.
4 Include the confounding traits as covariates in a regression model with SNB severity as the dependent trait. This is the method we used. The QTL detected when the corrected values are analysed are assumed to capture the true residual, genetic resistance to SNB. The corrected disease scores were annotated as 'corrected SNB severity'.

The effect of plant height on SNB development varied from year to year. As described in Section 2.1, the pathogen has certain rainfall and relative humidity requirements in order to sporulate, spread by rain-splash and successfully infect new leaves (Eyal et al., 1987). In years with moderate rainfalls during mid-June to July, the effect of plant height on disease severity was usually significant in our field trials. In years with low correlation between plant height and SNB development, extreme rainfalls (i.e. 76 mm in 24 h in 2015) seemed to reduce the differences between tall and short plants, by spreading the spores to the flag leaf even in tall cultivars.

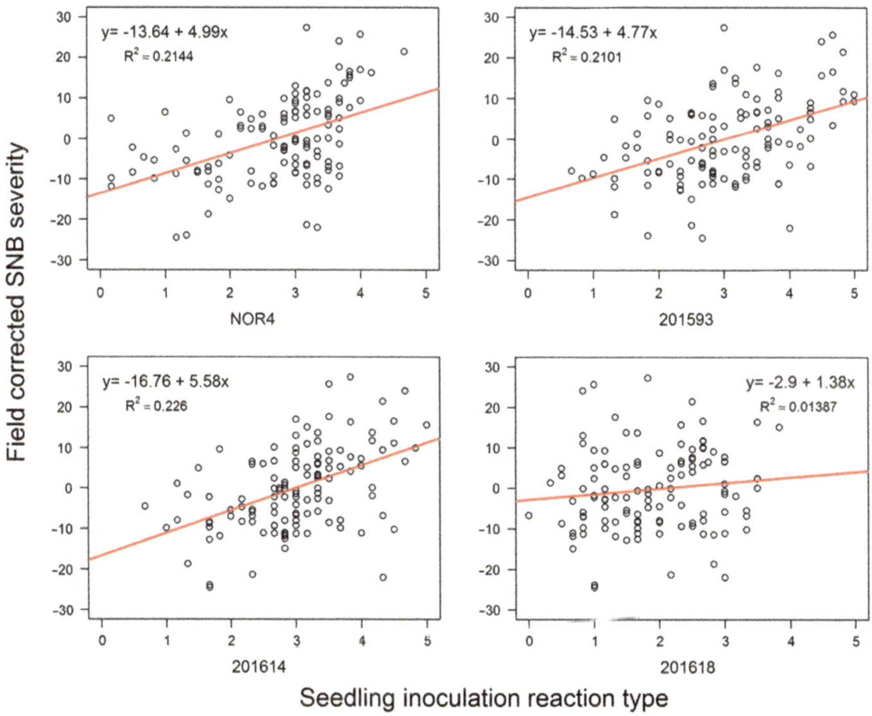

Figure 7 Correlation between seedling inoculation disease reaction type and the mean corrected SNB severity from seven years of field trials in a Nordic spring wheat collection (Ruud et al., 2018a,b).

Temperature affects plant development, such as the timing of heading. Although spring wheat normally does not require vernalization, that is, a cooler period after germination to induce heading and flowering, some of the lines we tested harbour the *Vrn-A1* gene (Yan et al., 2003). This gives them a weak vernalization requirement. In particular, this applies to several lines and cultivars originating from CIMMYT. When the spring is warm, the requirement may not be met, and the induction of heading and flowering is significantly delayed. Thus, the effects of the individual vernalization genes vary between years, depending on temperature.

6.2.2 How can we use the correlation between field and seedling experiments to select resistant genotypes?

To be relevant for breeding purposes, the genetic resistance needs to act at the adult plant stage in the field. However, disease evaluation in the field is resource demanding and is often complicated by genotype by environment interactions and sometimes by confounding traits. Thus, if sufficiently representative results are possible to score at the seedling stage, it would save time and money. A typical example of the relationships that can be obtained in mapping populations and germplasm collections is shown in Fig. 7. Significant correlations can be obtained by the use of representative isolates that reflect the natural pathogen population, but R^2 values rarely exceed 20–25% when field data are based on natural infection. When less relevant isolates are used, such as 201 618, there might not be any relationship at all. Therefore, good care should be taken in screening isolates beforehand if seedling inoculations are to be used as a selection tool in breeding, and results should always be confirmed by field testing.

As will be shown later in this section, we have also detected some QTL that were significant both at the seedling and adult plant stages, in particular, the *Snn3-B1* locus in SHA3/CBRD × Naxos (Ruud et al., 2017) and the QTL on 4B and 7A in our AM panel (Ruud et al., 2018a). We also have evidence that the *Tsn1* locus has an effect across developmental stages in our germplasm (Ruud et al., 2018b).

On the other hand, many of the QTL we have detected are *either* seedling stage *or* adult plant stage QTL, giving a strong indication that resistances at the two growth stages are only partially controlled by the same genes.

6.3 The role of known effector sensitivities in field resistance

A crucial issue for resistance breeding was to investigate the relationship between sensitivity to the cloned effectors SnToxA, SnTox1 and SnTox3 and their potential effect on disease severity under field conditions. A Nordic collection of spring wheat cultivars, landraces and breeding lines was used for this purpose. The plants were screened in mist-irrigated field trials from 2010 to 2016, and phenotyped for plant height, earliness (days from sowing to heading) and leaf blotch severity. Sensitivity to SnToxA and SnTox3 was common in the material (45% and 55%, respectively), while sensitivity to SnTox1 was only present in 12% of the lines. For SnTox3 sensitivity, we identified two distinct reaction types in our germplasm. While full necrosis and tissue collapse (reaction type 3) was common in CIMMYT germplasm and some of the Norwegian lines, most lines of Nordic origin showed weaker, chlorotic reaction to SnTox3 (reaction type 2) (Ruud et al., 2018b). Sensitivity to SnToxA was significantly correlated to higher SNB severity in the field, while sensitivity to SnTox1 and SnTox3 was not significant in the field in this material (Ruud et al., 2018b). Since the SnToxA-*Tsn1* interaction masks the SnTox3-*Snn3* interaction, it may have contributed to the lack of significance of SnTox3-sensitivity in this material. In a QTL mapping population that did not segregate for *Tsn1* and *Snn1* (both parents were SnToxA and SnTox1 insensitive), we detected a large and significant effect of the SnTox3-*Snn3* interaction in naturally infected field trials (Ruud et al., 2017, see below).

6.4 Understanding the genetics of resistance in relevant germplasm

In our research efforts, several mapping populations have been screened over several years under natural infection of *P. nodorum* in mist-irrigated field nurseries. The isolates used for seedling inoculations were selected based on SnToxA, SnTox1 and SnTox3-profile from PCR, virulence on differential lines and subsets of the mapping populations. In the two cases that will be reported here, we also inoculated and infiltrated with the North Dakotan isolate Sn4, which has been sequenced (Syme et al., 2013) and used in previous NE-*Snn*-characterization studies (Liu et al., 2015; Zhang et al., 2011). In both studies, the plants were also infiltrated with CFs from the isolates used for inoculation in order to unveil potential new NE-*Snn*-interactions and correlation between sensitivity to the CF, seedling inoculation and field susceptibility.

6.4.1 Case 1: QTL mapping

Initial field testing identified the line SHA3/CBRD ('Shanghai3/Catbird') from CIMMYT to be a promising source of SNB resistance. From previous projects, we already had a mapping population available of RILs from a cross between SHA3/CBRD and the German spring wheat cultivar Naxos that we previously had successfully used for mapping powdery mildew (Lu et al., 2012) and Fusarium head blight resistance (Lu et al., 2013).

The population was tested for four years with our established field testing methodology, and we also screened the parents for sensitivity to the known NEs SnToxA, SnTox1 and SnTox3. The population was found to segregate for SnTox3 sensitivity, and the existing genetic map of 567 SSR and DArT markers was used to perform a QTL analysis (Lu and Lillemo, 2014). Several QTL for field resistance were identified, but only a few of them were significant across years. However, we failed to identify the *Snn3-B1* locus, which is known to be the causal locus for SnTox3 sensitivity in bread wheat.

In parallel efforts to fine-map and validate powdery mildew resistance loci (Windju et al., 2017), the same mapping population was later genotyped with the Illumina iSelect 90K wheat SNP chip (Wang et al., 2014). With the improved linkage map of 4177 markers in total, we could map the *Snn3-B1* locus with tight linkage to a small group of SNP markers at the telomeric end of 5BS (Ruud et al., 2017). The corrected SNB severity data from the field trials were re-analysed with the new map, and the SnTox3-*Snn3* interaction was identified as a major determinant of susceptibility in the field, explaining up to 24% of the phenotypic variation. This is the first report of the significance of the SnTox3-*Snn3* interaction in the field. We also inoculated the population with four *P. nodorum* isolates at the seedling stage and infiltrated with filter-sterilized CFs from the same isolates. The locus explained up to 51% of the phenotypic variation after seedling inoculations, and was also the major determinant of sensitivity to CFs (Ruud et al., 2017).

6.4.2 Case 2: Genome-wide association mapping

A genome-wide association mapping study (GWAS) was carried out on 121 lines from the same spring wheat collection as used for investigating the role of NEs in field resistance (Ruud et al., 2018b). The aims were to investigate whether SnTox-*Snn*-interactions could be detected with significant MTAs in GWAS analysis, to identify stable adult plant SNB resistance in the Norwegian spring wheat material and to determine the degree of overlap between seedling and adult plant resistance.

The population was genotyped with the 90K SNP chip and the data combined with previously genotyped SSR and other gene-specific markers. A total of 22 031 polymorphic markers were included in the study. The population was also inoculated with four *P. nodorum* isolates and infiltrated with filter-sterilized CF from the same isolates, at the seedling stage in the greenhouse. GWAS was performed on the corrected SNB severities from the field trials, infiltrations with the purified SnToxA, SnTox1 and SnTox3, infiltrations with CFs from and inoculations with single isolates.

Markers associated with *Tsn1* conferring sensitivity to SnToxA were highly significant at the seedling stage, but were only detected below the significance threshold at the adult plant stage. Significant QTL for seedling resistance were located on 1A, 1B, 3A, 4B, 5B, 6B, 7A and 7B. At the adult plant stage, the most robust QTL were located on 2B, 2D, 4A, 4B, 5A, 6B, 7A and 7B. The most stable QTL in the field was located on 2DL and was significant in all years except 2012. QTL on 4B and 7A were significant both after seedling inoculations in the greenhouse and at the adult plant stage in the field (Ruud et al., 2018a). The QTL on 7A corresponded to the same QTL detected in the SHA3/CBRD × Naxos population (Ruud et al., 2017). This study highlighted the quantitative nature of field resistance to SNB in wheat, and that many QTL need to be combined in order to breed cultivars with high levels of field resistance.

6.5 Case study conclusions

As outlined in this case study, with relatively limited use of resources at a national level, we have been able to identify crucial information that will be useful in disease management and resistance breeding. The main results can be summarized as follows:

1 We have shown that *P. nodorum* is the most important pathogen causing leaf blotch on wheat in Norway – hence, information on SNB resistance in current cultivars is important for the farmers, and leaf blotch resistance breeding should focus on this pathogen.
2 Isolates were identified with reduced sensitivity to important fungicides – hence there is a need to diversify disease management strategies with a stronger focus on disease resistance and sustainable use of fungicides based on alternative active ingredients.
3 A reliable field testing methodology was established by mist irrigation in combination with straw inoculum in hillplots – this has provided important data on field resistance in relevant cultivars and breeding lines.
4 Seedling inoculation results showed only a weak association with field resistance – hence, resistance breeding should rely mostly on disease evaluations in the field.
5 Our field results together with leaf infiltrations showed that SnToxA and SnTox3 can contribute significantly to increased disease severities under field conditions

– hence, breeding efforts should focus on eliminating sensitivities to these NEs in the germplasm.

6 The frequencies of SnToxA- and SnTox3-producing *P. nodorum* isolates are high in the local pathogen population – hence, cultivars with sensitivity to these NEs should be avoided by farmers.

7 Important QTL for field resistance have been identified in relevant germplasm – these could be utilized in resistance breeding in combination with the elimination of NE sensitivities to build up higher levels of quantitative resistance.

7 Conclusion and future trends

The potential loss of fungicide sensitivity and concern about negative impacts of fungicides on the environment and human health provide strong incentives for breeding and growing SNB-resistant wheat cultivars. The discovery of the importance of NE sensitivities on host resistance led to a paradigm shift in the way of thinking about resistance breeding. Rather than searching for sources of resistance, there is a focus on eliminating susceptibility by use of either markers (Zhang et al., 2009) or infiltration assays (Vleeshouwers and Oliver, 2014). This approach is based on available polymorphisms in the breeding germplasm. With the recent developments in genome editing technologies (Gil-Humanes et al., 2017), we anticipate that breeders in the future will be able to eliminate NE sensitivities by editing or knocking out host receptors. The genomic resources now available, with a newly released, improved version of the wheat genome (Clavijo et al., 2017), allow for putative identification of gene functions and simplify the steps to candidate gene validation. The genetic background for potential *Snn* genes and the two reaction types for SnTox3 sensitivity observed in the Nordic spring wheat collection could be further investigated in this respect. While clearly some of the NE-*Snn* interactions show effect and are important at the adult plant stage, there are likely also many other mechanisms at play which are still poorly understood. It is expected that the available reference genome sequences of both the host and the pathogen will be useful in unravelling these mechanisms as well. Deeper insights into the biological mechanisms behind plant–pathogen interactions will likely come from pan-genome studies where the whole genome sequences of hundreds of wheat lines with various levels of disease resistance can be compared using bioinformatics pipelines to identify host resistance/sensitivity loci. Likewise, similar comparisons of the genome sequences of hundreds of isolates with various levels of virulence and/or aggressiveness on host cultivars will provide power to identify virulence factors on the pathogen side (Gao et al., 2016).

As we have shown in this chapter, the genetics of host resistance under field conditions at the adult plant stage is quite complex and involves many genes. Even if all the most important genetic factors can be identified, it will still be a challenge to handle more than a handful QTL by MAS or through effector screening. Further advancements in breeding methodology will therefore be needed. A quantitative trait like SNB resistance will be a good case for genomic selection. As genomic selection methodology is still in its infancy when it comes to wheat breeding, the coming years will likely see big advancements in the methodology. A challenge for resistance breeding will be to develop reliable genomic prediction models and integrate those with other agronomically important traits into a cost-effective breeding programme. In all cases, it will serve as a necessary fundament to have a reliable field testing methodology in place in order to make progress in resistance breeding to this challenging wheat disease.

8 Where to look for further information

Below are a few recommended papers for further reading on different topics related to SNB in wheat. The relatively short review paper by Solomon et al. (2006) gives an overview of the disease from the pathogen side. For more detailed information about the *P. nodorum* reference genome at the molecular level, the recent publication by Syme et al. (2016) can be recommended. A comprehensive review of SNB resistance in wheat from a breeding perspective is given by Francki (2013), while Oliver et al. (2012) summarized the recent molecular understanding of host–pathogen interactions for the disease. Friesen and Faris (2012) provide detailed protocols for dissecting plant–pathogen interactions and identification of NEs in this pathosystem, and examples of their use in breeding are given by Vleeshouwers and Oliver (2014).

9 References

Abeysekara, N. S., Friesen, T. L., Keller, B. and Faris, J. D. (2009). Identification and characterization of a novel host-toxin interaction in the wheat-*Stagonospora nodorum* pathosystem. *Theoretical and Applied Genetics*, 120(1), 117–26. doi:10.1007/s00122-009-1163-6

Abrahamsen, U. (2013). Spring wheat cultivars and fungicide resistance (In Norwegian). In: Strand, E. (Ed.), *Jord- og plantekultur 2013/Bioforsk FOKUS 8*, Volume 8. Norway: Bioforsk, pp. 124–9.

Adhikari, T. B., Jackson, E. W., Gurung, S., Hansen, J. M. and Bonman, J. M. (2011). Association mapping of quantitative resistance to *Phaeosphaeria nodorum* in spring wheat landraces from the USDA National Small Grains collection. *Phytopathology*, 101(11), 1301–10. doi:10.1094/phyto-03-11-0076

Aguilar, V., Stamp, P., Winzeler, M., Winzeler, H., Schachermayr, G., Keller, B., Zanetti, S. and Messmer, M. M. (2005). Inheritance of field resistance to *Stagonospora nodorum* leaf and glume blotch and correlations with other morphological traits in hexaploid wheat (*Triticum aestivum* L.). *Theoretical and Applied Genetics*, 111(2), 325–36.

Ali, S. and Adhikari, T. B. (2008). Variation in aggressiveness of *Stagonospora nodorum* isolates in North Dakota. *Journal of Phytopathology*, 156(3), 140–5. doi:10.1111/j.1439-0434.2007.01358.x

Alkan, C., Coe, B. P. and Eichler, E. E. (2011). Genome structural variation discovery and genotyping. *Nature Reviews. Genetics*, 12(5), 363–76. doi:10.1038/nrg2958

Allen, A. M., Winfield, M. O., Burridge, A. J., Downie, R. C., Benbow, H. R., Barker, G. L. A., Wilkinson, P. A., Coghill, J., Waterfall, C., Davassi, A., Scopes, G., Pirani, A., Webster, T., Brew, F., Bloor, C., Griffiths, S., Bentley, A. R., Alda, M., Jack, P., Phillips, A. L. and Edwards, K. J. (2017). Characterization of a wheat breeders' array suitable for high-throughput SNP genotyping of global accessions of hexaploid bread wheat (*Triticum aestivum*). *Plant Biotechnology Journal*, 15(3), 390–401. doi:10.1111/pbi.12635

Arseniuk, E., Czembor, P. C., Czaplicki, A., Song, Q. J., Cregan, P. B., Hoffman, D. L. and Ueng, P. P. (2004). QTL controlling partial resistance to *Stagonospora nodorum* leaf blotch in winter wheat cultivar Alba. *Euphytica*, 137(2), 225–31.

Barrus, M. (1942). Yellow-spot disease of wheat in New York State. *Plant Disease Reporter*, 26(11), 246.

Bartlett, D. W., Clough, J. M., Godwin, J. R., Hall, A. A., Hamer, M. and Parr-Dobrzanski, B. (2002). The strobilurin fungicides. *Pest Management Science*, 58(7), 649–62. doi:10.1002/ps.520

Bathgate, J. A. and Loughman, R. (2001). Ascospores are a source of inoculum of *Phaeosphaeria nodorum, P. avenaria* f. sp. *avenaria* and *Mycosphaerella graminicola* in Western Australia. *Australasian Plant Pathology*, 30(4), 317–22. doi:10.1071/ap01043

Bayles, R. A. (1991). Research note: Varietal resistance as a factor contributing to the increased importance of *Septoria tritici* Rob. and Desm. in the UK wheat crop. *Plant Varieties & Seeds*, 4(3), 177–83.

Bearchell, S. J., Fraaije, B. A., Shaw, M. W. and Fitt, B. D. L. (2005). Wheat archive links long-term fungal pathogen population dynamics to air pollution. *Proceedings of the National Academy of Sciences of the United States of America*, 102(15), 5438–42.

Bennett, R. S., Milgroom, M. G., Sainudiin, R., Cunfer, B. M. and Bergstrom, G. C. (2007). Relative contribution of seed-transmitted inoculum to foliar populations of *Phaeosphaeria nodorum*. *Phytopathology*, 97(5), 584–91.

Bernardo, R. (2014). Genomewide selection when major genes are known. *Crop Science*, 54(1), 68–75.

Bhathal, J., Loughman, R. and Speijers, J. (2003). Yield reduction in wheat in relation to leaf disease from yellow (Tan) spot and *Septoria nodorum* blotch. *European Journal of Plant Pathology*, 109(5), 435–43.

Bird, P. M. and Ride, J. P. (1981). The resistance of wheat to *Septoria nodorum*: Fungal development in relation to host lignification. *Physiological Plant Pathology*, 19(3), 289–99. doi:http://dx.doi.org/10.1016/S0048-4059(81)80063-X

Blixt, E., Olson, Å., Högberg, N., Djurle, A. and Yuen, J. (2008). Mating type distribution and genetic structure are consistent with sexual recombination in the Swedish population of *Phaeosphaeria nodorum*. *Plant Pathology*, 57(4), 634–41. doi:10.1111/j.1365-3059.2008.01826.x

Blixt, E., Djurle, A., Yuen, J. and Olson, Å. (2009). Fungicide sensitivity in Swedish isolates of *Phaeosphaeria nodorum*. *Plant Pathology*, 58(4), 655–64. doi:10.1111/j.1365-3059.2009.02041.x

Blixt, E., Olson, Å., Lindahl, B., Djurle, A. and Yuen, J. (2010). Spatiotemporal variation in the fungal community associated with wheat leaves showing symptoms similar to *Stagonospora nodorum* blotch. *European Journal Of Plant Pathology*, 126, 373–86.

Borlaug, N. E. (1966). Basic concepts which influence the choice of methods for use in breeding for disease resistance in cross-pollinated and self-pollinated crop plants. In: Gerhold, H. D., Schreiner, E. J., McDermott, R. E. and Wlnieski, J. A. (Eds), *Breeding Pest-Resistant Trees*:. Oxford, UK: Pergamon, pp. 327–44.

Bostwick, D. E., Ohm, H. W. and Shaner, G. (1993). Inheritance of *Septoria nodorum* glume blotch resistance in wheat. *Crop Science*, 33(3), 439–43.

Caldwell, R. M. (1968). Breeding for general and/or specific plant disease resistance. Paper presented at the *Proceedings of Third International Wheat Genetics Symposium*, Canberra, Australia.

Carmona, M., Ferrazini, M. and Barreto, D. (2006). Tan spot of wheat caused by *Drechslera tritici-repentis:* Detection, transmission, and control in wheat seed. *Cereal Research Communications*, 34(2–3), 1043–9.

Clavijo, B. J., Venturini, L., Schudoma, C., Accinelli, G. G., Kaithakottil, G., Wright, J., Borrill, P., Kettleborough, G., Heavens, D., Chapman, H., Lipscombe, J., Barker, T., Lu, F. H., McKenzie, N., Raats, D., Ramirez-Gonzalez, R. H., Coince, A., Peel, N., Percival-Alwyn, L., Duncan, O., Trosch, J., Yu, G., Bolser, D. M., Namaati, G., Kerhornou, A., Spannagl, M., Gundlach, H., Haberer, G., Davey, R. P., Fosker, C., Palma, F. D., Phillips, A. L., Millar, A. H., Kersey, P. J., Uauy, C., Krasileva, K. V., Swarbreck, D., Bevan, M. W. and Clark, M. D. (2017). An improved assembly and annotation of the allohexaploid wheat genome identifies complete families of agronomic genes and provides genomic evidence for chromosomal translocations. *Genome Research*, 27(5), 885–96. doi:10.1101/gr.217117.116

Collard, B. C. Y. and Mackill, D. J. (2008). Marker-assisted selection: An approach for precision plant breeding in the twenty-first century. *Philosophical Transactions of the Royal Society B: Biological Sciences*, 363(1491), 557–72. doi:10.1098/rstb.2007.2170

Collard, B. C. Y., Jahufer, M. Z. Z., Brouwer, J. B. and Pang, E. C. K. (2005). An introduction to markers, quantitative trait loci (QTL) mapping and marker-assisted selection for crop improvement: The basic concepts. *Euphytica*, 142(1), 169–96. doi:10.1007/s10681-005-1681-5

Cools, H. J., Mullins, J. G., Fraaije, B. A., Parker, J. E., Kelly, D. E., Lucas, J. A. and Kelly, S. L. (2011). Impact of recently emerged sterol 14α-demethylase (CYP51) variants of *Mycosphaerella graminicola* on azole fungicide sensitivity. *Applied and Environmental Microbiology*, 77(11), 3830–7. doi:10.1128/aem.00027-11

Cowger, C. and Murphy, J. P. (2007). Artificial inoculation of wheat for selecting resistance to *Stagonospora nodorum* blotch. *Plant Disease*, 91(5), 539–45. doi:10.1094/PDIS-91-5-0539

Cowger, C. and Silva-Rojas, H. V. (2006). Frequency of *Phaeosphaeria nodorum*, the sexual stage of *Stagonospora nodorum*, on winter wheat in North Carolina. *Phytopathology*, 96(8), 860–6. doi:10.1094/PHYTO-96-0860

Crossa, J., de Los Campos, G., Pérez, P., Gianola, D., Burgueño, J., Araus, J. L., Makumbi, D., Singh, R. P., Dreisigacker, S. and Yan, J. (2010). Prediction of genetic values of quantitative traits in plant breeding using pedigree and molecular markers. *Genetics*, 186(2), 713–24.

Cunfer, B. M. (1998). Seasonal availability of inoculum of *Stagonospora nodorum* in the field in the Southeastern US. *Cereal Research Communications*, 26, 259–63.

Cunfer, B. M. (1999). *Stagonospora* and *Septoria* pathogens of cereals: The infection process. Paper presented at the *Septoria and Stagonospora Diseases of Cereals: A Compilation of Global Research*, Mexico.

Czembor, P. C., Arseniuk, E., Czaplicki, A., Song, Q., Cregan, P. B. and Ueng, P. P. (2003). QTL mapping of partial resistance in winter wheat to *Stagonospora nodorum* blotch. *Genome*, 46, 546–54.

DePauw, R. (1995). Wheat improvement in Canada. PBI Bulletin. National Research Council/Plant Biotechnology Institute, Saskatoon, Canada.

Dodds, P. N. and Rathjen, J. P. (2010). Plant immunity: Towards an integrated view of plant–pathogen interactions. *Nature Reviews. Genetics*, 11(8), 539–48.

Elshire, R. J., Glaubitz, J. C., Sun, Q., Poland, J. A., Kawamoto, K., Buckler, E. S. and Mitchell, S. E. (2011). A robust, simple genotyping-by-sequencing (GBS) approach for high diversity species. *PLOS ONE*, 6(5), e19379. doi:10.1371/journal.pone.0019379

Engle, J. S., Madden, L. V. and Lipps, P. E. (2006). Distribution and pathogenic characterization of *Pyrenophora tritici-repentis* and *Stagonospora nodorum* in Ohio. *Phytopathology*, 96(12), 1355–62. doi:10.1094/phyto-96-1355

Eyal, Z. (1981). Integrated control of *Septoria* diseases of wheat. *Plant Disease*, 65(9), 763–8.

Eyal, Z. (1999). The *Septoria tritici* and *Stagonospora nodorum* blotch diseases of wheat. *European Journal of Plant Pathology*, 105(7), 629–41.

Eyal, Z. and Scharen, A. L. (1977). A quantitative method for the inoculation of wheat seedlings with pycnidiospores of *Septoria nodorum*. *Phytopathology*, 67, 712–14.

Eyal, Z., Scharen, A. L., Prescott, J. M. and van Ginkel, M. (1987). *The Septoria Diseases of Wheat: Concepts and Methods of Disease Management*. Mexico: CIMMYT.

FAO. (2017). FAOSTAT. Retrieved from: http://faostat.fao.org

Faris, J. D., Anderson, J. A., Francl, L. J. and Jordahl, J. G. (1996). Chromosomal location of a gene conditioning insensitivity in wheat to a necrosis-inducing culture filtrate from *Pyrenophora tritici-repentis*. *Phytopathology*, 86(5), 459–63.

Faris, J. D., Zhang, Z. C., Lu, H. J., Lu, S. W., Reddy, L., Cloutier, S., Fellers, J. P., Meinhardt, S. W., Rasmussen, J. B., Xu, S. S., Oliver, R. P., Simons, K. J. and Friesen, T. L. (2010). A unique wheat disease resistance-like gene governs effector-triggered susceptibility to necrotrophic pathogens. *Proceedings of the National Academy of Sciences of the United States of America*, 107(30), 13544–9. doi:10.1073/pnas.1004090107

Ficke, A., Abrahamsen, U. and Elen, O. (2011a). Fungicide resistance in diseases on cereals in Norway (in Norwegian). *Bioforsk Fokus*, 6(2), 96.

Ficke, A., Abrahamsen, U. and Elen, O. (2011b). Importance of the leaf blotch disease complex in Norwegian wheat (in Norwegian). *Bioforsk Fokus*, 6(1), 64–7.

Flor, H. H. (1956). The complementary genic systems in flax and flax rust. In: Demerec, M. (Ed.), *Advances in Genetics*, Volume 8. New York, NY: Academic Press, pp. 29–54.

Flor, H. H. (1971). Current status of the gene-for-gene concept. *Annual Review of Phytopathology*, 9(1), 275–96. doi:doi:10.1146/annurev.py.09.090171.001423

Fones, H. and Gurr, S. (2015). The impact of *Septoria tritici* Blotch disease on wheat: An EU perspective. *Fungal Genetics and Biology*, 79, 3–7. doi:http://dx.doi.org/10.1016/j.fgb.2015.04.004

Francki, M. G. (2013). Improving *Stagonospora nodorum* resistance in wheat: A review. *Crop Science*, 53(2), 355–65. doi:10.2135/cropsci2012.06.0347

Francki, M. G., Shankar, M., Walker, E., Loughman, R., Golzar, H. and Ohm, H. (2011). New quantitative trait loci in wheat for flag leaf resistance to *Stagonospora nodorum* blotch. *Phytopathology*, 101(11), 1278–84. doi:10.1094/phyto-02-11-0054

Fraser, D. E., Murphy, J. P., Leath, S. and Van Sanford, D. A. (2003). Effect of inoculation with selected isolates of *Stagonospora nodorum* on field evaluations of host resistance in winter wheat. *Plant Disease*, 87(10), 1213–20. doi:10.1094/pdis.2003.87.10.1213

Fried, P. M. and Meister, E. (1987). Inheritance of leaf and head resistance of winter wheat to *Septoria nodorum* in a diallel cross. *Phytopathology*, 77(10), 1371–5. doi:10.1094/Phyto-77-1371

Friesen, T. L. and Faris, J. D. (2012). Characterization of plant-fungal interactions involving necrotrophic effector-producing plant pathogens. *Methods in Molecular Biology*, 835, 191–207.

Friesen, T. L., Stukenbrock, E. H., Liu, Z., Meinhardt, S., Ling, H., Faris, J. D., Rasmussen, J. B., Solomon, P. S., McDonald, B. A. and Oliver, R. P. (2006). Emergence of a new disease as a result of interspecific virulence gene transfer. *Nature Genetics*, 38(8), 953–6. doi:http://www.nature.com/ng/journal/v38/n8/suppinfo/ng1839_S1.html

Friesen, T. L., Meinhardt, S. W. and Faris, J. D. (2007). The *Stagonospora nodorum*-wheat pathosystem involves multiple proteinaceous host-selective toxins and corresponding host sensitivity genes that interact in an inverse gene-for-gene manner. *Plant Journal*, 51(4), 681–92.

Friesen, T. L., Zhang, Z. C., Solomon, P. S., Oliver, R. P. and Faris, J. D. (2008). Characterization of the interaction of a novel *Stagonospora nodorum* host-selective toxin with a wheat susceptibility gene. *Plant Physiology*, 146(2), 682–93.

Friesen, T. L., Chu, C. G., Liu, Z. H., Xu, S. S., Halley, S. and Faris, J. D. (2009). Host-selective toxins produced by *Stagonospora nodorum* confer disease susceptibility in adult wheat plants under field conditions. *Theoretical and Applied Genetics*, 118(8), 1489–97. doi:10.1007/s00122-009-0997-2

Friesen, T. L., Chu, C., Xu, S. S. and Faris, J. D. (2012). SnTox5–*Snn5*: A novel *Stagonospora nodorum* effector–wheat gene interaction and its relationship with the SnToxA–*Tsn1* and SnTox3–*Snn3–B1* interactions. *Molecular Plant Pathology*, 13(9), 1101–9. doi:10.1111/j.1364-3703.2012.00819.x

Gao, Y., Faris, J. D., Liu, Z., Kim, Y. M., Syme, R. A., Oliver, R. P., Xu, S. S. and Friesen, T. L. (2015). Identification and characterization of the SnTox6-*Snn6* interaction in the *Parastagonospora nodorum*-wheat pathosystem. *Molecular Plant-Microbe Interactions*, 28(5), 615–25.

Gao, Y. Y., Liu, Z. H., Faris, J. D., Richards, J., Brueggeman, R. S., Li, X. H., Oliver, R. P., McDonald, B. A. and Friesen, T. L. (2016). Validation of genome-wide association studies as a tool to identify virulence factors in *Parastagonospora nodorum*. *Phytopathology*, 106(10), 1177–85. doi:10.1094/phyto-02-16-0113-fi

Gil-Humanes, J., Wang, Y. P., Liang, Z., Shan, Q. W., Ozuna, C. V., Sanchez-Leon, S., Baltes, N. J., Starker, C., Barro, F., Gao, C. and Voytas, D. F. (2017). High-efficiency gene targeting in hexaploid wheat using DNA replicons and CRISPR/Cas9. *Plant Journal*, 89(6), 1251–62. doi:10.1111/tpj.13446

Gisi, U., Sierotzki, H., Cook, A. and McCaffery, A. (2002). Mechanisms influencing the evolution of resistance to Qo inhibitor fungicides. *Pest Management Science*, 58(9), 859–67. doi:10.1002/ps.565

Gonzalez-Hernandez, J. L., Singh, P. K., Mergoum, M., Adhikari, T. B., Kianian, S. F., Simsek, S. and Elias, E. M. (2009). A quantitative trait locus on chromosome 5B controls resistance of *Triticum turgidum* (L.) var. *diccocoides* to *Stagonospora nodorum* blotch. *Euphytica*, 166(2), 199–206. doi:10.1007/s10681-008-9825-z

Goodwin, S., McPherson, J. D. and McCombie, W. R. (2016). Coming of age: Ten years of next-generation sequencing technologies. *Nature Reviews. Genetics*, 17(6), 333–51. doi:10.1038/nrg.2016.49

Gupta, P. K., Roy, J. K. and Prasad, M. (2001). Single nucleotide polymorphisms: A new paradigm for molecular marker technology and DNA polymorphism detection with emphasis on their use in plants. *Current Science*, 80(4), 524–35.

Gupta, P. K., Kulwal, P. L. and Jaiswal, V. (2014). Association mapping in crop plants: Opportunities and challenges. *Advances in Genetics*, 85, 109–47. doi:10.1016/b978-0-12-800271-1.00002-0

Gurung, S., Mamidi, S., Bonman, J. M., Xiong, M., Brown-Guedira, G. and Adhikari, T. B. (2014). Genome-wide association study reveals novel quantitative trait Loci associated with resistance to multiple leaf spot diseases of spring wheat. *PLOS ONE*, 9(9), e108179. doi:10.1371/journal.pone.0108179

Halama, P. and Lacoste, L. (1991). Déterminisme de la reproduction sexuée de *Phaeosphaeria (Leptosphaeria) nodorum*, agent de la septoriose du blé. I. Hétérothallisme et rôle des microspores. *Canadian Journal of Botany*, 69(1), 95–9. doi:10.1139/b91-013

Hane, J. K., Lowe, R. G., Solomon, P. S., Tan, K. C., Schoch, C. L., Spatafora, J. W., Crous, P. W., Kodira, C., Birren, B. W., Galagan, J. E., Torriani, S. F., McDonald, B. A. and Oliver, R. P. (2007). Dothideomycete plant interactions illuminated by genome sequencing and EST analysis of the wheat pathogen *Stagonospora nodorum*. *Plant Cell*, 19(11), 3347–68. doi:10.1105/tpc.107.052829

Hosford, R. M. (1982). Tan Spot of Wheat and Related Diseases Workshop, 14–15 July 1981, North Dakota State University, Fargo, North Dakota: North Dakota Agricultural Experiment Station, North Dakota State University.

Jalli, M., Laitinen, P. and Latvala, S. (2011). The emergence of cereal fungal diseases and the incidence of leaf spot diseases in Finland. *Agricultural and Food Science*, 20(1), 62–73.

Jighly, A., Alagu, M., Makdis, F., Singh, M., Singh, S., Emebiri, L. C. and Ogbonnaya, F. C. (2016). Genomic regions conferring resistance to multiple fungal pathogens in synthetic hexaploid wheat. *Molecular Breeding*, 36(9), 127. doi:10.1007/s11032-016-0541-4

Johnson, A. (1942). *Helminthosporium tritici-vulgaris* on wheat in Maryland. *Plant Disease Reporter*, 26, 246–7.

Jones, D. (1985). Partial resistance, cultivar mixture, and epidemic development in the *Septoria nodorum*: Wheat association. ARS US Department of Agriculture, Agricultural Research Service.

Jones, D. A. and Jones, J. D. G. (1997). The role of leucine-rich repeat proteins in plant defences. In: Andrews, J. H., Tommerup, I. C. and Callow, J. A. (Eds), *Advances in Botanical Research*, Volume 24. London, UK: Academic Press, pp. 89–167.

Jönsson, J. (1985). Evaluation of leaf resistance to *Septoria nodorum* in winter wheat at seedling and adult plant stage. *Agri Hortique Genetica*, 43, 52–68.

Juliana, P., Singh, R. P., Singh, P. K., Crossa, J., Rutkoski, J. E., Poland, J. A., Bergstrom, G. C. and Sorrells, M. E. (2017). Comparison of models and whole-genome profiling approaches for genomic-enabled prediction of *Septoria tritici* blotch, *Stagonospora nodorum* blotch, and tan spot resistance in wheat. *Plant Genome*, 10(2), 1–16. doi:10.3835/plantgenome2016.3808.0082

Karjalainen, R. (1985). Host-pathogen interaction between spring wheat and *Septoria nodorum* with reference to resistance breeding. *Journal of Agricultural Science in Finland*, 57(1), 1–66.

Kim, Y. K. and Bockus, W. W. (2003). Temperature-sensitive reaction of winter wheat cultivar AGSECO 7853 to *Stagonospora nodorum*. *Plant Disease*, 87(9), 1125–8. doi:10.1094/pdis.2003.87.9.1125

King, J. E., Cook, R. J. and Melville, S. C. (1983). A review of *Septoria* diseases of wheat and barley. *Annals of Applied Biology*, 103(2), 345–73. doi:10.1111/j.1744-7348.1983.tb02773.x

Kleijer, G., Bronnimann, A. and Fossati, A. (1977). Chromosomal location of a dominant gene for resistance at the seedling stage to *Septoria nodorum* Berk. in the wheat variety 'Atlas 66'. *Zeitschrift für Pflanzenzüchtung*, 78(2), 170–3.

Krattinger, S. G., Lagudah, E. S., Spielmeyer, W., Singh, R. P., Huerta-Espino, J., McFadden, H., Bossolini, E., Selter, L. L. and Keller, B. (2009). A putative ABC transporter confers durable resistance to multiple fungal pathogens in wheat. *Science*, 323(5919), 1360–3. doi:10.1126/science.1166453

Krupinsky, J. (1999). Influence of cultural practices on *Septoria/Stagonospora* diseases. Paper presented at the 5. *Proceedings of the International Septoria Workshop*, Mexico, DF (Mexico), 20–24 September 1999.

Lamari, L., Strelkov, S., Yahyaoui, A., Amedov, M., Saidov, M., Djunusova, M. and Koichibayev, M. (2005). Virulence of *Pyrenophora tritici-repentis* in the countries of the Silk Road. *Canadian Journal of Plant Pathology*, 27(3), 383–8.

Langridge, P., Lagudah, E. S., Holton, T. A., Appels, R., Sharp, P. J. and Chalmers, K. J. (2001). Trends in genetic and genome analyses in wheat: A review. *Australian Journal of Agricultural Research*, 52(12), 1043–77. doi:https://doi.org/10.1071/AR01082

Lehtinen, U. (1993). Plant cell wall degrading enzymes of *Septoria nodorum*. *Physiological and Molecular Plant Pathology*, 43(2), 121–34. doi:http://dx.doi.org/10.1006/pmpp.1993.1045

Leroux, P., Albertini, C., Gautier, A., Gredt, M. and Walker, A. S. (2007). Mutations in the CYP51 gene correlated with changes in sensitivity to sterol 14 alpha-demethylation inhibitors in field isolates of *Mycosphaerelia graminicola*. *Pest Management Science*, 63(7), 688–98. doi:10.1002/ps.1390

Lillemo, M. and Dieseth, J. A. (2011). Wheat breeding in Norway. In: Bonjean, A. P., Angus, W. J. and van Ginkel, M. (Eds), *The World Wheat Book. A History of Wheat Breeding*, Volume 2. Paris, France: Lavoisier Publishing, pp. 45–79.

Lillemo, M., Asalf, B., Singh, R. P., Huerta-Espino, J., Chen, X. M., He, Z. H. and Bjornstad, A. (2008). The adult plant rust resistance loci *Lr34/Yr18* and *Lr46/Yr29* are important determinants of partial resistance to powdery mildew in bread wheat line Saar. *Theoretical and Applied Genetics*, 116(8), 1155–66. doi:10.1007/s00122-008-0743-1

Liu, Z. H., Faris, J. D., Meinhardt, S. W., Ali, S., Rasmussen, J. B. and Friesen, T. L. (2004a). Genetic and physical mapping of a gene conditioning sensitivity in wheat to a partially purified host-selective toxin produced by *Stagonospora nodorum*. *Phytopathology*, 94(10), 1056–60.

Liu, Z. H., Friesen, T. L., Rasmussen, J. B., Ali, S., Meinhardt, S. W. and Faris, J. D. (2004b). Quantitative trait loci analysis and mapping of seedling resistance to *Stagonospora nodorum* leaf blotch in wheat. *Phytopathology*, 94(10), 1061–7.

Liu, Z. H., Friesen, T. L., Ling, H., Meinhardt, S. W., Oliver, R. P., Rasmussen, J. B. and Faris, J. D. (2006). The Tsn1-ToxA interaction in the wheat-*Stagonospora nodorum* pathosystem parallels that of the wheat-tan spot system. *Genome*, 49(10), 1265–73.

Liu, Z. H., Faris, J. D., Oliver, R. P., Tan, K. C., Solomon, P. S., McDonald, M. C., McDonald, B. A., Nunez, A., Lu, S. W., Rasmussen, J. B. and Friesen, T. L. (2009). SnTox3 acts in effector triggered susceptibility to induce disease on wheat carrying the *Snn3* gene. *PLOS ONE Pathogens*, 5(9), e1000581. doi:10.1371/journal.ppat.1000581

Liu, Z. H., Zhang, Z. C., Faris, J. D., Oliver, R. P., Syme, R., McDonald, M. C., McDonald, B. A., Solomon, P. S., Lu, S. W., Shelver, W. L., Xu, S. and Friesen, T. L. (2012). The cysteine rich necrotrophic effector SnTox1 produced by *Stagonospora nodorum* triggers susceptibility of wheat lines harboring *Snn1*. *PLOS ONE Pathogens*, 8(1), e1002467. doi:10.1371/journal.ppat.1002467

Liu, Z. H., El-Basyoni, I., Kariyawasam, G., Zhang, G. R., Fritz, A., Hansen, J., Marais, F., Friskop, A., Chao, S. M., Akhunov, E. and Baenziger, P. S. (2015). Evaluation and association mapping of resistance to tan spot and *Stagonospora nodorum* blotch in adapted winter wheat germplasm. *Plant Disease*, 99(10), 1333–41. doi:10.1094/pdis-11-14-1131-re

Liu, Z. H., Gao, Y. Y., Kim, Y. M., Faris, J. D., Shelver, W. L., de Wit, P., Xu, S. S. and Friesen, T. L. (2016). SnTox1, a *Parastagonospora nodorum* necrotrophic effector, is a dual-function protein that facilitates infection while protecting from wheat-produced chitinases. *New Phytologist*, 211(3), 1052–64. doi:10.1111/nph.13959

Lorang, J. M., Sweat, T. A. and Wolpert, T. J. (2007). Plant disease susceptibility conferred by a 'resistance' gene. *Proceedings of the National Academy of Sciences*, 104(37), 14861–6. doi:10.1073/pnas.0702572104

Lovell, D., Parker, S., Hunter, T., Royle, D. and Coker, R. (1997). Influence of crop growth and structure on the risk of epidemics by Mycosphaerella graminicola (*Septoria tritici*) in winter wheat. *Plant Pathology*, 46(1), 126–38.

Lu, Q. and Lillemo, M (2014). Molecular mapping of adult plant resistance to *Parastagonospora nodorum* leaf blotch in bread wheat lines 'Shanghai-3/Catbird' and 'Naxos'. *Theoretical and Applied Genetics*, 127(12), 2635–44. doi:10.1007/s00122-014-2404-x

Lu, Q., Bjørnstad, Å., Ren, Y., Asad, M., Xia, X., Chen, X., Ji, F., Shi, J. and Lillemo, M. (2012). Partial resistance to powdery mildew in German spring wheat 'Naxos' is based on multiple genes with stable effects in diverse environments. *Theoretical and Applied Genetics*, 125, 297–309. doi:10.1007/s00122-012-1834-6

Lu, Q. X., Lillemo, M., Skinnes, H., He, X. Y., Shi, J. R., Ji, F., Dong, Y. and Bjørnstad, Å. (2013). Anther extrusion and plant height are associated with Type I resistance to Fusarium head blight resistance in bread wheat line 'Shanghai-3/Catbird'. *Theoretical and Applied Genetics*, 126(2), 317–34

Ma, H. (1993). Genetic and cytogenetic studies of resistance to *Septoria nodorum* in tetraploid and hexaploid wheat. Ph.D. thesis, University of Saskatchewan, Saskatoon, Canada.

Ma, H. and Hughes, G. R. (1993). Resistance to *Septoria nodorum* blotch in several *Triticum* species. *Euphytica*, 70(1), 151–7. doi:10.1007/bf00029652

Ma, H. and Hughes, G. R. (1995). Genetic control and chromosomal location of *Triticum timopheevii*-derived resistance to *Septoria nodorum* blotch in durum wheat. *Genome*, 38(2), 332–8.

Magro, P. (1984). Production of polysaccharide-degrading enzymes by *Septoria Nodorum* in culture and during pathogenesis. *Plant Science Letters*, 37(1–2), 63–8. doi:http://dx.doi.org/10.1016/0304-4211(84)90204-9

Mammadov, J., Aggarwal, R., Buyyarapu, R. and Kumpatla, S. (2012). SNP markers and their impact on plant breeding. *International Journal of Plant Genomics*, 2012, 728398. doi:10.1155/2012/728398

Manickavelu, A., Kawaura, K., Oishi, K., Shin-I, T., Kohara, Y., Yahiaoui, N., Keller, B., Abe, R., Suzuki, A., Nagayama, T., Yano, K. and Ogihara, Y. (2012). Comprehensive functional analyses of expressed sequence tags in common wheat (*Triticum aestivum*). *DNA Research*, 19(2), 165–77. doi:10.1093/dnares/dss001

Manning, V. A., Chu, A. L., Steeves, J. E., Wolpert, T. J. and Ciuffetti, L. M. (2009). A host-selective toxin of *Pyrenophora tritici-repentis*, Ptr ToxA, induces photosystem changes and reactive oxygen species accumulation in sensitive wheat. *Molecular Plant-Microbe Interactions*, 22(6), 665–76. doi:10.1094/mpmi-22-6-0665

McDonald, M. C., Razavi, M., Friesen, T. L., Brunner, P. C. and McDonald, B. A. (2012). Phylogenetic and population genetic analyses of *Phaeosphaeria nodorum* and its close relatives indicate cryptic species and an origin in the Fertile Crescent. *Fungal Genetics and Biology: FG & B*, 49(11), 882–95. doi:10.1016/j.fgb.2012.08.001

McDonald, M. C., Oliver, R. P., Friesen, T. L., Brunner, P. C. and McDonald, B. A. (2013). Global diversity and distribution of three necrotrophic effectors in *Phaeosphaeria nodorum* and related species. *New Phytologist*, 199(1), 241–51. doi:10.1111/nph.12257

McDonald, M. C., Ahren, D., Simpfendorfer, S., Milgate, A. and Solomon, P. S. (2017). The discovery of the virulence gene ToxA in the wheat and barley pathogen *Bipolaris sorokiniana*. *Molecular Plant Pathology*, 19, 432–9. doi:10.1111/mpp.12535. doi:10.1111/mpp.12535

McMullen, M. and Adhikari, T. (2009). Fungal leaf spot diseases of wheat: Tan spot, *Stagonospora nodorum* blotch and *Septoria tritici* blotch. Retrieved from: http://www.ag.ndsu.edu/pubs/plantsci/pests/pp1249.pdf

Meehan, F. and Murphy, H. C. (1947). Differential phytotoxicity of metabolic by-products of *Helminthosporium victoriae*. *Science*, 106(2751), 270–1. doi:10.1126/science.106.2751.270-a

Meien-Vogeler, F., Bartels, G. and Fehrmann, H. (1994). Distribution of *S. tritici* on wheat in Germany and its regional importance. Paper presented at the *Proceedings of the 4th International Septoria of Cereals Workshop*. IHAR, Radzikow, Poland.

Melchinger, A. E., Utz, H. F. and Schon, C. C. (1998). Quantitative trait locus (QTL) mapping using different testers and independent population samples in maize reveals low power of QTL detection and large bias in estimates of QTL effects. *Genetics*, 149(1), 383–403.

Meuwissen, T. H., Hayes, B. J. and Goddard, M. E. (2001). Prediction of total genetic value using genome-wide dense marker maps. *Genetics*, 157(4), 1819–29.

Miedaner, T. and Korzun, V. (2012). Marker-assisted selection for disease resistance in wheat and barley breeding. *Phytopathology*, 102(6), 560–6. doi:10.1094/PHYTO-05-11-0157

Mitra, M. (1934). A leaf spot disease of wheat caused by *Helminthosporium tritici-repentis* Died. *Indian Journal of Agricultural Sciences*, 4, 692–700.

Mittelstädt, A. and Fehrmann, H. (1987). The occurrence of the perfect stage of *Septoria nodorum* in the Federal Republic of Germany. *Zeitschrift fuer Pflanzenkrankheiten und Pflanzenschutz (Germany, FR)*, 94, 380–5.

Moore, J. W., Herrera-Foessel, S., Lan, C. X., Schnippenkoetter, W., Ayliffe, M., Huerta-Espino, J., Lillemo, M., Viccars, L., Milne, R., Periyannan, S., Kong, X. Y., Spielmeyer, W., Talbot, M., Bariana, H., Patrick, J. W., Dodds, P., Singh, R. and Lagudah, E. (2015). A recently evolved hexose transporter variant confers resistance to multiple pathogens in wheat. *Nature Genetics*, 47(12), 1494–8. doi:10.1038/ng.3439

Mullaney, E. J., Martin, J. M. and Scharen, A. L. (1982). Generation mean analysis to identify and partition the components of genetic resistance to *Septoria nodorum* in wheat. *Euphytica*, 31(2), 539–45. doi:10.1007/bf00021674

Murray, G. M. and Brennan, J. P. (2009). Estimating disease losses to the Australian wheat industry. *Australasian Plant Pathology*, 38(6), 558–70. doi:10.1071/ap09053

Nagy, E. D. and Bennetzen, J. L. (2008). Pathogen corruption and site-directed recombination at a plant disease resistance gene cluster. *Genome Research*, 18(12), 1918–23. doi:10.1101/gr.078766.108

Oliver, R. P. and Solomon, P. S. (2010). New developments in pathogenicity and virulence of necrotrophs. *Current Opinion in Plant Biology*, 13(4), 415–19. doi:http://dx.doi.org/10.1016/j.pbi.2010.05.003

Oliver, R. P., Lord, M., Rybak, K., Faris, J. and Solomon, P. (2008). Emergence of tan spot disease caused by toxigenic *Pyrenophora tritici-repentis* in Australia is not associated with increased deployment of toxin-sensitive cultivars. *Phytopathology*, 98(5), 488–91.

Oliver, R. P., Rybak, K., Solomon, P. S. and Ferguson-Hunt, M. (2009). Prevalence of ToxA-sensitive alleles of the wheat gene *Tsn1* in Australian and Chinese wheat cultivars. *Crop and Pasture Science*, 60(4), 348–52. doi:http://dx.doi.org/10.1071/CP08259

Oliver, R. P., Friesen, T. L., Faris, J. D. and Solomon, P. S. (2012). *Stagonospora nodorum*: From pathology to genomics and host resistance. *Annual Review of Phytopathology*, 50(1), 23–43. doi:10.1146/annurev-phyto-081211-173019

Ornella, L., Singh, S., Perez, P., Burgueño, J., Singh, R., Tapia, E., Bhavani, S., Dreisigacker, S., Braun, H.-J. and Mathews, K. (2012). Genomic prediction of genetic values for resistance to wheat rusts. *The Plant Genome*, 5(3), 136–48.

Parlevliet, J. E. (1977). Plant pathosystems: An attempt to elucidate horizontal resistance. *Euphytica*, 26(3), 553–6. doi:10.1007/bf00021680

Paterson, A. H., Lander, E. S., Hewitt, J. D., Peterson, S., Lincoln, S. E. and Tanksley, S. D. (1988). Resolution of quantitative traits into Mendelian factors by using a complete linkage map of restriction fragment length polymorphisms. *Nature*, 335(6192), 721–6. doi:10.1038/335721a0

Pereira, D. A. S., McDonald, B. A. and Brunner, P. C. (2017). Mutations in the CYP51 gene reduce DMI sensitivity in *Parastagonospora nodorum* populations in Europe and China. *Pest Management Science*, 73(7), 1503–10. doi:10.1002/ps.4486

Phan, H. T. T., Rybak, K., Furuki, E., Breen, S., Solomon, P. S., Oliver, R. P. and Tan, K. C. (2016). Differential effector gene expression underpins epistasis in a plant fungal disease. *Plant Journal*, 87(4), 343–54. doi:10.1111/tpj.13203

Polley, R. and Thomas, M. (1991). Surveys of diseases of winter wheat in England and Wales, 1976–1988. *Annals of Applied Biology*, 119(1), 1–20.

Rosielle, A. A. and Brown, A. G. P. (1980). Selection for resistance to *Septoria nodorum* in wheat. *Euphytica*, 29(2), 337–46. doi:10.1007/bf00025132

Royle, D., Shaw, M. and Cook, R. (1986). Patterns of development of *Septoria nodorum* and *S. tritici* in some winter wheat crops in Western Europe, 1981–83. *Plant Pathology*, 35, 466–76.

Ruud, A. K., Windju, S., Belova, T., Friesen, T. L. and Lillemo, M. (2017). Mapping of SnTox3–*Snn3* as a major determinant of field susceptibility to *Septoria nodorum* leaf blotch in the SHA3/CBRD × Naxos population. *Theoretical and Applied Genetics*, 130(7), 1361–74. doi:10.1007/s00122-017-2893-5

Ruud, A. K., Dieseth, J. A., Ficke, A., Furuki, E., Phan, H. N. T., Oliver, R. P., Tan, K. C. and Lillemo, M. (2018a). Genome-wide association mapping of seedling and adult plant resistance to *Septoria nodorum* blotch in a Nordic spring wheat collection (*in progress*).

Ruud, A. K., Dieseth, J. A. and Lillemo, M. (2018b). Effects of three *Parastagonospora nodorum* necrotrophic effectors on spring wheat under Norwegian field conditions. *Crop Science*, 58, 159–68. doi:10.2135/cropsci2017.05.0281

Rybak, K., See, P. T., Phan, H. T. T., Syme, R. A., Moffat, C. S., Oliver, R. P. and Tan, K. C. (2017). A functionally conserved Zn(2)Cys(6) binuclear cluster transcription factor class regulates necrotrophic effector gene expression and host-specific virulence of two major Pleosporales fungal pathogens of wheat. *Molecular Plant Pathology*, 18(3), 420–34. doi:10.1111/mpp.12511

Scharen, A. L. (1999, September 1999). Biology of *Septoria/Stagonospora* pathogens: An overview. Paper presented at the *Proceedings of the Fifth International Septoria Workshop*, CIMMYT, Mexico.

Scharen, A. L. and Krupinsky, J. M. (1978). Detection and Manipulation of Resistance to *Septoria nodorum* in Wheat. *Phytopathology*, 68(2), 245–8.

Schnurbusch, T., Paillard, S., Fossati, D., Messmer, M., Schachermayr, G., Winzeler, M. and Keller, B. (2003). Detection of QTLs for *Stagonospora* glume blotch resistance in Swiss winter wheat. *Theoretical and Applied Genetics*, 107(7), 1226–34.

Scott, P. R., Benedikz, P. W. and Cox, C. J. (1982). A genetic study of the relationship between height, time of ear emergence and resistance to *Septoria nodorum* in wheat. *Plant Pathology*, 31(1), 45–60. doi:10.1111/j.1365-3059.1982.tb02811.x

Semagn, K., Babu, R., Hearne, S. and Olsen, M. (2014). Single nucleotide polymorphism genotyping using Kompetitive Allele Specific PCR (KASP): Overview of the technology and its application in crop improvement. *Molecular Breeding*, 33(1), 1–14. doi:10.1007/s11032-013-9917-x

Shankar, M., Walker, E., Golzar, H., Loughman, R., Wilson, R. E. and Francki, M. G. (2008). Quantitative trait loci for seedling and adult plant resistance to *Stagonospora nodorum* in wheat. *Phytopathology*, 98(8), 886–93. doi:10.1094/phyto-98-8-0886

Shatalina, M., Messmer, M., Feuillet, C., Mascher, F., Paux, E., Choulet, F., Wicker, T. and Keller, B. (2014). High-resolution analysis of a QTL for resistance to *Stagonospora nodorum* glume blotch in wheat reveals presence of two distinct resistance loci in the target interval. *Theoretical and Applied Genetics*, 127(3), 573–86. doi:10.1007/s00122-013-2240-4

Shaw, M. (1999). Population dynamics of *Septoria* in the crop ecosystem. In: Lucas, J. A., Bowyer, P. and Anderson, H. M. (Eds), *Septoria on Cereals: A Study of Pathosystems*. Wallingford, UK: CAB International, pp. 82–95.

Shi, G. J., Friesen, T. L., Saini, J., Xu, S. S., Rasmussen, J. B. and Faris, J. D. (2015). The wheat *Snn7* gene confers susceptibility on recognition of the *Parastagonospora nodorum* necrotrophic effector SnTox7. *Plant Genome*, 8(2), 39. doi:10.3835/plantgenome2015.3802.0007

Shi, G. J., Zhang, Z. C., Friesen, T. L., Bansal, U., Cloutier, S., Wicker, T., Rasmussen, J. B. and Faris, J. D. (2016a). Marker development, saturation mapping, and high-resolution mapping of the *Septoria nodorum* blotch susceptibility gene *Snn3-B1* in wheat. *Molecular Genetics and Genomics*, 291(1), 107–19. doi:10.1007/s00438-015-1091-x

Shi, G. J., Zhang, Z. C., Friesen, T. L., Raats, D., Fahima, T., Brueggeman, R. S., Lu, S. W., Trick, H. N., Liu, Z. H., Chao, W., Frenkel, Z., Xu, S. S., Rasmussen, J. B. and Faris, J. D. (2016b). The hijacking of a receptor kinase-driven pathway by a wheat fungal pathogen leads to disease. *Science Advances*, 2(10), e1600822. doi:10.1126/sciadv.1600822

Siegel, M. R. (1981). Sterol-inhibiting fungicides - effects on sterol biosynthesis and sites of action. *Plant Disease*, 65(12), 986–9.

Singh, R. P., Huerta-Espino, J., Bhavani, S., Herrera-Foessel, S. A., Singh, D., Singh, P. K., Velu, G., Mason, R. E., Jin, Y., Njau, P. and Crossa, J. (2011). Race non-specific resistance to rust diseases in CIMMYT spring wheats. *Euphytica*, 179(1), 175–86. doi:10.1007/s10681-010-0322-9

Solomon, P. S., Lowe, R. G. T., Tan, K.-C., Waters, O. D. C. and Oliver, R. P. (2006). *Stagonospora nodorum*: Cause of *Stagonospora nodorum* blotch of wheat. *Molecular Plant Pathology*, 7(3), 147–56. doi:10.1111/j.1364-3703.2006.00326.x

Sommerhalder, R. J., McDonald, B. A., Mascher, F. and Zhan, J. (2010). Sexual recombinants make a significant contribution to epidemics caused by the wheat pathogen *Phaeosphaeria nodorum*. *Phytopathology*, 100(9), 855–62. doi:10.1094/phyto-100-9-0855

Sommerhalder, R. J., McDonald, B. A., Mascher, F. and Zhan, J. (2011). Effect of hosts on competition among clones and evidence of differential selection between pathogenic and saprophytic phases in experimental populations of the wheat pathogen *Phaeosphaeria nodorum*. *BMC Evolutionary Biology*, 11, 188. doi:10.1186/1471-2148-11-188

Storlie, E. and Charmet, G. (2013). Genomic selection accuracy using historical data generated in a wheat breeding program. *The Plant Genome*, 6(1), 1–9. doi:10.3835/plantgenome2013.3801.0001

Stukenbrock, E. H., Banke, S. and McDonald, B. A. (2006). Global migration patterns in the fungal wheat pathogen *Phaeosphaeria nodorum*. *Molecular Ecology*, 15(10), 2895–904. doi:10.1111/j.1365-294X.2006.02986.x

Sutton, J. and Vyn, T. (1990). Crop sequences and tillage practices in relation to diseases of winter wheat in Ontario. *Canadian Journal of Plant Pathology*, 12(4), 358–68.

Syme, R. A., Hane, J. K., Friesen, T. L. and Oliver, R. P. (2013). Resequencing and comparative genomics of *Stagonospora nodorum*: Sectional gene absence and effector discovery. *G3: Genes|Genomes|Genetics*, 3(6), 959–69. doi:10.1534/g3.112.004994

Syme, R. A., Tan, K.-C., Hane, J. K., Dodhia, K., Stoll, T., Hastie, M., Furuki, E., Ellwood, S. R., Williams, A. H., Tan, Y.-F., Testa, A. C., Gorman, J. J. and Oliver, R. P. (2016). Comprehensive annotation of the *Parastagonospora nodorum* reference genome using next-generation genomics, transcriptomics and proteogenomics. *PLOS ONE*, 11(2), e0147221. doi:10.1371/journal.pone.0147221

Tan, K. C., Phan, H. T. T., Rybak, K., John, E., Chooi, Y. H., Solomon, P. S. and Oliver, R. P. (2015). Functional redundancy of necrotrophic effectors – consequences for exploitation for breeding. *Frontiers in Plant Science*, 6, Article 501. doi:10.3389/fpls.2015.00501

Tanaka, S. (1933). Studies on black spot disease of the Japanese pear (*Pirus serotina* Rehd.). Memoirs of the College of Agriculture, Kyoto Imperial University, No. 28 (Phytopathology series no. 6), 31pp.

Tanksley, S. D. (1983). Molecular markers in plant breeding. *Plant Molecular Biology Reporter*, 1(1), 3–8. doi:10.1007/bf02680255

Tomas, A. and Bockus, W. (1987). Cultivar-specific toxicity of culture filtrates of *Pyrenophora tritici-repentis*. *Phytopathology*, 77(9), 1337–40.

Tomas, A., Feng, G. H., Reeck, G. R., Bockus, W. W. and Leach, J. E. (1990). Purification of a cultivar-specific toxin from *Pyrenophora tritici-repentis*, causal agent of tan spot of wheat. *Molecular Plant-Microbe Interactions*, 3(4), 221–4. doi:10.1094/Mpmi-3-221

Tommasini, L., Schnurbusch, T., Fossati, D., Mascher, F. and Keller, B. (2007). Association mapping of *Stagonospora nodorum* blotch resistance in modern European winter wheat varieties. *Theoretical and Applied Genetics*, 115(5), 697–708.

Torriani, S. F. F., Melichar, J. P. E., Mills, C., Pain, N., Sierotzki, H. and Courbot, M. (2015). *Zymoseptoria tritici*: A major threat to wheat production, integrated approaches to control. *Fungal Genetics and Biology*, 79, 8–12. doi:http://dx.doi.org/10.1016/j.fgb.2015.04.010

Treangen, T. J. and Salzberg, S. L. (2011). Repetitive DNA and next-generation sequencing: Computational challenges and solutions. *Nature Reviews. Genetics*, 13(1), 36–46. doi:10.1038/nrg3117

Uphaus, J., Walker, E., Shankar, M., Golzar, H., Loughman, R., Francki, M. and Ohm, H. (2007). Quantitative trait loci identified for resistance to *Stagonospora glume* blotch in wheat in the USA and Australia. *Crop Science*, 47(5), 1813–22.

van't Slot, K. A. E., van den Burg, H. A., Kloks, C. P. A. M., Hilbers, C. W., Knogge, W. and Papavoine, C. H. M. (2003). Solution Structure of the plant disease resistance-triggering protein NIP1 from the Fungus *Rhynchosporium secalis* shows a novel β-sheet fold. *Journal of Biological Chemistry*, 278(46), 45730–6. doi:10.1074/jbc.M308304200

Vanden Bossche, H., Koymans, L. and Moereels, H. (1995). P450 inhibitors of use in medical treatment: Focus on mechanisms of action. *Pharmacology & Therapeutics*, 67(1), 79–100.

Virdi, S. K., Liu, Z. H., Overlander, M. E., Zhang, Z. C., Xu, S. S., Friesen, T. L. and Faris, J. D. (2016). New insights into the roles of host gene-necrotrophic effector interactions in governing susceptibility of durum wheat to tan spot and *Septoria nodorum* blotch. *G3-Genes Genomes Genetics*, 6(12), 4139–50. doi:10.1534/g3.116.036525

Vleeshouwers, V. and Oliver, R. P. (2014). Effectors as tools in disease resistance breeding against biotrophic, hemibiotrophic, and necrotrophic plant pathogens. *Molecular Plant-Microbe Interactions*, 27(3), 196–206. doi:10.1094/mpmi-10-13-0313-ia

Wang, S., Wong, D., Forrest, K., Allen, A., Chao, S., Huang, B. E., Maccaferri, M., Salvi, S., Milner, S. G., Cattivelli, L., Mastrangelo, A. M., Whan, A., Stephen, S., Barker, G., Wieseke, R., Plieske, J., International Wheat Genome Sequencing, C., Lillemo, M., Mather, D., Appels, R., Dolferus, R., Brown-Guedira, G., Korol, A., Akhunova, A. R., Feuillet, C., Salse, J., Morgante, M., Pozniak, C., Luo, M.-C., Dvorak, J., Morell, M., Dubcovsky, J., Ganal, M., Tuberosa, R., Lawley, C., Mikoulitch, I., Cavanagh, C., Edwards, K. J., Hayden, M. and Akhunov, E. (2014). Characterization of polyploid wheat genomic diversity using a high-density 90 000 single nucleotide polymorphism array. *Plant Biotechnology Journal*, 12(6), 787–96. doi:10.1111/pbi.12183

Waters, O. D., Lichtenzveig, J., Rybak, K., Friesen, T. L. and Oliver, R. P. (2011). Prevalence and importance of sensitivity to the *Stagonospora nodorum* necrotrophic effector SnTox3 in current Western Australian wheat cultivars. *Crop and Pasture Science*, 62(7), 556–62.

Wicki, W., Winzeler, M., Schmid, J. E., Stamp, P. and Messmer, M. (1999). Inheritance of resistance to leaf and glume blotch caused by *Septoria nodorum* Berk. in winter wheat. *Theoretical and Applied Genetics*, 99(7–8), 1265–72. doi:10.1007/s001220051332

Windju, S. S., Malla, K., Belova, T., Wilson, R. C., Dieseth, J. A., Alsheikh, M. K. and Lillemo, M. (2017). Mapping and validation of powdery mildew resistance loci from spring wheat cv. Naxos with SNP markers. *Molecular Breeding*, 37(5), 61. doi:10.1007/s11032-017-0655-3

Wolpert, T. J., Dunkle, L. D. and Ciuffetti, L. M. (2002). Host-selective toxins and avirulence determinants: What's in a name? *Annual Review of Phytopathology*, 40, 251–85. doi:10.1146/annurev.phyto.40.011402.114210

Yan, L., Loukoianov, A., Tranquilli, G., Helguera, M., Fahima, T. and Dubcovsky, J. (2003). Positional cloning of the wheat vernalization gene *VRN1*. *Proceedings of the National Academy of Sciences*, 100(10), 6263–8. doi:10.1073/pnas.0937399100

Zadoks, J. C., Chang, T. T. and Konzak, C. F. (1974). A decimal code for the growth stages of cereals. *Weed Research*, 14(6), 415–21. doi:10.1111/j.1365-3180.1974.tb01084.x

Zhang, Z. C., Friesen, T. L., Simons, K. J., Xu, S. S. and Faris, J. D. (2009). Development, identification, and validation of markers for marker-assisted selection against the *Stagonospora nodorum* toxin sensitivity genes *Tsn1* and *Snn2* in wheat. *Molecular Breeding*, 23(1), 35–49. doi:10.1007/s11032-008-9211-5

Zhang, Z. C., Friesen, T. L., Xu, S. S., Shi, G. J., Liu, Z. H., Rasmussen, J. B. and Faris, J. D. (2011). Two putatively homoeologous wheat genes mediate recognition of SnTox3 to confer effector-triggered susceptibility to *Stagonospora nodorum*. *Plant Journal*, 65(1), 27–38. doi:10.1111/j.1365-313X.2010.04407.x

Chapter 7

A lesser-known pathogen of wheat: *Bipolaris sorokiniana*

E. Hill and P. Solomon, The Australian National University, Australia

1 Introduction

Both spot blotch (SB) and common root rot (CRR) are responsible for significant yield reductions during severe disease outbreaks. In Asian and especially in Indian growing regions, *Bipolaris sorokiniana* is considered the foliar and root pathogen of greatest concern. SB symptoms appear on leaf tissue as small dark brown necrotic lesions. Ovular in shape and 1–3 mm in length with tapered ends, lesions appear in the early stages of infection and coalesce into broad chlorotic regions (Poloni et al., 2009; Singh et al., 1998; Duveiller, 2004). This reduction in photosynthetically active tissue limits plant growth and grain production, leading to overall reductions in yield quality, as measured by thousand kernel weight (TKW) and quantity (Joshi et al., 2007a). Disease severity in the field is largely assessed by measuring the intensity and development of tissue damage over a defined time period, to generate an area under the disease progress curve (AUDPC) or AUDPC/day. This quantification allows comparisons between annual disease outbreaks and a standardised method of assessing susceptibility among wheat cultivars (Sharma et al., 2007; Jeger, 2004; Duveiller and Sharma, 2012). CRR symptoms are characterised by the browning and necrosis of root tissue. Tissue damage resulting from infection is suggested to compromise nutrient and water uptake, resulting in stunted vegetative and reproductive growth (Duveiller and Sharma, 2009). Wheat infected with *B. sorokiniana* may also show blackening of the grain. High incidences are associated with high precipitation during reproductive growth (Viedma and Kohli, 1998).

http://dx.doi.org/10.19103/AS.2018.0039.07

2 Geographic distribution of disease

B. sorokiniana is found globally in wheat- and barley-growing regions, but disease epidemics are concentrated in the warmer, nontraditional wheat growing regions in South America, South-East Asia and the Indian subcontinent. Disease outbreaks affect 25 million hectares of cropping land, of which more than 30% resides in the rice–wheat cropping regions of the Indo-Gangetic Plains, including central and eastern India, Bangladesh and Nepal (Duveiller et al., 2005; Duveiller and Sharma, 2012). *B. sorokiniana* remains a major barrier to secure wheat production in these regions, particularly for small producers who contribute the main body of wheat production in Southern Asia and growers with limited resources (Joshi et al., 2007b; Duveiller, 2004; Sharma and Duveiller, 2006). Foliar outbreaks are promoted where mean minimum temperatures remain above 17°C, and high relative humidity and moisture levels cause prolonged periods of leaf wetness (Duveiller et al., 2005; Sharma et al., 2007). CRR is more prominent in drier conditions, posing problems in northern Australia, the United States, Canada and Eastern Europe (Duveiller and Sharma, 2009; Murray and Brennan, 2009; Chowdhury et al., 2013).

Saari et al. (1998) estimated 20% losses across these regions from SLB or Helminthosporium leaf blight (HLB), with potential for 100% losses under favourable environmental conditions. Duveiller et al. (2005) more recently reported 20–30% losses from these blights under both field and experimental field stations in Nepal in the plains bordering the Himalayas, whilst Joshi et al. (2012) estimated 17–19% losses from HLB across Southern Asia. Losses from CRR are poorly documented in these regions, but they are estimated to vary between 5% and 20% (Mehta, 1998). However, actual yield losses probably deviate, given yield records are inconsistent in many regions (Duveiller et al., 2007). Disease severity and total yield losses are also heavily dependent on environmental conditions, cultivar resistance and crop husbandry practices (Mehta, 1998; Sharma and Duveiller, 2004; Gurung et al., 2012; Duveiller and Altamirana, 2000). Consequently, Duveiller et al. (2007) noted that only a proportion of *B. sorokiniana*-affected regions will suffer ~20% losses each season depending on the combination of genotype and environmental and cultural factors at play (Mehta, 1998). The impacts of the disease in Southern Asia are not to be underestimated. Rates of wheat production in these regions have slowed since the 1990s, likely due to the spread of intensive wheat–rice cropping systems and their associated environmental impacts, alongside the pressures of a warming climate. Together, these factors hold serious implications for future agronomic productivity and food security (Sharma et al., 2007; Joshi et al., 2007b; Duveiller et al., 2007).

The current literature reflects the impact of *B. sorokiniana* in southern parts of Asia. However, SB and CRR have been an increasing problem in China since the late 1950s, spreading across the country from the north-eastern growing regions (Chang and Wu, 1998). Yield losses in the most severely affected northern regions are reported between 10% and 30% from the foliar disease, where high temperatures and heavy rainfall align with the reproductive stages of crop growth (Chang and Wu, 1998; Xiao et al., 1998).

In South America, the nontraditional wheat growing areas of Paraguay, Brazil and eastern lowlands of Bolivia are affected by SB (Duveiller and Sharma, 2012). These nontraditional areas have mild climates, with variable precipitation levels and the potential for high rainfall during vegetative and reproductive growth periods of winter and early spring, respectively (Viedma and Kohli, 1998). Growing seasons experiencing high precipitation levels and winter temperatures exceeding 18°C are prone to disease outbreaks (Toledo and Guzman,

1998; Duveiller et al., 2005). Yield losses from SB and HLB are largely undefined but have exceeded 70% when cropping regions received unusually high levels of rainfall (Mehta et al., 1992). These severe outbreaks continue to significantly limit wheat production in Brazil and Latin America (Mehta et al., 1992).

3 How does *B. sorokiniana* cause disease in wheat?

B. sorokiniana is a filamentous, hemibiotrophic pathogen capable of infecting all parts of its host. Its capacity to cause disease is well documented, but the mechanisms underlying its infection of wheat and how they differ in root and foliar and grain tissues are unclear. The mechanisms governing the pathogen's broad host specificity are also undetermined. Here we discuss the current status of understanding of the foliar infection process and the pathogenic lifestyle of *B. sorokiniana* in the field. The mechanisms of CRR infection in wheat are less well studied and will not be discussed here.

In the field, *B. sorokiniana* pathogenesis is driven by the asexual lifecycle of the heterothallic fungus. Conidia are asexually produced from necrotic lesions on root or foliar tissue, dispersing to new hosts by wind or water droplets during periods of high precipitation (Shoemaker, 1955; Duveiller and Sharma, 2012). Infected seed and soil-borne conidia are also likely contributors to pathogen persistence and spread (Duveiller, 2004). The hydrophobic surface of conidia allows adhesion to host tissue surfaces, where moist conditions promote germination. Germination is associated with the development of a proteinaceous, β-glucan-rich extracellular matrix (ECM), suggested to maintain adhesion and provide a protective coating for the growing germ tube (Apoga et al., 2001; Jansson and Akesson, 2003; Apoga and Jansson, 2000). Esterase and other cellulolytic enzymes localised at the ECM are implicated in degrading the waxy cuticle surface of wheat leaves and compromising the structural integrity of the epidermal layer (Jansson and Akesson, 2003; Han et al., 2010; Geimba et al., 1999). Helminthosporol toxins, a class of non-host specific secondary metabolites, are also secreted by germlings and infection hyphae to facilitate the weakening of host tissues. The diffusion of these cytotoxic compounds is also linked to the development of local and distal necrotic and chlorotic lesions (Carlson et al., 1991; Briquet et al., 1998; Apoga et al., 2002; Apoga and Jansson, 2000). These biochemical mechanisms aid *B. sorokiniana* in mechanically penetrating epidermal cell walls or junctions using appressoria structures developed after germination (Han et al., 2010). The initial colonisation of epidermal cells is proposed to mark a short biotrophic phase (Kumar et al., 2002); however, *B. sorokiniana* appears to predominantly adopt a necrotrophic lifestyle, aggressively spreading through the epidermal and mesophyll tissues and rapidly inducing widespread cell death (Huang and Tinline, 1976; Yadav, 1981; Ibeagha et al., 2005).

4 Pathogenic variability

B. sorokiniana is a highly variable pathogen at the morphological, physiological and genetic levels (Poloni et al., 2009; Gurung et al., 2013). A range of morphologies and growth patterns are seen across populations; isolates vary between constrained, black mycelial growth to grey-white, cottony growth with several intermediate phenotypes (Poloni et al., 2009; Mahto et al., 2012). Variations in the fungus' aggressiveness on wheat are also complex. *B. sorokiniana* isolates from across Southern Asia and Latin America show no distinct patterns in virulence like those seen in other wheat pathogens, such as the race structures of the necrotroph *P. tritici-repentis* and the biotrophic cereal rusts (*Puccinia* spp.) (Duveiller and Altamirana, 2000; Webb and Fellers, 2006). This variability has been described as a continuum of aggressiveness (Duveiller, 2004; Duveiller and Sharma, 2012), with high levels of genetic and pathological diversity occurring within and between populations of isolates from wheat (Ruckstuhl, 1998; Maraite et al., 1998; Pandey et al., 2008; Gurung et al., 2013). Recent studies of Brazilian, Nepalese and Indian *B. sorokiniana* isolates have sought to define a correlation between morphological types and aggressiveness. No consensus on the grouping of isolates by pathogenicity and morphology has been reached, partly due to inconsistencies in the pathogen isolates, wheat cultivars and molecular analyses used by these independent studies (Mahto et al., 2012; Gurung et al., 2013; Jaiswal et al., 2007; Pandey et al., 2008; Kumar et al., 2015a). Comprehensive population-level analyses of molecular diversity in Southern Asian and South American isolates are clearly needed. Regardless of this, it is evident that this genetic diversity grants the adaptive potential of the pathogen, which stands as an important consideration for resistance breeding (Duveiller and Altamirana, 2000; Pandey et al., 2008; Gurung et al., 2013).

The influence of histochemical and physiological traits has also been investigated in relation to pathogenic variation. Apoga et al. (2002) analysed differential helminthosporol toxin production between isolates to determine its relationship with aggressiveness and the role in pathogenicity. Isolates that were more aggressive on barley seedling roots produced more of the most bioactive toxin, prehelminthosporol (PHL), during growth *in vitro* than isolates that were less affecting barley seedling roots. However, this correlation was not verified *in planta* in barley or wheat, so the contribution of PHL production to variable pathogenicity in wheat remains unclear (Apoga et al., 2002). Given our limited understanding of the pathogenicity factors driving disease in wheat, there may be other secondary metabolites, degradative enzymes or proteinaceous effectors influencing variations in pathogenicity. For example, the recent discovery of the proteinaceous effector gene *ToxA* in some Australian isolates of *B. sorokiniana* is associated with enhanced virulence on wheat cultivars containing the corresponding susceptibility gene, *Tsn1* (McDonald et al., 2017; Friesen et al., 2008). Initial analyses suggest that the effector gene is inconsistently dispersed among Australian isolates, but more comprehensive population-level studies are required (McDonald et al., 2017). The presence of *ToxA* among international populations is also undetermined. Given the rich effector and secondary metabolite-encoding capacity of related filamentous phytopathogens, including other *Bipolaris* cereal pathogens, it is plausible that other virulence factors may be at play in the wheat–*B. sorokiniana* interaction and account for some degree of pathogenic diversity (Ellis et al., 2009; Stergiopoulos et al., 2013; Condon et al., 2013).

5 Disease management

Exploiting genetic resistance offers an economical and environmentally friendly approach to SB and HLB control. However, while genetics has had a significant impact on managing other wheat diseases, genetic resistance to date has not proven particularly effective at controlling SB or HLB in wheat (Sharma and Duveiller, 2006). Concerted efforts for sourcing germplasm resistant to *B. sorokiniana* were only initiated in the 1980s when the impact of the disease on wheat was fully realised (Duveiller and Gilchrist, 1994). Since then, international efforts have been made to identify and exploit germplasm harbouring resistance to *B. sorokiniana*. Initial efforts focused on germplasm from affected areas in Brazil, Zambia and China with some lines identified expressing reasonable levels of resistance. CIMMYT later used a variety of wide cross derivatives to identify several genotypes with good SB resistance (Duveiller and Sharma, 2012; Dubin et al., 1998). Additionally, further sources were identified in India and South Asia which have been previously reviewed by Duveiller and Sharma (2009b, 2012). Subsequent studies have identified further sources of resistant germplasm and also markers linked to SB and HLB resistance (Adhikari et al., 2012; Kumar et al., 2015b,c, 2016; Lu et al., 2016; Meena et al., 2014; Osman et al., 2016; Singh et al., 2016; Zhu et al., 2014; Duveiller and Sharma, 2009). It is reasonable to suggest that despite these efforts, our understanding of the interaction of *B. sorokiniana* and wheat at the genetic level is poor compared to other pathosystems. A review of the literature reveals some confusion surrounding these genetics. The interaction is controlled by a variety of mechanisms from single gene qualitative resistance to polygenic quantitative traits with little consistency in the loci being identified. Whilst issues exist in terms of consistency of disease screening (e.g. comparing seedling and adult plant resistance) (Duveiller and Sharma, 2012), SB and HLB are clearly complex diseases that require a concerted effort to truly understand their genetic basis.

Farming practices also play a significant role in managing disease development (Duveiller and Sharma, 2009). Relative humidity is a key variable dictating the severity of SB, and the timing of seeding is critical (Sharma et al., 2003; Sharma and Duveiller, 2004; Duveiller et al., 2005). Growers are typically advised to plant early in their respective growing seasons to enable the crop to outgrow conditions that are conducive to disease development (i.e. avoiding heat stress and high humidity). Indeed late seeding has been reported to lead to 25% yield losses due to SB and HLB, even with the use of more resistant genotypes, highlighting the importance of early seeding (Duveiller and Sharma, 2009).

It has been demonstrated that lower seeding rates and increased row spacing restrict the spread of SB and HLB due to the lower relative humidity (Sentelhas et al., 1995). Adequate soil fertilisation is also important in minimising the onset and development of SB. Wheat grown in soils deficient in nitrogen (Sharma and Duveiller, 2004; Wall et al., 1990) or potassium (Sharma et al., 2005; Regmi et al., 2002) appears more susceptible to SB and HLB infection. Crop rotations are also important in minimising the impact of disease. Rotations with non-hosts such as oat, vetch or canola play a role in reducing primary inoculum, particularly when zero-tillage practices are applied (Reis et al., 1997). In South Asia, rice is often rotated with wheat and it was a long-held view that this practice led to increased *B. sorokiniana* diseases. However, it has been reported by Dubbin and Bimb (1991) that this rotation did not lead to increased pathogen build-up or infection. Indeed it was suggested that the flooding process used for irrigated rice could actually play a sanitising role in reducing the impact of *B. sorokiniana* (Dubin and Bimb, 1991; Duveiller and Sharma, 2009).

Fungicides are also used to control SB and HLB in affected regions, and multiple chemistries have been proven effective in protecting germinating seeds and seedlings from early infection. Recently, Sharma-Poudyal et al. (2016) published a comprehensive report on the efficacy of chemical control for HLB. They reported that a single spray of propiconazole significantly reduced foliar disease with increases in grain yields of 14% and 15% reported over consecutive years. This finding supported a previous demonstration of the efficacy of propiconazole, confirming its use in controlling HLB in warm growing regions (Sharma-Poudyal et al., 2005). In the same study, Sharma-Poudyal et al. (2016) also examined the effect of different fungicides as a seed treatment option for managing HLB. Other fungicide treatments have been shown to consistently reduce the number of infected seedlings in trials over consecutive years but did not result in reduced disease. Interestingly though, some fungicide combinations have resulted in the enhanced seed germination, a greater number of effective tillers and increased grain yield. As such, growers have been recommended to use a fungicide seed treatment as well as a single application of propiconazole to maximise yield and minimise the impact of SB and HLB.

However, like all diseases, management of SB or HLB is not achieved through a single approach and integrated management is recommended. The appropriate choice of cultivar, timing of seeding, rotations and optimal use of chemical control is recommended for reducing losses due to SB and HLB. However, in terms of improving management to minimise the input for growers, further advances need to be made in dissecting the genetic basis of these diseases and to use this knowledge in generating genotypes with significantly increased disease resistance.

6 Conclusion and future trends

It is evident from the above review that both *B. sorokiniana* and the diseases it causes are poorly understood. The pathogenic machinery required to facilitate disease is unknown and a detailed understanding of the genetic basis of resistance in wheat is lacking. In contrast, the knowledge base of comparable diseases such as tan spot and septoria nodorum blotch is far more advanced with both pathogen and host components identified and being used for disease management. The reason for this lack of progress in understanding *B. sorokiniana* is likely due to the global distribution of the pathogen. SB and HLB predominantly affect warmer wheat growing areas in developing countries that are unlikely to have the resources to dissect these diseases to the detail required to enhance management strategies. However, the generation of at least partially resistant varieties, the use of improved farming practices and chemical control have helped reduce the impact of disease in affected growing areas. However, further resources, particularly from developed countries, are now needed in order to advance our understanding of these diseases and provide a meaningful contribution to their management.

7 References

Adhikari, T. B., Gurung, S., Hansen, J. M., Jackson, E. W. and Bonman, J. M. (2012). Association mapping of quantitative trait loci in spring wheat landraces conferring resistance to bacterial leaf streak and spot blotch. *Plant Genome*, 5, 1–16.

Apoga, D. and Jansson, H.-B. (2000). Visualization and characterization of the extracellular matrix of *Bipolaris sorokiniana*. *Mycological Research*, 104, 564–75.

Apoga, D., Jansson, H.-B. and Tunlid, A. (2001). Adhesion of conidia and germlings of the plant pathogenic fungus *Bipolaris sorokiniana* to solid surfaces. *Mycological Research*, 105, 1251–60.

Apoga, D., Akesson, H., Jansson, H. B. and Odham, G. (2002). Relationship between production of the phytotoxin prehelminthosporol and virulence in isolates of the plant pathogenic fungus *Bipolaris sorokiniana*. *European Journal of Plant Pathology*, 108, 519–26.

Briquet, M., Vilret, D., Goblet, P., Mesa, M. and Eloy, M. C. (1998). Plant cell membranes as biochemical targets of the phytotoxin helminthosporol. *Journal of Bioenergetics and Biomembranes*, 30, 285–95.

Carlson, H., Nilsson, P., Jansson, H. B. and Odham, G. (1991). Characterization and determination of prehelminthosporol, a toxin from the plant pathogenic fungus *Bipolaris sorokiniana*, using liquid-chromatography mass spectrometry. *Journal of Microbiological Methods*, 13, 259–69.

Chang, N. and Wu, Y. (1998). Incidence and current management of spot blotch of wheat in China. In: *Helminthosporium Blights of Wheat: Spot Blotch and Tan Spot*. Duveiller, E., Dubin, H. J., Reeves, J. and McNab, A. (Eds). Mexico, DF: CIMMYT, pp. 119–25.

Chowdhury, A. K., Singh, G., Tyagi, B. S., Ojha, A., Dhar, T. and Bhattacharya, P. M. (2013). Spot blotch disease of wheat - a new thrust area for sustaining productivity. *Journal of Wheat Research*, 5, 1–11.

Condon, B. J., Leng, Y., Wu, D., Bushley, K. E., Ohm, R. A., Otillar, R., Martin, J., Schackwitz, W., Grimwood, J., MohdZainudin, N., Xue, C., Wang, R., Manning, V. A., Dhillon, B., Tu, Z. J., Steffenson, B. J., Salamov, A., Sun, H., Lowry, S., LaButti, K., Han, J., Copeland, A., Lindquist, E., Barry, K., Schmutz, J., Baker, S. E., Ciuffetti, L. M., Grigoriev, I. V., Zhong, S. and Turgeon, B. G. (2013). Comparative genome structure, secondary metabolite, and effector coding capacity across Cochliobolus pathogens. *PLoS Genetics*, 9, e1003233.

da Luz, W. C. and Bergstrom, G. C. (1987). Interactions between *Cochliobolus sativus* and *Pyrenophera triciti-repentis* on wheat leaves. *Phytopathology*, 77, 1355–60.

Dubin, H. J. and Bimb, H. P. (1991). Effects of soil and foliar treatments on yields and diseases of wheat in lowland Nepal. In: *Wheat for the Nontraditional Warm Areas*. Saunders, D. A. (Ed.). Mexico, DF: CIMMYT, pp. 484–5.

Dubin, H. J., Arun, B., Begum, S. N., Bhatta, M., Dhari, R., Goel, L. B., Joshi, A. K., Khanna, B. N., Malaker, P. K., Pokhrel, D. R., Rahman, M. M., Saha, N. K., Shaheed, M. A., Sharma, R. C., Singh, A. K., Singh, R. M., Singh, R. V., Vargas, M. and Verma, P. C. (1998). Results of the South Asia regional Helminthosporium leaf blight and yield experiment, 1993–94. In: *Helminthosophism Diseases of Wheat: Spot Blotch and Tan Spot*. Duveiller, E., Dubin, H. J., Reeves, J. and McNab, A. (Eds). Mexico, DF: CIMMYT, pp. 182–7.

Duveiller, E. (2004). Controlling foliar blights of wheat in the rice-wheat systems of Asia. *Plant Disease*, 88, 552–6.

Duveiller, E. and Gilchrist, L. (1994). Production constraints due to *Bipolaris sorokiniana* in wheat: Current situation and future prospects. In: *Wheat in Heat Stressed Environments: Irrigated Dry Areas and Rice-Wheating Farming Systems*. Saunders, D. A. and Hettel, G. P. (Eds). Mexico, DF: CIMMYT, pp. 343–52.

Duveiller, E. and Altamirana, G. (2000). Pathogenicity of *Bipolaris sorokiniana* isolates from wheat roots, leaves and grains in Mexico. *Plant Pathology*, 49, 235–42.

Duveiller, E. M. and Sharma, R. C. (2009). Genetic improvement and crop management strategies to minimise yield losses in warm non-traditional wheat growing areas due to spot blotch pathogen *Cochliobolus sativus*. *Journal of Phytopathology*, 157, 521–34.

Duveiller, E. and Sharma, R. C. (2012) Wheat resistance to spot blotch or foliar blight. In: *Disease Resistance in Wheat*. Sharma, I. (Ed.). Wallingford, UK: CABI, pp. 120–35.

Duveiller, E., Kandel, Y. R., Sharma, R. C. and Shrestha, S. M. (2005). Epidemiology of foliar blights (spot blotch and tan spot) of wheat in the plains bordering the Himalayas. *Phytopathology*, 95, 248–56.

Duveiller, E., Singh, R. P. and Nicol, J. M. (2007). The challenges of maintaining wheat productivity: Pests, diseases, and potential epidemics. *Euphytica*, 157, 417–30.

Ellis, J. G., Rafiqi, M., Gan, P., Chakrabarti, A. and Dodds, P. N. (2009). Recent progress in discovery and functional analysis of effector proteins of fungal and oomycete plant pathogens. *Current Opinion in Plant Biology*, 12, 399–405.

Friesen, T. L., Zhang, Z., Solomon, P. S., Oliver, R. P. and Faris, J. D. (2008). Characterization of the interaction of a novel *Stagonospora nodorum* host-selective toxin with a wheat susceptibility gene. *Plant Physiology*, 146, 682–93.

Geimba, M. P., Riffel, A., Agostini, V. and Brandelli, A. (1999). Characterisation of cellulose-hydrolysing enzymes from the fungus *Bipolaris sorokiniana*. *Journal of the Science of Food and Agriculture*, 79, 1849–54.

Gurung, S., Sharma, R. C., Duveiller, E. and Shrestha, S. M. (2012). Comparative analyses of spot blotch and tan spot epidemics on wheat under optimum and late sowing period in South Asia. *European Journal of Plant Pathology*, 134, 257–66.

Gurung, S., Mahto, B. N., Gyawali, S. and Adhikari, T. B. (2013). Phenotypic and molecular diversity of *Cochliobolus sativus* populations from wheat. *Plant Disease*, 97, 62–73.

Han, Q., Huang, L., Buchenauer, H., Wang, C. and Kang, Z. (2010). Cytological study of wheat spike infection by *Bipolaris sorokiniana*. *Journal of Phytopathology*, 158, 22–9.

Huang, H. D. and Tinline, R. D. (1976). Histology of *Cochliobolus sativus* infection in subcrown internodes of wheat and barley. *Canadian Journal of Botany*, 54, 1344–54.

Ibeagha, A. E., Huckelhoven, R., Schafer, P., Singh, D. P. and Kogel, K. H. (2005). Model wheat genotypes as tools to uncover effective defence mechanisms against the hemibiotrophic fungus *Bipolaris sorokiniana*. *Phytopathology*, 95, 528–32.

Jaiswal, S. K., Sweta, Prasad, L. C., Sharma, S., Kumar, S., Prasad, R., Pandey, S. P., Chand, R. and Joshi, A. K. (2007). Identification of molecular marker and aggressiveness for different groups of *Bipolaris sorokiniana* isolates causing spot blotch disease in wheat (*Triticum aestivum* L.). *Current Microbiology*, 55, 135–41.

Jansson, H.-B. and Akesson, H. (2003). Extracellular matrix, esterase and the phytotoxin prehelminthosporol in infection of barley leaves by *Bipolaris sorokiniana*. *European Journal of Plant Pathology*, 109, 599–605.

Jeger, M. J. (2004). Analysis of disease progress as a basis for evaluating disease management practices. *Annual Review of Phytopathology*, 42, 61–82.

Joshi, A., Kumari, M., Singh, V., Reddy, C., Kumar, S., Rane, J. and Chand, R. (2007a). Stay green trait: Variation, inheritance and its association with spot blotch resistance in spring wheat (*Triticum aestivum* L.). *Euphytica*, 153, 59–71.

Joshi, A. K., Mishra, B., Chatrath, R., Ortiz Ferrara, G. and Singh, R. P. (2007b). Wheat improvement in India: Present status, emerging challenges and future prospects. *Euphytica*, 157, 431–46.

Kumar, J., Schäfer, P., Hückelhoven, R., Langen, G., Baltruschat, H., Stein, E., Nagarajan, S. and Kogel, K. H. (2002). *Bipolaris sorokiniana*, a cereal pathogen of global concern: Cytological and molecular approaches towards better control. *Molecular Plant Pathology*, 3, 185–95.

Kumar, S., Prasad, L. C., Tripathi, R., Singh, A. K., Singh, R., Chand, R. and Joshi, A. K. (2015a). Genetic diversity in the isolates of spot blotch of wheat caused by *Bipolaris sorokiniana* using RAPD markers. *Biotech Today: An International Journal of Biological Sciences*, 5, 63.

Kumar, S., Röder, M. S., Tripathi, S. B., Kumar, S., Chand, R., Joshi, A. K. and Kumar, U. (2015b). Mendelization and fine mapping of a bread wheat spot blotch disease resistance QTL. *Molecular Breeding*, 35, 218.

Kumar, S., Tripathi, S. B. and Kumar, U. (2015c). Dissection of wheat spot blotch disease resistance QTLs into single Mendelian genes. *Indian Journal of Genetics and Plant Breeding*, 75, 434–9.

Kumar, S., Röder, M. S., Singh, R. P., Kumar, S., Chand, R., Joshi, A. K. and Kumar, U. (2016). Mapping of spot blotch disease resistance using NDVI as a substitute to visual observation in wheat (Triticum aestivum L.). *Molecular Breeding*, 36, 95.

Lu, P., Liang, Y., Li, D., Wang, Z., Li, W., Wang, G., Wang, Y., Zhou, S., Wu, Q., Xie, J., Zhang, D., Chen, Y., Li, M., Zhang, Y., Sun, Q., Han, C. and Liu, Z. (2016). Fine genetic mapping of spot blotch resistance gene Sb3 in wheat (Triticum aestivum). *Theoretical and Applied Genetics*, 129, 577–89.

Mahto, B. N., Gurung, S., Nepal, A. and Adhikari, T. B. (2012). Morphological, pathological and genetic variations among isolates of *Cochliobolus sativus* from Nepal. *European Journal of Plant Pathology*, 133, 405–17.

Maraite, H., Zinno, T. D., Longree, H., Daumeria, V. and Duveiller, E. (1998). Fungi associated with foliar blight of wheat in warm areas. In: *Helminthosporium Blights of Wheat: Spot Blotch and Tan Spot*. Duveiller, E., Dubin, H. J., Reeves, J. and McNab, A. (Eds). Mexico, CF: CIMMYT.

McDonald, M. C., Ahren, D., Simpfendorfer, S., Milgate, A. and Solomon, P. S. (2017). The discovery of the virulence gene *ToxA* in the wheat and barley pathogen *Bipolaris sorokiniana*. *Molecular Plant Pathology*.

Meena, N., Mishra, V. K., Baranwal, D. K., Singh, A. K., Rai, V. P., Prasad, R., Arun, B. and Chand, R. (2014). Genetic evaluation of spring wheat (Triticum aestivum L.) recombinant inbred lines for spot blotch (*Bipolaris sorokiniana*) resistance and yield components under natural conditions for South Asia. *Journal of Agricultural Science and Technology*, 16, 1429–40.

Mehta, Y. R. (1998). Constraints on the integrated management of spot blotch of wheat. In: *Helminthosporium Blights of Wheat: Spot Blotch and Tan Spot*. Duveiller, E., Dubin, H. J., Reeves, J. and McNab, A. (Eds). Mexico, CF: CIMMYT.

Mehta, Y. R., Riede, C. R., Campos, L. A. C. and Kohli, M. M. (1992). Integrated management of major wheat diseases in Brazil: An example for the Southern Cone region of Latin America. *Crop Protection*, 11, 517–25.

Murray, G. M. and Brennan, J. P. (2009). The current and potential costs from diseases of wheat in Australia. *Australasian Plant Pathology*, 38, 558–70.

Osman, M., He, X., Benedettelli, S., Ali, S. and Singh, P. K. (2016). Identification of new sources of resistance to fungal leaf and head blight diseases of wheat. *European Journal of Plant Pathology*, 145, 305–20.

Pandey, S. P., Sharma, S., Chand, R., Shahi, P. and Joshi, A. K. (2008). Clonal variability and its relevance in generation of new pathotypes in the spot blotch pathogen, *Bipolaris sorokiniana*. *Current Microbiology*, 56, 33–41.

Poloni, A., Pessi, I. S., Frazzon, A. P. and Van der Sand, S. T. (2009). Morphology, physiology, and virulence of *Bipolaris sorokiniana* isolates. *Current Microbiology*, 59, 267–73.

Regmi, A. P., Ladha, J. K., Pasuquin, E., Pathak, H., Hobbs, P. R., Shrestha, L. L., Gharti, D. B. and Duveiller, E. (2002). The role of potassium in sustaining yields in a long-term rice-wheat experiment in the Indo-Gangetic Plains of Nepal. *Biology and Fertility of Soils*, 36, 240–7.

Reis, E. M., Madeiros, C. and Casa, R. T. (1997). Control of leaf blights of wheat by the elimination of the inoculum source. In: *Proceedings of the International Workshop Helminthosporium Diseases of Wheat: Spot Blotch and Tan Spot, CIMMYT, El Batan, Mexico, 9–14 February 1997*. Duveiller, E., Dubin, H. J., Reeves, J. and McNab, A. (Eds). Mexico, CF: CIMMYT, pp. 327–32.

Ruckstuhl, M. (1998). Population structure and epidemiology of *Bipolaris sorokiniana* in the rice-wheat cropping pattern of Nepal. In: *Helminthosporium Blights of Wheat: Spot Blotch and Tan Spot*. Duveiller, E., Dubin, H. J., Reeves, J. and McNab, A. (Eds). Mexico, CF: CIMMYT, pp. 88–105.

Saari, E. E. (1998). Leaf blight diseases and associated soilborne fungal pathogens of wheat in South and Southeast Asia. In: *Helminthosporium Blights of Wheat: Spot Blotch and Tan Spot*. Duveiller, E., Dubin, H. J., Reeves, J. and McNab, A. (Eds). Mexico, CF: CIMMYT, pp. 37–51.

Sentelhas, P. C., Pedro, M. J. J. and Felicio, J. C. (1995). Effects of different conditions of irrigation and crop density on microclimate and occurrence of spot blotch and powdery mildew. *Bragantia*, 52, 45–52.

Sharma, R. C. and Duveiller, E. (2004). Effect of helminthosporium leaf blight on performance of timely and late-seeded wheat under optimal and stressed levels of soil fertility and moisture. *Field Crops Research*, 89, 205–18.

Sharma, R. C. and Duveiller, E. (2006). Spot blotch continues to cause substantial grain yield reductions under resource-limited farming conditions. *Journal of Phytopathology*, 154, 482–8.

Sharma, R. C., Shrestha, S. M. and Duveiller, E. (2003). Incidence of *Bipolaris sorokiniana* and *Pyrenophora tritici-repentis* on wheat in the lowlands of Nepal. In: *Proceedings of the Fourth International Wheat Tan Spot and Spot Blotch Workshop*. Rasmussen, J. B., Friesen, T. L. and Ali, S. (Eds). Fargo and Bemidji, MN: North Dakota State University, pp. 122–7.

Sharma, S., Duveiller, E., Basnet, R., Karki, C. B. and Sharma, R. C. (2005). Effect of Potash fertilization on Helminthosporium leaf blight severity in wheat, and associated increases in grain yield and kernel weight. *Field Crops Research*, 93, 142–50.

Sharma-Poudyal, D., Duveiller, E. and Sharma, R. C. (2005). Effects of seed treatment and foliar fungicides on Helminthosporium leaf blight and on performance of wheat in warmer growing conditions. *Journal of Phytopathology*, 153, 401–8.

Sharma, R. C., Duveiller, E. and Ortiz-Ferrara, G. (2007). Progress and challenge towards reducing wheat spot blotch threat in the Eastern Gangetic Plains of South Asia: Is climate change already taking its toll? *Field Crops Research*, 103, 109–18.

Sharma-Poudyal, D., Sharma, R. C. and Duveiller, E. (2016). Control of Helminthosporium leaf blight of spring wheat using seed treatments and single foliar spray in Indo-Gangetic Plains of Nepal. *Crop Protection*, 88, 161–6.

Shoemaker, R. A. (1955). Biology, cytology and taxonomy of *Cochliobolus sativus*. *Canadian Journal of Botany*, 33, 562–76.

Singh, R. V., Singh, A. K. and Singh, S. P. (1998). Distribution of pathogens causing foliar blight of wheat in India and Neighbouring countries. In: *Helminthosporium Blights of Wheat: Spot Blotch and Tan Spot*. Duveiller, E., Dubin, H. J., Reeves, J. and McNab, A. (Eds). Mexico, CF: CIMMYT.

Singh, V., Singh, G., Chaudhury, A., Ojha, A., Tyagi, B. S., Chowdhary, A. K. and Sheoran, S. (2016). Phenotyping at hot spots and tagging of QTLs conferring spot blotch resistance in bread wheat. *Molecular Biology Reports*, 43, 1293–303.

Stergiopoulos, I., Collemare, J., Mehrabi, R. and De Wit, P. J. (2013). Phytotoxic secondary metabolites and peptides produced by plant pathogenic Dothideomycete fungi. *FEMS Microbiology Reviews*, 37, 67–93.

Toledo, B. J. and Guzman, A. E. (1998). Importance of Spot Blotch caused by *Bipolaris sorokiniana* in Bolivia. In: *Helminthosporium Blights of Wheat: Spot Blotch and Tan Spot*. Duveiller, E., Dubin, H. J., Reeves, J. and McNab, A. (Eds). Mexico, CF: CIMMYT, pp. 146–9.

Viedma, L. Q. and Kohli, M. M. (1998). Spot Blotch and Tan Spot of wheat in Paraguay. In: *Helminthosporium Blights of Wheat: Spot Blotch and Tan Spot*. Duveiller, E., Dubin, H. J., Reeves, J. and McNab, A. (Eds). Mexico, CF: CIMMYT.

Wall, P. C., Hobbs, P. R., Saunders, D. A., Sayre, K. D. and Tanner, D. G. (1990) Wheat crop management in warmer areas: A review of issues and advances. In: *Wheat for the Nontraditional, Warm Areas*. Saunders, D. A. (Ed.). Mexico, CF, CIMMYT, pp. 225–41.

Webb, C. A. and Fellers, J. P. (2006). Cereal rust fungi genomics and the pursuit of virulence and avirulence factors. *FEMS Microbiology Letters*, 264, 1–7.

Xiao, Z., Sun, L. and Xin, W. (1998) Breeding for foliar blight resistance in Heilongjiang province, China. In: *Helminthosporium Blights of Wheat: Spot Blotch and Tan Spot*. Duveiller, E., Dubin, H. J., Reeves, J. and McNab, A. (Eds). Mexico, CF: CIMMYT, pp. 114–18.

Yadav, B. S. (1981). Behavior of *Cochliobolus sativus* during its infection of barley and wheat leaves. *Australian Journal of Botany*, 29, 71–9.

Zhu, Z., Bonnett, D., Ellis, M., Singh, P., Heslot, N., Dreisigacker, S., Gao, C. and Mujeeb-Kazi, A. (2014). Mapping resistance to spot blotch in a CIMMYT synthetic-derived bread wheat. *Molecular Breeding*, 34, 1215–28.

Chapter 8

Diseases affecting wheat: wheat blast

J. L. N. Maciel, Embrapa Wheat, Brazil

1 Introduction

Pyricularia oryzae Cavara (syn. *Magnaporthe oryzae* B. C. Couch) (Couch and Kohn 2002) is a fungus that causes blast disease in more than 50 species of the Poaceae family (the grass family), including agricultural crops such as rice, wheat, barley, millet and oat (Urashima and Kato 1998; Couch and Kohn 2002; Takabayashi et al. 2002; Murakami et al. 2003; Couch et al. 2005; Prabhu and Filippi 2006; Castroagúdin et al. 2016). In Brazil, in addition to rice, the wheat crop is severely affected by outbreaks of the disease, especially in the states of Paraná, Minas Gerais, Mato Grosso do Sul, Mato Grosso, São Paulo and Goiás (Igarashi et al. 1986; Goulart and Paiva 1992; Anjos et al. 1996; Goulart and Paiva 2000). Wheat blast was first reported in Londrina, Paraná, Brazil, in 1985 (Igarashi et al. 1986). Reports of the rapid spread of the pathogen in other Brazilian states demonstrated the rapid dispersal of the pathogen from its original focus (Picinini and Fernandes 1990; Goulart et al. 1990; Goulart and Paiva 1992; Prabhu et al. 1992; Urashima and Kato 1994). Measures adopted for controlling wheat blast have not been effective, with no resistant wheat cultivars available and low efficiency of fungicides, conditions that are reflected in the losses caused by the disease, which may reach up to 100% (Goulart and Paiva 2000; Torres et al. 2009; Castroagúdin et al. 2015; Maciel 2011).

Although there have been reports of wheat blast occurrence in Argentina (Cabrera and Gutiérrez 2007), until the beginning of 2016 reports of outbreaks of wheat blast were restricted to three countries: Brazil, Bolivia and Paraguay (Duveiller et al. 2010; Maciel 2011). However,

http://dx.doi.org/10.19103/AS.2018.0039.08
Published by Burleigh Dodds Science Publishing Limited, 2019.

in March 2016, a very severe wheat blast epidemic was reported in Bangladesh (Callaway 2016). More recently, in February 2017 there was a report about blast disease in Indian fields near the Bangladeshi border causing great damage to wheat grain production (Bhattacharya and Pal 2017). This situation has generated enormous concern within Bangladesh, and the concerned authorities have adopted extreme measures, such as burning affected wheat fields, in an attempt to eradicate the disease. The extreme severity and rapid spread demonstrated by wheat blast in Brazil since its occurrence in the country, in the mid-1980s, contrast with the condition that may have prevailed in India and Pakistan. Diekmann and Putter (1995) mentioned that in 1922 and 1943 there were reports of the natural occurrence of wheat blast in those two countries, respectively. However, the disease never reached epidemic levels. Based on phylogenetic analyses, Islam et al. (2016) proposed that wheat blast arrived in Bangladesh from wheat grain imported from South America.

2 Symptoms and conditions for wheat blast development

Wheat blast symptoms vary greatly in severity on leaves and heads according to weather conditions and wheat genotype (Urashima et al. 2009; Goulart 2005). The most favourable conditions for wheat blast development are a combination of temperature ranging from 23° to 28°C, excessive rain, long and frequent leaf wetness and poor fungicide efficacy (Goulart 2005). Danelli (2015) determined that the concentration of conidia of *Pyricularia oryzae* in the air in the municipality of Passo Fundo, southern Brazil, is enormous, although there may be some variation in the concentration (conidia/m^3 of air) throughout the year.

Pyricularia oryzae affects both leaves and heads of wheat, but the most significant damage occurs on the heads (Cruz et al. 2015; Igarashi 1990; Urashima et al. 2009). The typical symptoms of wheat blast on the leaves, even though rare before the heading stage, are the presence of elliptical to elongate lesions with white to light brown centres and dark grey to reddish-brown borders. The main problem comes from the infection in the rachis, causing partial or total sterility of the heads. All the spikelets above the infection point in the rachis turn white. In general, symptoms on heads may vary from elliptical lesions with bleached centres on glumes to partial or total spike bleaching, sterility and empty grains (Goulart and Paiva 1992; Urashima et al. 2009). The infection in the rachis interrupts the translocation of photosynthates to upper parts of the spike. Blast-infected grains from highly susceptible cultivars are usually small, wrinkled, deformed and have low-test weight (Goulart 2005).

3 Origin of wheat blast

Several hypotheses were made with regard to the origin of wheat blast in Brazil after its first report in the mid-1980s. The most discussed hypothesis emphasized the relationship between rice and wheat crops, indicating that *Pyricularia oryzae* of wheat came from rice plants (Mehta et al. 1992). However, molecular and virulence assays detected important differences between isolates of *Pyricularia oryzae* coming from rice and wheat (Couch and Kohn et al. 2002; Maciel et al. 2014). New hypotheses emerged based on modern evolutionary theories. In this context, evidence showed that sympatric populations of *Pyricularia oryzae* adapted to rice rarely migrated to wheat, suggesting that the population that infects wheat originated from a population infecting an unknown weed Poaceae, distinct from rice (Ceresini et al. 2011). Considering all these aspects, the most plausible hypothesis for the origin of wheat blast is from a clonal propagation of a well-adapted and epidemic lineage of *Pyricularia oryzae* belonging to populations of other host grasses. The bottleneck effect in the original *Pyricularia oryzae* population and intensification of wheat fields in southern Brazil are also part of this scenario (Ceresini et al. 2011; Maciel et al. 2014). Besides, it is likely that *Pyricularia oryzae* of wheat presents a mixed reproductive system in which sexual reproduction is followed by the dispersal of clones (Maciel et al. 2014), and that invasive plants of the Poaceae family influence the development of wheat blast (Maciel et al. 2013).

4 Reproductive biology of *Magnaporthe oryzae*

Pyricularia oryzae of rice is used as model fungal pathogen species (Talbot 2003). The teleomorph or sexual phase *of Pyricularia oryzae* is called *Magnaporthe oryzae* and, due to its self-incompatibility, it is formed only when there is a crossing between two sexually compatible and fertile individuals. This occurs when the female receptive structure called ascogonium is able to receive the compatible nucleus or nuclei of the male donor by means of conidia or receptor hyphae (Kang et al. 1994; Moreira et al. 2015).

What we know about reproductive biology of *Magnaporthe oryzae* in the literature was generated, mainly, with controlled crosses conducted in laboratories because the sexual phase of the pathogen is rarely found in the field (Saleh et al. 2012). Mating-type genes regulate the recognition of sexually compatible individuals (Turgeon 1998; Moreira et al. 2015). The gene that controls the compatibility in *M. oryzae* has two idiomorphs, *Mat1-1* and *Mat1-2* (Nottéghem and Sulué 1992; Peixoto 2014). The detection of gametic balance in populations of the pathogen is an indirect way of determining whether the reproductive mode of the pathogen in the field is mainly sexual or clonal (Maciel et al. 2014).

5 The new causal agent of wheat blast: *Pyricularia graminis-tritici*

Based on host specificity, mating capacity and genetic similarity, isolates of *Pyricularia oryzae* are characterized as belonging to several pathotypes (Urashima et al. 1993; Kato et al. 2000; Tosa et al. 2004; Tosa and Chuma 2014). The following is the list of the pathotypes of *Pyricularia oryzae* already described and the hosts in which it is pathogenic: Oryza, *Oryza sativa* (rice); Setaria, *Setaria italica* (foxtail millet); Panicum, *Panicum miliaceum* (millet); Eleusine, *Eleusine coracana* (finger millet); Triticum, *Triticum aestivum* (wheat); Oat, *Avena sativa*, (oat); and Lolium, *Lolium perenne* (ryegrass).

Pyricularia oryzae Triticum is the causal agent of wheat blast in South America, including Brazil, and is responsible for blast in barley, rye, triticale and *Brachiaria* (*Urochloa* spp., ex *Brachiaria* spp.) (Lima and Minella 2003; Verzignassi et al. 2012). As the wheat blast arose in an area in southern Brazil close to areas where rice fields were cultivated and occurrence of rice blast on those fields was common, the initial proposition was that *Pyricularia oryzae* Oryza had undergone some type of mutation and evolved virulence on wheat (Igarashi et al. 1986). In southern Brazil, wheat and rice crops are grown in relatively close regions, although for reasons related to agricultural aptitude, the same areas are not used for cultivating the two crops. However, based on evaluations of pathogenicity, reproductive isolation and genetic analysis, Urashima et al. (1993) provided evidence that there were two distinct groups of *Pyricularia oryzae* that cause blast diseases in Brazil, one that infects wheat and another that infects rice. In these evaluations, it was verified that the isolates obtained from wheat plants were able to infect plants from six tribes of the family Poaceae. Urashima et al. (1993) also verified that *Pyricularia oryzae* Triticum isolates were able to recombine sexually, producing perithecia with viable ascospores, when crossed with isolates from *Eleusine coracana*, *Urochloa plantaginea* (ex *Brachiaria plantaginea*) and *Setaria italica*. However, when crosses were carried out between wheat and rice isolates, a complete absence of fertility was obtained, with no perithecia or ascospores.

Specific comparisons between isolates of *Pyricularia oryzae* belonging to the Oryza and Triticum pathotypes have already been made using different approaches. Using the RFLP technique, with the probes MGR563 and MGR586, Farman (2002) determined the existence of a substantial difference between the two pathotypes. According to analyses using 11 microsatellite loci, Maciel et al. (2014) showed a significant difference between *Pyricularia oryzae* Triticum and *Pyricularia oryzae* Oryza (FCT = 0.896, $P \leq 0.001$; FCT, variance between the two populations compared). Besides, Maciel et al. (2014) also determined that none of the 69 isolates of *Pyricularia oryzae* Triticum used in inoculation procedures were able to infect rice plants.

The genus *Pyricularia*, which has approximately 50 species, is quite complex, albeit it is believed to be monophyletic (Hirata et al. 2007; Choi et al. 2013; Klaubauf et al. 2014). Species of this genus have increasingly been subject to reclassifications and/or subdivisions, which have been mainly based on multilocus analyses of housekeeping genes. An important reclassification scheme divided *Pyricularia grisea* into two species, *Pyricularia grisea* and *Pyricularia oryzae* (Couch and Kohn 2002). Castroagúdin et al. (2016) proposed the existence of a new species *Pyricularia graminis-tritici*. According to the authors of the work, this species, together with the pathotype Triticum of *Pyricularia oryzae*, is the causative agent of wheat blast.

6 Alternatives for survival of the causal agent

Pyricularia oryzae is known to survive from one season to another on seeds and live plants or, as saprophytic mycelium, on straw (Ou 1987; Kimati et al. 2005). However, the relative extent and importance of these survival alternatives is unknown. Castroagúdin et al. (2016) indicated that plants of the genus *Urochloa* play an important role in the survival and generation of the pathogen variability. It is critical that we get more information about the characteristics of each one of the alternatives of survival of the causal agents of wheat blast in order to manage and control the disease.

Even though the proposal for the new species *Pyricularia graminis-tritici* is recent, this fungus also needs to be included in the discussion about its alternatives for the survival in the absence of wheat plants. However, this situation should be contextualized. Information on *Pyricularia graminis-tritici* is scarce and is restricted to the publication in which the proposal of this new species was presented (Castroagudin et al. 2016). In this sense, one of the most important factors reported by the authors is that both species *Pyricularia oryzae* and

Pyricularia graminis-tritici are widely distributed in the Brazilian wheat fields. In fact, there is little or no knowledge about the characteristics of *Pyricularia graminis-tritici*, especially in relation to phenotypic aspects, for example, its ability to survive in environments such as straw, seeds or alternative hosts. From now on, it seems most reasonable to imagine that the new studies on wheat blast, especially those conducted in Brazil, should consider the existence of the two species, *Pyricularia oryzae* and *Pyricularia graminis-tritici*, as the causal agent of wheat blast. Thus, the currently available knowledge on the alternatives of survival of the causative agent of wheat blast refers almost exclusively to the species *Pyricularia oryzae*, which represents the main content of this chapter. Currently the main source on *Pyricularia graminis-tritici* is the work of Castroagúdin et al. (2016) who isolated the pathogen in a number of grass species.

6.1 Infective activity in alternative hosts

The large number of plant species on which *Pyricularia oryzae* may survive represents a great difficulty in advancing the knowledge about the role that alternative hosts play in wheat blast outbreaks occurring in South America (Castroagúdin et al. 2016; Couch et al. 2005; Couch and Kohn 2002; Murakami et al. 2003; Prabhu and Filippi 2006; Takabayashi et al. 2002; Urashima and Kato 1998). The situation became even more complex with the new proposal that *Pyricularia graminis-tritici* was also a causal agent of wheat blast (Castroagúdin et al. 2016). This species has been reported as being capable of infecting *Urochloa* spp., *Echinochloa* spp., *Cenchrus echinatus*, *Rhynchelytrum repens*, *Eleusine indica*, *Avena sativa*, *Cynodon dactylon*, among others (Castroagúdin et al. 2016). Directly associated with this aspect is the problem of the lack of consistent information on the virulence of *Pyricularia graminis-tritici* on wheat and, much less, on wheat cultivars. One of the few pieces of information we have about the characteristics of *Pyricularia graminis-tritici* was demonstrated by Maciel et al. (2013) using isolates obtained from *Urochloa* spp. to inoculate leaves and heads of a set of wheat genotypes. They verified that *Pyricularia graminis-tritici* was as virulent as some isolates of *Pyricularia oryzae* Triticum used in the experiment.

6.2 Infection of wheat seeds

The first report of transmission of *Pyricularia oryzae* through seed was in rice seeds in Japan (Kuribayashi 1928). In relation to wheat, most of the studies on the detection of *Pyricularia oryzae* in wheat seeds were carried out in Brazil during the 1990s. The most important study is probably the one conducted by Goulart et al. (1995). In this work, carried out for 6 years (1987–92), 2238 samples of grain and wheat seeds, produced in 11 municipalities in the Brazilian state of Mato Grosso do Sul of several cultivars, were analysed. The presence of *Pyricularia oryzae* was detected in 16.5% of the samples and the average incidence of the fungus in the grains and/or seeds was 3% in 1990 and 0.2% in 1991 and 1992. The efficiency of the transmission of the pathogen by the seeds is high, capable of reaching 28% (Goulart and Paiva 1993).

Reis et al. (1995) determined that *Pyricularia oryzae* may survive for a relatively long period in wheat seeds. They tested seeds of two wheat cultivars, Anahuac and Tapejara, both harvested in the 1987 season, in the state of Paraná. Both samples had an average incidence of the fungal infection of around 45%, and it was possible to verify that *Pyricularia oryzae* survived in the seeds for 19 and 22 months, respectively, with a viability of 60% after 8 months of storage.

The presence of *Pyricularia oryzae* in seeds, besides damaging the plant that will emerge from this infected seed, may introduce the fungus in areas where it was not present. Considering the potential for dissemination of the pathogen through wheat seed, the use of healthy seeds, free from the pathogen, is an important measure in the management of blast. Currently, the seed cleanliness testing for *Pyricularia oryzae* is uncommon in South America, although this practice is becoming more prevalent in international seed trading (International Seed Testing Association 1996).

Detection of pathogenic fungi in cereal seeds may be difficult, especially when infection rates are very low. For a diagnostic test to be adequate, it should be sensitive and specific, avoiding cross-reactions with other fungal species (Konstantinova et al. 2002). The most commonly used seed tests for the detection of *Pyricularia oryzae* have been made on rice seeds using the Blotter test or growth on Petri dishes with culture medium (Neergaard et al. 1970). These procedures usually involve steps such as the isolation and cultivation of pathogens, as well as identification by the morphology of reproductive structures and biochemical tests. The test based on molecular methods, especially using PCR, is a viable alternative that has been shown to be very efficient for the detection of many fungi (Vandemark et al. 2000; Konstantinova et al. 2002).

The use of molecular techniques to detect *Pyricularia oryzae* in complex environments and under low concentrations has been demonstrated by Villari et al. (2017). In this study, a pathogen-specific quantitative-loop-mediated isothermal amplification (*qLAMP*) assay coupled with a spore trap system was developed for the rapid detection and quantification of airborne inoculums of *Pyricularia oryzae* from perennial ryegrass. In this sense, although the application of molecular strategies for the detection and differentiation of *Pyricularia oryzae* and *Pyricularia graminis-tritici* in seeds has not yet been demonstrated, it is likely that this procedure is achievable. The next step is the selection of specific primers and validation of the methodology.

6.3 Saprophytic activity in wheat straw

The great adaptability of the fungus *Pyricularia oryzae* is closely related to its nutritional requirements. The fungus is classified as a hemibiotroph, combining aspects of biotrophy and necrotrophy at different points during infection. *Pyricularia oryzae* can survive on dead plant tissue, but the timespan for viability of the pathogen in these circumstances is unknown. Soil type and tillage methods in addition to climatic conditions would affect viability. Most of the wheat fields in South America are in no-tillage areas, where plant residues remain on the surface of the soil (van Capelle et al. 2012). It is unknown if the slower degradation of wheat straw in no-tillage also means a longer survival period of *Pyricularia oryzae*. A great challenge is to relate the use of no-tillage with the survival of *Pyricularia oryzae* in straw and, more importantly, to establish the importance of this inoculum in the structuring of wheat blast outbreaks.

7 Genetic control

In Brazil, since the first report of the disease in the mid-1980s, the search for resistance to wheat blast has been intense, but the results are erratic and inconsistent (Goulart et al. 1995; Cruz et al. 2010). Wheat cultivars, apparently resistant in some localities, are susceptible in other regions of the country, indicating interaction between cultivar and environment (Urashima et al. 2004). Considering that *Pyricularia oryzae* Triticum and *Pyricularia graminis-tritici* are present in all Brazilian wheat agroecosystems (Duveiller et al. 2010; Maciel et al. 2014; Castroagúdin et al. 2016), it is probable that both incidence and severity of blast are dependent on the fungal variant prevalent in that agroecosystem. Maciel et al. (2014) detected both complete and partial resistance in seven Brazilian wheat cultivars. Based on those results, it was possible to classify the wheat blast isolates used on the experiments in eight and fourteen virulence groups according to response in terms of complete and partial resistance, respectively. The high phenotypic variability of the pathogen will probably result in low durability of the resistance of wheat cultivars.

Selected wheat genotypes, which displayed high resistance to blast in early trials, have been evaluated in more detail, either under controlled conditions or in the field environment. The cultivar BR 18-Terena is widely used in wheat breeding programmes to obtain durable resistance to wheat blast. Results obtained in the field trials, under high incidence of the disease, determined that the bread wheat BR 18-Terena had head blast incidences varying from 8.2% to 24%, rates relatively low for wheat blast (Goulart et al. 1991; Goulart and Paiva 1992; Sousa 2002). Another wheat cultivar, classified as moderately resistant, that has demonstrated a good degree of resistance to wheat blast is BRS 229. In a field assay, conducted in Londrina, Paraná, Brunetta et al. (2006) evaluated BRS 229 for blast resistance and determined a mean score of 2 for severity on its heads, in a scale ranging from 0 to 9, with 9 as the highest intensity of the disease.

The winter wheat cultivar Renan was reported as having low disease rates (Danelli 2015; Tufan et al. 2009). In work conducted under controlled conditions, Danelli (2015) evaluated the reaction of Renan heads to individual inoculations of 144 isolates of *Pyricularia oryzae* obtained from wheat plants. There was considerable variability in the response of the cultivar, but overall the cultivar showed high resistance. In the evaluation of the cellular responses in Renan, the formation of papilla-like structures at the infection sites was observed, and the analysis of the transcription of response signals showed, 24 hours after inoculation, a significant number of transcripts associated with defence mechanisms (Tufan et al. 2009).

In Bolivia, many wheat growers use the cultivars Motacú, Urubo and San Pablo, which are classified as moderately resistant. More recently, cultivars, such as USDA 8 and CMT-50, with higher resistance levels to the disease have been available (Baldelomar et al. 2017).

Wheat cultivars currently available to South American growers are very susceptible to wheat blast with the exception of two or three cultivars, which at best may be classified as moderately resistant. This is thought to be due to the narrow genetic basis of wheat cultivars in Brazil. A solution to this situation is to use wide crosses to the progenitors of wheat (Cruz et al. 2010). It has already been shown that some *T. tauschii* accessions (synonyms *Aegilops squarrosa*, *Aegilops tauschii*) are resistant to *Pyricularia oryzae* (Urashima and Kato 1994). Bockus et al. (2012) evaluated 10 different genotypes of this species, with each one coming from a different Asian country, and showed a great deal of variability in blast responses in heads. The mean severity ranged from 5.1% to 99.9% depending on the genotype. Cruz et al. (2010), studying partial resistance to wheat blast, verified that synthetic wheat genotypes, developed by Moraes-Fernandes et al. (2000) from the crossing of *T. durum* and *T. tauschii*, showed lower blast severity than cultivars of wheat traditionally used in Brazil. This lower severity was observed in both leaf and head reactions.

7.1 Genetic resistance and molecular markers

The presence of *Avr1-CO39* avirulence gene in *Pyricularia oryzae* Triticum and *Pyricularia graminis-tritici* prevents their infection on rice (Maciel et al. 2014) as this plant species carries the *Pi-CO39 (t)* resistance gene. The *Pi-CO39 (t)* gene is rare or absent in plant species belonging to the genera *Avena*, *Urochloa*, *Eleusine*, *Eragrostis*, *Lolium*, *Panicum*, *Setaria*, *Triticum* and also some species of the genus *Oryza* (Tosa et al. 2005). The corresponding *Avr1-CO39* avirulence gene occurs generally in pathogen populations that do not infect rice, which include isolates of *Pyricularia* spp. of most of the above-mentioned hosts (Couch et al. 2005). These data indicate that the presence of *Pi-CO39* in wheat genotypes, which is still unknown, appears as highly relevant scientific information to be obtained in the future, considering the possibility that this gene may be transferred to wheat genotypes using, for example, transgenic strategy since *Pi-CO39* is cloned (Chauhan et al. 2002) .

Approximately 100 rice blast-resistant genes have already been identified and mapped (Sharma et al. 2012; Wang et al. 2014). Of the genes mapped, 22 have already been cloned (Liu et al. 2010; Sharma et al. 2012; Singh et al. 2015; Wang et al. 2014; Xiao et al. 2015). Because it is a relatively recent disease, many aspects related to the molecular interaction between *Pyricularia oryzae* and wheat remain unknown. Eight resistance genes to wheat blast have already been identified, known as *Rmg*1 to *Rmg*8 (Anh et al. 2015; Nga et al. 2009; Tagle et al. 2015; Takabayashi et al. 2002; Vy et al. 2014; Zhan et al. 2008). In addition, transcriptome studies have also been conducted (Tufan et al. 2009; Ha et al. 2016). In a study published in 2016, evaluating differential expression in wheat in response to blast, the authors recommended pyramiding a large number of resistance genes to wheat blast (Ha et al. 2016).

Another study published in 2016 reported the association between the presence of the 2NS chromosomal translocation segment of *Aegilops ventricosa* and resistance to wheat blast (Cruz et al. 2016). The resistance of *Aegilops ventricosa* to wheat diseases including rust was first mentioned in the 1960s (Maia 1967). A long chromosomal segment of 25 to 38 cM located on the 2NS chromosome was transferred to chromosome 2AS of *T. aestivum* (McIntosh et al. 1995). The segment 2NS/AS carries the following genes: *Rkn3*, for gall nematode resistance (*Meloidogyne* spp.) (Williamson et al. 2013); *Cre5*, for resistance to the French pathotype Ha12 of the cereal cyst nematode (*Heterodera avenae*) (Jahier et al. 2001); and the genes *Lr37*, *Sr38* and *Yr17*, which act against *Puccinia* spp., causative agents of leaf, stem and stripe rusts, respectively (Bariana and McIntosh 1993; McIntosh et al. 1995). The 2NS/AS fragment was initially transferred to the VPM1 wheat cultivar and then to other cultivars such as Madsen and Thatcher (Helguera et al. 2003). Helguera et al. (2003) developed molecular markers to confirm the presence of the 2NS/AS segment in wheat genotypes, and, based on them, Cruz et al. (2016) reported the association between the presence of this segment and the reduction of blast severity on heads of American cultivars of spring and winter wheat. According to Cruz et al. (2016), wheat genotypes with the 2NS/AS segment, when inoculated with wheat *Pyricularia oryzae* isolates, showed 64% to 80.5% less disease on the heads than those that did not have the segment. Nonetheless it is clear that other genes confer resistance in wheat against *Pyricularia oryzae* infection.

8 Chemical control

Chemical control is a major aspect part of wheat blast management and it is used in two situations. The first one is the seed treatment to prevent infection of seedlings after germination. Tests show that fungicidal treatment of wheat seeds can satisfactorily control seedling infections caused by wheat blast. In Brazil, there are five commercial fungicides officially registered for wheat blast seed treatment (Anon. 2017). The active ingredients are iprodione, fluazinam, difeconazol, carboxin and thiram. The list of these products has not been subject to major changes since when the disease appeared and began to spread in the 1980s. Among them, the commercial mixture carboxin + thiram has been the most cost-effective product to control wheat blast via seed treatment (Goulart and Paiva 1991).

Leaves and heads can also be protected by applying fungicide one or more times, but results have been disappointing (Maciel 2011; Pagani et al. 2014; Rocha et al. 2014). This situation may be due to insensitivity of the pathogen to fungicides, use of too low a dose, poor coverage of the fungicide on the heads and inappropriate timing of the spray (Maciel 2011; Panisson et al. 2014; Castroagúdin et al. 2015; Oliveira et al. 2015). To remedy this situation, an important initiative started in Brazil in 2010, when a collaborative action was established between some Brazilian universities and wheat research companies (public and private). The 'Cooperative Assays Network' (CAN) conducts field trials using a standard protocol to evaluate the efficiency of fungicides to control wheat blast on heads. In 5 years, 39 field trials were held in the main wheat-producing regions of the country with a history of wheat blast. It was shown that there is no economic return for any fungicide sprayings if the incidence of head wheat blast in the field is more than 75%. Economic return was only guaranteed when the incidence of blast on the heads reached a maximum of 25%. When the incidence of the disease ranged from 25% to 75%, there was variation in relation to the economic viability, which was quite dependent on the yield of grains promoted by aerial application of fungicides. Another important result obtained in the scope of the CAN is that the products that contain mancozeb performed consistently well.

A question that still needs to be addressed is whether leaf infection of either *Pyricularia oryzae* Triticum or *Pyricularia graminis-tritici* leads directly to head blast. We think any analysis or speculations on the subject must take into account how South American growers, especially the Brazilian ones, cultivate their wheat crop. The occurrence of leaf blast is a rare event, and one reason that may contribute much to the difficulty in finding leaves with blast lesions is the increasing of the number of fungicide sprayings at early stages of development of the plants. In addition to mancozeb, the active ingredients of fungicide action most used in Brazil for the control of wheat blast belong to triazole or strobilurin chemical groups, which are registered and allowed to be sold in the country. Their performance to control the development of the causal agent of wheat blast in tests conducted *in vitro* has been considered satisfactory. Wheat growers use these products even before the heading stage with the aim of controlling foliar diseases such as mildew, leaf rust, tan spot, among others. In addition, Silva et al. (2015) showed that chemical control of blast on the wheat leaf is much more feasible than on the heads. In fact, the problem is that the applied fungicide does not fully cover the rachis adequately. The question of fungicide application technology is one of the factors that most needs to be investigated in relation to the chemical control of the blast. It is important to relate efficiency of the application with environmental aspects that occur during the crop season such as temperature and precipitation.

Evolved resistance to fungicides is a major concern. *In vitro* tests determined the sensitivity to the triazole tebuconazole of *Pyricularia oryzae* and *Pyricularia graminis-tritici* isolates obtained from the Federal District and from six Brazilian wheat-producing states.

Boaretto (2016) determined that 35% of the tested isolates had reduced sensitivity. In the case of trifloxystrobin (strobilurin chemical group), total insensitivity is widespread in the Brazilian population of *Pyricularia oryzae* Triticum and *Pyricularia graminis-tritici*, with 80% of the isolates demonstrating some kind of insensitivity to this fungicide (Castroagúdin et al. 2015). Examination of the cytochrome b (*Cyt-b*) gene, which is the target site, confirmed the results obtained *in vitro*. Of 234 tested isolates of *Pyricularia oryzae* Triticum and *Pyricularia graminis-tritici*, obtained from the major Brazilian wheat-producing regions in 2005 and 2012, 82% of them had the resistant G143A *Cyt-b* gene mutation. Castroagúdin et al. (2015) verified that the frequency of mutation had increased, from 36% in 2005 to 90% in 2012. The results indicate that, in Brazil, monitoring of populations of *Pyricularia oryzae* Triticum and *Pyricularia graminis-tritici* for sensitivity losses to triazole fungicides should be a priority and that alternatives to strobilurin fungicides are urgently needed.

9 Conclusion and future trends

The occurrence of wheat blast epidemics in Asia will bring a new perspective to this disease on the world stage. It is expected that there will be an increase in the number of people interested in developing research actions aimed at managing or controlling this disease, which has already demonstrated its harmful effects on the South American continent.

Regardless of who will do what and where it will be carried out, specific studies on wheat blast should be considered as priorities. Among these actions, we highlight the following that should be intensified: molecular and virulence characterization of the causative agents of wheat blast; molecular studies of the interaction of the pathogens *Pyricularia oryzae* Triticum and *Pyricularia graminis-tritici* in relation to the host *T. aestivum*; the search for new sources and resistance genes, considering the related species of *T. aestivum*, especially the species *Aegilops squarrosa* with a great potential to be exploited; development and evaluation of new molecules or alternative substances of fungicidal action to control wheat blast; evaluation of new application technologies that allow the fungicide to have a good covering action on the head, especially of the rachis; and better dimensioning of the role of alternative hosts, seeds and wheat straw as sources of inoculum for wheat blast.

10 References

Anh, V. L., Anh, N. T., Tagle, A. G., Vy, T. T. P., Inoue, Y., Takumi, S., Chuma, I. and Tosa, Y. (2015). Rmg8, a new gene for resistance to *Triticum* isolates of *Pyricularia oryzae* in hexaploid wheat. *Phytopathology* 105, 1568–72.

Anjos, J. R. N., Silva, D. B., Charcar, M. J. D. and Rodrigues, G. C. (1996). Ocorrência de brusone (*Pyricularia grisea*) em trigo e centeio na região dos cerrados do Brasil Central. *Pesqui. Agropecu. Bras.* 31, 79–82.

Anon. (2017). Ministério da Agricultura, Pecuária e Abastecimento. Agrofit, Sistema de Agrotóxicos Fitossanitários. http://agrofit.agricultura.gov.br/agrofit_cons/principal_agrofit_cons/ (accessed 20 August 2017).

Baldelomar, D., Cardenas, S. and Quispe, K. (2017). Caracterización de genotipos de trigo (*Triticum aestivum*) con resistencia a Piricularia durante el Verano 2014/2015 en Quirusillas, Santa Cruz, Bolivia. *Info INIAF* 1, 17–21.

Published by Burleigh Dodds Science Publishing Limited, 2019.

Bariana, H. S. and McIntosh, R. A. (1993). Cytogenetic studies in wheat. XV. Location of rust resistance genes in VPM1 and their genetic linkage with other disease resistance genes in chromosome 2A. *Genome* 36, 476–82.

Bhattacharya, R. and Pal, S. (2017). Deadly wheat blast symptoms enters India through the Bangladesh border. *Hindustan Times*, Updated: 05 March 2017, 13:52 IST. http://www.hindustantimes.com/kolkata/deadly-wheat-blast-symptoms-enters-india-through-the-bangladesh-border-bengal-govt-burning-crops-on-war-footing/story-3zoWQ0H7sdMU4HxQyzWUsN.html.

Boaretto, C. (2016). *Magnaporthe oryzae*: Genetic variability and chemical control. Master's Thesis. Brazil: University of Passo Fundo. Retrieved from: https://secure.upf.br/pdf/2016CristinaBoaretto.pdf.

Bockus, W. W., Cruz, C. D., Kalia, B., Gill, B. S., Stack, J. P., Pedley, K. F., Peterson, G. L. and Valent, B. (2012). Reaction of selected accessions of *Aegilops tauschii* to wheat blast, 2011. *Plant Dis. Manage. Rep.* 6, CF005.

Brunetta, D., Bassoi, M. C., Dotto, S. R., Scheeren, P. L., Miranda, M. Z. D., Tavares, L. C. V. and Miranda, L. C. (2006). Novas cultivares: Características e desempenho agronômico da cultivar de trigo BRS 229 no Paraná. *Pesqui. Agropecu. Bras.* 41, 889–992.

Cabrera, M. G. and Gutiérrez, S. A. (2007). Primer registro de *Pyricularia grisea* en cultivos de trigo del noreste de Argentina. In: *Jornada de actualización en enfermedades de trigo*. 1, *Resúmenes*... Buenos Aires: La vallol., 1 CD-ROM.

Callaway, E. (2016). Devastating wheat fungus appears in Asia for first time. *Nature* 532, 421–2.

Castroagúdin, V. L., Ceresini, P. C., Oliveira, S. C., Reges, J. T., Maciel, J. L. N., Bonato, A. L., Dorigan, A. F. and McDonald, B. A. (2015). Resistance to QoI fungicides is widespread in Brazilian populations of the wheat blast pathogen *Magnaporthe oryzae*. *Phytopathology* 105, 284–94.

Castroagúdin, V. L., Moreira, S. I., Pereira, D. A. S., Moreira, S. S., Brunner, P. C., Maciel, J. L. N., Crous, P., McDonald, B. A. and Ceresini, P. C. (2016). *Pyricularia graminis-tritici*, a new Pyricularia species causing wheat blast. *Persoonia* 37, 199–216.

Ceresini, P. C., Maciel, J. L. N., Kohn, L., Levy, M. and McDonald, B. A. (2011). A 'de novo' origin for the wheat – adapted populations of *Magnaporthe oryzae* in Southern Brazil and levels of gene flow 20 years after the first epidemics. *Phytopathology* 101(Suppl.), S221 (Abstract).

Chauhan, R. S., Farman, M. L., Zhang, H.-B. and Leong, A. S. (2002). Genetic and physical mapping of a rice blast resistance locus, *Pi-CO39(t)*, that corresponds to the avirulence gene *AVR1-CO39* of *Magnaporthe grisea*. *Mol. Genet. Genomics* 267, 603–12.

Choi, J., Park, S.-Y., Kim, B.-R., Roh, J.-H., Oh, I.-S., Han, S.-S. and Lee, Y.-H. (2013). Comparative Analysis of pathogenicity and phylogenetic relationship in *Magnaporthe grisea* species complex. *PLoS ONE* 8(2), e57196. http://doi.org/10.1371/journal.pone.0057196

Couch, B. C. and Kohn, L. M. (2002). A multilocus gene genealogy concordant with host preference indicates segregation of a new species, *Magnaporthe oryzae*, from *M. grisea*. *Mycologia* 94, 683–93.

Couch, B. C., Fudal, I., Lebrun, M.-H., Tharreau, D., Valent, B., van Kim, P. and Kohn, L. M. (2005). Origins of host-specific populations of the blast pathogen *Magnaporthe oryzae* in crop domestication with subsequent expansion of pandemic clones on rice and weeds of rice. *Genetics* 170, 613–30.

Cruz, M. F. A., Prestes, A. M., Maciel, J. L. N. and Scheeren, P. L. (2010). Resistência parcial à brusone de genótipos de trigo comum e sintético nos estádios de planta jovem e de planta adulta. *Trop. Plant Pathol.* 35, 24–31.

Cruz, C. D., Kiyuna, J., Bockus, W. W., Todd, T. C., Stack, J. P. and Valent, B. (2015). *Magnaporthe oryzae* conidia on basal wheat leaves as a potential source of wheat blast inoculum. *Plant Pathol.* 64, 1491–8.

Cruz, C. D., Peterson, G. L., Bockus, W. W., Kankanala, P., Dubcovsky, J., Jordan, K. W., Akhunov, E., Chumley, F., Baldelomar, F. D. and Valent, B. (2016). The 2NS translocation from *Aegilops ventricosa* confers resistance to the Triticum pathotype of *Magnaporthe oryzae*. *Crop Sci.* 56, 990–1000.

Danelli, A. L. D. (2015). *Pyricularia oryzae*: Virulence of isolates, conidia density in the air and nitrogen effects on wheat susceptibility. Doctoral dissertation. Brazil: University of Passo Fundo. Retrieved from: https://secure.upf.br/pdf/2015AndersonDaneli.pdf.

Diekmann, M. and Putter, C. A. J. (1995). Small grain temperate cereals. In: *FAO/IPGRI Technical Guidelines for the Safe Movement of Germplasm (FAO/IPGRI), no. 14 / FAO, Rome (Italy)*. Rome, Italy: Plant Production and Protection Div.; International Plant Genetic Resources Inst., 67pp.

Duveiller, E., Hodson, D. and Van Tiedmann, A. (2010). Wheat blast caused by *Magnaporthe grisea*: A reality and new challenge for wheat research. In: *International Wheat Conference*, 8. 2010, St. Petersburg. *Abstracts* ... St. Petersburg: Vavilov Research Institute of Plant Industry, pp. 247–8.

Farman, M. L. (2002). *Pyricularia grisea* isolates causing gray leaf spot on perennial ryegrass (*Lolium perenne*) in the United States: Relationship to *P. grisea* isolates from other host plants. *Phytopathology* 92, 245–54.

Goulart, A. C. P. (2005). Perdas em trigo causada pela brusone. In: *Workshop de epidemiologia de doenças de plantas. Anais* Viçosa, Brasil: Universidade Federal de Viçosa, pp. 123–30.

Goulart, A. C. P. and Paiva, F. A. (1991). Controle de *Pyricularia oryzae* e *Helminthosporium sativum* pelo tratamento de sementes de trigo com fungicidas. *Pesq. Agropec. Bras.* 26, 1983–8.

Goulart, A. C. P. and Paiva, F. A. (1992). Incidência da brusone (*Pyricularia oryzae*) em diferentes cultivares de trigo (*Triticum aestivum*) em condições de campo. *Fitopatol. Bras.* 17, 321–5.

Goulart, A. C. P. and Paiva, F. A. (1993). Eficiência do tratamento químico de sementes de trigo no controle de *Helminthosporium sativum* e *Pyricularia oryzae*. *Summa Phytopathol.* 19, 199–202.

Goulart, A. C. P. and Paiva, F. A. (2000). Perdas no rendimento de grãos de trigo causada por *Pyricularia grisea*, nos anos de 1991 e 1992, no Mato Grosso do Sul. *Summa Phytopathol.* 26, 279–82.

Goulart, A. C. P., Paiva, F. A. and Mesquita, A. N. (1990). Ocorrência da brusone (*Pyricularia orysae*) do trigo (*Triticum aestivum*) em Mato Grosso do Sul. *Fitopatol. Bras.* 15, 112–14.

Goulart, A. C. P., Paiva, F. A. and Colman, O. P. (1991). Reação de cultivares de trigo (*Triticum aestivum* L.) à brusone (*Pyricularia oryzae* Cav) em condições de campo. In: *Reunião da Comissão Centro-Sul Brasileira de Pesquisa de Trigo*, 7, 1991, Curitiba. *Anais*... Dourados: Embrapa-Uepae, Dourados, pp. 118–20.

Goulart, A. C. P., Paiva, F. A. and Andrade, P. J. M. (1995). Qualidade sanitária de sementes de trigo produzidas no Mato Grosso do Sul, safras 1987 a 1992. *Summa Phytopathol.* 21, 235–8.

Ha, X., Koopmann, B. and Von Tiedemann, A. (2016). Wheat blast and fusarium head blight display contrasting interaction patterns on ears of wheat genotypes differing in resistance. *Phytopathology* 106, 270–81.

Helguera, M., Khan, I. A., Kolmer, J., Lijavetzki, D., Zhong-Qi, L. and Dubcovsky, J. (2003). PCR assays for the *Lr37-Yr17-Sr38* cluster of rust resistance genes and their use to develop isogenic hard red spring wheat lines. *Crop Sci.* 43, 1839–47.

Hirata, K., Kusaba, M., Chuma, I., Osue, J., Nakayashiki, H., Mayama, S. and Tosa, Y. (2007). Speciation in Pyricularia inferred from multilocus phylogenetic analysis. *Mycol. Res.* 111, 799–808

Igarashi, S. (1990). Update on wheat blast (*Pyricularia oryzae*) in Brazil. In: Saunders, D. (Ed.), *Proceedings of the International Conference – Wheat for the Nontraditional Warm Areas*. Mexico DF, Mexico: CIMMYT, pp. 480–3.

Igarashi, S., Utiamada, C. M., Igarashi, L. C., Kazuma, A. H. and Lopes, R. S. (1986). *Pyricularia* sp. em trigo. I. Ocorrência de *Pyricularia* sp. no estado do Paraná. In: *Reunião Nacional de Pesquisa de Trigo*, 14, 1986, Londrina. *Resumos* ... Londrina: IAPAR, p. 57.

International Seed Testing Association. (1996). International rules for testing. *Proc. Int. Seed Testing Assoc.* 31, 1–152.

Islam, T., Croll, D., Gladieux, P., Soanes, D., Persoons, A., Bhattacharjee, P., Hossain, S., Gupta, D., Rahman, Md. M., Mahboob, M. G., Cook, N., Salam, M., Bueno Sancho, V., Maciel, J. L. N., Nhani, A., Castroagúdin, V., Reges, J., Ceresini, P., Ravel, S., Kellner, R., Fournier, E., Tharreau, D., Lebrun, M. H., McDonald, B., Stitt, T., Swan, D., Talbot, N., Saunders, D., Win, J. and Kamoun, S. (2016). Emergence of wheat blast in Bangladesh was caused by a South American lineage of *Magnaporthe oryzae*. *BMC Biology* 14, 1–11.

Jahier, J., Abelard, P. A. M., Tanguy, F., Dedryver, F., Rivoal, R., Khatkar, S. and Bariana, H. S. (2001). The *Aegilops ventricosa* segment on chromosome 2AS of the wheat cultivar 'VPM1' carries the cereal cyst nematode resistance gene *Cre5*. *Plant Breed.*, 120, 125–8.

Kang, S., Chumley, F. G. and Valent, B. (1994). Isolation of the mating-type genes of the phytopathogenic fungus *Magnaporthe grisea* using genomic subtraction. *Genetics* 138, 289–96.

Kato, H., Yamamoto, M., Yamaguchi-Ozaki, T., Kadouchi, H., Iwamoto, Y., Nakayashiki, H., Tosa, Y., Mayama, S. and Mori, N. (2000). Pathogenicity, mating ability and DNA restriction fragment length polymorphisms of *Pyricularia* populations isolated from Gramineae, Bambusideae and Zingiberaceae plants. *J. Gen. Plant Pathol.* 66, 30–47.

Kimati, H., Amorin, L., Bergamim Filho, A., Camargo, L. E. A. and Rezende, J. A. M. (2005). *Manual de fitopatologia: Doenças das plantas cultivadas.* 3 ed., v. 2. Editora Agronômica Ceres, São Paulo, SP, Brasil, 663pp.

Klaubauf, S., Tharreau, D., Fournier, E., Groenewald, J. Z., Crous, P. W., Vries, R. P. de and Lebrun, M.-H. (2014). Resolving the polyphyletic nature of *Pyricularia* (*Pyriculariaceae*). *Stud. Mycol.* 79, 85–120.

Konstantinova, P., Bonants, P. J. M., Pelzer, V. G., Zouwen, P. V. D. and Bulk, R. V. D. (2002). Development of specific primers for detection and identification of *Alternaria* spp. in carrot material by PCR and comparison with blotter and plating assay. *Mycol. Res.* 106, 23–33.

Kuribayashi, K. (1928). Studies on overwintering, primary infection and control of rice blast fungus, *Piricularia oryzae*. *Ann. Phytopathol. Soc. Japan* 2, 99–117.

Lima, M. I. P. and Minella, E. (2003). Occurrence of head blast in barley. *Fitopatol. Bras.* 28, 207.

Liu, J., Wang, X., Mitchell, T., Hu, Y., Liu, X., Dai, L. and Wang, G. L. (2010). Recent progress and understanding of the molecular mechanisms of the rice–*Magnaporthe oryzae* interaction. *Mol. Plant pathol.* 11, 419–27.

Maciel, J. L. N. (2011). *Magnaporthe oryzae*, the blast pathogen: Current status and options for its control. *CABI Rev.* 6, 1–28.

Maciel, J. L. N., Danelli, A. L. D., Boaretto, C. and Forcelini, C. A. (2013). Virulência de isolados de *Magnaporthe oryzae* do trigo e de Poáceas invasoras. In: *VII Reunião da Comissão Brasileira de Pesquisa de Trigo e Triticale*, 2013, Londrina. *Anais* Londrina: Fundação Meridional/Iapar.

Maciel, J. L. N., Ceresini, P. C., Castroagúdin, V. L., Zala, M., Kema, G. H. J. and McDonald, B. A. (2014). Population structure and pathotype diversity of the wheat blast pathogen *Magnaporthe oryzae* 25 years after its emergence in Brazil. *Phytopathology* 104, 95–107.

Maia, N. (1967). Obtention des bles tendres resistants au pietin-verse par croisements interspecifiques bles X Aegilops. *Comptes Rendus de l'Academie d'Agriculture de France* 53, 149–54.

McIntosh, R. A., Wellings, C. R. and Park, R. F. (1995). *Wheat Rusts: An Atlas of Resistance Genes.* Australia: CSIRO, 200pp.

Mehta, Y. R., Riede, C. R., Campos, L. A. C. and Kohli, M. M. (1992). Integrated management of major wheat diseases in Brazil: An example for the Southern Cone region of Latin America. *Crop Prot.* 11, 517–24.

Moraes-Fernandes M. I. B., Zanatta A. C. A., Prestes A. M., Caetano V. R., Barcellos A. L., Angra D. C. and Pandolfi V. (2000). Cytogenetics and immature embryo culture at Embrapa Trigo breeding program: Transfer of disease resistance from related species by artificial resynthesis of hexaploid wheat (*Triticum aestivum* L. em. Thell). *Genet. Mol. Biol.* 23, 1051–62.

Moreira, S. I., Ceresini, P. C. and Alves, E. (2015). Reprodução sexuada em *Pyricularia oryzae*. *Summa Phytopathol.* 41, 175–82.

Murakami, J., Tomita, R., Kataoka, T., Nakayashiki, H., Tosa, Y. and Mayama, S. (2003). Analysis of host species specificity of *Magnaporthe grisea* toward foxtail millet using a genetic cross between isolates from wheat and foxtail millet. *Phytopathology* 93, 42–5.

Neergaard, P., Lambat, A. K. and Mathur, S. B. (1970). Seed health testing of rice III: Testing procedure for detection of *Pyricularia oryzae* Cav. *Proc. Intl. Seed Testing Assoc.* 35, 165–8.

Nga, N. T. T., Hau, V. T. B. and Tosa, Y. (2009). Identification of genes for resistance to a *Digitaria* isolate of *Magnaporthe grisea* in common wheat cultivars. *Genome* 52, 801–9.

Nottéguem, J.-L. and Sulué, D. (1992). Distribution of mating type alleles in Magnaporthe grisea populations pathogenic on rice. *Phytopathology* 82, 421–4.

Oliveira, S. C., Castroagudín, V. L., Maciel, J. L. N., Pereira, D. A. S. and Ceresini, P. C. 2015. Resistência cruzada aos fungicidas IQo azoxistrobina e piraclostrobina no patógeno da brusone do trigo *Pyricularia oryzae* no Brasil. *Summa Phytopathol.* 41, 298–304.

Ou, S. H. (1987). *Rice Diseases.* Kew: Commonwealth Mycological Institute.

Pagani, A. P. S., Dianese, A. C. and Café-Filho, A. C. (2014). Management of wheat blast with synthetic fungicides, partial resistance and silicate and phosphite minerals. *Phytoparasitica* 42, 609–17.

Panisson, E., Boller, W. and Reis, E. M. (2014). Avaliação da deposição de calda em anteras de trigo, para o estudo do controle químico de giberela (*Gibberella zeae*). *Eng. Agric.* 24, 111–20.

Peixoto, L. F. (2014). Mating type, sexuality, fertility and *AVR1*-CO39 identification of *Magnaporthe oryzae* populations collected from rice fields in Brazil. Master's Thesis. Goiânia, Brazil: Goiás Federal University. Retrieved from: http://docplayer.com.br/14721409-Lorena-ferreira-peixoto.html.

Picinini, E. C. and Fernandes, J. M. C. (1990). Ocorrência da brusone (*Pyricularia oryzae*) em lavouras comerciais de trigo (*Triticum aestivum*) no estado do Rio Grande do Sul. *Fitopatol. Bras.* 15, 83–4.

Prabhu, A. S. and Filippi, M. C. C. (2006). *Brusone em arroz: Controle genético, progresso e perspectivas.* Santo Antônio de Goiás: Embrapa Arroz e Feijão, 387pp.

Prabhu, A. S., Filippi, M. C. C. and Castro, N. (1992). Pathogenic variation among isolates of *Pyricularia grisea* affecting rice, wheat and grass in Brazil. *Trop. Pest Manage.* 38, 367–71.

Reis, E. M., Blum, M. C. and Forcelini, C. A. (1995). Sobrevivência de *Pyricularia oryzae*, associado a sementes de trigo. *Summa Phytopathol.* 21, 43–4.

Rocha, J., Pimentel, A., Ribeiro, G. and de Souza, M. (2014). Eficiência de fungicidas no controle da brusone em trigo. *Summa Phytopathol.* 40, 347–52

Saleh, D., XU, P., Shen, Y., Li, C., Adreit, H., Millazo, J. L., Ravigné, V., Bazin, E., Nottéguem, J.-L., Fournier, E. and Tharreau, D. (2012). Sex at the origin: An Asian population of the rice blast fungus *Magnaporthe oryzae* reproduces sexually. *Mol. Ecol.*, 21, 1330–44.

Sharma, T. R., Rai, A. K., Gupta, S. K., Vijayan, J., Devanna, B. N. and Ray, S. (2012). Rice blast management through host-plant resistance: Retrospect and prospects. *Agr. Res.* 1, 37–52.

Silva, E. V. L., Sussel, A. A. B., Perozzo, J. and Santos, F. E. (2015). Cobertura da ráquis do trigo com corante marcador em função de diferentes pontas de pulverização e adjuvantes. In: *IX Reunião da Comissão Brasileira de Pesquisa de Trigo e Triticale*, 2015, Passo Fundo, RS. *Anais...* Passo Fundo: Embrapa Trigo.

Singh, A. K., Singh, P. K., Arya, M., Singh, N. K. and Singh, U. S. (2015). Molecular screening of blast resistance genes in rice using SSR markers. *Plant Pathol. J.* 31, 12–24.

Sousa, P. G. (2002). Novas cultivares: BR 18-Terena, cultivar de trigo para o Brasil. *Pesqui. Agropecu. Bras.* 37, 1039–43.

Tagle, A. G., Chuma, I. and Tosa, Y. (2015). *Rmg7*, a new gene for resistance to *Triticum* isolates of *Pyricularia oryzae* identified in tetraploid wheat. *Phytopathology* 105, 495–9.

Takabayashi, N., Tosa, Y., Oh, H. S. and Mayama, S. (2002). A gene-for-gene relationship underlying the species-specific parasitism of Avena/Triticum isolates of *Magnaporthe grisea* on wheat cultivars. *Phytopathology* 92, 1182–8.

Talbot, N. On the trail of a cereal killer: Exploring the biology of *Magnaporthe grisea*. (2003). *Annu. Rev. Microbiol.* 57, 177–202.

Torres, G. A. M., Santana, F. M., Fernandes, J. M. C. and Só e Silva, M. (2009). *Doenças da espiga causam perda de rendimento em trigo nos estados do Paraná, São Paulo e Mato Grosso do Sul, em 2009.* Passo Fundo: Embrapa Trigo – CNPT, p. 10 (EMBRAPA - CNPT Comunicado Técnico online, 255).

Tosa, Y. and Chuma, I. (2014). Classification and parasitic specialization of blast fungi. *J. Gen. Plant Pathol.* 80, 202–9.

Tosa, Y., Hirata, K., Tamba, H., Nakagawa, S., Chuma, I., Isobe, C., Osue, J., Urashima, A. S., Don, L. D., Kusaba, M., Nakayashiki, H., Tanaka, A., Tani, T., Mori, N. and Mayama, S. (2004). Genetic constitution and pathogenicity of Lolium isolates of *Magnaporthe oryzae* in comparison with host species-specific pathotypes of the blast fungus. *Phytopathology* 94, 454–62.

Tosa, Y., Osue, J., Eto, Y., Oh, H.-S., Nakayashiki, H., Mayama, S. and Leong, S. A. (2005). Evolution of an avirulence gene, *AVR1-CO39*, concomitant with the evolution and differentiation of *Magnaporthe oryzae*. *Mol. Plant Microbe Interact.* 18, 1148–60.

Tufan, H. A., McGrann, G. R., Magusin, A., Morel, J. B., Miché, L. and Boyd, L. A. (2009). Wheat blast: Histopathology and transcriptome reprogramming in response to adapted and nonadapted *Magnaporthe* isolates. *New Phytol.* 184, 473–84.

Turgeon, B. G. (1998). Application of mating type gene technology to problems in fungal biology. *Annu. Rev. of Phytopathol.* 36, 115–37.

Urashima, A. S. and Kato, H. (1994). Varietal resistance and chemical control of wheat blast fungus. *Summa Phytopathol.* 20, 107–12.

Urashima, A. S. and Kato, H. (1998). Pathogenic relationship between isolates of *Pyricularia grisea* of wheat and other hosts at different host developmental stages. *Fitopatol. Bras.* 23, 30–5.

Urashima, A. S., Igarashi, S. and Kato, H. (1993). Host range, mating type, and fertility of *Pyricularia grisea* from wheat in Brazil. *Plant Dis.* 77, 1211–16.

Urashima, A. S., Lavorenti, N. A., Goulart, A. C. P. and Mehta, Y. R. (2004). Resistance spectra of wheat cultivars and virulence diversity of *Magnaporthe grisea* isolates in Brazil. *Fitopatol. Bras.* 29, 511–8.

Urashima, A. S., Grosso, C. R. F., Stabili, A., Freitas, E. G., Silva, C. P., Netto, D. C. S., Franco, I. and Merola Bottan, J. H. (2009). Effect of *Magnaporthe grisea* on seed germination, yield and quality of wheat. In: Wang, G. L. and Valent, B. (Eds), *Advances in Genetics, Genomics and Control of Rice Blast Disease*. Berlin: Springer, pp. 267–77.

Van Capelle, C., Schrader, S. and Brunotte, J. (2012). Tillage-induced changes in the functional diversity of soil biota – A review with a focus on German data. *Eur. J. Soil Biol.* 50, 165–81.

Vandemark, G. J., Kraft, J. M., Larsen, R. C., Gritsenko, M. A. and Boge, W. L. (2000). A PCR-based assay by sequence characterised DNA markers for the identification and detection of *Aphanomyces euteiches*. *Phytopathology* 90, 1137–44.

Verzignassi, J. R., Poltronieri, L. S., Benchimol, R. L., França, S. K. S. de, Carvalho, E. de A. and Fernandes, C. D. (2012). *Pyricularia grisea*: New pathogen on *Brachiaria brizantha* cv. *Marandu* in Pará. *Summa Phytopathol.* 38, 254.

Villari, C., Mahaffee, W. F., Mitchell, T. K., Pedley, K. F., Pieck, M. L., Hand, F. P. (2017). Early detection of airborne inoculum of *Magnaporthe oryzae* in turfgrass fields using a quantitative LAMP assay. *Plant Dis.* 101, 170–7.

Vy, T. T. P., Hyon, G. S., Nga, N. T. T., Inoue, Y., Chuma, I. and Tosa, Y. (2014). Genetic analysis of host–pathogen incompatibility between *Lolium* isolates of *Pyricularia oryzae* and wheat. *J. Gen. Plant Pathol.* 80, 59–65.

Wang, X., Lee, S., Wang, J., Ma, J., Bianco, T. and Jia, Y. (2014). Current advances on genetic resistance to rice blast disease. In: Yan, W. and Bao, J. (Eds), *Rice Germplasm, Genetics and Improvement*. Rijeka: InTech, pp. 195–217.

Williamson, V. M., Thomas, V., Ferris, H. and Dubcovsky, J. (2013). An *Aegilops ventricosa* translocation confers resistance against root-knot nematodes to common wheat. *Crop Sci.* 53, 1412–18.

Xiao, W., Yang, Q., Sun, D., Wang, H., Guo, T., Liu, Y., Zhu, X. and Chen, Z. (2015). Identification of three major R genes responsible for broad-spectrum blast resistance in an indica rice accession. *Mol. Breed.* 35, 1–11.

Zhan, S. W., Mayama, S. and Tosa, Y. (2008). Identification of two genes for resistance to *Triticum* isolates of *Magnaporthe oryzae* in wheat. *Genome* 51, 216–21.

Chapter 9

Diseases affecting barley: net blotches

Simon R. Ellwood, Curtin University, Australia; and Hugh Wallwork, South Australian Research and Development Institute, Australia

1 Introduction

Net blotch of barley (*Hordeum vulgare* L.) consists of two closely related but separate diseases that occur worldwide in barley growing areas. These are the net form of net blotch (NFNB), caused by *Pyrenophora teres* f. *teres* (PTT), and the spot form of net blotch (SFNB), caused by *P. teres* f. *maculata* (PTM) (Smedegård-Petersen 1971). As their names suggest, the net form has characteristic netting symptoms resulting from lesions along leaf veins with longitudinal and transverse dark striations. The spot form shows ovoid to round lesions often surrounded by a chlorotic border (Fig. 1).

Both NFNB and SFNB are damaging diseases with reported yield losses of over 30% and 40%, respectively (Jayasena et al. 2007; Khan 1987, 1989; Steffenson et al. 1991). These publications provide a record of losses, but observations by field pathologists suggest NFNB is more damaging than SFNB, as larger NFNB lesions can completely abscise leaves and weaken stems more effectively. In South Australia, following a breakdown in resistance in the cultivar Skiff in 1998, most crops untreated with fungicides recorded 100% losses after screening to remove undersized grain that is unsuitable for malting (Wallwork et al. 2016).

While PTT and PTM are morphologically similar and phylogenetically closely related (Ellwood et al. 2012), several studies in different countries have concluded these pathogens are genetically isolated populations (Akhavan et al. 2016b; Bakonyi and Justesen 2007; Bogacki et al. 2010; Lehmensiek et al. 2010; Leišová-Svobodová et al. 2014; Rau et al. 2003, 2007). Interestingly, PTT and PTM may be crossed in the laboratory (Campbell and Crous 2003; Jalli 2011; Louw et al. 1994; McDonald 1967; Smedegård-Petersen 1977),

http://dx.doi.org/10.19103/AS.2017.0039.09
Published by Burleigh Dodds Science Publishing Limited, 2019.

Figure 1 Images illustrating typical symptoms of net form of net blotch (left) and spot form of net blotch (right) on susceptible barley lines. The photographs were taken early during infection before more widespread and severe symptoms developed.

but only a few natural hybrids have been inferred though not conclusively confirmed by shared DNA bands in AFLP and RAPD marker studies (Campbell et al. 2002; Leišova et al. 2005; McLean et al. 2014). A recent study by Poudel et al. (2017), using expressed regions unique to each form and based on whole-genome assembly comparisons, was able to confirm the existence of field hybrids. Their relative rarity, however, suggests that there is a natural barrier or fitness penalty restricting survival, but such events potentially provide a means for novel virulence gene combinations and fungicide-resistant variants to arise (Campbell et al. 1999; Jalli 2011).

The biological characteristics of PTT and PTM make them challenging pathogens from several perspectives. PTT is readily dispersed on infected seed material (Hampton 1980; Jordan 1981) and although there are no publications on PTM in this respect, low levels of seed-borne contamination has potentially played a role in long-distance dispersal that has enabled this pathogen to spread globally since the 1970s. Both are wind and splash dispersed and persist in the field on stubble from the previous year's crop, with no-till practices appearing to have contributed to an increased prevalence (Jordan and Allen 1984; McLean et al. 2009). They are heterothallic, meaning different mating type loci or sexes are present within individuals and are therefore obligate outcrossing fungi, leading to genetically diverse populations (Akhavan et al. 2016a,b; Liu et al. 2012; Tekauz 1990). The main means of reproduction during the growing season is through polycyclic asexual conidia, allowing virulent clones to spread and increase rapidly within a susceptible crop. In PTT, regional adaption to resistance genes present in local cultivars occurs while at the same time populations may retain virulences to historically grown cultivars, indicating prolonged survival of virulence gene combinations (Fowler et al. 2017).

Published by Burleigh Dodds Science Publishing Limited, 2019.

The diverse nature of the NFNB population supports the retention of a wide range of virulences and virulence combinations to a greater degree than might be the case where sexual recombination did not occur. This increases the chances for the 'breakdown' of resistance in new varieties that has frequently been observed with NFNB. There is much less evidence for variation in virulence in PTM for reasons that are not fully understood. PTM populations are less genetically diverse than PTT, perhaps reflecting the relatively recent colonisation of barley, with the shorter period of coexistence meaning the pathogen is yet to develop a wider range of alternative virulence mechanisms. Interestingly, the genome of PTT is larger than PTM, expanded by transposon repeats (unpublished data). In fungal crop pathogens, genes involved in host interactions are often present within repeat-rich regions that are associated with the emergence of new virulence traits (Raffaele and Kamoun 2012).

In common with other fungal pathogens, sustained use of fungicides has led to the evolution of resistant populations. In Europe, PTT has acquired resistance to the quinone outside inhibitor (QoI) and succinate dehydrogenase inhibitor fungicide classes through mutations in their respective target genes, cytochrome b and three of the four succinate dehydrogenase subunit genes (Rehfus et al. 2016; Semar et al. 2007; Sierotzki et al. 2007). In *P. teres* cytochrome b, the F129L mutation is common, while the G143A mutation, which confers a higher level of resistance than F129L in other fungi, is lethal due the presence of an intron directly after that codon (Sierotzki et al. 2007). In the absence of the G143A mutation or other modes of resistance, the weaker F129L mutation in isolation has so far allowed the continued efficacy of QoI fungicides. In Australia, which has a history of over-reliance on single class of fungicides known as the azole or demethylase inhibitor (DMI) group, resistance has emerged which is associated with both a target site mutation and enhanced induced expression of two Cyp51A paralogs and the Cyp51B gene (Mair et al. 2016).

Resistance to net blotch in barley is complex, with a continuous distribution of disease severity shown by different barley lines, indicating resistance is often quantitative and controlled by several loci (Beattie et al. 2007; Burlakoti et al. 2016; Robinson and Jalli 1997; Steffenson and Webster 1992). Seedling, adult and all stage resistance genes or quantitative trait loci (QTLs) have been reported across the barley genome to different pathotypes of both forms of the disease (reviewed by Liu et al. 2011). Pathogen epidemiology, genetic diversity and complex host–pathogen genetics in *P. teres* pose challenges in selecting stable genetic resistance in breeding programmes. The following sections examine these in more detail and discuss current approaches and practical cultural options for barley growers.

2 Control measures for net blotch

A range of measures may be deployed to reduce the severity of net blotch disease. Perhaps the most cost-effective and environmentally friendly factor is good cultivar resistance. However, the availability of resistant cultivars may be limited when selecting varieties adapted to a particular region. In Western Australia, for example, most varieties recommended by the Department of Agriculture and Food (2017) were susceptible to SFNB. Among export-accredited recommended malting varieties, two were intermediate in their response to NFNB with the remainder either moderately susceptible or susceptible.

Net blotch shares many of the cultural methods used to reduce severity of similar diseases such as spot blotch (*Cochliobolus sativus*) and scald (*Rhynchosporium commune*, previously *R. secalis*), reviewed by (McLean et al. 2009; Turkington et al. 2011). Crop rotation avoids the build-up of inoculum when susceptible cultivars are grown, while wind direction may be significant if a field is downwind from a site with high disease levels the previous year. Stubble management is also considered important in some regions, for example in Mediterranean climates, where barley is grown over winter and dry summers do not allow much stubble breakdown, although in the Middle East and North Africa heavy grazing removes the stubble, greatly reducing inoculum carryover. In regions with temperate climates such as Canada, trials did not find a relationship between no or minimum tillage and disease (Turkington et al. 2006) while in contrast, in the United Kingdom, the incidence of disease was more severe in direct-drilled plots compared to ploughed (Jordan and Allen 1984), with different straw removal methods (burning and baling leaving culm bases) reducing but not eliminating disease.

Additional cultural practices to control early net blotch include the use of disease-free seeds, fungicide seed treatments and foliar sprays. Foliar fungicides are registered for and regularly used to manage both forms of net blotch. Whilst some field trials have indicated that late-season sprays of the upper two leaves provide more effective control than early sprays (McLean et al. 2016; Turkington et al. 2004; Turkington et al. 2015), these trials were compromised by the very close proximity of treated and untreated plots. In these situations as the effect of the treatments wore off, there was a supply of high levels of inoculum in adjacent plots that negated much of the value of treatments in reducing inoculum levels. In practice, in South Australia at least, farmers have found that where early treatments are not applied to susceptible varieties the disease becomes hard to manage later in the season. Early season treatments have, therefore, become almost universal where a threat is perceived and follow-up sprays are applied when needed.

3 Host–pathogen genetic interactions

3.1 Spot form of net blotch

Genetic mapping studies since the 1990s have revealed the locations of genes and QTL for SFNB resistance (reviewed by Liu et al. 2011; McLean et al. 2009). These early studies were mainly conducted using segregating double haploid populations and were able to define separate resistance loci from NFNB, confirming phenotypic observations that indicated for breeding purposes the two pathosystems might be considered separate. More recently, association mapping has been used to survey resistance in larger germplasm samples. Association mapping is a technique that does not require time-consuming development of mapping populations and detects QTL across large numbers of landraces, cultivars and selected lines.

Wang et al. (Wang et al. 2015) conducted association mapping with 898 lines from four independent breeding populations in a breeding programme based in Queensland, Australia, at both seedling and adult stages. A large number of QTLs (29) varying in effect size were detected in 24 genomic regions, of which 22 functioned at both stages, two at the seedling stage only and five at the adult stage only. The authors were able to co-locate 21 QTLs identified in previous studies with five genomic regions and identified

Published by Burleigh Dodds Science Publishing Limited, 2019.

a complex region on chromosome 7 containing highly significant resistance effects. This region was composed of four separate QTLs over a 36-cM distance, one of which was co-located with a dominant resistance gene, *Rpt4*, which has a significant effect in seedlings (Williams et al. 1999) but little effect at later growth stages (Williams et al. 2003). Marker haplotype groupings suggested that such resistance QTLs alone do not provide high levels of resistance and that multiple minor-effect QTLs need to be incorporated in breeding programmes.

Subsequently, Tamang et al. (2015) used association mapping with four PTM isolates with diverse origins from Australia, Denmark, New Zealand and the United States to detect seedling resistance QTL among 1480 lines from a worldwide collection of barley. Twenty-one new resistance loci were detected in addition to the six previously identified loci. Similarly, Burlakoti et al. (2016) were able to detect ten seedling resistance QTL from 376 Upper Midwest USA breeding lines, with four being not reported previously. Dominant and recessive effects suggested six of the QTLs conditioned susceptibility, the remainder resistance. In contrast to PTT, no defined genetic studies involving the genetics of PTM virulence and avirulence have been reported. However, the presence in the host of dominant and recessive loci suggests that a similar complement of avirulence gene products recognised by dominant resistance genes and necrotrophic effectors targeting susceptibility loci may occur.

3.2 Net form of net blotch

Barley possesses both major gene resistance and minor gene resistance to NFNB, unlike SFNB. Resistance to NFNB also has a longer history of genetic analysis than SFNB, largely due to the fact that NFNB is the only form of net blotch reported before the 1970s. A large number of resistance loci have been determined (Liu et al. 2011). Richards et al. (2017) deployed association mapping to reveal the genomic distribution of seedling resistance genes in 957 accessions from a world barley core collection using three isolates. Sixteen separate loci were found, with five being unique to the study. Three loci were identified on chromosomes 6H and 7H that were shared by all three isolates. One of these loci, a centromeric region of barley chromosome 6H has been detected in several studies, including for SFNB resistance (Liu et al. 2011) and represents a complex locus containing both dominant resistance and susceptibility genes. Similarly, a region on 4H was shared by two isolates but has also previously been implicated in SFNB resistance (Grewal et al. 2008; Tamang et al. 2015) as well as NFNB resistance. Islamovic et al. (2017) were able to show this region, or one closely associated, provided varying levels of resistance to four geographically diverse isolates.

In contrast to SFNB, several studies on NFNB have examined the pathogen virulence/avirulence genes, based on crosses between isolates with reciprocal virulence or avirulence phenotypes on different cultivars. Weiland et al. (1999) first demonstrated in a cross between isolates 01 and 15A, virulent and avirulent on the barley cultivar (cv.) Harbin respectively, that avirulence was controlled by a single major gene, *avrHar*. The authors suggested additional genes may modify the expression of the major virulence gene as many of the progeny had phenotypes different to the parents or that some variation was inherent in the disease phenotyping. Lai et al. (2007) used the same two isolates to show that single genes also controlled 01 virulence on the cvs. Tifang and Canadian Lake Shore, while virulence on a third, cv. Prato, was controlled by two epistatic genes in 15A. One of these genes mapped to the same location as the *avrHar* but with the opposite phenotype.

Beattie et al. (2007) and Afanasenko et al. (2007) conducted similar studies showing that 1–2 avirulence genes define the interaction in a gene-for-gene manner (Flor 1971). However, preselection of isolates with strongly contrasting phenotypes may have biased how commonly this occurs, given that between three and five loci often provide resistance in the host (Liu et al. 2011). These studies also lacked corresponding data on the location of host resistance loci in each interaction and the virulence genes involved. Shjerve et al. (2014) addressed this to some extent by demonstrating PTT progeny harbouring one of three genetically defined virulence loci interact with dominant susceptibility genes in the same region of barley chromosome 6H.

4 Case study: breeding for durable resistance to the net form of net blotch in South Australia

NFNB is a potentially devastating disease of barley in South Australia. A crop of a susceptible cultivar Maritime in the Yorke Peninsula that was left untreated was noted as having lost at least 70% of its yield (unpublished data) through infection of the leaves, heads and stems, leading to severe lodging and grain shrivelling (Fig. 2). That losses are normally only slight is largely due to the use of genetic resistance in current and

Figure 2 Severe net form of net blotch disease on an untreated crop of cultivar Maritime in the Yorke Peninsula, South Australia. The crop lost over 70% of its yield through infection of the leaves, heads and stems, leading to severe lodging and grain shrivelling.

past varieties and, where that has failed, control with foliar fungicides. Partial control by delaying epidemics is also provided through the removal of infected volunteer barley plants growing through the summer by grazing and use of herbicides. The disrupting force that makes disease management an ongoing problem comes from the variable nature of the pathogen which undergoes frequent mutation and recombination allowing new virulences and virulence combinations to emerge in the pathogen population, leading to situations where repeated doses of chemical control becomes essential. Changes in virulence are more frequent where pathogen populations are high so that widespread use of effective resistance and good management techniques not only reduce yield losses in the year of application but also reduce the rate of pathogen evolution and thus losses in future years. A sustainable system for the control of NFNB, therefore, requires as its basis an effective level of durable resistance that eliminates or at least reduces the need for chemical control and other forms of disease management.

The search for durable resistance has largely been an exercise in historical observations. Durable resistance after all is defined as resistance in a variety that has held up over a long period of time when that variety, or closely related varieties, has been grown over a wide area that is prone to the disease in question. With increasing knowledge of the genetic control of resistance, the historical approach can be bypassed where specific resistance genes or gene combinations are known to provide the durable resistance.

In South Australia, NFNB was almost absent from the state for at least two decades prior to 1993. During this time the variety Clipper (Released in 1968) and subsequently Schooner (1983) covered very large areas of the state and had good resistance to NFNB. It is likely that the cultivation of these two varieties in particular was responsible for the effective control of NFNB for over 20 years. Both varieties are susceptible as seedlings so the resistance behind them is adult plant or minor gene resistance. Other varieties that have shown durable resistance, albeit not exposed to infection for as long are Chebec, a backcross derivative of Schooner, Sloop and two backcross derivatives of this variety, Sloop SA and Sloop Vic. Each of these varieties is also susceptible as seedlings to NFNB. Varieties whose resistance has not endured after their initial release are Skiff, Maritime, Fleet and Oxford, all of whose resistances appeared to be based on a single major seedling resistance gene such that when novel virulences emerged, these varieties went from being rated resistant to very susceptible in a single step. It should be noted that adult plant resistance alone is not a determinant of durable resistance. Keel, which for many years was rated moderately resistant, and which has adult plant resistance, became moderately susceptible to susceptible from 2007.

Experience from resistance studies against other cereal crop diseases suggests that durable resistance can come from single or minor genes conditioning adult plant resistance. For example, *mlo* resistance to barley powdery mildew and Lr34/Yr18 and Lr46/Yr29 partial resistance to leaf and stripe rust in wheat. In other cases durable resistance is likely to come from the complexity of resistance where multiple genes are providing the resistance. In the case of NFNB it is likely the complexity of the resistance is providing the durable resistance. In adult plant tests under controlled environment conditions some variations in resistance have been noted when large numbers of isolates are tested against the lines with durable resistance (Wallwork et al. 2016). However in no tests has a susceptible reaction been observed. The genetic basis for the durable resistance in Clipper, Schooner and SloopSA is being investigated in research in SARDI in Adelaide. Preliminary results have identified QTL on 2H, 6H and 7H in SloopSA from a SloopSA × Fathom mapping population (unpublished data). Development of durable resistance donors for breeding programmes will come from the pyramiding of QTLs from sources such as SloopSA × Fathom, especially where transgressive segregants are selected from such populations. In the SloopSA × Fathom population, two QTLs on Fathom 5H were also detected and lines containing the five QTLs showed a high level of resistance (unpublished data).

5 Summary and future trends in research

The host–pathogen genetic interactions in both NFNB and SFNB portray a complex picture, certainly more complex than found in definitively necrotrophic pathogens such as *P. tritici-repentis*, the cause of tan spot on wheat, where just three necrotrophic effectors are the primary determinants of disease (Ciuffetti et al. 2010). NFNB host resistance appears to comprise elements of both biotrophic and necrotrophic genetic interactions. This may be a consequence of this pathogen's apparently hemi-biotrophic lifestyle with a short biotrophic or asymptomatic phase followed by a necrotrophic phase with characteristic necrotic lesions due to phytotoxin damage. Thus both dominant major resistance genes typical of biotrophic interactions and dominant susceptibility genes typical of necrotrophic interactions exist, together with quantitative (genes with a large or minor effect but not necessarily race non-specific) resistances. Resistance to SFNB by contrast appears to be conditioned only by quantitative genes.

In biotrophic pathosystems, dominant resistance is usually provided by single major resistance genes which readily breakdown and are therefore effective only in the short term. Such genes may be responsible for the dramatic breakdowns in resistance to NFNB. In necrotrophic systems, the products of dominant susceptibility genes are targets for phytotoxic effectors. Loss or alteration of these genes provides effective recessive resistance; however, direct selection requires the isolation of individual effectors. In this regard, the completion of high-quality reference genomes assemblies for PTT and PTM will assist effector isolation strategies (EBI project accession PRJEB18107). Until a detailed view of the numbers of effectors in both net blotch diseases and their corresponding insensitivity loci are determined, multigene resistance is preferable as this is inherently more likely to be durable simply because of the greater number of genes and the complexity of the resistance mechanisms involved.

Researchers have already identified QTLs shared between NFNB and SFNB that may simplify the incorporation of dual resistance regions into breeding programmes. Although the exact nature of such regions has not been dissected, they may provide only a limited number of genes to breeders and therefore combining them with other QTLs using marker-assisted selection may provide more effective long-term resistance.

6 Where to look for further information

A wealth of crop disease and management information is available to growers, consultants and researchers, aided by Internet access and the growing use of mobile phone applications and social media. The information is available at several levels, from local field pathologists and agricultural offices to regional and national government departments, breeder and agrochemical companies, academic publishers and researchers.

The Centre for Agricultural Bioscience International (CABI) hosts the Plantwise Knowledge Bank website which publishes plant disease factsheets and links to useful external management guides and factsheets (http://www.plantwise.org/KnowledgeBank/home.aspx). The site also provides pest alerts and has a comprehensive country-based disease and pest identification portal.

The American Phytopathological Society (APS, http://www.apsnet.org/Pages/default.aspx) publishes Plant Disease Management Reports covering fungicides and pesticides, variety performance and integrated disease management, though mostly related to the United States. The site also has links to external crop protection and management news reports and the APS Press Bookstore provides a large number of titles related to plant diseases.

In the United Kingdom the Agriculture and Horticulture Development Board hosts an encyclopaedia that describes the symptoms and life cycle of both common and less frequently found diseases of cereal crops and a barley disease management guide (https://cereals.ahdb.org.uk/cereal-disease-encyclopedia.aspx). Fungicide and disease management guides are available at https://cereals.ahdb.org.uk/publications/2016/february/23/barley-disease-management-guide.aspx. The site also contains links to disease monitoring and forecast sites and fungicide performance trials. In Australia disease activity reports and variety guides are available from state agricultural departments. The Grains Research and Development Corporation (GRDC, https://grdc.com.au/) publishes regional variety trial results, manuals, articles and factsheets and grower alerts.

An international conference which often features presentations on barley net blotch is the International Workshop on Barley Leaf Diseases. Regional conferences include the American Malting Barely Association's Barley Improvement Conference, the Australian Barley Technical Symposium and the combined Canadian Barley Symposium and North American Barley Researchers Workshop.

7 References

2017 *Barley Variety Sowing Guide for WA*. Department of Agriculture and Food, Western Australia, Perth, pp. 1–36.

Afanasenko, O., Mironenko, N., Filatova, O., Kopahnke, D., Krämer, I. and Ordon, F. (2007) Genetics of host-pathogen interactions in the *Pyrenophora teres* f. *teres* (net form) – barley (*Hordeum vulgare*) pathosystem. *Eur. J. Plant Pathol.* 117:267–80.

Akhavan, A., Turkington, T. K., Askarian, H., Tekauz, A., Xi, K., Tucker, J. R., Kutcher, H. R. and Strelkov, S. E. (2016a) Virulence of *Pyrenophora teres* populations in western Canada. *Can J. Plant Pathol.* 38:183–96.

Akhavan, A., Turkington, T. K., Kebede, B., Xi, K., Kumar, K., Tekauz, A., Kutcher, H. R., Tucker, J. R. and Strelkov, S. E. (2016b) Genetic structure of *Pyrenophora teres* f. *teres* and P. *teres* f. *maculata* populations from western Canada. *Eur. J. Plant Pathol.* 146:325–35.

Bakonyi, J. and Justesen, A. F. (2007) Genetic relationship of *Pyrenophora graminea, P. teres* f. *maculata* and *P. teres* f. *teres* assessed by RAPD analysis. *J. Phytopathol.* 155:76–83.

Beattie, A. D., Scoles, G. J. and Rossnagel, B. G. (2007) Identification of molecular markers linked to a *Pyrenophora teres* avirulence gene. *Phytopathology* 97:842–9.

Bogacki, P., Keiper, F. J. and Oldach, K. H. (2010) Genetic structure of South Australian *Pyrenophora teres* populations as revealed by microsatellite analyses. *Fungal Biol.* 114:834–41.

Burlakoti, R. R., Gyawali, S., Chao, S., Smith, K. P., Horsley, R. D., Cooper, B., Muehlbauer, G. J. and Neate, S. M. (2016) Genome-wide association study of spot form of net blotch resistance in the Upper Midwest barley breeding programs. *Phytopathology* 107:100–8.

Campbell, G. F. and Crous, P. W. (2003) Genetic stability of net × spot hybrid progeny of the barley pathogen *Pyrenophora teres*. *Australas. Plant Pathol.* 32:283–7.

Campbell, G. F., Crous, P. W. and Lucas, J. A. (1999) *Pyrenophora teres* f. *maculata*, the cause of *Pyrenophora* leaf spot of barley in South Africa. *Mycol. Res.* 103:257–67.

Campbell, G. F., Lucas, J. A. and Crous, P. W. (2002) Evidence of recombination between net- and spot-type populations of *Pyrenophora teres* as determined by RAPD analysis. *Mycol. Res.* 106:602–8.

Ciuffetti, L. M., Manning, V. A., Pandelova, I., Betts, M. F. and Martinez, J. P. (2010) Host-selective toxins, Ptr ToxA and Ptr ToxB, as necrotrophic effectors in the *Pyrenophora tritici-repentis*–wheat interaction. *New Phytol.* 187:911–19.

Ellwood, S. R., Syme, R. A., Moffat, C. S. and Oliver, R. P. (2012) Evolution of three *Pyrenophora* cereal pathogens: Recent divergence, speciation and evolution of non-coding DNA. *Fungal Genet. Biol.* 49:825–9.

Flor, H. H. (1971) Current status of the gene-for-gene concept. *Annu. Rev. Phytopathol.* 9:275–96.

Fowler, R. A., Platz, G. J., Bell, K. L., Fletcher, S. E. H., Franckowiak, J. D. and Hickey, L. T. (2017) Pathogenic variation of *Pyrenophora teres* f. *teres* in Australia. *Australas. Plant Pathol.* 46:115–28.

Grewal, T. S., Rossnagel, B. G., Pozniak, C. J. and Scoles, G. J. (2008) Mapping quantitative trait loci associated with barley net blotch resistance. *Theor. Appl. Genet.* 116:529–39.

Hampton J. G. (1980) The role of seed-borne inoculum in the epidemiology of net blotch of barley in New Zealand. *N. Z. J. Exp. Agric.* 8:297–9.

Islamovic, E., Bregitzer, P. and Friesen, T. L. (2017) Barley 4H QTL confers NFNB resistance to a global set of *P. teres* f. *teres* isolates. *Mol. Breed.* 37:29.

Jalli, M. (2011) Sexual reproduction and soil tillage effects on virulence of *Pyrenophora teres* in Finland. *Ann. Appl. Biol.* 158:95–105.

Jayasena, K. W., Van Burgel, A., Tanaka, K., Majewski, J. and Loughman, R. (2007) Yield reduction in barley in relation to spot-type net blotch. *Australas. Plant Pathol.* 36:429–33.

Jordan, V. W.L. (1981) Aetiology of barley net blotch caused by *Pyrenophora teres* and some effects on yield. *Plant Pathol.* 30:77–87.

Jordan, V. W.L. and Allen, E. C. (1984) Barley net blotch: Influence of straw disposal and cultivation methods on inoculum potential, and on incidence and severity of autumn disease. *Plant Pathol.* 33:547–59.

Khan, T. N. (1987) Relationship between net blotch (*Drechslera teres*) and losses in grain yield of barley in Western Australia. *Aust. J. Agric. Res.* 38:671–9.

Khan, T. N. (1989) Effect of spot-type net blotch (*Drechslera teres* (Sacc.) Shoem) infection on barley yield in short season environment of northern cereal belt of Western Australia. *Aust. J. Agric. Res.* 40:745–52.

Lai, Z., Faris, J. D., Weiland, J. J., Steffenson, B. J. and Friesen, T. L. (2007) Genetic mapping of *Pyrenophora teres* f. *teres* genes conferring avirulence on barley. *Fungal Genet. Biol.* 44:323–9.

Lehmensiek, A., Bester-van der Merwe, A. E., Sutherland, M. W., Platz, G., Kriel, W. M., Potgieter, G. F. and Prins, R. (2010) Population structure of South African and Australian *Pyrenophora teres* isolates. *Plant Pathol.* 59:504–15.

Leišová-Svobodová, L., Minaříková, V., Matušinsky, P., Hudcovicová, M., Ondreičková, K. and Gubiš, J. (2014) Genetic structure of *Pyrenophora teres* net and spot populations as revealed by microsatellite analysis. *Fungal Biol.* 118:180–92.

Leišova, L., Minarı̌kova, V., Kučera, L. and Ovesná, J. (2005) Genetic diversity of *Pyrenophora teres* isolates as detected by AFLP analysis. *J. Phytopathol.* 153:569–78.

Liu, Z., Ellwood, S. R., Oliver, R. P. and Friesen, T. L. (2011) *Pyrenophora teres*: Profile of an increasingly damaging barley pathogen. *Mol. Plant Pathol.* 12:1–19.

Liu, Z. H., Zhong, S., Stasko, A. K., Edwards, M. C. and Friesen, T. L. (2012) Virulence profile and genetic structure of a North Dakota population of *Pyrenophora teres* f. *teres*, the causal agent of net form net blotch of barley. *Phytopathology* 102:539–46.

Louw, J. P.J., Crous, P. W. and Holz, G. (1994) *Pyrenophora japonica* occurring on barley in South Africa. *Plant Pathol.* 43:420–3.

Published by Burleigh Dodds Science Publishing Limited, 2019.

Mair, W. J., Deng, W., Mullins, J. G. L., West, S., Wang, P., Besharat, N., Ellwood, S. R., Oliver, R. P. and Lopez-Ruiz, F. J. (2016) Demethylase inhibitor fungicide resistance in *Pyrenophora teres* f. sp. *teres* associated with target site modification and inducible overexpression of *Cyp51*. *Front. Microbiol.* 7:1279.

McDonald, W. C. (1967) Variability and the inheritance of morphological mutants in *Pyrenophora teres*. *Phytopathology* 57:747–55.

McLean, M. S., Howlett, B. J. and Hollaway, G. J. (2009) Epidemiology and control of spot form of net blotch (*Pyrenophora teres* f. *maculata*) of barley: A review. *Crop Past. Sci.* 60:303–15.

McLean, M. S., Martin, A., Gupta, S., Sutherland, M. W., Hollaway, G. J. and Platz, G. J. (2014) Validation of a new spot form of net blotch differential set and evidence for hybridisation between the spot and net forms of net blotch in Australia. *Australas. Plant Pathol.* 43:223–33.

McLean, M. S., Weppler, R., Howlett, B. J. and Hollaway, G. J. (2016) Spot form of net blotch suppression and yield of barley in response to fungicide application in the Wimmera region of Victoria, Australia. *Australas. Plant Pathol.* 45:37–43.

Poudel, B., Ellwood, S. R., Testa, A. C., McLean, M., Sutherland, M. W. and Martin, A. (2017) Rare *Pyrenophora teres* hybridization events revealed by development of sequence-specific PCR markers. *Phytopathology* 107(7): 878–84.

Raffaele, S. and Kamoun, S. (2012) Genome evolution in filamentous plant pathogens: Why bigger can be better. *Nat. Rev. Micro.* 10:417–30.

Rau, D., Attene, G., Brown, A., Nanni, L., Maier, F., Balmas, V., Saba, E., Schäfer, W. and Papa, R. (2007) Phylogeny and evolution of mating-type genes from *Pyrenophora teres*, the causal agent of barley 'net blotch' disease. *Curr. Genet.* 51:377–92.

Rau, D., Brown, A. H. D., Brubaker, C. L., Attene, G., Balmas, V., Saba, E. and Papa, R. (2003) Population genetic structure of *Pyrenophora teres* Drechs. the causal agent of net blotch in Sardinian landraces of barley (*Hordeum vulgare* L.). *Theor. Appl. Genet.* 106:947–59.

Rehfus, A., Miessner, S., Achenbach, J., Strobel, D., Bryson, R. and Stammler, G. (2016) Emergence of succinate dehydrogenase inhibitor resistance of *Pyrenophora teres* in Europe. *Pest Manag. Sci.* 72:1977–88.

Richards, J. K., Friesen, T. L. and Brueggeman, R. S. (2017) Association mapping utilizing diverse barley lines reveals net form net blotch seedling resistance/susceptibility loci. *Theor. Appl. Genet.* 130:915–27.

Robinson, J. and Jalli, M. (1997) Quantitative resistance to *Pyrenophora teres* in six Nordic spring barley accessions. *Euphytica* 94:201–8.

Semar, M., Strobel, D., Koch, A., Klappach, K. and Stammler, G. (2007) Field efficacy of pyraclostrobin against populations of *Pyrenophora teres* containing the F129L mutation in the cytochrome b gene. *J. Plant Dis. Prot.* 114:117–19.

Shjerve, R. A., Faris, J. D., Brueggeman, R. S., Yan, C., Zhu, Y., Koladia, V. and Friesen, T. L. (2014) Evaluation of a *Pyrenophora teres* f. *teres* mapping population reveals multiple independent interactions with a region of barley chromosome 6H. *Fungal Genet. Biol.* 70:104–12.

Sierotzki, H., Frey, R., Wullschleger, J., Palermo, S., Karlin, S., Godwin, J. and Gisi, U. (2007) Cytochrome b gene sequence and structure of *Pyrenophora teres* and *P. tritici-repentis* and implications for QoI resistance. *Pest Manag. Sci.* 63:225–33.

Smedegård-Petersen, V. (1971) *Pyrenophora teres* f. *maculata* f. nov. and *Pyrenophora teres* f. *teres* on barley in Denmark. *Kgl. Vet. Landbohojsk. Arsskr.* 1971:124–44.

Smedegård-Petersen, V. (1977) Inheritance of genetic factors for symptoms and pathogenicity in hybrids of *Pyrenophora teres* and *Pyrenophora graminea*. *J. Phytopathol.* 89:193–202.

Steffenson, B., Webster, R. K. and Jackson, L. F. (1991) Reduction in yield loss using incomplete resistance to *Pyrenophora teres* f. *teres* in barley. *Plant Dis.* 75:96–100.

Steffenson, B. J. and Webster, R. K. (1992) Pathotype diversity of *Pyrenophora teres f. teres* on barley. *Phytopathology* 82:170–7.

Tamang, P., Neupane, A., Mamidi, S., Friesen, T. and Brueggeman, R. (2015) Association mapping of seedling resistance to spot form net blotch in a worldwide collection of barley. *Phytopathology* 105:500–8.

Tekauz, A. (1990) Characterisation and distribution of pathogenic variation in *Pyrenophora teres* f. *teres* and *P. teres* f. *maculata* from western Canada. *Can. J. Plant Pathol.* 12:141–8.

Turkington, T. K., Clayton, G. W., Harker, K. N., Kutcher, H. R., O'Donovan, J. T., Johnston, A. M., Xi, K. and Stevenson, F. C. (2004) Impact of seedbed utilization and fungicide application on severity of net blotch [*Pyrenophora teres*] and production of barley. *Can J. Plant Pathol.* 26:533–47.

Turkington, T. K., Clayton, G. W., Klein-Gebbinck, H. W., Lupwayi, N. Z., Harker, K. N., O'Donovan, J. T., Burnett, P. A. and Xi, K. (2006) Impact of crop management on leaf diseases in Alberta barley fields, 1995–1997. *Can. J. Plant Pathol.* 28:441–9.

Turkington, T. K., Ó'Donovan, J. T., Harker, K. N., Xi, K., Blackshaw, R. E., Johnson, E. N., Peng, G., Kutcher, H. R., May, W. E., Lafond, G. P., Mohr, R. M., Irvine, R. B. and Stevenson, C. (2015) The impact of fungicide and herbicide timing on foliar disease severity, and barley productivity and quality. *Can. J. Plant Sci.* 95:525–37.

Turkington, T. K., Tekauz, A., Xi, K. and Kutcher, H. R. (2011) Foliar diseases of barley: Don't rely on a single strategy from the disease management toolbox. Prairie Soils and Crops. *Prairie Soils Crops* 4:142–50. Saskatchewan Soil Conservation Association.

Wallwork, H., Butt, M. and Capio, E. (2016) Pathogen diversity and screening for minor gene resistance to *Pyrenophora teres* f. *teres* in barley and its use for plant breeding. *Australas. Plant Pathol.* 45:527–31.

Wang, X., Mace, E. S., Platz, G. J., Hunt, C. H., Hickey, L. T., Franckowiak, J. D. and Jordan, D. R. (2015) Spot form of net blotch resistance in barley is under complex genetic control. *Theor. Appl. Genet.* 128:489–99.

Weiland, J. J., Steffenson, B. J., Cartwright, R. D. and Webster, R. K. (1999) Identification of molecular genetic markers in Pyrenophora teres f. teres associated with low virulence on 'Harbin' barley. *Phytopathology* 89:176–81.

Williams, K. J., Lichon, A., Gianquitto, P., Kretschmer, J. M., Karakousis, A., Manning, S., Langridge, P. and Wallwork, H. (1999) Identification and mapping of a gene conferring resistance to the spot form of net blotch (*Pyrenophora teres* f *maculata*) in barley. Theor. Appl. Genet.99:323–7.

Williams, K. J., Platz, G. J., Barr, A. R., Cheong, J., Willsmore, K., Cakir, M. and Wallwork, H. (2003) A comparison of the genetics of seedling and adult plant resistance to the spot form of net blotch *Pyrenophora teres* f. *maculata*. *Aust. J. Agric. Res.* 54:1387–94.

Published by Burleigh Dodds Science Publishing Limited, 2019.

Chapter 10

Diseases affecting barley: scald

Wolfgang Knogge, Leibniz Institute of Plant Biochemistry, Germany

1 Introduction

Scald (leaf blotch) is a common disease of barley (*Hordeum vulgare*; Fig. 1). It is caused by the fungal pathogen *Rhynchosporium commune*, which occurs worldwide due to its high genetic variability and climatic adaptability (Stefansson et al. 2013), but develops best in barley-growing regions with longer periods of cool, semi-humid weather. *R. commune* is a member of a fungal genus, of which five species with differing host specificities have been described to date (Caldwell 1937, Zaffarano et al. 2011; King et al. 2013). A previous review article (Shipton et al. 1974) mentioned yield losses of up to 40%, a number which, since then, has been cited frequently. However, the average yield losses in Europe and North America are rather closer to 5–10%, even if the frequency of symptom occurrence is higher. Nevertheless, regional losses as high as 65% can occur locally in epidemic years (Beigi et al. 2013). Furthermore, from areas where barley is continuously grown virtually without crop rotation such as in some parts of North Africa (Tunisia, Morocco, Eritrea and Ethiopia; Bouajila et al. 2006), South America (Peru, Colombia, Bolivia and Ecuador) and Central and Western Asia (Turkey, Syria and Iran; Bouajila et al. 2006; Arabi et al. 2008; Beigi et al. 2013), yield losses close to 100% have been reported (Abang et al. 2006).

http://dx.doi.org/10.19103/AS.2017.0039.10

Figure 1 Severe scald symptoms on plants in a barley field.

2 History, origins and general characteristics of scald

2.1 History, host specialization, taxonomy and phylogeny

The fungus was first recorded in 1897 in the Netherlands on rye (*Secale cereale*) under the name *Marsonia secalis* Oud. (Oudemans 1897), later renamed *Marssonina secalis* (Oud.) Magnus (Magnus 1906). In the same year, it was recorded on rye and barley (*H. vulgare*) from several parts of Germany (Frank 1897). The peculiarly beaked, two-celled (uniseptate) conidiospores (*cf.* Fig. 3) led to the establishment of a new fungal genus, *Rhynchosporium* (from Greek *rhynchos*, beak), and the fungus was designated *R. graminicola* Heinsen (Frank 1897; Heinsen 1900). Eventually, the name was changed to *R. secalis* (Oud.) Davis (Davis 1919; 1922) and the disease it causes is referred to as 'scald' or 'leaf blotch' (Caldwell 1937). By the 1920s, *R. secalis* was recorded from the United States, Canada, the United Kingdom and Australia (Brooks 1928), and by the 1930s from Tunisia, Argentina and Peru, with severe yield losses from the United States and Germany (Caldwell 1937). Today, the fungus has been reported from five dozen countries across all continents except Antarctica (www.cabi.org/isc/datasheet/47480).

Since the early days of research on *Rhynchosporium*, host specialization has been under debate. Cross-inoculation experiments with fungal isolates from barley, rye and other grasses (briefly reviewed in Zaffarano et al., 2011) gave varying results which are indicative of strict, weak or even no specialization (*cf.* Caldwell 1937; Sarasola and Campi 1947; Owen 1958; Kay and Owen 1973a; Ali and Boyd 1973; Wilkins 1973; Jackson and Webster 1976c; Cromey and Mulholland 1987; Brown 1990; Lyngs Jørgensen and Smedegaard-Petersen 1995; Robinson et al. 1996; Kilby and Robinson 2001; Lebedeva and Tvarůžek 2006; Kiros-Meles et al. 2011). Finally, as a consequence of phylogenetic analyses based on the sequences of four housekeeping genes (ITS, EF-1α, α-tubulin and β-tubulin), *R. secalis* was dissolved into a complex consisting of three species that are specialized on different grass hosts. The original name *R. secalis* was kept for the fungus infecting rye and triticale, whereas the species grown on common couch grass (*Agropyron repens*, syn. *Elymus repens*) and bearded couch grass (*Agropyron caninum*, syn. *Elymus caninus*) was designated as *R. agropyri*. The fungal species pathogenic to barley, to other *Hordeum* species and to ripgut brome (*Bromus diandrus*), was named *R. commune* (Zaffarano et al. 2011). These three species produce the typical beak-shaped conidia (beaked conidia group, BCG) and form one branch of the *Rhynchosporium* phylogeny (Fig. 2). The other branch contains another new species, *R. lolii*, from perennial ryegrass (*Lolium perenne*; King et al. 2013) along with *R. orthosporum*, a species attacking cocksfoot (orchard grass, *Dactylis glomerata*) that was already described 80 years ago (Caldwell 1937). The latter species are characterized by straight cylindrical conidia (cylindrical conidia group, CCG), which reveals an inconsistency in the species name for *R. orthosporum* (from Greek *orthos*, straight). By far, most studies have been carried out with the barley pathogen, *R. commune*, due to its economic importance.

The genus *Rhynchosporium* belongs to the Leotiomycetes class of Ascomycota (Goodwin 2002, Penselin et al. 2016). This class harbours other plant pathogens such as *Sclerotinia sclerotiorum* and *Botrytis cinerea*, causing white mould and grey mould, respectively, on a wide variety of plants, as well as *Blumeria graminis*, the powdery mildew pathogen of grasses. It should be noted here that two other fungal species, the rice leaf scald pathogen *R. oryzae* and *R. alismatis*, a pathogen of different arrowhead genera (Alismataceae),

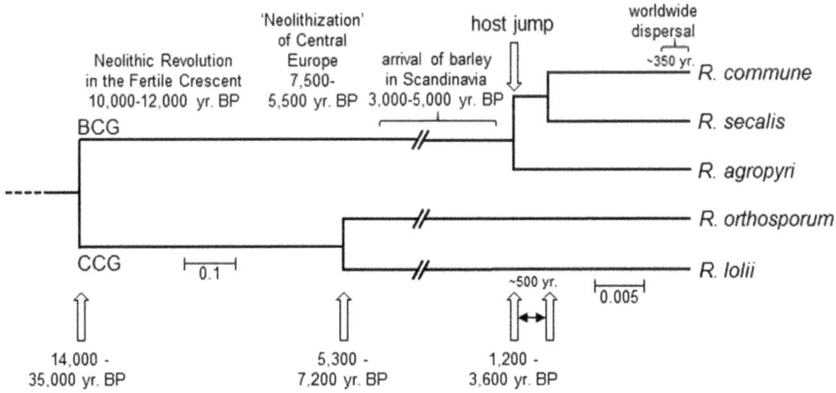

Figure 2 *Rhynchosporium* evolution with estimated divergence times. The BCG and CCG branches of the genus diverged already before barley and rye were domesticated 10 000–12 000 yr. BP, years before present (Salamini et al., 2002). From the Fertile Crescent the domesticated cereals were brought to Europe. Supposedly in Scandinavia, the common ancestor of the *Rhynchosporium* BCG species 'jumped' to the newcomer species from an unknown host plant not longer than 3600 yr. ago. After another divergence of the lineages about 500 y later *Rhynchosporium* speciation led to the recent members of the genus. Scale: number of substitutions per site and number of single nucleotide polymorphisms (SNPs) per position.

group into a different class of Ascomycota, the Sordariomycetes. Consequently, they were renamed *Microdochium oryzae* (teleomorph: *Monographella albescens*; Parkinson et al. 1981) and *Plectosphaerella alismatis* (Carlucci et al. 2012), respectively.

Several phylogenetic studies based on DNA sequence data (α-tubulin, β-tubulin and ITS), on rDNA structures and on the analysis of whole-genome single nucleotide polymorphisms (Penselin et al. 2016) combined with divergence time estimation (Tamura et al. 2012) dealt with the relationship of the five currently accepted *Rhynchosporium* species. The present understanding is that the lineages leading to the three BCG species on the one hand and to the CCG species on the other hand split about 14 500–35 000 years ago (Zaffarano et al. 2008, 2011). The most recent common ancestor (TMRCA) of *R. orthosporum* and *R. lolii* lived between 5300 and 7200 years ago (King et al. 2013). Time to TMCRA of the three BCG species was estimated to be ~1200–3600 years (Zaffarano et al. 2008), when *R. agropyri* branched off about 500 years before *R. secalis* and *R. commune* diverged (Penselin et al. 2016).

2.2 Origin and global migration

Origin and genetic history of the barley pathogen *R. commune* were reconstructed by studying populations collected from different parts of the world. To this end, neutral restriction fragment length polymorphism (RFLP) markers (McDonald et al. 1999; Salamati et al. 2000; Zaffarano et al. 2006; 2009), sequences of the avirulence gene *NIP1* (Brunner et al. 2007) and of several housekeeping genes (Zaffarano et al. 2008), of RFLP loci and the mating-type idiomorphs (Zaffarano et al. 2006, 2009) and microsatellite loci (Linde et al. 2009) were used. The highest genetic diversity was found in fungal populations from Scandinavia, not in those from the Fertile Crescent of the Middle East or Ethiopia as expected if the fungus originated at the primary or secondary centre of origin of its host plant, barley (Salamini et al. 2002). This is underpinned by the recent finding that fungal population structures from wild (*H. vulgare* ssp. *spontaneum*) or cultivated barley in Syria and Jordan are very similar showing low levels of genotypic diversity. This suggests that the fungus was introduced in the Middle East with barley and did not originate from wild barley (Kiros-Meles et al. 2011). Instead, an origin in northern Europe is proposed, where the fungus made a host shift not later than 3600 years ago, that is, around the time when the CCG species diverged, probably from a wild grass to barley after the introduction of this crop species to Scandinavia ~5000–3000 years ago (Fig. 2; Zaffarano et al. 2008). A demographic bottleneck occurred in all species most likely as a result of the host jump, before ancient fungal founder populations spread to barley-growing areas in Central Europe (Brunner et al. 2007). Worldwide dispersal of *R. commune* over the last 350 years is probably anthropogenic, following the major colonial and international barley trading routes from Europe to California as well as to South Africa, Australia and New Zealand, and from North America to North Africa and the Near East (see global map in McDonald, 2015). In the arrival regions, more recent founder populations were established, which went through another demographic bottleneck before expanding again in the barley agricultural systems (Linde et al. 2009).

2.3 Epidemiology

Infected plant debris (stubble and straw) represents the major source of *R. commune* primary inoculum (Ayesu-Offei and Carter 1971; Kay and Owen 1973b; Mayfield and Clare 1984b).

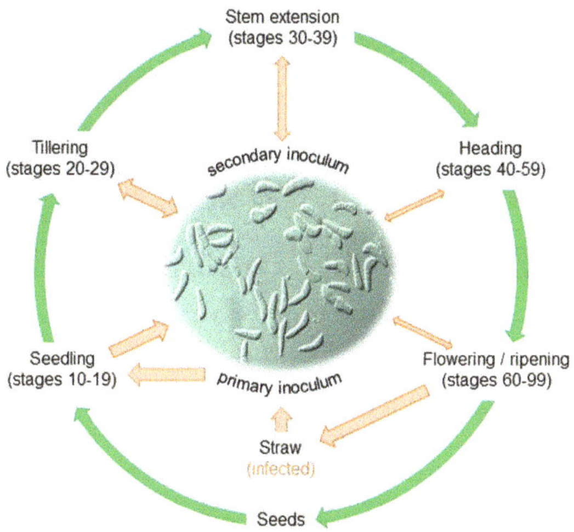

Figure 3 *R. commune* infection cycle. Infected plant debris (stubble, straw) is the major source of primary inoculum, whereas secondary inoculum may be generated on leaves throughout fungal development on the plant. The phenological development stages of barley are indicated using the BBCH scale (Lancashire et al. 1991; Meier et al. 2009).

In addition, the fungus can be introduced by seed-borne inoculum (Fig. 3; Skoropad 1959; Habgood 1971; Kay and Owen 1973b; Lee et al. 2001, 2002; Fountaine et al. 2010), which also enables long-distance dispersal through human activities. Upon germination and limited growth on the leaf surface of host plants, the germ tubes form more or less distinct appressoria, through which the fungus penetrates the leaf cuticle (Fig. 4; Ayesu-Offei and Clare 1970; Jones and Ayres 1974; Lehnackers and Knogge 1990; Lyngs Jørgensen et al. 1993; Xi et al.

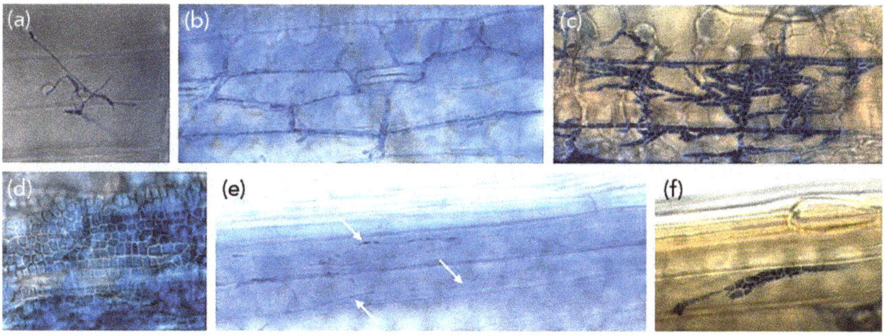

Figure 4 *R. commune* development *in planta*. (a) Germ tubes originating from fungal spores are penetrating the surface of a barley leaf. At penetration sites the plant responds with halo formation (appositions on the inner side of the epidermal wall). (b) Thin hyphae grow along the anticlinal walls of epidermis cells, at some sites also across a cell. (c) The occurrence of thicker, shorter septate hyphae initiates the formation of a dense subcuticular stroma (d). On a resistant barley cultivar fungal hyphae (e, arrows) and beginning stroma formation (f) can be observed usually on cells near the leaf margin. Fungal structures were stained with Trypan Blue.

Figure 5 Typical scald symptoms. Necrotic lesions on primary leaves about two weeks after inoculation of a susceptible barley cultivar with *R. commune* spores under conducive growth chamber conditions.

2000; Linsell et al. 2011; Horbach et al. 2011). Thin, wide-septate fungal hyphae grow mainly in the subcuticular region above the anticlinal walls between epidermal cells. After some time, hyphae with thicker and shorter cells are formed, and eventually a dense stroma is generated (Ayesu-Offei and Clare 1970; Lehnackers and Knogge 1990; Lyngs Jørgensen et al. 1993). First microscopic symptoms occur when single epidermal cells collapse. The typical necrotic disease symptoms, however, do not become visible before larger regions of the mesophyll underneath the stroma collapse (Fig. 5). Between inoculation and symptom development, 10–14 d may be required under artificial conducive conditions, but up to several months may pass by in the field (Zhan et al. 2008). Therefore, the reclassification of the fungus as a hemibiotroph rather than a necrotroph has been suggested (Zhan et al. 2008). New conidia are formed on the stroma in lesions, and also on the healthy looking (symptomless) areas of the leaves (Howlett and Cooke 1987, 1992; Davis and Fitt 1990). Spores are dispersed only over short distances by water droplets (rain, mist; Stedman 1980a,b, Fitt et al. 1986, 1988, 1989). The fungus is polycyclic, that is, several rounds of secondary infections can occur during a single growing season (Fig. 3). Normally, the lower leaves are infected before the stem extension. Later, infection in the upper parts of a host plant may occur again as a result of spore transmission, and not of fungal growth through the shoot. *R. commune* isolates were also obtained from several wild barley species (*H. spontaneum*, *H. murinum* and *H. leporinum*) as well as from brome grass (*B. diandrus*; Caldwell 1937; Wilkins 1973; Shipton et al. 1974; Zaffarano et al. 2011; King et al. 2013). These isolates showed strong genetic similarities with isolates from cultivated barley (Zaffarano et al. 2011). Furthermore, in California, wild rye (*Elymus triticoides*) was found to be susceptible to fungal isolates from barley (Jackson and Webster 1976c). Therefore, the question is immanent as to whether these uncultivated grass hosts may represent additional sources of inoculum for infecting cultivated barley. During pathogenesis, the formation of the large chlorotic and necrotic disease symptoms leads to a reduction in the foliar area available for photosynthesis. In particular, infection in the upper leaves that support heading (panicle and fruit formation) results in severe damage to further plant development. As a consequence, agricultural yield losses occur in the form of smaller and fewer kernels of lower quality.

3 Haplotypes, virulence variation and recombination

R. commune populations are characterized by a high degree of virulence variation, although to varying levels in different barley-growing regions. This variability is manifested within fields and even within single leaves (Brown 1985). Using differential barley cultivars containing known major resistance genes (Goodwin et al. 1990) and/or local cultivars, a high number of fungal pathotypes could be differentiated in numerous studies from barley-growing areas around the world, for example, South Eastern Australia (Ali et al. 1976; Brown 1985; 1990; Burdon et al. 1994), New Zealand (Cromey 1987), the United States (California; Jackson and Webster 1976b; McDermott et al. 1989; Zhang et al. 1992), Idaho and Oregon (Goodwin et al. 1992), Canada (Xue et al. 1991; Tekauz 1991; Xi et al. 2002, 2003), the United Kingdom (Newton et al. 2001), Denmark (Lyngs Jørgensen and Smedegaard-Petersen 1995), Norway (Salamati and Tronsmo 1997), Italy (Ceoloni 1980), Tunisia (Bouajila et al. 2006, 2007, 2010), West Asia and North Africa (von Korff et al. 2004), Turkey (Araz and Maden 2006), Syria (Kiros-Meles et al. 2005; Arabi et al. 2008; 82009), Iran (Beigi et al. 2013) and South Africa (Robbertse and Crous 2000; Robbertse et al. 2000). However, results from classical pathotype analyses which are carried out such a long period using varying differential cultivars and a visual disease scoring system that is prone to subjective influences are impossible to critically compare. Furthermore, the interaction is very sensitive to environmental conditions (Ali 1974). In addition to the always reported high level of virulence polymorphism, no conventional race structure could be established.

To assess the general genetic variability of *R. commune*, several neutral markers such as isozymes, DNA fingerprints, RFLPs, random amplified polymorphic DNAs, amplified fragment length polymorphisms and microsatellites have been employed over the years (McDermott et al. 1989; Goodwin et al. 1993; Burdon et al. 1994; Robbertse and Crous 2000; Salamati et al. 2000; Newton et al. 2001; Keiper et al. 2006; Bouajila et al. 2007; Linde et al. 2009; Beigi et al. 2013). To characterize the global genetic structure, a hierarchical sampling strategy was used and fungal isolates were assayed for several RFLP loci (Salamati et al. 2000; Zaffarano et al. 2006). The majority of gene diversity within regions was found within a barley field. In fungal field populations from five continents, 58% of the diversity was distributed within fields, 11% between fields within regions and 31% between regions (McDonald et al. 1999; Zaffarano et al. 2006). This suggests that the gene flow is substantial between regional populations, but decreases with increasing geographical distance leading to subdivision among populations separated by long distances (Zaffarano et al. 2006). Furthermore, the high within-field genetic diversity along with the substantial virulence variability suggests that *R. commune* populations are able to evolve relatively quickly and to adapt rapidly to newly introduced resistance genes.

The mechanism underlying the high degree of variability remains to be elucidated. The fungus is haploid and a teleomorph has never been observed. A somatic recombination was shown to occur (Newman and Owen 1985; Forgan et al. 2007). Nevertheless, its role in generating genetic variability appears to be minor (Goodwin et al. 1994). There are, however, several lines of evidence for the existence of a cryptic sexual stage. The majority of populations analysed for a multilocus association using RFLP markers turned out to be in gametic equilibrium (McDonald et al. 1999; Salamati et al. 2000). Furthermore, for most populations, the mating-type genes (Foster and Fitt 2003) were found in equal ratios (Linde et al. 2003; Zaffarano et al. 2006; King et al. 2015). Possible exceptions were reported for two of the three Iranian populations (Arzanlou et al. 2016). Finally, it was shown that core meiotic genes along with more than 140 genes required for recombination and meiosis in other fungi are present in the *Rhynchosporium* genome (Penselin et al. 2016). Taken together, these results strongly suggest that sexual recombination takes place and that a teleomorph exists. It should be noted, however, that the earlier results regarding gametic equilibrium were partially corrected when microsatellite markers were used to estimate more recent fungal migration events. In contrast to the RFLP-based estimates, a significant gametic disequilibrium was detected in the populations studied. Microsatellites evolve more rapidly than nucleotide sequence loci. Therefore, this result is interpreted such that most current *R. commune* populations are founder populations, which did not have enough time to recover from a significant population bottleneck over the last ~100 years. In addition, 'low migration rates and an excess of asexual compared to sexual reproduction cycles' were discussed in order to most likely contribute to this disequilibrium (Linde et al. 2009).

4 Resistance genetics and mechanisms

4.1 Resistance genetics

In the early 1930s, scald was considered 'the most destructive disease to barley in California' (Mackie 1931, cited in Riddle and Suneson, 1948). After the first report of host resistance in 1920 (Johnson and Mackie 1920), it was not until the 1930s and 1940s that several resistant barley varieties were identified and utilized in breeding programmes (Riddle and Suneson 1948; Riddle and Briggs 1950). The first example is cultivar 'Atlas 46', which resulted from the introgression of a major resistance gene from cultivar 'Turk' into the genetic background of cultivar 'Atlas' (Riddle and Suneson 1948; Riddle and Briggs 1950). In the 1960s and 1970s, work was undertaken in Australia, Canada, the United Kingdom, the United States and Argentina to extend the genetic knowledge of resistance (Schein 1958; Ali 1975a,b, Baker and Larter 1963; Riddle and Briggs 1950; Dyck and Schaller 1961a,b, Habgood 1972; 1974, 1977; Habgood and Hayes 1971; Bockelman et al. 1977). As a consequence, several dominant and recessive resistance genes (*Rh, Rh2, Rh3, Rh4, Rh5, rh6, rh7, rh8, Rh9, Rh10, rh11*) were described (Table 1). Furthermore, the genes *Rh3* and *Rh4* were found to be closely linked (Dyck and Schaller 1961b) and associated with chromosome 3H (Dyck and Schaller 1961a). Due to the fact that different fungal isolates and a symptom-based disease scoring system were used to define these resistances, uncertainties remained, however, regarding the identity of resistance genes/alleles in different host cultivars. In an attempt to elucidate this somewhat confusing situation, crossing studies with resistant and susceptible barley cultivars were carried out. As a consequence, the genetic constitution of 16 resistant barley cultivars was amended using revised *Rh* symbols for barley genes conferring resistance to *R. commune* (Habgood and Hayes 1971). In particular, the genes for resistance from several barley cultivars (*Rh*, Rh^2, Rh^3, Rh^4, rh^5) were localized into a single complex locus on chromosome 3H, which had been previously defined by the genes *Rh, Rh3* and *Rh4* (Dyck and Schaller 1961a).

Eventually, in a large effort based on marker and phenotypic analyses of near-isogenic barley lines, a more appropriate and updated terminology of genes for resistance to *R. commune* in barley was suggested (Bjørnstad et al. 2002). The major locus on chromosome 3H was now designated *Rrs1* (Lehnackers and Knogge 1990) according to the nomenclature suggestion for plant resistance genes (Moseman 1972). In addition, the name of the reference cultivar was included as a suffix in the gene designation to allow the differentiation between the 15 alleles described so far. The *Rrs1* locus was mapped close to the centromere on the long arm of chromosome 3 (3HL; Barua et al. 1993; Graner and Tekauz 1996). However, it is still under debate whether the complex *Rrs1* locus is multiallelic or rather harbours a series of tightly linked *R* genes. Support for the latter came from inoculation experiments with *R. commune* isolates that were able to cause

Table 1 Major *Rhynchosporium* resistance genes

Gene/allele designation	Alternative designation	Chromosome	Origin	Barley accession[a]	References and remarks
Rrs1$_{Brier}$	*Rh, Rh1*	3HS (3)	*H. vulgare* ssp. *vulgare*	CI 7157	Bryner 1957 (*Rh*); Dyck and Schaller 1961b (*Rh*: Brier - CI 7157); Habgood and Hayes 1971 (*Rh1*: Brier, Hudson - CI 8067); Graner and Tekauz 1996 (*Rh1*: Triton); Bjørnstad et al. 2002
Rrs1$_{Turk}$	*Rh3*	3HS (3)	*H. vulgare* ssp. *vulgare*	CI 14400[b]	Dyck and Schaller 1961b (Turk - CI 5611-2[b], Atlas 46 - CI 7323, possibly Brier); Wells and Skoropad 1963 (Bey, Osiris - CI 1622, Rivale, Turk, CI 3515, CI 8256, 36Ab1991); Baker and Larter 1963 (Turk); Starling et al., 1971 (Turk); Habgood and Hayes 1971 (Wisconsin Winter × Glabron - CI 8162); Bjørnstad et al. 2002
Rrs1$_{La\ Mesita}$	*Rh4*, *Rh*^{4}	3HS (3)	*H. vulgare* ssp. *vulgare*	CI 7565	Riddle and Briggs 1950 (La Mesita - CI 7565: no designation); Dyck and Schaller 1961b (*Rh4*: La Mesita, Trebi - CI 936, Osiris - CI 1622); Habgood and Hayes 1971 (*Rh*^{4}: Osiris, La Mesita, CI 3515, CI 8256, Gembloux 14 -CI 8286); Bjørnstad et al. 2002; Hofmann et al. 2013
Rrs1$_{Modoc}$	*Rh4*^{2}, *Rh*^{2}	3HS (3)	*H. vulgare* ssp. *vulgare*	CI 7566	Riddle and Briggs 1950 (Modoc; - CI 7566: no designation); Dyck and Schaller, 1961 (*Rh4*^{2}: Modoc); Habgood and Hayes 1971 (*Rh4*^{2}: Osiris, La Mesita); Habgood and Hayes 1971 (*Rh*^{2}: Modoc); Bjørnstad et al. 2002

(Continued)

Table 1 (*Continued*)

Gene/allele designation	Alternative designation	Chromosome	Origin	Barley accession[a]	References and remarks
rrs1$_{Jet}$	*rh7*, *rh*^{5}	3HS (3)	*H. vulgare* ssp. *vulgare*	CI 967	Baker and Larter 1963 (*rh7*: Jet - CI 967; Steudelli - CI 2266: complementary to *rh6*); Habgood and Hayes 1971 (Jet: *rh7* redesignated to *rh*^{5}); Bockelman et al. 1977 (trisomic analysis: Jet - *rrs1* on chr. 3H; *rrs6* on 4H); Bjørnstad et al. 2002
Rrs1$_{CI\ 8162}$	*Rh*^{3}	3HS (3)	*H. vulgare* ssp. *vulgare*	CI 8162	Habgood and Hayes 1971 (Wisconsin Winter × Glabron - CI 8162); Reitan et al. 2002; Bjørnstad et al. 2002
Rrs1$_{AB240}$	–	3HS (3)	*H. vulgare* ssp. *vulgare*	AB240	Garvin et al., 1997 (BC_3F_2 population; CPI 109853 accession from Turkey as donor parent); Bjørnstad et al. 2002; Genger et al. 2003a(failed to respond to fungal avirulence factor NIP1)
Rrs1$_{Abyssinian}$	–	3HS (3)	*H. vulgare* ssp. *vulgare*	CI 668	Grønnerød et al. 2002 (Abyssinian); Bjørnstad et al. 2002
Rrs1$_{Steudelli}$	*rh7*, *rh*^{5}	3HS (3)	*H. vulgare* ssp. *vulgare*	CI 967, CI 2266	Baker and Larter 1963 (Jet, Steudelli: *rh6*,*rh7*); Habgood and Hayes 1971 (Jet, Turk, Brier, Modoc, Osiris): *rh7* redesignated *rh*^{5}; Bjørnstad et al. 2002 (s.o.)
Rrs1$_{B87/14}$	–	3HS (3)	*H. vulgare* ssp. *vulgare*	B87/14	Williams et al. 2001 (*Rrs. B87*); Bjørnstad et al. 2002
Rrs1$_{CI\ 11549}$	–	3HS (3)	*H. vulgare* ssp. *vulgare*	CI 11549[c]	Patil et al. 2003; Bjørnstad et al. 2002
Rrs1$_{CIho3515}$	–	3HS (3)	*H. vulgare* ssp. *vulgare*	CI 3515	Hofmann 2014 (CIho3515: Spanish accession)
Rrs1$_{Triton}$	–	3HS (3)	*H. vulgare* ssp. *vulgare*		Graner and Tekauz 1996 (*Rh:* Triton; German winter barley)
Rrs1$_{Halcyon}$	–	3HS (3)	*H. vulgare* ssp. *vulgare*	PI 531062	Genger et al. 2003b (Halcyon)

(*Continued*)

Gene/allele designation	Alternative designation	Chromosome	Origin	Barley accession[a]	References and remarks
Rrs1$_{Sultan}$	–	3HS (3)	*H. vulgare* ssp. *vulgare*	PI 94886 (?)	Genger et al. 2003b (Sultan)
Rrs2$_{Atlas}$	*Rh2*	7HS (1)	*H. vulgare* ssp. *vulgare*	CI 4118	Dyck and Schaller, 1961b (Atlas - CI 4118, Atlas 46); Starling et al., 1971 (Atlas, Atlas 46); Schweizer et al. 1995; Penner et al. 1996 (Falcon); Patil et al. 2002; Bjørnstad et al. 2002; Hanemann et al. 2009
Rrs2$_{Steudelli}$	–	7HS (1)	*H. vulgare* ssp. *vulgare*	CI 2266	Bjørnstad et al. 2002, 2004
Rrs2$_{CI\ 2222}$	–	7HS (1)	*H. vulgare* ssp. *vulgare*	CI 2222[c]	Patil et al. 2003
Rrs3	*rh6*, *Rh9*, *rrs6*$_{Jet}$, *Rrs9*$_{Kitchin}$	4H (4)	*H. vulgare* ssp. *vulgare*	CI 668 CI 2266 CI 1296	Baker and Larter 1963 (*rh6*: Jet - CI 967, Steudelli - CI 2266: complementary to *rh7*), Habgood and Hayes 1971 (Turk, Brier, Osiris, Modoc, Jet: *rh6*); Bockelman et al. 1977 (trisomic analysis: *rh6* inJet - CI 967 redesignated *rrs6* on chr. 4H); Bjørnstad et al. 2004 (Jet)/Baker and Larter 1963 (*Rh9*: Abyssinian - CI 668, Kitchin - CI 1296); Bockelman et al. 1977 (trisomic analysis: *Rh9* in Kitchin CI 1296 redesignated *Rrs9* on chr. 4H); not detected neither by Patil et al, 2002, or Grønnerød et al. 2002; Bjørnstad et al. 2002 (*rh7*/*Rh9* redesignated *Rrs3*)
Rrs4$_{E224/3}$	*Rhy*	3HL (3)	*H. vulgare* ssp. *vulgare*		Barua et al. 1993 (E224/3: Scottish breeding line); Bjørnstad et al. 2002
Rrs4$_{CI\ 11549}$	*rrsx (rhx)*	3HL (3)	*H. vulgare* ssp. *vulgare*	CI 11549[c]	Patil et al. 2002, 2003; Bjørnstad et al. 2002
Rrs5	*Rh5*	?	*H. vulgare* ssp. *vulgare*	CI 5611-2	Dyck and Schaller 1961b (Turk)

(*Continued*)

Table 1 (*Continued*)

Gene/allele designation	Alternative designation	Chromosome	Origin	Barley accession[a]	References and remarks
rrs8	*rh8*	?	*H. vulgare* ssp. *vulgare*	CI 2222[c]	Wells and Skoropad, 1963 (Nigrinudum - CI 2222[c]); Habgood and Hayes 1971 (Nigrinudum - CI 11797[c]); not found by Patil et al. 2003
Rrs10	*Rh4, Rh10*	3HL (3)	*H. vulgare* ssp. *vulgare*		Dyck and Schaller 1961b (Osiris and La Mesita: *Rh4* distinct from *Rh3* on chr. 3H); Habgood and Hayes 1971 (CI 3515, CI 8256: *Rh4* complementary to *Rh*[4], *Rh10* separate gene), Hansen and Magnus, 1973 (Pioneer: *Rh10*); not detected by Patil et al. 2002; Wallwork and Grcic 2011 (Osiris, La Mesita: $Rrs1_{Rh4/Rh10}$ distinct from $Rrs1_{Rh/Rh3}$)
rrs11	*rh11*	?	*H. vulgare* ssp. *vulgare*	CI 4364	Habgood and Hayes 1971 (CI 4364, CI 4368)
Rrs12	*Rrs2*	7HS (1)	*H. vulgare* ssp. *spontaneum*		Abbott et al. 1992 (*Rrs12, Rrs13*); Bjørnstad et al. 2002; Genger et al. 2003b; (Genger et al, 2005: *Rrs12* maybe allelic to *Rrs2;* Wagner et al. 2008: *Rrs2/Rrs12* is only one probably multiallelic locus)
Rrs13		6HS (6)	*H. vulgare* ssp. *spontaneum*		Abbott et al. 1992, 1995; Bjørnstad et al. 2002; Genger et al. 2003b
Rrs14		1HS (5)	*H. vulgare* ssp. *spontaneum*		Garvin et al. 1997, 2000; Bjørnstad et al. 2002; Genger et al. 2003b
Rrs15		7HL (1)	*H. vulgare* ssp. *spontaneum*		Genger et al. 2005
Rrs16		4HS (4)	*H. bulbosum*		Pickering et al. 2006

(*Continued*)

Gene/allele designation	Alternative designation	Chromosome	Origin	Barley accession[a]	References and remarks
Rrs17	*Rrs15*$_{CI\ 8288}$	2HS (2)	*H. vulgare* ssp. *vulgare*	CI 8288	Schweizer et al. 2004 (*Rrs15*$_{CI\ 8288}$); Wagner et al. 2008 (*Rrs15*$_{CI\ 8288}$); Zhan et al. 2008 (*Rrs15*$_{CI\ 8288}$ redesignated *Rrs17*)
Rrs18	–	6HS (6)	*H. vulgare* ssp. *vulgare*	CI 3515	Hofmann 2014

[a] U.S. National Plant Germplasm System.
[b] Bjørnstad used Turk accession CI 14400 from Turkey instead of accession CI 5611-2 from Azerbaijan.
[c] Different Nigrinudum accessions (CI 2222 from Ethiopia, CI 11549 from Japan, CI 11797 from The Netherlands).

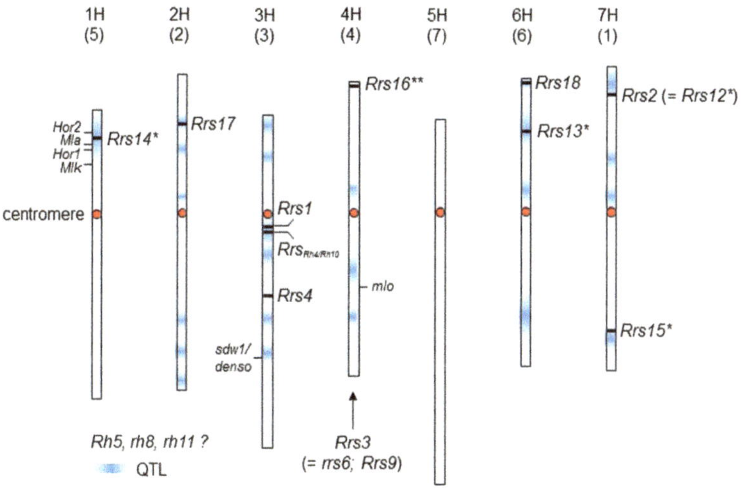

Figure 6 Barley genes conferring resistance to *R. commune*. The approximate position of major *R genes* and resistance QTLs on the barley chromosomes is depicted. Chromosome numbering corresponds to the *Triticeae* nomenclature (Linde-Laursen et al. 1997). In addition, the original numbers, which were mostly used in the older literature, are presented in brackets (Burnham and Hagberg 1956). * *H. spontaneum*, ** *H. bulbosum*; denso: prostrate growth (favours infection; Jensen et al. 2002), sdw1: semi-dwarfing (restriction of splash dispersal in tall cultivars; Looseley et al. 2015); *Mla, Mlk, mlo*: major *R* genes conferring resistance to powdery mildew, *Hor1, Hor2*: hordein genes.

distinguishable resistance patterns on plants carrying the genes *Rrs1* and *Rh4/Rh10* near the centromere of chromosome 3H (Wallwork and Grcic 2011). This suggests that *Rh4/Rh10* resistance is not caused by an *Rrs1* allele (Bjørnstad et al. 2002) but by a very closely linked distinct gene, *Rrs*$_{Rh4/Rh10}$.

In the meantime, genes conferring resistance to *R. commune* have been mapped to all barley chromosomes except 5H (Fig. 6). In addition to the *Rrs1*–*Rrs*$_{Rh4/Rh10}$ loci, the chromosome 3HL carries another *R* gene, *Rrs4*, mapping about 22 cM distally of *Rrs1* (Patil et al. 2003). Two resistance loci, *Rrs2* (Schweizer et al. 1995) and *Rrs15* (Genger et al.

2005), were assigned to chromosomes 7HS and 7HL, respectively. The same chromosome also harbours *Rrs12* (Abbott et al. 1992), which has been hypothesized to be located at the same probably multiallelic locus as *Rrs2* (Wagner et al. 2008). The gene *Rrs3* has a somewhat complicated history. Originally, the genes rh_6 and Rh_9 were identified in different barley cultivars (Baker and Larter 1963; Habgood and Hayes 1971). Renamed *rrs6* and *Rrs9*, they were assigned to chromosome 4H by a trisomic analysis (Bockelman et al. 1977) and later suggested to be at a single locus, which was redesignated as *Rrs3* (Bjørnstad et al. 2002). The genes *Rrs13*, *Rrs14* and *Rrs16* were mapped to chromosomes 6HS, 1HS and 4HS, respectively. *Rrs12*, *Rrs13* (Abbott et al. 1992), *Rrs14* (Garvin et al. 2000) and *Rrs15* (Genger et al. 2005) were identified in wild barley accessions (*H. vulgare* subsp. *spontaneum*). *Rrs16* was introgressed from another wild barley species, *H. bulbosum*, to *H. vulgare* (Pickering et al. 2006). The latest addenda to the *R* gene list are *Rrs17* and *Rrs18*, which were identified on chromosomes 2HS (Schweizer et al. 2004) and 6HS (Hofmann 2014), respectively. *Rrs17* was originally published under the designation *Rrs15*, but was recommended to be renamed (Zhan et al. 2008) due to the earlier referencing of *Rrs15* on chromosome 7HL (Dahleen 2006).

In addition to the major *R* genes, quantitative trait loci (QTLs) conferring resistance to *R. commune* were identified in the barley genome (Zhan et al. 2008). Again with the exception

Table 2 *Rhynchosporium* QTLs

	Chromosome						
Reference	1H	2H	3H	4H	5H	6H	7H
Backes et al. 1995		+	+			+, +	+
Thomas et al. 1995	+		+, +, +	+			+
Spaner et al. 1998			+	+		+	
Jensen et al. 2002			L (*denso*)	+		+	
Grønnerød et al. 2002			S, L, L				S
Genger et al. 2003b			L				
Bjørnstad et al. 2004	+		+	+		+	+ (*Rrs2*?)
Sayed et al. 2004		+	+, +				
von Korff et al. 2005		S	L	S			
Yun et al. 2005, 2006	+						
Shtaya et al. 2006			L	L		S, S	
Cheong et al. 2006			L			L	S, L
Wagner et al. 2008		S				S	S, S
Li and Zhou 2011	S	+	L, L				
Looseley et al. 2012		L	L (*sdw1*)				+
Looseley et al. 2015			S, L, L			S	
Hofmann et al. 2013		S	L	S			
Hofmann 2014			S, L			S	

+: QTL present, S: short chromosome arm, L: long chromosome arm.

of chromosome 5H, QTLs were mapped to all other chromosomes (Fig. 6; for details, see Table 2). In several barley mapping populations, QTLs were identified in chromosomal regions, where major *R* genes had been described. In particular in the centromeric region of chromosome 3H, which harbours the *Rrs1–Rrs*$_{Rh4/Rh10}$ loci, quantitative resistance could be mapped (Genger et al. 2003b; Bjørnstad et al. 2004; Shtaya et al. 2006; Li and Zhou 2011; Looseley et al. 2015; Hofmann 2014). In addition, QTLs were detected in the regions of *Rrs2/Rrs12* on chromosome 7HS (Cheong et al. 2006; Wagner et al. 2008; Hofmann et al. 2013), of *Rrs13* on chromosome 6HS (Shtaya et al. 2006; Wagner et al. 2008), of *Rrs14* on chromosome 1HS (Yun et al. 2005, 2006), of *Rrs15* on chromosome 7HL (Cheong et al. 2006), of *Rrs17* (=*Rrs15*$_{CI8288}$) on chromosome 2HS (Wagner et al. 2008; Hofmann et al. 2013) and of *Rrs18* (Jensen et al. 2002; Cheong et al. 2006; Hofmann 2014). In each of these cases, the QTL may represent a weaker allele of the respective resistance gene. However, conclusions need to be drawn with caution because results from the mapping experiments are often difficult to compare due to the use of different barley genotypes, different fungal isolates and above all different markers (marker systems). Consequently, the position on the genetic map (Fig. 6) and the identity of *R* genes and QTLs represent suggestions, which need to be confirmed by physical/sequencing data. Finally, it should be noted that several of the previously described resistance genes (*Rh5*, *rh8*, *rh11*) have not been confirmed and mapped yet.

4.2 Resistance mechanisms

Several histological analyses involving different barley cultivars and *R. commune* isolates were carried out to follow the infection process from conidial germination to sporulation and to identify putative resistance mechanisms (Ayesu-Offei and Clare 1970; Ayres and Owen 1971; Ryan and Grivell 1974; Hosemans and Branchard 1985; Howlett and Cooke 1987; Howlett and Cooke 1992; Lehnackers and Knogge 1990; Lyngs Jørgensen et al. 1993). The inhibition of spore germination on the leaf surface (Lehnackers and Knogge 1990) and the prevention of cuticular penetration by host cell wall modifications (Xi et al. 2000) were described to contribute to resistance. Penetration attempts are usually associated with the formation of papillae and haloes (Fig. 4; Ayesu-Offei and Clare 1970; Lehnackers and Knogge 1990). Furthermore, localized thickenings of the cell wall in the vicinity of subcuticular hyphae have been reported (Jones and Ayres 1974). However, fungal hyphae grow beneath the cuticle, whereas papillae are formed on the inner side of the epidermal cell walls, suggesting that resistance cannot be based on papillae as physical barriers. Nevertheless, in some studies, a correlation was found between the degree of resistance and the size of the haloes, suggesting papilla formation to be part of a primary resistance mechanism (Lyngs Jørgensen et al. 1993).

During infection, *R. commune* remains in the apoplast (Stotz et al. 2014). The resistance of barley is not mediated by a hypersensitive response and the pathogen is not usually killed (Mur et al. 2008). Barley cultivars are called resistant if they fail to generate scald lesions upon inoculation. However, infection can be symptomless (Davis and Fitt 1990) and fungal mycelial structures could be detected even in resistant cultivars, mostly near the leaf margins (Fig. 4; Ayres and Owen 1971; Ali 1974; Lehnackers and Knogge 1990; Lyngs Jørgensen et al. 1993). Meanwhile, molecular techniques (qPCR) were used to show symptomless colonization of a resistant winter barley cultivar by *R. commune* throughout the growing season, albeit fungal DNA levels stayed about 10-fold lower than in a susceptible cultivar (Fountaine et al. 2007). Nevertheless, the formation of spores was observed during symptomless infection (Thirugnanasambandam et al. 2011). Inoculation of two barley cultivars differing in the *Rrs1* gene (both carrying the *Rrs2* gene) with an *Rrs1*-incompatible fungal isolate resulted in a clear-cut difference in fungal growth morphology. In the incompatible interaction, the fungal isolate failed to establish a fully functional mycelial network in the leaf, growing less extensive and less regular than in the compatible interaction. It remains to be shown though which mechanisms provoke this 'disorientation of hyphal growth' (Thirugnanasambandam et al. 2011). However, these observations suggest that the resistance of barley to *R. commune* generally may be a quantitative rather than a qualitative feature.

To date, none of the *R* genes conferring resistance to *R. commune* has been identified and sequenced. However, for two loci, *Rrs1* and *Rrs2,* fine mapping approaches were carried out, both impeded, however, by significantly reduced recombination frequencies. This is typical for the centromeric regions of barley chromosomes such as the one on chromosome 3HL harbouring *Rrs1* (Graner and Tekauz 1996). Also, *Rrs2* resides in a distal region of chromosome 7HS that shows suppressed recombination in the mapping population (Hanemann et al. 2009). Thus, for both genes, map-based cloning strategies have been aggravated by insufficient recombination. Nevertheless, *Rrs1* was mapped to a region of 1.2 cM, which defined an interval of *c.* 8.9 Mb with an estimate of about 30 annotated genes (Hofmann et al. 2013).

In contrast to the host side, the fungal (avirulence) gene pivotal to *Rrs1*-based resistance was identified. *R. commune* secretes a set of small proteinaceous virulence effectors, which had been named necrosis-inducing peptides (NIP1, NIP2 and NIP3) due to their induction of necrosis upon injection into barley leaves (Wevelsiep et al. 1991). No biochemical function could be assigned yet to NIP2. In contrast, NIP1 and NIP3 were found to stimulate the activity of the plasma membrane-localized H^+-ATPase in the host (Wevelsiep et al. 1993). This enzyme generates a cellular membrane potential, which may affect nutrient accumulation and fungal membrane transport processes. In addition, the acidification of the apoplast may create more convenient growth conditions for the fungus (Elmore and Coaker 2011). Importantly, NIP1 is an effector protein with dual functions (van't Slot and Knogge 2002): in addition to its non-specific function in virulence (H^+-ATPase stimulation), it is able to induce defence genes specifically in *Rrs1* cultivars (Hahn et al. 1993; Rohe et al. 1995; Fiegen and Knogge 2002; Steiner-Lange et al. 2003; Kirsten et al. 2012). The *NIP1* gene was found in only ~55% of the 196 fungal isolates analysed (Schürch et al. 2004). Hence, gene deletion is the major mechanism employed by the fungus to re-establish virulence. In addition, *NIP1* alleles were identified, which code for protein haplotypes with few variable amino acid positions, some of which affecting both the virulence function (H^+-ATPase stimulation) and the avirulence function (defence gene induction; Fiegen and Knogge 2002). The protein was expressed *ex planta* (Gierlich et al. 1999) to solve its three-dimensional structure (van't Slot et al. 2003). A binding site was identified on plasma membranes of barley irrespective of the resistance genotype and of other cereals (van't Slot et al. 2007), suggesting that on the plant side, a putative membrane 'receptor' alone is not sufficient for defence induction. It remains to be shown whether it is one of the proteins encoded in the genomic *Rrs1* interval that is required to interact with NIP1 to stimulate host defence and plant resistance.

Rrs2 could be delimited to an interval of 0.08 cM (Hanemann et al. 2009). Sequence information was obtained from BACs spanning the interval, leaving, however, a gap of unknown size. Eleven putative genes were identified with homology to barley ESTs and rice genes. In addition to a putative serine–threonine protein kinase and a β-1,3-glucanase, five proteins containing a pectin esterase inhibitor (PEI) domain are encoded in this genomic region. Pectin is an important component of plant cell walls. Its precursors are synthesized and secreted in methyl esterified form and partially de-esterified in the plant cell wall by plant pectin methyl esterases (PMEs), which are regulated by plant PEIs. It is noteworthy in this context that subcuticular hyphae of *R. commune* grow in a region of the epidermal cell wall that is predominantly cellulosic, and is also rich in pectic substances (Jones and Ayres 1974). Furthermore, evidence for the involvement of fungal pectin-degrading enzymes has been reported (Jones and Ayres 1974; Linsell et al. 2011). Hence, by inhibiting plant PME activity, PEIs lead to highly methyl esterified pectin, which is less susceptible to the degradation by secreted pathogen enzymes, rendering the plant more resistant to pathogen attack (for review, see Lionetti et al., 2012).

Four of the five *PEI* genes (*PEI2*, *PEI3*, *PEI4* and *PEI6*) in the *Rrs2* region showed different haplotypes in *Rrs2* and *rrs2* barley cultivars (Marzin et al. 2016). To date, three of the genes (*PEI2*, *PEI3* and *PEI4*) were therefore used to stably transform barley cultivar 'Golden Promise' (*rrs2*). However, none of the genes conferred a high level of resistance to the transgenic plants, although an improved resistance level was found with four of the five *PEI4*-transgenic lines. It remains to be shown therefore whether the not yet tested *PEI6* or a combination of different *PEI* genes are causal for *Rrs2*-based resistance. Furthermore, it cannot be excluded that so far unknown *PEI* genes or totally different genes at this locus are required.

5 Host specificity, effector proteins and agricultural consequences of fungal biology

5.1 Host specificity and the role of effector proteins

Recently, the genomes of isolates from all currently defined *Rhynchosporium* species were sequenced, assembled and annotated *ab initio* (Penselin et al. 2016). This enabled a comparative genomic approach to identify genes, which may play a role in controlling virulence and host specificity. Fungal effector proteins affecting the host physiology are of prime interest in this context. The necrosis-inducing proteins NIP1, NIP2 and NIP3 had been the first effector proteins described for *R. commune* (Wevelsiep et al. 1991). The *NIP3* gene is missing in the genomes of the CCG isolates, the *NIP1* gene in the isolates of the two BCG species *R. secalis* and *R. agropyri*. It cannot be excluded, however, that both genes are present in other isolates of these fungal species. When present in a genome, *NIP1* and *NIP3* occur as single-copy genes. This is also true for *NIP2* in the CCG genomes. In contrast, however, the BCG species are characterized by *NIP2* families of 7–10 members, which appear to be the result of duplication events in a common ancestor before divergence of the BCG species (Penselin et al. 2016). Hence, this duplication process may have allowed or facilitated the host jump and the colonization of a new crop grass species.

The simplest model explains host specificity by the occurrence of effector genes in the genome of one *Rhynchosporium* species and their absence from other species. Alternatively, effector genes, albeit present in the genomes of several *Rhynchosporium* species, may be specifically expressed in the interaction of one species with its host species only, but not in the interactions of the other fungal species. Finally, host specificity may originate from structural differences of effector proteins rather than from the presence or expression of specific genes. As a starting point, a comparative analysis of the *Rhynchosporium* genomes resulted in the identification in *R. commune* of seven candidate genes coding for secreted putative effector proteins (*R. commune*-specific proteins, RcSPs; Penselin et al. 2016). Upon deletion of these genes, which are all expressed transiently early during the pathogenesis, the resulting mutants were compared to the wild-type isolate regarding disease phenotype and fungal development *in planta*. Surprisingly, all mutants showed a stronger development, that is, a higher fungal biomass; some also showed a more severe disease phenotype in susceptible barley leaves than the wild type. Or else, in the presence of these *R. commune*-specific effector proteins, fungal growth was decelerated. A similar increase in fungal biomass had been observed with several mutants of the grass endophyte *Epichloë festucae*, which had switched from mutualistic to pathogenic growth by the inactivation of different (non-effector) genes (Tanaka et al. 2006, 2008; Takemoto et al. 2006; Johnson et al. 2013). Therefore, the RcSP effectors may play a role in establishing an endophyte-like growth in the host leaves during the symptomless stage of pathogenesis. It is tempting to speculate that the hemibiotrophic apoplastic fungus *R. commune* adapts to its barley host by stabilizing the biotrophic stage in favour of the necrotrophic destructive stage.

In addition to proteins, small molecules such as host-specific toxins originating from fungal secondary metabolism may play a role in colonizing a particular host niche. Comparative *Rhynchosporium* genomics identified genes coding for key enzymes of fungal secondary metabolism such as polyketide synthases and non-ribosomal peptide synthetases, which show specific occurrence in the *Rhynchosporium* genus as well as in *Rhynchosporium* species and in isolates (Penselin et al. 2016). However, the analysis of *Rhynchosporium* secondary metabolism is still in its infancy and future research needs to shed light on the compounds and functions and their possible involvement in determining host specificity.

5.2 Agricultural consequences of fungal biology

To design strategies to avoid or minimize the damage caused by a disease and to decrease the severity of epidemics, it is important to understand the biology of the pathogen. The characteristics of the developmental cycle of *R. commune* provide the points of attack for useful agricultural practices. Fungal primary inoculum is derived from infected plant debris (stubble and straw), in particular if left under moist conditions conducive for sporulation (Mayfield and Clare 1984b). Deep tilling (ploughing) after harvest or stubble treatment (burning and grazing) may therefore help to reduce or remove this inoculum source (Elen 2003). Crop rotation with non-host plants aims into the same direction to avoid the establishment or deplete the soil of infectious material from the previous growing season, and it was shown to be efficient (Shipton et al. 1974; Mayfield and Clare 1984a; Arvidsson 1998; Zhan et al. 2008). The secondary inoculum of the polycyclic pathogen *R. commune* is spread from the infected lower leaves to the non-infected upper leaves through water droplets. The success of the leaf-to-leaf transfer may be impaired by using cultivar mixtures with differing resistance levels or by reducing the sowing density and, hence, the canopy density. The latter may also influence the duration of leaf wetness required for efficient spore germination (Davis and Fitt 1994; Zhan et al. 2008). Seeds from affected crops should be avoided and the proper sowing time should be selected to avoid high inoculum levels or weather conditions conducive to fungal development (Davis and Fitt 1992).

In addition to cultivation practices, barley genetic factors are able to restrict scald infection and epidemics. It is discussed, for instance, that the growth habit of a cultivar (e.g. prostrate *vs.* erect) as associated with the dwarfing genes *sdw1* and *ari-eGP* (Bingham et al. 2009; Looseley et al. 2012, 2015) may be helpful in limiting the spread of secondary inoculum; tall plants are able to grow away from the disease at the lower leaves (disease escape; Zhan et al. 2008; Avrova and Knogge 2012; Walters et al. 2012). This was substantiated by finding *sdw1* to be a resistance QTL (Looseley et al. 2012, 2015). Major *R* genes and/or QTLs are certainly classical genetic factors, which are deployed to avoid or restrict disease development. One of the first *R* genes utilized in breeding programmes for resistance to *R. commune* was $Rrs1_{Turk}$ (Table 1; Riddle and Suneson 1948). This gene was extracted from cultivar 'Turk' by backcrossing with highly susceptible malting barley 'Atlas' (Turk X Atlas[7]). The resulting cultivar was combined with a similarly backcross-derived, powdery mildew-resistant cultivar (Hanna X Atlas[6]) to yield the scald- and mildew-resistant cultivar 'Atlas 46'. This cultivar was released for the growers in California in 1947 (Schaller 1948). By 1953, scald was reported on it and by 1956, it had become extremely susceptible (Shipton et al. 1974). Therefore, the history of this cultivar highlights the general problem with deploying single major *R* genes, which usually are overcome by pathogens within a short time. Furthermore, *Rhynchosporium* populations turn out to be extremely variable and easily adapt to new situations (Jackson and Webster 1976a; Zhang et al. 1992; Newton et al. 2001; Xi et al. 2003; Stefansson et al. 2013). This extraordinary adaptiveness even led to the establishment of a new host species, the wheat-rye crossing product triticale (× *Triticosecale* Wittmack). About 30 years after this man-made crop had been introduced, it was found to be susceptible to a member of the *R. secalis* lineage (Oettler 2005; Welty and Metzger 1996).

To be successful, breeding programmes need to aim at reducing the likelihood of fungal resistance breaking through a single or very few mutations. To this end, several genetic tools need to be generated and/or improved. Extended barley germplasm has been successfully used already to identify novel resistances in wild barley (*H. spontaneum*, Table 1; Abbott et al. 1992; Garvin et al. 2000; Genger et al. 2005; Yun et al. 2005) and in bulbous barley (*H. bulbosum*, Table 1; Pickering et al. 2006). Identification and integration of novel resistances, that is, single *R* genes as well as genetically more complex and, hence, most likely more durable QTLs, into elite barley cultivars by marker-assisted selection need to be intensified. Today, adequate diagnostic markers for *R* genes and QTLs can be more efficiently obtained due to the availability of high-density maps and DNA sequence information of barley. Furthermore, the availability of the barley genome sequence (Mayer et al. 2011, 2012) facilitates map-based cloning of genes from resistance loci. The final goal may be the pyramiding of several *R* genes, preferentially located on different chromosomes, to 'simulate' the situation of cultivar mixtures and their protection benefits in a single cultivar (Walters et al. 2012). Furthermore, breeding programmes may want to combine *R* genes with strong/stable QTLs. This approach has been successfully carried out, for instance, with the resistance to barley stripe rust, where a major *R* gene was combined with several QTLs on different chromosomes (Castro et al. 2003). Quantitative resistance conferred by QTLs is, in general, not very well understood, if at all. Therefore, future research programmes should attempt to unravel the molecular basis of these resistances, including the question whether at least some of the *R* genes conferring resistance to *R. commune* are closely associated with QTLs or even very efficient QTL genes. A very promising goal may be, however, to identify old Scandinavian barley landraces and, in particular, the original host species of *Rhynchosporium*. As the result of a long selection period and of the plant–fungus coevolution, respectively, these plants can be expected to be rich sources of novel resistances (McDonald 2015).

6 Crop protection and management

6.1 Fungicide-based crop protection

As for most crop plants, fungicides are key players in protecting barley from fungal pathogens. However, to maintain their efficiency, fungicides should only be used when necessary. For instance, scald frequently occurs on the seedlings of winter barley during fall. Nevertheless, fungicides should only be applied in spring during tillering or at stem extension (Fig. 3). Provided that the growing season is wet, *R. commune* can also spread to the ears and awns of the host. In this case, the upper leaves should be protected with an accurately timed fungicide treatment (Walters et al. 2012). In addition, the selection of a particular product requires knowledge of the fungicide sensitivity levels of the local fungal population.

Since the 1970s, methyl benzimidazole carbamates (MBCs) and demethylation inhibitors (DMIs, triazoles) have been used, followed increasingly from the mid-1990s by the quinone outside inhibitors (QoIs, strobilurins). However, like other plant pathogenic fungi, *R. commune* has evolved resistances to several of these fungicides (Kendall et al. 1994, Locke and Phillips 1995; Wheeler et al. 1995; Taggart et al. 1999). MBCs bind to β-tubulin and interfere with mitosis and other cellular tubulin-requiring processes (Davidse and Flach 1978). Resistance is associated with amino acid alterations in the β-tubulin protein, which decrease MBC binding (Hollomon et al. 1998). Most fungi including *R. commune* have only one β-tubulin gene. Hence, a single gene controls resistance to MBC fungicides. Nevertheless, it took *R. commune* 17 years to develop resistance in the United Kingdom (Grimmer et al. 2014).

Ergosterol is a critical compound in cell membranes of fungi, but absent in plants and animals. Therefore, a number of fungicides affecting different steps of ergosterol biosynthesis were developed. The class of DMI fungicides (azoles) targets the P450 enzyme lanosterol 14α-demethylase, which is encoded by the *CYP51* gene (Becher and Wirsel 2012). Due to the extensive use of these fungicides (Azevedo et al. 2015), not surprisingly many fungi including *R. commune* evolved a reduced DMI sensitivity or even azole resistance within less than 10 years (Kendall et al. 1993; Robbertse et al. 2001; Grimmer et al. 2014). Three main mechanisms were identified to diminish the activity of DMI fungicides: mutations in the CYP51 enzyme, overexpression of the *CYP51* gene or reduction of intracellular azole accumulation *via* the activation of efflux pumps (Becher and Wirsel 2012; Cools et al. 2013; Omrane et al. 2015). Several ascomycetes such as *R. commune* contain multiple *CYP51* paralogues. In a UK study, it was demonstrated that the decrease in azole sensitivity of *R. commune* correlated with the emergence of the *CYP51A* paralogue in the *R. commune* population and with the intensification of azole use in agriculture (Hawkins et al. 2014). An analysis of *R. commune* isolates from 18 global field populations confirmed this result by detecting the *CYP51A* gene in all UK isolates. In contrast, the gene was present only in a fraction of the isolates from New Zealand (~85%), Switzerland (~76%) and Spain (~20%) or missing in isolates from Australia, Scandinavia, California and Ethiopia (Brunner et al. 2016). This demonstrates that the azole resistance situation varies drastically in different barley-growing regions. Furthermore, particular members of this fungicide group may remain very effective to *R. commune* (Cools et al. 2013). In a recent study (Mohd-Assaad et al. 2016), three additional azole resistance loci were identified which code for saccharopine dehydrogenase (SDH, lysine biosynthesis), vacuolar cation channel YVC1 (regulation of Ca^{2+} homeostasis) and a fungal-specific transcription activator (TA). Therefore, DMI fungicides appear to have pleiotropic effects. Nevertheless, the major contribution to resistance variation came from *CYP51A* presence–absence (62%), whereas SDH, YVC1 and TA contributed 21.9%, 15.2% and 0.7%, respectively. The mechanisms linking the three novel proteins with azole resistance need to be further elucidated.

The fungicides of the strobilurin-based class of QoIs target the mitochondrial cytochrome b encoded by the mitochondrial *cytb* gene. Several fungi quickly developed resistance to these fungicides by a single mutation in codon 143 of the *cytb* gene leading to a glycine-to-alanine exchange (G143A) (Grimmer et al. 2014). Furthermore, moderate resistance is inferred by a second mutation, a phenylalanine-to-leucine exchange at position 129 (F129L). However, in a global sample of *R. commune* collected between 1997 and 2004, the G143A mutation was not found (Torriani et al. 2009). *R. commune* does not possess the intron at codon 143 (Torriani et al. 2014), which protects QoI sensitivity in *Puccinia* spp. (Grasso et al. 2006) and *Pyrenophora teres* (Sierotzki et al. 2007). Eventually, the mutation was detected in a routine screening in 2008 in a few *R. commune* isolates in northern France (Walters et al. 2012), and occasionally, albeit at a very low frequency, in isolates from France, the United Kingdom and Spain (FRAG-UK 2016; Fountaine 2011). Also in Irish *R. commune* populations sampled between 2006 and 2014, only a low frequency of the mutation (2–18%) was found by targeted sequencing (Phelan et al. 2017). In contrast, according to a recent information sheet from the UK Agriculture and Horticulture Development Board (AHDB), disease control levels by different strobilurins have already become relatively low in the United Kingdom by now (AHDB 2017). It remains to be shown therefore why the rapid buildup of complete QoI resistance that was seen in other major pathogens did not occur in *R. commune* at least in some of the growing regions.

Another class of fungicides originally called carboxamide fungicides target complex II of fungal respiration. Since the early 2000s, a series of 'next-generation' succinate dehydrogenase inhibitors (SDHI) has been generated with activity towards a broad spectrum of fungi (Sierotzki and Scalliet 2013). Isolates of several fungi showed mutations in different SDH subunits, which affect SDHI binding and reduce the sensitivity to the fungicides (Stammler et al. 2015). However, no decrease in the sensitivity has been described to date for *R. commune* and according to the AHDB, SDHIs alone and in mixtures with other fungicides have the highest control levels towards *R. commune* (Oilseeds 2017).

Following pathogen infection, plants often develop broad-spectrum, systemic and long-lasting enhanced resistance to subsequent infection (systemic acquired resistance; Durrant and Dong 2004). In addition, non-pathogenic rhizobacteria are able to trigger plant defence reactions (induced systemic resistance; Pieterse et al., 2014). This response can also be induced by various biotic and abiotic agents such as bacterial or fungal cell wall fragments, proteins or various chemicals (Wiesel et al. 2014). Therefore, the practical applicability of these resistance elicitors (also called plant immunity inducers, *cf.* Dewen et al. 2017) and their protective effect towards *R. commune* were tested on barley under greenhouse and field conditions (Walters et al. 2014). Under greenhouse conditions, the combination of the resistance elicitors, namely acibenzolar-S-methyl, β-aminobutyric acid and *cis*-jasmone provided more protection to *R. commune* than the single substances. In field experiments, however, the effect of the elicitor combination proved to be inconsistent by varying drastically between significant control in one year and infectivity in the other. Better disease control and consistency were obtained when the elicitor treatment was combined with the fungicide treatment, even at a lower rate. As a consequence, elicitor/

fungicide combinations may help to reduce fungicide use and to prolong the effectivity period of particular fungicides.

6.2 Integrated crop protection management

In particular in northwestern Europe, fungicides play a dominant role in protecting plants from infection by pathogenic fungi. However, the development of resistance to fungicides in *R. commune* threatens the concept of chemical scald control. Currently, there is still sufficient, although regionally varying, sensitivity even to single members of fungicide families to which resistance has been described already. Nevertheless, the use of mixtures of compounds that belong to different fungicide families is recommended (AHDB 2017). More importantly, however, an integrated crop protection management is required to keep fungicides effective as long as possible. Briefly, this management should bring together a careful risk assessment, adequate agricultural practices, intelligent breeding and deployment of resistant/less susceptible barley cultivars, avoidance of gene flow and educated usage of fungicides. Risk assessment should be based on information regarding the host (cultivar, sowing time, growth habit), the environment (soil condition, weather) and the fungus (disease monitoring including molecular methods, prognostic system). Agricultural practices include crop rotation and cultivation (stubble treatment, deep tilling). Breeding strategies should select and combine different resistances (*R* genes and QTLs) in valuable barley germplasm depending on the regional needs. Seeds as well as other plant material should only cautiously, if at all, be transmitted over longer distances to avoid the translocation of fungal virulence and fungicide resistance genes. Finally, the knowledge of the fungicide resistances prevalent in the local *R. commune* population should be regarded as the *sine qua non* for fungicide selection. Mixtures of different fungicide classes, possibly complemented with resistance elicitors, should be used with careful timing and quantity.

7 Conclusion

Fungi of the *Rhynchosporium* genus are serious pathogens of several sweet grass species including cereals. In particular, *R. commune* poses a severe threat to barley production worldwide. However, research in recent years on fungal virulence and fungicide resistance and in particular on fungal population biology furnished several key points for disease management. These include barley breeding strategies aiming at quantitative rather than qualitative resistance to *R. commune* because of the mostly quantitative nature of this pathogen's virulence. When major resistance genes are provided, they should be pyramided or used in cultivar mixtures to pose an, as costly as possible, obstacle to the pathogen. The supra-regional movement of the pathogen in particular through seeds should be controlled to avoid the spread of unfavourable traits such as fungicide resistance. Fungicides should be selected based on the local resistance prevalence and compounds should be alternated based on their mode of action. Furthermore, the simple removal of plant straw and stubble treatment are well suited to reduce fungal inoculum. Finally, to counteract variability and adaptability of the fungus, future studies need to find out where and how sexual recombination takes place. All in all, the fungus still offers a tremendous potential for basic research on several aspects of its biology, which will be facilitated now by the availability of the *Rhynchosporium* genome sequences.

8 Where to look for further information

The following articles provide overviews and more detailed insight, respectively, into some of the aspects covered by the present review on scald.

- Reviews on *Rhynchosporium*: Shipton et al. 1974, Zhan et al. 2008, Avrova and Knogge 2012.
- Infection strategies of phytopathogenic fungi including *Rhynchosporium*: Horbach et al. 2011.
- *Rhynchosporium* genomics: Penselin et al. 2016, Torriani et al. 2014.
- Deduction of disease management strategies from *Rhynchosporium* population biology: McDonald 2015.
- Integrated crop protection strategies: Walters et al. 2012.
- Resistance elicitors: Wiesel et al. 2014.
- Fungicide resistance and management: Chapters 12–14; van den Bosch et al. 2014.
- Emergence and evolution of new pathogens in agro-ecosystems: McDonald and Stukenbrock 2016.

9 References

Abang, M. M., Baum, M., Ceccarelli, S., Grando, S., Linde, C. C., Yahyaoui, A., Zhan, J. S. and McDonald, B. A. (2006) 'Differential selection on *Rhynchosporium secalis* during parasitic and saprophytic phases in the barley scald disease cycle', *Phytopathology*, 96(11), 1214–22.

Abbott, D. C., Brown, A. H. D. and Burdon, J. J. (1992) 'Genes for scald resistance from wild barley (*Hordeum vulgare* ssp. *spontaneum*) and their linkage to isozyme markers', *Euphytica*, 61(3), 225–32.

AHDB Cereals & Oilseeds (2017) Agriculture and Horticulture Development Board, Stoneleigh Park, Kenilworth, Warwickshire, CV8 2TLUK.

Ali, S. M. (1974) 'Factors influencing infection, colonization and symptom expression in barley by *Rhynchosporium secalis*', *Aust. J. Agric. Res.*, 25(1), 9–20.

Ali, S. M. (1975a) 'Inheritance of scald resistance in barley. I. Resistance genes of group A barley cultivars', *Aust. J. Agric. Res.*, 26(2), 243–50.

Ali, S. M. (1975b) 'Inheritance of scald resistance in barley. II. Resistance genes of group B barley cultivars', *Aust. J. Agric. Res.*, 26(2), 251–7.

Ali, S. M. and Boyd, W. J. R. (1973) 'Host range and physiologic specialization in *Rhynchosporium secalis*', *Aust. J. Agric. Res.*, 25(1), 21–31.

Ali, S. M., Mayfield, A. H. and Clare, B. G. (1976) 'Pathogenicity of 203 isolates of *Rhynchosporium secalis* on 21 barley cultivars', *Physiol. Plant Pathol.*, 9(2), 135–43.

Arabi, M. I. E., Al-Shehadah, E. and Jawhar, M. (2009) 'Virulence spectrum to barley (*Hordeum vulgare* L.) in isolates of *Rhynchosporium secalis* from Syria', *J. Plant Dis. Protect.*, 116(6), 274–7.

Arabi, M. I. E., Jawhar, M. and Al-Shehadah, E. (2008) 'Molecular and pathogenic variation identified among isolates of *Rhynchosporium secalis* from Syria', *J. Plant Pathol.*, 90(2), 179–84.

Araz, A. and Maden, S. (2006) 'Pathogenic variation among isolates of *Rhynchosporium secalis* from cultivated barley growing in Central Anatolia, Turkey', *Plant Pathology J.*, 5(2), 244–7.

Arvidsson, J. (1998) 'Effects of cultivation depth in reduced tillage on soil physical properties, crop yield and plant pathogens', *Eur. J. Agron.*, 9(2–3), 79–85.

Arzanlou, M., Karimi, K. and Mirabi, F. (2016) 'Some evidence for skewed mating type distribution in Iranian populations of *Rhynchosporium commune*, the cause of barley scald disease', *J. Plant Prot. Res.*, 56(3), 237–43.

Avrova, A. and Knogge, W. (2012) '*Rhynchosporium commune*: a persistent threat to barley cultivation', *Mol. Plant Pathol.*, 13(9), 986–97.

Ayesu-Offei, E. N. and Carter, M. V. (1971) 'Epidemiology of leaf scald of barley', *Aust. J. Agric. Res.*, 22(3), 383–90.

Ayesu-Offei, E. N. and Clare, B. G. (1970) 'Processes in the infection of barley leaves by *Rhynchosporium secalis*', *Aust. J. Biol. Sci.*, 23(2), 299–307.

Ayres, P. G. and Owen, H. (1971) 'Resistance of barley varieties to establishment of subcuticular mycelia by *Rhynchosporium secalis*', *Trans. Br. Mycol. Soc.*, 57(2), 233–40.

Azevedo, M. M., Faria-Ramos, I., Cruz, L. C., Pina-Vaz, C. and Rodrigues, A. G. (2015) 'Genesis of azole antifungal resistance from agriculture to clinical settings', *J. Agric. Food Chem.*, 63(34), 7463–8.

Baker, R. J. and Larter, E. N. (1963) 'The inheritance of scald resistance in barley', *Can. J. Genet. Cytol.*, 5(4), 445–9.

Barua, U. M., Chalmers, K. J., Hackett, C. A., Thomas, W. T. B., Powell, W. and Waugh, R. (1993) 'Identification of RAPD markers linked to a *Rhynchosporium secalis* resistance locus in barley using near-isogenic lines and bulked segregant analysis', *Heredity*, 71(2), 177–84.

Becher, R. and Wirsel, S. G. R. (2012) 'Fungal cytochrome P450 sterol 14α-demethylase (CYP51) and azole resistance in plant and human pathogens', *Appl. Microbiol. Biotechnol.*, 95(4), 825–40.

Beigi, S., Zamanizadeh, H., Razavi, M. and Zare, R. (2013) 'Genetic diversity of Iranian isolates of barley scald pathogen (*Rhynchosporium secalis*) making use of molecular markers', *J. Agr. Sci. Tech.*, 15(4), 843–54.

Bingham, I. J., Walters, D. R., Foulkes, M. J. and Paveley, N. D. (2009) 'Crop traits and the tolerance of wheat and barley to foliar disease', *Ann. Appl. Biol.*, 154(2), 159–73.

Bjørnstad, A., Grønnerod, S., Key, J. M., Tekauz, A., Crossa, J. and Martens, H. (2004) 'Resistance to barley scald (*Rhynchosporium secalis*) in the Ethiopian donor lines 'Steudelli' and 'Jet', analysed by partial least squares regression and interval mapping', *Hereditas*, 141(2), 166–79.

Bjørnstad, A., Patil, V., Tekauz, A., Maroy, A. G., Skinnes, H., Jensen, A., Magnus, H. and MacKey, J. (2002) 'Resistance to scald (*Rhynchosporium secalis*) in barley (*Hordeum vulgare*) studied by near-isogenic lines: I. Markers and differential isolates', *Phytopathology*, 92(7), 710–20.

Bockelman, H. E., Sharp, E. L. and Eslick, R. F. (1977) 'Trisomic analysis of genes for resistance to scald and net blotch in several barley cultivars', *Can. J. Botany*, 55(15), 2142–8.

Bouajila, A., Abang, M. M., Haouas, S., Udupa, S., Rezgui, S., Baum, M. and Yahyaoui, A. (2007) 'Genetic diversity of *Rhynchosporium secalis* in Tunisia as revealed by pathotype, AFLP, and microsatellite analyses', *Mycopathologia*, 163(5), 281–94.

Bouajila, A., Haouas, S., Fakhfakh, M., Rezgui, S., El Ahmed, M. and Yahyaoui, A. (2006) 'Pathotypic diversity of *Rhynchosporium secalis* (Oudem) in Tunisia', *Afr. J. Biotechnol.*, 5(8), 570–9.

Bouajila, A., Zoghlami, N., Ghorbel, A., Rezgui, S. and Yahyaoui, A. (2010) 'Pathotype and microsatellite analyses reveal new sources of resistance to barley scald in Tunisia', *FEMS Microbiol. Lett.*, 305(1), 35–41.

Brooks, F. T. (1928) 'Observations on *Rhynchosporium secalis* (Oud.) Davis, leaf blotch of barley and rye', *New Phytol.*, 27(4), 215–19.

Brown, J. S. (1985) 'Pathogenic variation among isolates of *Rhynchosporium secalis* from cultivated barley growing in Victoria, Australia', *Euphytica*, 34(1), 129–33.

Brown, J. S. (1990) 'Pathogenic variation among isolates of *Rhynchosporium secalis* from barley grass growing in South Eastern Australia', *Euphytica*, 50(1), 81–9.

Brunner, P. C., Schürch, S. and McDonald, B. A. (2007) 'The origin and colonization history of the barley scald pathogen *Rhynchosporium secalis*', *J. Evol. Biol.*, 20(4), 1311–21.

Brunner, P. C., Stefansson, T. S., Fountaine, J., Richina, V. and McDonald, B. A. (2016) 'A global analysis of *CYP51* diversity and azole sensitivity in *Rhynchosporium commune*', *Phytopathology*, 106(4), 355–61.

Burdon, J. J., Abbott, D. C., Brown, A. H. D. and Brown, J. S. (1994) 'Genetic structure of the scald pathogen (*Rhynchosporium secalis*) in South East Australia: Implications for control strategies', *Aust. J. Agric. Res.*, 45(7), 1445–54.

Burnham, C. R. and Hagberg, A. (1956) 'Cytogenetic notes on chromosomal interchanges in barley', *Hereditas*, 42(3–4), 467–482.
Caldwell, R. M. (1937) '*Rhynchosporium* scald of barley, rye, and other grasses', *J. Agric. Res.*, 55(3), 175–98.
Carlucci, A., Raimondo, M. L., Santos, J. and Phillips, A. J. L. (2012 June) '*Plectosphaerella* species associated with root and collar rots of horticultural crops in southern Italy', *Persoonia*, 28, 34–48.
Castro, A. J., Capettini, F., Corey, A. E., Filichkina, T., Hayes, P. M., Kleinhofs, A., Kudrna, D., Richardson, K., Sandoval-Islas, S., Rossi, C. and Vivar, H. (2003) 'Mapping and pyramiding of qualitative and quantitative resistance to stripe rust in barley', *Theor. Appl. Genet.*, 107(5), 922–30.
Ceoloni, C. (1980) 'Race differentiation and search for sources of resistance to *Rhynchosporium secalis* in barley in Italy', *Euphytica*, 29(3), 547–53.
Cheong, J., Williams, K. and Wallwork, H. (2006) 'The identification of QTLs for adult plant resistance to leaf scald in barley', *Aust. J. Agric. Res.*, 57(9), 961–5.
Cools, H. J., Hawkins, N. J. and Fraaije, B. A. (2013) 'Constraints on the evolution of azole resistance in plant pathogenic fungi', *Plant Pathol.*, 62(S1), 36–42.
Cromey, M. G. (1987) 'Pathogenic variation in *Rhynchosporium secalis* on barley in New Zealand', *New Zeal. J. Agr. Res.*, 30(1), 95–9.
Cromey, M. G. and Mulholland, R. I. (1987) 'Host specialization of *Rhynchosporium secalis* in New Zealand', *New Zeal. J. Agr. Res.*, 30(3), 345–8.
Dahleen, L. S. (2006) 'Coordinator's report: chromosome 7H', *Barley Genet. Newslett.*, 36, 63–5.
Davidse, L. C. and Flach, W. (1978) 'Interaction of thiabendazole with fungal tubulin', *Biochim. Biophys. Acta*, 543(1), 82–90.
Davis, H. and Fitt, B. D. (1990) 'Symptomless infection of *Rhynchosporium secalis* on leaves of winter barley', *Mycol. Res.*, 94(4), 557–60.
Davis, H. and Fitt, B. D. L. (1992) 'Seasonal changes in primary and secondary inoculum during epidemics of leaf blotch (*Rhynchosporium secalis*) on winter barley', *Ann. Appl. Biol.*, 121(1), 39–49.
Davis, H. and Fitt, B. D. L. (1994) 'Effects of temperature and leaf wetness on the latent period of *Rhynchosporium secalis* (leaf blotch) on leaves of winter barley', *J. Phytopathol.*, 140(3), 269–79.
Davis, J. J. (1919) 'Notes on parasitic fungi in Wisconsin. VI', *Trans. Wisc. Acad. Sci. Arts. Lett.*, 19(2), 705–27.
Davis, J. J. (1922) 'Notes on parasitic fungi in Wisconsin. VIII', *Trans. Wisc. Acad. Sci. Arts. Lett.*, 20, 413–31.
Dewen, Q., Yijie, D., Yi, Z., Shupeng, L. and Fachao, S. (2017) 'Plant immunity inducer development and application', *Mol. Plant-Microbe Interact.*, 30(5), 355–60.
Durrant, W. E. and Dong, X. (2004) 'Systemic acquired resistance', *Annu. Rev. Phytopathol.*, 42, 185–209.
Dyck, P. L. and Schaller, C. W. (1961a) 'Association of two genes for scald resistance with a specific barley chromosome', *Can. J. Genet. Cytol.*, 3(2), 165–9.
Dyck, P. L. and Schaller, C. W. (1961b) 'Inheritance of resistance in barley to several physiologic races of the scald fungus', *Can. J. Genet. Cytol.*, 3(2), 153–64.
Elen, O. (2003) 'Long-term experiments with reduced tillage in spring cereals. III. Development of leaf diseases', *Crop Prot.*, 22(1), 65–71.
Elmore, J. M. and Coaker, G. (2011) 'The role of the plasma membrane H^+-ATPase in plant–microbe interactions', *Mol. Plant*, 4(3), 416–27.
Fiegen, M. and Knogge, W. (2002) 'Amino acid alterations in isoforms of the effector protein NIP1 from *Rhynchosporium secalis* have similar effects on its avirulence- and virulence-associated activities on barley', *Physiol. Mol. Plant Pathol.*, 61(5), 299–302.
Fitt, B. D. L., Creighton, N. F., Lacey, M. E. and McCartney, H. A. (1986) 'Effects of rainfall intensity and duration on dispersal of *Rhynchosporium secalis* conidia from infected barley leaves', *Trans. Br. Mycol. Soc.*, 86(4), 611–18.

Fitt, B. D. L., McCartney, H. A., Creighton, N. F., Lacey, M. E. and Walklate, P. J. (1988) 'Dispersal of *Rhynchosporium secalis* conidia from infected barley leaves or straw by simulated rain', *Ann. Appl. Biol.*, 112(1), 49–59.

Fitt, B. D. L., McCartney, H. A. and Walklate, P. J. (1989) 'The role of rain in dispersal of pathogen inoculum', *Annu. Rev. Phytopathol.*, 27, 241–70.

Forgan, A. H., Knogge, W. and Anderson, P. A. (2007) 'Asexual genetic exchange in the barley pathogen *Rhynchosporium secalis*', *Phytopathology*, 97(5), 650–4.

Foster, S. J. and Fitt, B. D. L. (2003) 'Isolation and characterisation of the mating-type (*MAT*) locus from *Rhynchosporium secalis*', *Curr. Genet.*, 44(5), 277–86.

Fountaine, J. A., Shaw, M. W., Napier, B., Ward, E. and Fraaije, B. A. (2007) 'Application of real-time and multiplex polymerase chain reaction assays to study leaf blotch epidemics in barley', *Phytopathology*, 97(3), 297–303.

Fountaine, J. M. (2011) 'Screening for QoI resistance in UK populations of *Rhynchosporium secalis*', *HGCA project report*.

Fountaine, J. M., Shaw, M. W., Ward, E. and Fraaije, B. A. (2010) 'The role of seeds and airborne inoculum in the initiation of leaf blotch (*Rhynchosporium secalis*) epidemics in winter barley', *Plant Pathol.*, 59(2), 330-7.

FRAG (Fungicide Resistance Action Group)-UK (2016) Fungicide Resistance Management in Cereals, pp. 16–17.

Frank, A. B. (1897) 'Über die Zerstörung der Gerste durch einen neuen Getreidepilz', *Wochenschr. Brauerei*, 14(42), 518–20.

Garvin, D. F., Brown, A. H. D., Raman, H. and Read, B. J. (2000) 'Genetic mapping of the barley *Rrs14* scald resistance gene with RFLP, isozyme and seed storage protein markers', *Plant Breed.*, 119(3), 193–6.

Genger, R. K., Nesbitt, K., Brown, A. H. D., Abbott, D. C. and Burdon, J. J. (2005) 'A novel barley scald resistance gene: genetic mapping of the *Rrs15* scald resistance gene derived from wild barley, *Hordeum vulgare* ssp. *spontaneum*', *Plant Breed.*, 124(2), 137–41.

Genger, R. K., Brown, A. H. D., Knogge, W., Nesbitt, K., and Burdon, J. J. (2003a) 'Development of SCAR markers linked to a scald resistance gene derived from wild barley', *Euphytica*, 134(2), 149–59, doi: 10.1023/B:EUPH.0000003833.63547.78.

Genger, R. K., Williams, K. J., Raman, H., Read, B. J., Wallwork, H., Burdon, J. J. and Brown, A. H. D. (2003b) 'Leaf scald resistance genes in *Hordeum vulgare* and *Hordeum vulgare* ssp. *spontaneum*: parallels between cultivated and wild barley', *Aust. J. Agric. Res.*, 54(11–12), 1335–42.

Gierlich, A., van't Slot, K. A. E., Li, V. M., Marie, C., Hermann, H. and Knogge, W. (1999) 'Heterologous expression of the avirulence gene product, NIP1, from the barley pathogen *Rhynchosporium secalis*', *Protein Expr. Purif.*, 17(1), 64–73.

Goodwin, S. B. (2002) 'The barley scald pathogen *Rhynchosporium secalis* is closely related to the discomycetes *Tapesia* and *Pyrenopeziza*', *Mycol. Res.*, 106(6), 645–54.

Goodwin, S. B., Allard, R. W., Hardy, S. A. and Webster, R. K. (1992) 'Hierarchical structure of pathogenic variation among *Rhynchosporium secalis* populations in Idaho and Oregon', *Can. J. Botany*, 70(4), 810–17.

Goodwin, S. B., Allard, R. W. and Webster, R. K. (1990) 'A nomenclature for *Rhynchosporium secalis* pathotypes', *Phytopathology*, 80(12), 1330–6.

Goodwin, S. B., Saghai Maroof, M. A., Allard, R. W. and Webster, R. K. (1993) 'Isozyme variation within and among populations of *Rhynchosporium secalis* in Europe, Australia, and the United States', *Mycol. Res.*, 97(1), 49–58.

Goodwin, S. B., Webster, R. K. and Allard, R. W. (1994) 'Evidence for mutation and migration as sources of genetic variation in populations of *Rhynchosporium secalis*', *Phytopathology*, 84(10), 1047–53.

Graner, A. and Tekauz, A. (1996) 'RFLP mapping in barley of a dominant gene conferring resistance to scald (*Rhynchosporium secalis*)', *Theor. Appl. Genet.*, 93(3), 421–5.

Grasso, V., Palermo, S., Sierotzki, H., Garibaldi, A. and Gisi, U. (2006) 'Cytochrome b gene structure and consequences for resistance to Qo inhibitor fungicides in plant pathogens', *Pest Manag. Sci.*, 62(6), 465–72.

Grimmer, M. K., van den Bosch, F., Powers, S. J. and Paveley, N. D. (2014) 'Evaluation of a matrix to calculate fungicide resistance risk', *Pest Manag. Sci.*, 70(6), 1008–16.

Habgood, R. M. (1971) 'The transmission of *Rhynchosporium secalis* by infected barley seed', *Plant Pathol.*, 20(2), 80–1.

Habgood, R. M. (1972) 'Resistance to *Rhynchosporium secalis* in the winter barley cultivar Vulcan', *Ann. Appl. Biol.*, 72, 265-71.

Habgood, R. M. (1974) 'The inheritance of resistance to *Rhynchosporium secalis* in some European spring barley cultivars', *Ann. Appl. Biol.*, 77, 191–200.

Habgood, R. M. (1977) 'Resistance of barley cultivars to *Rhynchosporium secalis*', *Trans. Br. Mycol. Soc.*, 69(2), 281-6.

Habgood, R. M. and Hayes, J. D. (1971) 'The inheritance of resistance to *Rhynchosporium secalis* in barley', *Heredity*, 27(1), 25–37.

Hahn, M., Jüngling, S. and Knogge, W. (1993) 'Cultivar-specific elicitation of barley defense reactions by the phytotoxic peptide NIP1 from *Rhynchosporium secalis*', *Mol. Plant-Microbe Interact.*, 6(6), 745–54.

Hanemann, A., Schweizer, G. F., Cossu, R., Wicker, T. and Röder, M. S. (2009) 'Fine mapping, physical mapping and development of diagnostic markers for the *Rrs2* scald resistance gene in barley', *Theor. Appl. Genet.*, 119(8), 1507–22.

Hawkins, N. J., Cools, H. J., Sierotzki, H., Shaw, M. W., Knogge, W., Kelly, S. L., Kelly, D. E. and Fraaije, B. A. (2014) 'Paralog re-emergence: A novel, historically contingent mechanism in the evolution of antimicrobial resistance', *Mol. Biol. Evol.*, 31(7), 1793–802.

Heinsen, E. (1900) 'Beobachtungen über den neuen Getreidepilz *Rhynchosporium graminicola*', *Jahrb. d. Hamburg. wissenschaftl. Anstalt.*, 18, 43–55.

Hofmann, K. (2014) Phenotypic assessment and genetic mapping of genes conferring resistance to leaf scald (*Rhynchosporium commune*) in barley (*Hordeum vulgare*), unpublished thesis (PhD thesis), Justus-Liebig-University Giessen.

Hofmann, K., Silvar, C., Casas, A. M., Herz, M., Buttner, B., Gracia, M. P., Contreras-Moreira, B., Wallwork, H., Igartua, E. and Schweizer, G. (2013) 'Fine mapping of the *Rrs1* resistance locus against scald in two large populations derived from Spanish barley landraces', *Theor. Appl. Genet.*, 126(12), 3091–102.

Hollomon, D. W., Butters, J. A., Barker, H. and Hall, L. (1998) 'Fungal β-tubulin, expressed as a fusion protein, binds benzimidazole and phenylcarbamate fungicides', *Antimicrobial Agents & Chemotherapy*, 42(9), 2171–3.

Horbach, R., Navarro-Quesada, A. R., Knogge, W. and Deising, H. B. (2011) 'When and how to kill a plant cell: infection strategies of plant pathogenic fungi', *J. Plant Physiol.*, 168(1), 51–62.

Hosemans, D. and Branchard, M. (1985) 'Étude *in vitro* de la Rhynchosporiose de l'orge: Cycle de reproduction du parasite - histopathologie de l'hôte', *Phytopath. Z.*, 112(2), 127–42.

Howlett, S. G. and Cooke, B. M. (1987) 'Scanning electron microscopy of sporulation in *Rhynchosporium secalis*', *Trans. Br. Mycol. Soc.*, 88(4), 547–9.

Howlett, S. G. and Cooke, B. M. (1992) 'More scanning electron micrographs of conidial production in *Rhynchosporium secalis* on barley', *Mycologist*, 6(1), 16–17.

Jackson, L. F. and Webster, R. K. (1976a) 'The dynamics of a controlled population of *Rhynchosporium secalis*, changes in race composition and frequencies', *Phytopathology*, 66(6), 726–8.

Jackson, L. F. and Webster, R. K. (1976b) 'Race differentiation, distribution, and frequency of *Rhynchosporium secalis* in California', *Phytopathology*, 66(6), 719–25.

Jackson, L. F. and Webster, R. K. (1976c) 'Seed and grasses as possible sources of *Rhynchosporium secalis* for barley in California', *Plant Dis. Rep.*, 60(3), 233–6.

Jensen, J., Backes, G., Skinnes, H. and Giese, H. (2002) 'Quantitative trait loci for scald resistance in barley localized by a non-interval mapping procedure', *Plant Breed.*, 121(2), 124-8.

Johnson, A. G. and Mackie, W. W. (1920) 'Evidence of disease resistance in barley to attacks of *Rhynchosporium*. (Abstract)', *Phytopathology*, 10, 54.

Johnson, L. J., Koulman, A., Christensen, M., Lane, G. A., Fraser, K., Forester, N., Johnson, R. D., Bryan, G. T. and Rasmussen, S. (2013) 'An extracellular siderophore is required to maintain the mutualistic interaction of *Epichloë festucae* with *Lolium perenne*', *PLoS Pathogens*, 9(5), e1003332.

Jones, P. and Ayres, P. G. (1974) '*Rhynchosporium* leaf blotch of barley studied during the subcuticular phase by electron microscopy', *Physiol. Plant Pathol.*, 4(2), 229–33.

Kay, J. G. and Owen, H. (1973a) 'Host range of *Rhynchosporium secalis*', *Trans. Br. Mycol. Soc.*, 60(3), 413–22.

Kay, J. G. and Owen, H. (1973b) 'Transmission of *Rhynchosporium secalis* on barley grain', *Trans. Br. Mycol. Soc.*, 60(3), 405–11.

Keiper, F. J., Hayden, M. J. and Wallwork, H. (2006) 'Development of sequence tagged microsatellites (STMs) for the barley scald pathogen *Rhynchosporium secalis*', *Mol. Ecol. Notes*, 6(2), 543–6.

Kendall, S., Hollomon, D. W., Ishii, H. and Heaney, S. P. (1994) 'Characterization of benzimidazole-resistant strains of *Rhynchosporium secalis*', *Pestic. Sci.*, 40(3), 175–81.

Kendall, S. J., Hollomon, D. W., Cooke, L. R. and Jones, D. R. (1993) 'Changes in sensitivity to DMI fungicides in *Rhynchosporium secalis*', *Crop Prot.*, 12(5), 357–62.

Kilby, N. J. and Robinson, J. (2001) 'Pathotypes and *NIP1* gene sequences of Finnish *Rhynchosporium secalis* isolates from barley, couch grass and rye', *Euphytica*, 120(2), 265-72.

King, K. M., West, J. S., Brunner, P. C., Dyer, P. S. and Fitt, B. D. L. (2013) 'Evolutionary relationships between *Rhynchosporium lolii* sp. nov. and other *Rhynchosporium* species on grasses', *PLoS ONE*, 8(10), e72536.

King, K. M., West, J. S., Fitt, B. D. L. and Dyer, P. S. (2015) 'Differences in MAT gene distribution and expression between *Rhynchosporium* species on grasses', *Plant Pathol.*, 64(2), 344–54.

Kiros-Meles, A., Gomez, D., McDonald, B. A., Yahyaoui, A. and Linde, C. C. (2011) 'Invasion of *Rhynchosporium commune* onto wild barley in the Middle East', *Biol. Invasions*, 13(2), 321–30.

Kiros-Meles, A., Udupa, S., Abang, M. M., Abu-Blan, H., Baum, M., Ceccarelli, S. and Yahyaoui, A. H. (2005) 'Amplified fragment length polymorphism among *Rhynchosporium secalis* isolates collected from a single barley field in Syria', *Ann. Appl. Biol.*, 146(3), 389–94.

Kirsten, S., Navarro-Quezada, A., Penselin, D., Wenzel, C., Matern, A., Leitner, A., Baum, T., Seiffert, U. and Knogge, W. (2012) 'Necrosis-inducing proteins of *Rhynchosporium commune*, effectors in quantitative disease resistance', *Mol. Plant-Microbe Interact.*, 25(10), 1314-25.

Lancashire, P. D., Bleiholder, H., Vandenboom, T., Langeluddeke, P., Stauss, R., Weber, E. and Witzenberger, A. (1991) 'A uniform decimal code for growth stages of crops and weeds', *Ann. Appl. Biol.*, 119(3), 561–601.

Lebedeva, L. and Tvarůžek, L. (2006) 'Specialisation of *Rhynchosporium secalis* (Oud.) J.J. Davis infecting barley and rye', *Plant Prot. Sci.*, 42(3), 85–93.

Lee, H. K., Tewari, J. P. and Turkington, T. K. (2001) 'Symptomless infection of barley seed by *Rhynchosporium secalis*', *Can. J. Plant Pathol.*, 23(3), 315–17.

Lee, H. K., Tewari, J. P. and Turkington, T. K. (2002) 'Quantification of seedborne infection by *Rhynchosporium secalis* in barley using competitive PCR', *Plant Pathol.*, 51(2), 217–24.

Lehnackers, H. and Knogge, W. (1990) 'Cytological studies on the infection of barley cultivars with known resistance genotypes by *Rhynchosporium secalis*', *Can. J. Botany*, 68(9), 1953–61.

Li, H. B. and Zhou, M. X. (2011) 'Quantitative trait loci controlling barley powdery mildew and scald resistances in two different barley doubled haploid populations', *Mol. Breed.*, 27(4), 479–90.

Linde, C. C., Zala, M., Ceccarelli, S. and McDonald, B. A. (2003) 'Further evidence for sexual reproduction in *Rhynchosporium secalis* based on distribution and frequency of mating-type alleles', *Fungal Genet. Biol.*, 40(2), 115–25.

Linde, C. C., Zala, M. and McDonald, B. A. (2009) 'Molecular evidence for recent founder populations and human-mediated migration in the barley scald pathogen *Rhynchosporium secalis*', *Mol. Phylogenet. Ebol.*, 51(3), 454–64.

Linde-Laursen, I., Heslop-Harrison, J. S., Shepherd, K. W. and Taketa, S. (1997) 'The barley genome and its relationship with the wheat genomes. A survey with an internationally agreed recommendation for barley chromosome nomenclature', *Hereditas*, 126(1), 1–16.

Linsell, K. J., Keiper, F. J., Forgan, A. and Oldach, K. H. (2011) 'New insights into the infection process of *Rhynchosporium secalis* in barley using GFP', *Fungal Genet. Biol.*, 48(2), 124–31.

Lionetti, V., Cervone, F. and Bellincampi, D. (2012) 'Methyl esterification of pectin plays a role during plant-pathogen interactions and affects plant resistance to diseases', *J. Plant Physiol.*, 169(16), 1623–30.

Locke, T. and Phillips, A. N. (1995) 'The occurrence of carbendazim resistance in *Rhynchosporium secalis* on winter barley in England and Wales in 1992 and 1993', *Plant Pathol.*, 44(2), 294-300.

Looseley, M. E., Keith, R., Guy, D., Barral-Baron, G., Thirugnanasambandam, A., Harrap, D., Werner, P. and Newton, A. C. (2015) 'Genetic mapping of resistance to *Rhynchosporium commune* and characterisation of early infection in a winter barley mapping population', *Euphytica*, 203(2), 337-47.

Looseley, M. E., Newton, A. C., Atkins, S. D., Fitt, B. D. L., Fraaije, B. A., Thomas, W. T. B., Keith, R., Macaulay, M., Lynott, J. and Harrap, D. (2012) 'Genetic basis of control of *Rhynchosporium secalis* infection and symptom expression in barley', *Euphytica*, 184(1), 47–56.

Lyngs Jørgensen, H. J., de Neergaard, E. and Smedegaard-Petersen, V. (1993) 'Histological examination of the interaction between *Rhynchosporium secalis* and susceptible and resistant cultivars of barley', *Physiol. Mol. Plant Pathol.*, 42(5), 345–58.

Lyngs Jørgensen, H. J. and Smedegaard-Petersen, V. (1995) 'Pathogenic variation of *Rhynchosporium* secalis in Denmark and sources of resistance in barley', *Plant Dis.*, 79(3), 297–301.

Mackie, W. W. (1931) 'Diseases of grain and their control', *Calif. AES Bull.*, 511.

Magnus, P. (1906) 'Notwendige Umänderung des Namens der Pilzgattung *Marssonia* Fisch', *Hedwigia*, 45(2), 88–91.

Marzin, S., Hanemann, A., Sharma, S., Hensel, G., Kumlehn, J., Schweizer, G. and Roder, M. S. (2016) 'Are *PECTIN ESTERASE INHIBITOR* genes involved in mediating resistance to *Rhynchosporium commune* in barley?', *PLoS ONE*, 11(3), e0150485.

Mayer, K. F. X., Martis, M., Hedley, P. E., Simkova, H., Liu, H., Morris, J. A., Steuernagel, B., Taudien, S., Roessner, S., Gundlach, H., Kubalakova, M., Suchankova, P., Murat, F., Felder, M., Nussbaumer, T., Graner, A., Salse, J., Endo, T., Sakai, H., Tanaka, T., Itoh, T., Sato, K., Platzer, M., Matsumoto, T., Scholz, U., Dolezel, J., Waugh, R. and Stein, N. (2011) 'Unlocking the barley genome by chromosomal and comparative genomics', *Plant Cell*, 23(4), 1249–63.

Mayer, K. F. X., Waugh, R., Langridge, P., Close, T. J., Wise, R. P., Graner, A., Matsumoto, T., Sato, K., Schulman, A., Muehlbauer, G. J., Stein, N., Ariyadasa, R., Schulte, D., Poursarebani, N., Zhou, R. N., Steuernagel, B., Mascher, M., Scholz, U., Shi, B. J., Langridge, P., Madishetty, K., Svensson, J. T., Bhat, P., Moscou, M., Resnik, J., Close, T. J., Muehlbauer, G. J., Hedley, P., Liu, H., Morris, J., Waugh, R., Frenkel, Z., Korol, A., Berges, H., Graner, A., Stein, N., Steuernagel, B., Taudien, S., Groth, M., Felder, M., Lonardi, S., Duma, D., Alpert, M., Cordero, F., Beccuti, M., Ciardo, G., Ma, Y., Wanamaker, S., Stein, N., Close, T. J., Platzer, M., Brown, J. W. S., Schulman, A., Platzer, M., Fincher, G. B., Muehlbauer, G. J., Sato, K., Taudien, S., Sampath, D., Swarbreck, D., Scalabrin, S., Zuccolo, A., Vendramin, V., Morgante, M., Schulman, A. and Cons, I. B. G. S. (2012) 'A physical, genetic and functional sequence assembly of the barley genome', *Nature*, 491(7426), 711–17.

Mayfield, A. H. and Clare, B. G. (1984a) 'Effects of common stubble treatments and sowing sequences on scald disease (*Rhynchosporium secalis*) in barley crops', *Aust. J. Agric. Res.*, 35(6), 799–805.

Mayfield, A. H. and Clare, B. G. (1984b) 'Survival over summer of *Rhynchosporium secalis* in host debris in the field', *Aust. J. Agric. Res.*, 35(6), 789–97.

McDermott, J. M., McDonald, B. A., Allard, R. W. and Webster, R. K. (1989) 'Genetic variability for pathogenicity, isozyme, ribosomal DNA and colony color variants in populations of *Rhynchosporium secalis*', *Genetics*, 122(3), 561–5.

McDonald, B. A. (2015) 'How can research on pathogen population biology suggest disease management strategies? The example of barley scald (*Rhynchosporium commune*)', *Plant Pathol.*, 64(5), 1005–13.

McDonald, B. A. and Stukenbrock, E. H. (2016) 'Rapid emergence of pathogens in agro-ecosystems: Global threats to agricultural sustainability and food security', Phil. Trans. R. Soc. B. , 371(1709).

McDonald, B. A., Zhan, J. and Burdon, J. J. (1999) 'Genetic structure of *Rhynchosporium secalis* in Australia', *Phytopathology*, 89(8), 639–45.

Meier, U., Bleiholder, H., Buhr, L., Feller, C., Hack, H., Heß, M., Lancashire, P. D., Schnock, U., Stauß, R., van den Boom, T., Weber, E., and Zwerger, P. (2009) 'The BBCH system to coding the phenological growth stages of plants – history and publications', *J. Kulturpfl.*, 61(2), 41–52.

Mohd-Assaad, N., McDonald, B. A. and Croll, D. (2016) 'Multilocus resistance evolution to azole fungicides in fungal plant pathogen populations', *Mol. Ecol.*, 25(24), 6124–42.

Moseman, J. G. (1972) 'Report on genes for resistance to pests', *Barley Genet. Newslett.*, 2, 145–7.

Mur, L. A. J., Kenton, P., Lloyd, A. J., Ougham, H. and Prats, E. (2008) 'The hypersensitive response; the centenary is upon us but how much do we know?', *J. Exp. Botany*, 59(3), 501–20.

Newman, P. L. and Owen, H. (1985) 'Evidence of asexual recombination in *Rhynchosporium secalis*', *Plant Pathol.*, 34(3), 338–40.

Newton, A. C., Searle, J., Guy, D. C., Hackett, C. A. and Cooke, D. E. L. (2001) 'Variability in pathotype, aggressiveness, RAPD profile, and rDNA ITS1 sequences of UK isolates of *Rhynchosporium secalis*', *J. Plant Dis. Protect.*, 108(5), 446–58.

Oettler, G. (2005) 'The fortune of a botanical curiosity – Triticale: Past, present and future', *J. Agric. Sci.*, 143(5), 329–46.

Omrane, S., Sghyer, H., Audeon, C., Lanen, C., Duplaix, C., Walker, A. S. and Fillinger, S. (2015) 'Fungicide efflux and the MgMFS1 transporter contribute to the multidrug resistance phenotype in *Zymoseptoria tritici* field isolates', *Environ. Microbiol.*, 17(8), 2805–23.

Oudemans, C. A. J. A. (1897) 'Observations mycologiques', *Verslag. Wis- en Natuurk. Afd. K. Akad. Wetensch. Amsterdam*, 6, 86–92.

Owen, H. (1958) 'Physiologic specialisation in *Rhynchosporium secalis*', *Trans. Br. Mycol. Soc.*, 41(1), 99–108.

Parkinson, V. O., Sivanesan, A. and Booth, C. (1981) 'The perfect state of the rice leaf scald fungus and the taxonomy of both the perfect and imperfect states', *Trans. Br. Mycol. Soc.*, 76(1), 59–69.

Patil, V., Bjørnstad, A. and Mackey, J. (2003) 'Molecular mapping of a new gene $Rrs4_{CI11549}$ for resistance to barley scald (*Rhynchosporium secalis*)', *Mol. Breed.*, 12(2), 169–83.

Penselin, D., Münsterkotter, M., Kirsten, S., Felder, M., Taudien, S., Platzer, M., Ashelford, K., Paskiewicz, K. H., Harrison, R. J., Hughes, D. J., Wolf, T., Shelest, E., Graap, J., Hoffmann, J., Wenzel, C., Wöltje, N., King, K. M., Fitt, B. D. L., Güldener, U., Avrova, A. and Knogge, W. (2016) 'Comparative genomics to explore phylogenetic relationship, cryptic sexual potential and host specificity of *Rhynchosporium species* on grasses', *BMC Genomics*, 17(1), 953.

Phelan, S., Barthe, M. S., Tobie, C. and Kildea, S. (2017) 'Detection of the cytochrome b mutation G143A in Irish *Rhynchosporium commune* populations using targeted 454 sequencing', *Pest Manag. Sci.*, 73(6), 1154–60.

Pickering, R., Ruge-Wehling, B., Johnston, P. A., Schweizer, G., Ackermann, P. and Wehling, P. (2006) 'The transfer of a gene conferring resistance to scald (*Rhynchosporium secalis*) from *Hordeum bulbosum* into *H. vulgare* chromosome 4HS', *Plant Breed.*, 125(6), 576–9.

Pieterse, C. M. J., Zamioudis, C., Berendsen, R. L., Weller, D. M., Van Wees, S. C. M. and Bakker, P. A. H. M. (2014) 'Induced systemic resistance by beneficial microbes' in VanAlfen, N. K., ed. *Annu. Rev. Phytopathol.*, 52, 347–75.

Riddle, O. C. and Briggs, F. N. (1950) 'Inheritance of resistance to scald in barley', *Hilgardia*, 20(2), 19–27.

Riddle, O. C. and Suneson, C. A. (1948) 'Sources and use of scald resistance in barley', *Agron. J.*, 40, 926–8.

Robbertse, B. and Crous, P. W. (2000) 'Genotypic variation in *Rhynchosporium secalis* pathotypes collected in the Western Cape Province of South Africa', *S. Afr. J. Sci.*, 96(7), 391–5.

Robbertse, B., Lennox, C. L., van Jaarsveld, A. B., Crous, P. W. and van der Rijst, M. (2000) 'Pathogenicity of the *Rhynchosporium secalis* population in the Western Cape province of South Africa', *Euphytica*, 115(1), 75–82.
Robbertse, B., van der Rijst, M., van Aarde, I. M. R., Lennox, C. and Crous, P. W. (2001) 'DMI sensitivity and cross-resistance patterns of *Rhynchosporium secalis* isolates from South Africa', *Crop Prot.*, 20(2), 97–102.
Robinson, J., Lindqvist, H. and Jalli, M. (1996) 'Genes for resistance in barley to Finnish isolates of *Rhynchosporium secalis*', *Euphytica*, 92(3), 295–300.
Rohe, M., Gierlich, A., Hermann, H., Hahn, M., Schmidt, B., Rosahl, S. and Knogge, W. (1995) 'The race-specific elicitor, NIP1, from the barley pathogen, *Rhynchosporium secalis*, determines avirulence on host plants of the *Rrs1* resistance genotype', *EMBO J.*, 14(17), 4168–77.
Ryan, C. C. and Grivell, C. J. (1974) 'An electron microscope study of the outer layers of barley leaves infected with *Rhynchosporium secalis*', *Aust. J. Plant Physiol.*, 1, 313–17.
Salamati, S. and Tronsmo, A. M. (1997) 'Pathogenicity of *Rhynchosporium secalis* isolates from Norway on 30 cultivars of barley', *Plant Pathol.*, 46(3), 416–24.
Salamati, S., Zhan, J., Burdon, J. J. and McDonald, B. A. (2000) 'The genetic structure of field populations of *Rhynchosporium secalis* from three continents suggests moderate gene flow and regular recombination', *Phytopathology*, 90(8), 901–8.
Salamini, F., Ozkan, H., Brandolini, A., Schafer-Pregl, R. and Martin, W. (2002) 'Genetics and geography of wild cereal domestication in the Near East', *Nat. Rev. Genet.*, 3(6), 429–41.
Sarasola, J. A. and Campi, M. D. (1947) 'Reacción de algunas cebadas con respecto a '*Rhynchosporium secalis*' en Argentina', *Rev. Invest. agric.*, 1, 243–60.
Schaller, C. W. (1948) 'Atlas 46, a new barley resistant to mildew and scald', *Cailf. Agr.*, 2(5), 15.
Schein, R. D. (1958) 'Pathogenic specialization in *Rhynchosporium secalis*', *Phytopathology*, 48, 477–80.
Schürch, S., Linde, C. C., Knogge, W., Jackson, L. F. and McDonald, B. A. (2004) 'Molecular population genetic analysis differentiates two virulence mechanisms of the fungal avirulence gene *NIP1*', *Mol. Plant-Microbe Interact.*, 17(10), 1114–25.
Schweizer, G., Herz, M., Mikolajewski, S., Brenner, M., Hartl, L. and Baumer, M. (2004) 'Genetic mapping of a novel scald resistance gene $Rrs15_{CI\ 8288}$ in barley'. In: *9th International Barley Genetics Symposium*, Brno, Czech Republic, 20–26 June, pp. 258–65.
Schweizer, G. F., Baumer, M., Daniel, G., Rugel, H. and Röder, M. S. (1995) 'RFLP markers linked to scald (*Rhynchosporium secalis*) resistance gene *Rh2* in barley', *Theor. Appl. Genet.*, 90(7), 920–4.
Shipton, W. A., Boyd, W. J. R. and Ali, S. M. (1974) 'Scald of barley', *Rev. Plant Pathol.*, 53, 839–61.
Shtaya, M. J. Y., Marcel, T. C., Sillero, J. C., Niks, R. E. and Rubiales, D. (2006) 'Identification of QTLs for powdery mildew and scald resistance in barley', *Euphytica*, 151(3), 421–9.
Sierotzki, H., Frey, R., Wullschleger, J., Palermo, S., Karlin, S., Godwin, J. and Gisi, U. (2007) 'Cytochrome b gene sequence and structure of *Pyrenophora teres* and *P. tritici-repentis* and implications for QoI resistance', *Pest Manag. Sci.*, 63(3), 225–33.
Sierotzki, H. and Scalliet, G. (2013) 'A review of current knowledge of resistance aspects for the next-generation succinate dehydrogenase inhibitor fungicides', *Phytopathology*, 103(9), 880–7.
Skoropad, W. P. (1959) 'Seed and seedling infection of barley by *Rhynchosporium secalis*', *Phytopathology*, 49(10), 623-6.
Stammler, G., Rehfus, A., Bryson, R. and Strobel, D. (2015) 'Fungicide resistance in cereal pathogens'. In: *XVIII. International Plant Protection Congress (IPPC)*, Berlin, 24–27 August, 2015, p. 179.
Stedman, O. J. (1980a October) 'Dispersal of spores of *Rhynchosporium secalis* from a dry surface by water drops', *Trans. Br. Mycol. Soc.*, 75, 339–40.
Stedman, O. J. (1980b) 'Observations on the production and dispersal of spores, and infection by *Rhynchosporium secalis*', *Ann. Appl. Biol.*, 95(2), 163–75.
Stefansson, T. S., McDonald, B. A. and Willi, Y. (2013) 'Local adaptation and evolutionary potential along a temperature gradient in the fungal pathogen *Rhynchosporium commune*', *Evol. Appl.*, 6(3), 524–34.

Steiner-Lange, S., Fischer, A., Böttcher, A., Rouhara, I., Liedgens, H., Schmelzer, E. and Knogge, W. (2003) 'Differential defense reactions in leaf tissues of barley in response to infection by *Rhynchosporium secalis* and to treatment with a fungal avirulence gene product', *Mol. Plant-Microbe Interact.*, 16(10), 893–902.

Stotz, H. U., Mitrousia, G. K., de Wit, P. J. G. M. and Fitt, B. D. L. (2014) 'Effector-triggered defence against apoplastic fungal pathogens', *Trends Plant Sci.*, 19(8), 491–500.

Taggart, P. J., Locke, T., Phillips, A. N., Pask, N., Hollomon, D. W., Kendall, S. J., Cooke, L. R. and Mercer, P. C. (1999) 'Benzimidazole resistance in *Rhynchosporium secalis* and its effect on barley leaf blotch control in the UK', *Crop Prot.*, 18(4), 239–43.

Takemoto, D., Tanaka, A. and Scott, B. (2006) 'A p67(Phox)-like regulator is recruited to control hyphal branching in a fungal-grass mutualistic symbiosis', *Plant Cell*, 18(10), 2807–21.

Tamura, K., Battistuzzi, F. U., Billing-Ross, P., Murillo, O., Filipski, A. and Kumar, S. (2012) 'Estimating divergence times in large molecular phylogenies', *Proc. Natl. Acad. Sci. USA*, 109(47), 19333–8.

Tanaka, A., Christensen, M. J., Takemoto, D., Park, P. and Scott, B. (2006) 'Reactive oxygen species play a role in regulating a fungus-perennial ryegrass mutualistic interaction', *Plant Cell*, 18(4), 1052–66.

Tanaka, A., Takemoto, D., Hyon, G. S., Park, P. and Scott, B. (2008) 'NoxA activation by the small GTPase RacA is required to maintain a mutualistic symbiotic association between *Epichloë festucae* and perennial ryegrass', *Mol. Microbiol.*, 68(5), 1165–78.

Tekauz, A. (1991) 'Pathogenic variation in *Rhynchosporium secalis* on barley in Canada', *Can. J. Plant Pathol.*, 13(4), 298–304.

Thirugnanasambandam, A., Wright, K. M., Atkins, S. D., Whisson, S. C. and Newton, A. C. (2011) 'Infection of *Rrs1* barley by an incompatible race of the fungus *Rhynchosporium secalis* expressing the green fluorescent protein', *Plant Pathol.*, 60(3), 513–21.

Torriani, S. F. F., Linde, C. C. and McDonald, B. A. (2009) 'Sequence conservation in the mitochondrial cytochrome b gene and lack of G143A QoI resistance allele in a global sample of *Rhynchosporium secalis*', *Australas. Plant Pathol.*, 38(2), 202–7.

Torriani, S. F. F., Penselin, D., Knogge, W., Felder, M., Taudien, S., Platzer, M., McDonald, B. A. and Brunner, P. C. (2014) 'Comparative analysis of mitochondrial genomes from closely related *Rhynchosporium* species reveals extensive intron invasion initiated by recent lateral intron transfer', *Fungal Genet. Biol.*, 62(1), 34–42.

van't Slot, K. A. E., Gierlich, A. and Knogge, W. (2007) 'A single binding site mediates resistance- and disease-associated activities of the effector protein NIP1 from the barley pathogen *Rhynchosporium secalis*', *Plant Physiol.*, 144(3), 1654–66.

van't Slot, K. A. E. and Knogge, W. (2002) 'A dual role of microbial pathogen-derived proteins in plant disease and resistance', *Crit. Rev. Plant Sci.*, 21(3), 229–71.

van't Slot, K. A. E., van den Burg, H. A., Kloks, C. P. A. M., Hilbers, C. W., Knogge, W. and Papavoine, C. H. M. (2003) 'Solution structure of the plant disease resistance-triggering protein NIP1 from the fungus *Rhynchosporium secalis* shows a novel β-sheet fold', *J. Biol. Chem.*, 278(46), 45730–6.

van den Bosch, F., Oliver, R., van den Berg, F. and Paveley, N. (2014) 'Governing principles can guide fungicide-resistance management tactics' in VanAlfen, N. K., ed. *Annu. Rev. Phytopathol.*, 175–95.

von Korff, M., Udupa, S. M., Yahyaoui, A. and Baum, M. (2004) 'Genetic variation among *Rhynchosporium secalis* populations of West Asia and North Africa as revealed by RAPD and AFLP analysis', *J. Phytopathol.*, 152(2), 106–13.

Wagner, C., Schweizer, G., Kramer, M., Dehmer-Badani, A., Ordon, F. and Friedt, W. (2008) 'The complex quantitative barley-*Rhynchosporium secalis* interaction: Newly identified QTL may represent already known resistance genes', *Theor. Appl. Genet.*, 118(1), 113–22.

Wallwork, H. and Grcic, M. (2011) 'The use of differential isolates of *Rhynchosporium secalis* to identify resistance to leaf scald in barley', *Australas. Plant Pathol.*, 40(5), 490–6.

Walters, D. R., Avrova, A., Bingham, I. J., Burnett, F. J., Fountaine, J., Havis, N. D., Hoad, S. P., Hughes, G., Looseley, M., Oxley, S. J. P., Renwick, A., Topp, C. F. E. and Newton, A. C. (2012) 'Control of foliar diseases in barley: Towards an integrated approach', *Eur. J. Plant Pathol.*, 133(1), 33–73.

Walters, D. R., Havis, N. D., Paterson, L., Taylor, J., Walsh, D. J. and Sablou, C. (2014) 'Control of foliar pathogens of spring barley using a combination of resistance elicitors', *Front. Plant Sci.*, 5, 241.

Welty, R. E. and Metzger, R. J. (1996) 'First report of scald of triticale caused by *Rhynchosporium secalis* in North America', *Plant Dis.*, 80(11), 1220–3.

Wevelsiep, L., Kogel, K.-H. and Knogge, W. (1991) 'Purification and characterization of peptides from *Rhynchosporium secalis* inducing necrosis in barley', *Physiol. Mol. Plant Pathol.*, 39(6), 471–82.

Wevelsiep, L., Rüpping, E. and Knogge, W. (1993) 'Stimulation of barley plasmalemma H^+-ATPase by phytotoxic peptides from the fungal pathogen *Rhynchosporium secalis*', *Plant Physiol.*, 101(1), 297–301.

Wheeler, J. E., Kendall, S. J., Butters, J., Hollomon, D. W. and Hall, L. (1995) 'Using allele-specific oligonucleotide probes to characterize benzimidazole resistance in *Rhynchosporium secalis*', *Pestic. Sci.*, 43(3), 201-9.

Wiesel, L., Newton, A. C., Elliott, I., Booty, D., Gilroy, E. M., Birch, P. R. J. and Hein, I. (2014) 'Molecular effects of resistance elicitors from biological origin and their potential for crop protection', *Front. Plant Sci.*, 5, 655.

Wilkins, P. (1973) 'Infection of *Lolium multiflorum* with *Rhynchosporium* species', *Plant Pathol.*, 22(3), 107–11.

Xi, K., Burnett, P. A., Tewari, J. P., Chen, M. H., Turkington, T. K. and Helm, J. H. (2000) 'Histopathological study of barley cultivars resistant and susceptible to *Rhynchosporium secalis*', *Phytopathology*, 90(1), 94–102.

Xi, K., Turkington, T. K., Helm, J. H. and Bos, C. (2002) 'Pathogenic variation of *Rhynchosporium secalis* in Alberta', *Can. J. Plant Pathol.*, 24(2), 176-83.

Xi, K., Turkington, T. K., Helm, J. H., Briggs, K. G., Tewari, J. P., Ferguson, T. and Kharbanda, P. D. (2003) 'Distribution of pathotypes of *Rhynchosporium secalis* and cultivar reaction on barley in Alberta', *Plant Dis.*, 87(4), 391–6.

Xue, G., Hall, R. and Falk, D. (1991) 'Pathogenic variation in *Rhynchosporium secalis* from Southern Ontario', *Plant Dis.*, 75(9), 934–8.

Yun, S. J., Gyenis, L., Bossolini, E., Hayes, P. M., Matus, I., Smith, K. P., Steffenson, B. J., Tuberosa, R. and Muehlbauer, G. J. (2006) 'Validation of quantitative trait loci for multiple disease resistance in barley using advanced backcross lines developed with a wild barley', *Crop Sci.*, 46(3), 1179–86.

Yun, S. J., Gyenis, L., Hayes, P. M., Matus, I., Smith, K. P., Steffenson, B. J. and Muehlbauer, G. J. (2005) 'Quantitative trait loci for multiple disease resistance in wild barley', *Crop Sci.*, 45(6), 2563–72.

Zaffarano, P. L., McDonald, B. A. and Linde, C. C. (2008) 'Rapid speciation following recent host shifts in the plant pathogenic fungus *Rhynchosporium*', *Evolution*, 62(6), 1418–36.

Zaffarano, P. L., McDonald, B. A. and Linde, C. C. (2009) 'Phylogeographical analyses reveal global migration patterns of the barley scald pathogen *Rhynchosporium secalis*', *Mol. Ecol.*, 18(2), 279–93.

Zaffarano, P. L., McDonald, B. A. and Linde, C. C. (2011) 'Two new species of *Rhynchosporium*', *Mycologia*, 103(1), 195–202.

Zaffarano, P. L., McDonald, B. A., Zala, M. and Linde, C. C. (2006) 'Global hierarchical gene diversity analysis suggests the Fertile Crescent is not the center of origin of the barley scald pathogen *Rhynchosporium secalis*', *Phytopathology*, 96(9), 941–50.

Zhan, J., Fitt, B. D. L., Pinnschmidt, H. O., Oxley, S. J. P. and Newton, A. C. (2008) 'Resistance, epidemiology and sustainable management of *Rhynchosporium secalis* populations on barley', *Plant Pathol.*, 57(1), 1–14.

Zhang, Q., Webster, R. K., Crandall, B. A., Jackson, L. F. and Saghai Maroof, M. A. (1992) 'Race composition and pathogenicity associations of *Rhynchosporium secalis* in California', *Phytopathology*, 82(7), 798–803.

Part 2

Key challenges in integrated disease management of cereals

Chapter 11

Challenges and prospects for fungicidal control of wheat diseases

R. J. Bryson and H-D. Brix, BASF SE, Germany

1 Introduction

> *It's the 25th of April, 683 BC in the time of Numa Pompilius, the second King of Rome and successor of Romulus. The Robigalia games have just started in the rural outskirts of Rome and the blood and entrails of an unweaned puppy have been consumed in a communal meal. The local farmers hope that Robigus, the Roman God of Rust, will be pleased with their sacrifice and will bless them with a healthy crop for the coming season and good yields so that they can feed their families next year.*

Fortunately, farmers in the twenty-first century do not have to resort to such drastic measures to ensure that their crops stay free from disease. However, understanding the causes of crop disease and the development of effective solutions is relatively recent in historic terms.

It was only in the nineteenth century that the concept of 'spontaneous generation', the view that symptoms on plants were a result of bad environmental conditions, was superseded by the concept of 'germs' being the cause of plant diseases. The early protagonists of this new concept are now accepted as some of the very early pioneers of plant pathology, such as Tillet who described bunt for the first time in 1755, and Berkeley and De Bary who described late blight fungus of potatoes in the 1850s (Russell, 2005), even preceding Pasteur and Koch who pioneered the concept that germs cause animal diseases. However, this was not until after the Irish potato famine, between 1840 and 1845, which resulted in the death of over 1 million people in Ireland with over half the population migrating to the United States. Early attempts at

http://dx.doi.org/10.19103/AS.2018.0039.14

chemical control of crop disease, such as the suggestion in 1807 by Prevost that bunt of wheat caused by *Tilletia caries* could be controlled to a certain extent by copper sulphate, fell on deaf ears for over 50 years (Prevost, 1807). It was not until the work of Millardet in 1882 on downy mildew on vines (*Plasmopara viticola*) that it was discovered a liquid made from lime and copper sulphate, which was initially applied to vines to stop thieves from stealing the grapes, also reduced the levels of downy mildew (Russell, 2005); hence, Bordeaux mixture was born. Throughout the nineteenth century and early part of the twentieth century, fungicide products were largely 'do-it-yourself' or fairly basic mixtures made by small commercial companies. The big leap in the development of new chemistry occurred after the Second World War, when great advances in the chemical industry were made and the war efforts were redirected, to some degree, towards agricultural products.

It is tempting to review at length the history of mycology and plant pathology, as well as the parallel developments of the early evolution of fungicides, but these are well documented in some excellent reviews including: Orlob (1973), Kelman and Peterson (2002), McCallan (1967), Horsfall (1975) and Russell (2005, 2006), among many others. It is interesting that many plant pathologists seem to have a fascination for the history of their science. This may be due to the fact that although agriculture has become more intensive and sophisticated in many parts of the world compared with the past, even in the twenty-first century, plant pathologists are still battling against the complexity of plant pathogens which our forefathers confronted. The consequences of which can still result in devastating crop loss, poisonous toxin production and an impressive ability to adapt to their environment often thwarting the best efforts of plant breeders and development chemists all over the world.

As pathologists, we can now smile wryly at the comment from Biffen in 1931 who, buoyed by early successes in breeding for yellow rust resistance with the new-found knowledge of Mendelian genetics, predicted that 'the solutions of the most important problems of rust control are in sight' (Biffen, 1931; Manners, 1969). Indeed, in 1992, Johnson reviewed the statement from Biffen and others with similar views saying that in relation to the management of rust in cereals 'although one can still optimistically look forward to the final solution, it is rather seen through the wrong end of the telescope' (Johnson, 1992). As plant pathologists in the twenty-first century, we have learnt many lessons from the past and are bravely fighting the battle against disease control but have not, and may never, win the war. This chapter therefore seeks to examine the present-day status of the threat of fungal diseases and our ability to control them in wheat, which is an economically important crop. We look at the challenges and opportunities of maintaining and developing an effective and environmentally safe fungicide pipeline whilst acknowledging the need for complementary technologies and innovation.

2 Global wheat production

It is believed that wheat was first cultivated around 10 000 years ago, most likely originating from the south-east region of Turkey. The spread of wheat from its site of origin across the world has been described by Feldman (2001) and it appears that the main route into Europe was via Greece and then northwards to the Balkans and the Danube, across to

Italy, France and Spain, and finally reaching the United Kingdom and Scandinavia. To the east, wheat reached China via Iran, and Africa via Egypt.

Wheat is one of the 'big three' cereal crops due to the range of geographies in which it can be cultivated (Feldman, 1995). It is a vital food source and is the dominant crop in temperate environments, being used for both human food and animal feed. Its success depends partly on its adaptability and high yield potential but also on the gluten protein fraction, which confers viscoelastic properties allowing dough to be processed into bread, pasta, noodles and other food products (Shewry, 2009).

Global wheat production has been increasing year-on-year and currently stands at approximately 735 million tonnes, with the top five producing countries or regions being the European Union (EU), China, India, Russia and the United States (Table 1 and Fig. 1), which account for over 65% of the total global production. However, the levels of productivity differ significantly across regions, as shown in Table 1, highlighting that the EU 28 is the most intensely productive area with an average of 6.04 tonnes/ha. China has a similar level of productivity with an average yield of 5.39 t/ha; however, Russia, India and the United States have much more extensive rather than intensive production systems with yield only averaging between 2.41 and 2.93 t/ha.

It is clear that the EU 28 is a major contributor to global wheat production and it is the most important food crop in this region, and second globally only to rice (Fones and Gurr, 2015). Within Europe, the leading country for cereal production in 2014 was France with 25% of the total for the EU 28, followed by Germany (18.5%) and the United Kingdom (11.1%), with the EU currently exporting up to 15% of its annual harvest. The levels of production and distribution of the wheat crop in the EU 28 are commensurate with the growing conditions, that is, soil type, weather conditions and water availability. This high level of production inevitably results from intensive cropping in key countries such as France, Germany, the United Kingdom and Poland. It also follows that where conditions for wheat production are good, so too are conditions for weeds, pests and disease development. Pest and disease risk is exacerbated by the monoculture nature of wheat production and by the relatively low number of wheat cultivars grown, thereby reducing the level of genetic diversity. This level of intensity is reflected in the high levels of fungicide input to the wheat crop in key countries with over 95% of the crop area treated in France, Germany and the United Kingdom and nearly 60% (and increasing year-on-year) in Poland (BASF internal data). The frequency of these treatments is also high compared with that of less intensive production areas, such as the United States, with the average number of treatments per season being 3.73 (UK), 2.35 (FR), 2.28 (DE) and 1.96 (PL) in 2015, which is equivalent to over 38 million product-treated hectares (Jorgensen et al., 2017).

Table 1 Top five global wheat producers (FAO, 2014) – % global wheat production, harvest area and yield (2014/15 summary)

	% of global wheat production	Harvest area (million ha)	Average yield (tonnes/ha)
EU 28	21.82	26.53	6.04
China	17.7	24.14	5.39
India	11.76	31.46	2.75
Russia	8.4	25.58	2.41
USA	7.63	19.14	2.93

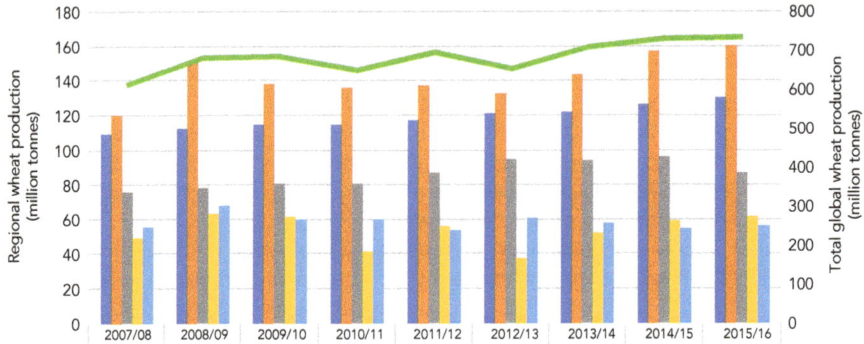

	2007/08	2008/09	2009/10	2010/11	2011/12	2012/13	2013/14	2014/15	2015/16
China	109.29	112.45	115.11	115.18	117.4	120.84	121.92	126.2	130.18
EU 28	119.99	150.54	138.6	136.13	137.62	132.57	143.56	157.1	160.5
India	75.81	78.57	80.68	80.8	86.87	94.88	93.51	95.85	86.53
Russia	49.36	63.76	61.73	41.5	56.23	37.72	52.09	59.71	61.78
USA	55.82	68.01	60.36	60.06	54.24	61.29	58.1	55.14	56.11
World	611.64	681.32	684.01	651.47	697.46	654.56	711.37	730.74	735.64

Wheat production per global region (million tonnes – FAO Stats 2017)

China EU 28 India Russia USA World

Figure 1 Changes in global wheat production over time with 2015–16 data which are estimates of yield (FAO).

3 Crop loss, plant health and the value of fungicide applications

Over the past few decades, the concept of food security has been the subject of considerable debate and discussion (Oerke, 2006). It can be defined as the balance between the growing food demand of the world population and global agricultural output. However, this equation glosses over the obvious discrepancies between supply and demand at regional, national and local scales (Ingram, 2011). This imbalance is not new but has the potential to lead to food crises, as witnessed in 2008 which resulted in political and social unrest in some regions (Smil, 2000; Savary et al., 2012). Plant protection, and specifically the protection of crops against plant diseases, clearly has a role to play in meeting the growing demand for food both in terms of quantity and quality (Strange and Scott, 2005). The issue of food security tends to be the focus of the developing world, whilst food safety is the more dominant concern of the developed world. However, it could be argued that the developed world holds the trump cards to address the issue of food security by having the most efficient and productive areas, Europe in particular. However, the developing world poses the biggest threat to food safety due to poor crop sanitation and protection, inadequate storage and poor regulation of the agricultural industry.

In the agricultural industry, we tend to focus on crop loss and its economic effect at the farmer level, but the true 'cost' of crop loss impacts the consumer, public health, societies, the environment as well as the economic fabric of the industry. Crop losses due to disease and the implication for food production loss, as well as food security, have been reviewed by several authors (Oerke, 2006; Zadoks, 2008; Savary et al., 2012). At a European level, it is very rare for a disease epidemic to cause significant yield loss in a wheat crop on a large scale, although localised, uncontrolled epidemics may occur at a local level from time to time, causing significant economic loss to an individual business or enterprise. Within the European wheat production environment, it is currently perhaps more appropriate to talk about maintaining 'plant health' rather than managing significant 'crop losses' due to diseases. The former term is widely used within the European legislative framework and is increasingly being applied to evaluate the need (or otherwise) for plant protection products (Regulation (EC) 1107/2009).

In a review on the concepts of plant health by Doring et al. (2012), the authors attempted to explore the issue in the context of an analogy with human and animal health. Indeed, for both humans and plants: (1) health varies between individuals (e.g. as a result of age), (2) health and disease are dynamic: within individuals they change over time, (3) health is subject to geographic patterns in pathogen occurrence and severity and (4) pathogens can develop resistance. However, perhaps not surprisingly, there are also some fundamental differences. For example, ill-health or even death of a plant can be tolerated if this is compensated for by other healthier individuals within a population. Additionally, in many cases, humans and animals can be cured by treatment, whereas plants are rarely 'cured' instead they are mostly prevented from becoming (more) diseased or the disease is kept under control long enough for the 'marketable yield' to be obtained.

Boorse (1977) defined plant health very simply as 'the absence of disease'; however, this is perhaps too simplistic a view. A better description was provided by Agrios (2005) when he stated that, 'a plant is healthy, or normal, when it can carry out its physiological functions to the best of its genetic potential', that is, when photosynthesis, growth and reproduction can be optimised within the given environment and agronomic conditions. This approach is best suited to a crop production situation where, for broad acre crops, monoculture of generally genetically homogeneous plants are grown for their marketable yield such as in the case of wheat production. Although it has been concluded by authors such as Schlösser (1997), Lucas (1998) and Agrios (2005) that a clear definition of health (and disease) in plants is difficult to achieve, there are approaches that can help to quantify the value of plant health by demonstrating the benefits of fungicide applications even in an intensive wheat production environment like Europe.

In a recent paper, by Fones and Gurr (2015), the authors estimated the economic advantage of a fungicide spray programme in a realistic agronomic scenario for the three leading wheat-producing countries in Europe – France, Germany and the United Kingdom (Table 2). They calculated the value of the wheat harvest to the individual economies of the three countries as France: €7.2 billion, Germany: €4.9 billion and the United Kingdom: €2.4 billion (Table 2).

Table 2 Economic advantage of a standard fungicide programme in France, Germany and the UK

Wheat (2013 harvest)	France	Germany	UK
Hectares planted (million)	4.95	3.1	1.63
Av. yield per hectare	7.4	8	7.4
Harvest total (million tonnes)	37	24.9	12.1
Value per tonne (€)	195	195	195
Value to country economy (million €)	7200	4900	2400

Reproduced from Fones and Gurr (2015).
Data from: Analyst Agritel; Agri.eu/wheat market; Farming statistics@defra; Federal Ministry of Food and Agriculture; International grain council wheat index; Agrimoney.

They assumed disease mitigation measures which would normally be used in a commercial farming situation, that is, the planting of a moderately susceptible wheat variety to *Septoria tritici*, and the application of a robust three-spray fungicide programme. Even under good production conditions, they still assumed a yield loss of between 5% and 10%, which was considered realistic under the growing and weather conditions in the three countries under consideration (Brown et al., 2015; Torriani et al., 2015). The result of this calculation is shown in Table 3. In summary, the cost to the economy of each country was two-fold, that is (1) direct cost of the loss of wheat yield due to the disease and (2) the cost of fungicide application, with total cost reduced due to enhanced yields. As a result, the 5–10% yield losses in France, Germany and the United Kingdom resulted in direct costs of between €120 and €700 million. Fungicide treatments cost the farmer between €160 and €500 million across the three countries based on the cost of the chemical and area treated. However, the costs were counterbalanced by the increase in treated yield (estimated to be 2.5 tonnes/ha – Torriani et al., 2015), leading to a positive financial return of between €800 and €2400 million. Such calculations, although only estimates based on one scenario, illustrate both the high return on investment for the farmer from his input based on effective chemistry but also the value of an effective fungicide solution being available to the farmer to maintain the health of the plant in an intensive environment.

Many similar studies have been carried out either focused on specific fungicide groups or individual active ingredients. These studies generally come to similar conclusions, that is, that there is a clear and quantifiable benefit of controlling plant diseases of wheat to maintain plant health, thereby ensuring a measurable return on investment whilst also producing safe, wholesome food (Teng et al., 1993; Cook and Veseth, 1992; Hughes et al., 1999; Leadbeater et al., 2000; Jorgensen et al., 2008).

4 Evolution of the pesticide regulatory environment

Up until the 1940s, chemical disease control mainly relied on inorganic chemical formulations often prepared by the user. The post-war decade saw a great expansion of the chemical industry, but the 1960s and 1970s were the halcyon days of rapid expansion of research and development and market growth. Further developments occurred in the 1980s and 1990s, but issues around fungicide resistance and the effect of an increasingly regulated industry also started to impact (Russell, 2005, 2006). Now in the twenty-first century,

Table 3 Estimation of economic loss in France, Germany and the United Kingdom based on poor control of *Septoria tritici* and showing the benefit of a robust fungicide programme

Losses to *Septoria tritici*[a]	France	Germany	UK
Value of 5–10% harvest loss due to *Septoria* (million €)	350–700	250–500	120–240
Spraying costs @ 100€/ha[b]	459×10^6	310×10^6	163×10^6
Added value (million €) (assuming 2.5 t/ha advantage after spraying)[c]	2400	1500	790

Reproduced from Fones and Gurr (2015).

[a] Figures are given in Euros.

[b] Spraying costs equate to three fungicide sprays per season.

[c] Yield enhancements following fungicide sprays (see Torriani et al., 2015).

fungicide development is a very complex and expensive business which has resulted in significant consolidation of the main companies involved with product development, a situation still ongoing today (Copping, 2003; Russell, 2005).

Alongside the evolution of fungicide resistance, it was inevitable that over time, there would be a need for increased management and regulation of pesticide use. Initially, regulation focused on the protection of the operator but over time developed to also consider the potential risks to consumers of treated crops.

Early regulatory schemes were voluntary such as the Pesticide Safety Precaution Scheme in the United Kingdom whereas today, pesticides are strictly controlled by legal obligations (Russell, 2005). Pesticide legislation varies greatly worldwide because countries have different requirements, guidelines and legal limits for plant protection. Generally, developed nations have more stringent regulations than developing countries, with the latter generally lacking resources and expertise to adequately implement and enforce legislation (Handford et al., 2015). These global differences can cause trade barriers, and there have been attempts to harmonise pesticide legislation by, for example, harmonising maximum residue limits. Despite some improvements in key markets, these limits still remain variable at a global level.

International and national legal frameworks governing pesticide management have undergone significant changes over the past 25 years. The *International Code of Conduct on the Distribution and Use of Pesticides* was first adopted by the FAO in 1985. Since the first adoption of the code, there have been several international instruments dealing directly or indirectly with pesticides and pesticide management. As a result of many of these international instruments, the Code of Conduct was updated in 2013, and in 2014 the WHO also adopted the Code of Conduct as its reference framework for international guidance on pesticide management. The purpose of the Code of Conduct is to provide guidance to governments that seek to review, update or design national pesticide legislation. The complete *International Code of Conduct on Pesticide Management: Guidelines on Pesticide Legislation* is available on the FAO website (www.fao.org) and the WHO website (www.who.int). In addition, the FAO maintains a repository of national legislation relevant to agriculture called FAO-LEX which can be found at http://faolex.fao.org/. A recent review of global pesticide legislation summarises the present-day situation (Handford, 2015).

5 Status of the global fungicide market

In 2015, the global market for crop protection products (CPPs) was valued at $51 210 million, with fungicides accounting for 26.7% ($15 348) of that market (Phillips McDougall, 2016). This sector is led by the developed markets, particularly Western Europe, Japan and North America. The fungicide market has grown at the fastest rate since 2010 and experienced a lesser decline in 2015 than other major market sectors. For fungicide products, some of this growth has come from increasing usage in order to control Asian soybean rust, particularly in Brazil where this is a devastating disease and one that is facing challenges due to fungicide resistance. The cereal crop sector accounts for the largest percentage of total global sales at 19.1%, followed by fruit and vegetable crops (15.8%) and soybean (14.8%). In terms of chemical class, strobilurins, triazoles and other azoles plus the Succinate DeHydrogenase Inhibitors (SDHis) account for a total of $9414 million of global sales which is over 60% of the total fungicide sector, demonstrating the very high reliance of the industry on just three modes of action.

In Fig. 2 the progression of the life cycle of a theoretical active ingredient is shown. Once an active ingredient is launched on to the market, sales will increase over time as new markets and products are launched and added to the portfolio [green dotted line (a)]. The duration of this life cycle will vary depending on the active ingredient, but there are many actives on the market today that have had very long lifespans (Russell 2005). However, in today's agronomic and legislative environment, there are two major threats to this longevity, particularly, although not exclusively, in the European market. The two main threats are: (1) the impact of resistance by a pathogen or number of pathogens to the active ingredients, thereby reducing its value or usefulness in a market [red dotted line (a)] and (2) the impact of legislation which may either restrict the market for an active ingredient or, in a worst-case scenario, meaning that it is removed totally from the market [grey box – (b) sales]. The reality in Europe can be that both legislation and resistance could impact upon the life cycle of the active ingredient severely reducing a financial return on investment for the company. The differences in the area under the curve in both (a) and (b) (Fig. 2) show the potential loss of sales and hence return on investment that may occur for the company. With these potential threats, it is very important that a company evaluates the risk versus the value in developing any new active substance.

5.1 Fungicide resistance and its impact on the active ingredient life cycle

In Europe, some of the first incidents of fungicide resistance were in the 1980s when the key wheat pathogen, *Septoria tritici*, was found to have mutation E198A, which conferred resistance to MBCs (methyl benzimidazoles carbamates) (Lucas and Fraaije, 2008; Leroux et al., 2007). Since those early findings, understanding the risks and management of fungicide resistance by industry has increased significantly. However, it is a constant challenge to try to reduce and mitigate the risk of fungicide resistance in pathogens such as *Septoria tritici* whilst at the same time ensuring effective disease control (Kuck

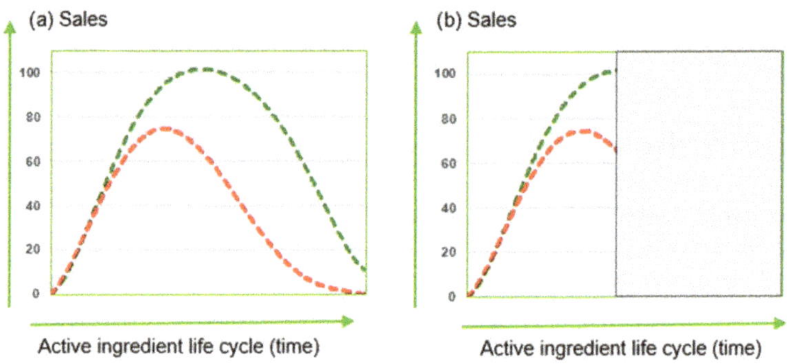

Figure 2 Return on investment from the launch of an active ingredient (a) when resistance alone impacts the life cycle (based on sales) and (b) when resistance and legislation impact the life cycle (based on sales). Green dotted line [charts (a) and (b)] – normal assumed life cycle of an active ingredient with no negative impacts. Red dotted line [charts (a) and (b)] – theoretical life cycle of an active ingredient with the onset of resistance against a target pathogen. Grey shading (chart (b) only) – premature loss of an active ingredient due to loss of approval as a result of 'cut-off criteria' according to Dir. EC 1107/2009.

and Russell, 2006; Grimmer et al., 2015). The development of fungicide resistance and the mechanisms involved are well documented and have been reviewed by Lucas et al. (2015). In addition, the assessment of the possible risk of resistance to either an existing or new active ingredient is a fundamental part of European legislation (EC 1107/2009) and requires agrochemical companies to invest significant resources in monitoring programmes, genetic studies and stewardship measures to maintain the effectiveness of active ingredients for as long as possible (Russell, 2001). This is, of course, in the best interests of both the market and the manufacturer, as shown in Fig. 2.

5.2 The European regulatory environment and its impact on the active ingredient life cycle

The EU is renowned for having one of the strictest regulatory systems in the world and is considered a key player internationally with its harmonised approach among member states to pesticide standards. It is suggested by many that the EU provides a good example for global harmonisation and it is true that it is the most harmonised of all approaches on a global scale (Erlacher and Wang, 2011; Handford, 2015). However, many who work closely within the European legislation environment recognise that despite harmonised regulation, national interpretations by member states and political interference can lead to inconsistent evaluations and unpredictable outcomes. For example, despite a harmonised approach to pesticide regulation, some countries have imposed additional restrictions at a national level. In Denmark, where ground water contamination is of particular concern, several active ingredients that are commonly found in other European countries are not registered and those on the market are subject to a pesticide tax. This tax is based on a pesticide load index (PLI) – the higher the PLI, the higher the tax. In France, the government has taken a different approach, starting with the 'Grenelle' programme, which led to nearly 40 active ingredients being banned. Ecophyto I and, now, Ecophyto II, which aims to reduce pesticide use by 25% by 2020 and 50% by 2025, supersede this programme.

In 1993, EU Directive 91/414/EEC was implemented to harmonise pesticide approvals across Europe. The net result of this process was the loss of nearly 75% of active ingredients from the European market. However, many were lost simply because they were not supported by the approval holder as they had been superseded by technically superior actives or were uneconomic rather than judged unsafe under the EU Directive (King, 2014). The active ingredients that remained on the market and were approved under Dir. 91/414/EEC were given approval for 10 years with a requirement for a second review to be completed before the end of the 10-year period, ensuring that risk assessments were kept up to date. EU Regulation EC 1107/2009, which came into force in 2011, replaced Directive 91/414/EEC and it is under this regulation that the present renewal process is taking place (European Parliament and Council, 2009). In the new regulation, there was a key change from a risk-based to a hazard-based assessment, whereby the intrinsic properties of the active substance itself became key criteria for whether or not an approval could be supported. These defined properties, referred to as 'cut-off criteria', now mean that if they are applicable to the active ingredient under evaluation, no further assessment will take place and approval will not be granted. These criteria include potential impacts on human health if a substance has carcinogenic, mutagenic and/or reproductive toxicity properties (CMRs 1a/1b). In the future, endocrine-disrupting properties are also likely to fulfil the 'cut-off criteria' and those active ingredients deemed to meet this criterion will also not be approved (Anon., 2014). In addition, the 'cut-off criteria' include factors such as being a persistent organic pollutant, persistent bioaccumulative toxin and very persistent bioaccumulative toxin.

In addition to this renewal process, further assessments to reduce risks are also included, with the main one being Comparative Assessment (CA) (Article 50 and Annex IV of EU Reg. EC 1107/2009) (Williams, 2011). This assessment differs from the renewal process in that it focuses on products containing active ingredients that are on a candidate for substitution (CfS) list and are evaluated at a national rather than zonal level. The CfS list, published in 2015, currently contains 78 active substances (Traon, 2013). It is estimated that nearly 70% of all registered fungicide products in Europe either contain an active substance which meets the cut-off criteria or contains a CfS (BASF internal data). It is hard to predict how many active substances and products will be lost in the European market over the next 5–10 years, but it is clear that the wheat fungicide market has become increasingly complex and unpredictable, and the commitment to future development will be challenged in Europe in the future (King, 2014).

5.3 The cost of developing new fungicide active ingredients

The search for new active ingredient molecules capable of becoming profitable products is becoming increasingly complex, expensive and unpredictable, particularly in a European setting. Finding an active molecule is only the first step in the process. The real challenge is in producing an active molecule that has all the necessary attributes to allow the product to compete against existing products in the market, comply with a myriad of regulatory hurdles at regional and global levels, as well as having a broad applicability to a range of crops and pathogenic species. Considerable thought goes into defining the required properties that the new molecule should ideally have but there is no blueprint and it is often the case that along with desirable properties from a biological efficacy approach to target pathogens comes a range of contraindications that should be avoided. In reality, most molecules and the resulting products that come on to the market are a compromise such that they allow the molecule to be used in its most effective way without posing undue regulatory risks to the user, the consumer or the environment (Leadbeater et al., 2000; Russell, 2006) (Table 4).

In a recent survey of leading crop protection companies, Phillips McDougall (2016) evaluated the expenditure necessary for the discovery and development of a new crop protection active ingredient looking at the change in costs over time from 1995, 2000

Table 4 Desirable properties for new fungicides and concerns that could arise from those properties

Desirable	Contraindications for regulatory reasons
Multisite of action to reduce resistance risk	Most likely a general toxophore with a non-specific mode of action; raises issues with possible non-target toxicity effects
Long-term efficacy	Persistency in crop residues, harvested produce and the environment
Systemicity	Accumulation in non-treated but harvested produce, for example foliar applications moving to fruit
Broad spectrum	Non-target effects
Defined mode of action	Generally associated with a single site of action, increased resistance risk
High efficacy, low dose rate	Increased hazard resulting from the intrinsically increased activity

Reproduced from Russell (2006).

and 2005 to 2008 and 2010 to 2014. In addition to examining the development costs the report also looked at the number of molecules that needed to be screened in order to develop one through to registration plus the time to market.

The survey looked at three main areas necessary to achieve the registration of a new active ingredient:

1 Research: chemistry, biology, toxicology and environmental chemistry
2 Development: chemistry, field trials, toxicology and environmental chemistry
3 Registration

As shown in Fig. 3, costs in all three areas have increased from 1995 to 2014. It is not surprising that costs would have increased over time due to impacts of inflation and wage increase; however, the scale of the increase in costs cannot be accounted for by inflation and wage costs alone. The overall costs of discovery and development increased by 21.1% between 1995 and 2000, 39.1% from 2000 to 2005–8, and 11.7% from 2005–8 to 2010–14. Within the development phase, the biggest leap in costs occurred between 2000 and 2005–8 where costs rose by 84.8%. Although costs in all areas increased, environmental chemistry costs increased significantly which is probably attributable to the increasing regulatory demand being placed on the industry. The biggest expenditure within the development phase has always been the area of field trials. These costs are hard to reduce as legal requirements are set out in order to demonstrate the effectiveness of a molecule in the field. However, new technologies in digital farming and remote measurements of symptom expression may help to support reductions in costs and increases in efficiency in the future, but these savings have yet to be realised. More recent figures for the overall cost of the development of an active ingredient have not been published but are now estimated to be in the region of $300 million (approx. €250 million) (BASF internal data).

Within the research phase, the cost of biology (screening) is now the biggest area for expenditure having risen from $32 to $51 million (equivalent to 59.4%) between 2005–8 and 2010–14, a rapid rise over a short period of time. These increases in costs are probably

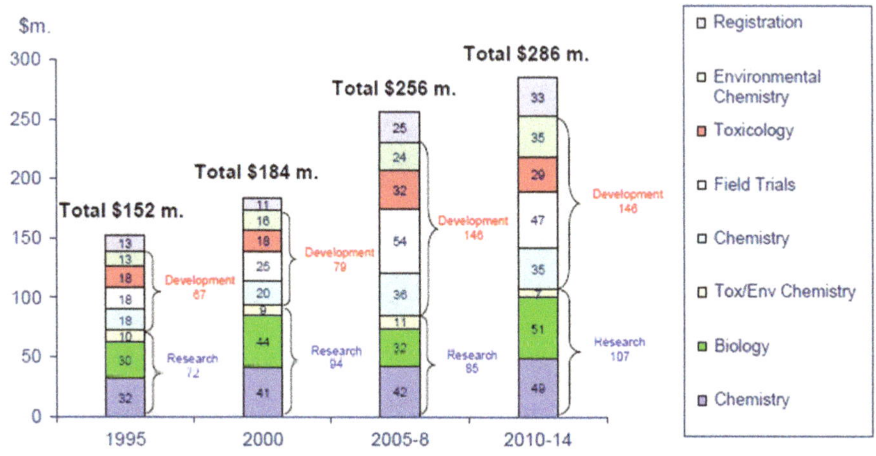

Figure 3 Discovery and development costs of a new crop protection active ingredient.

due to the higher number of candidates to be screened and the increased number of tests that need to be performed in order to look for a suitable candidate.

Clearly the costs of bringing a new active to the market have increased significantly, but another challenge for the agrochemical manufacturers is the time it takes for that active ingredient to be approved. Table 5 shows the increase in the number of candidate molecules that have to be screened before a single candidate can be registered and in addition, the time to market has increased from 8.3 to 11.3 years.

In summary, the cost of discovering and developing a new active ingredient has increased from $152 to $286 million in 20 years, but at the same time it also takes 3 years to even start to see a return on the investment. If this investment is further threatened by the factors of resistance and regulatory changes, as shown in Fig. 3, it is perhaps not surprising that the industry is further consolidating in order to withstand both the costs and risks.

The economics of active ingredient discovery and development are having a particularly significant impact in Europe. Between 2003 and 2011, Europe was the leading regional agrochemical market worldwide; in 2012, it was overtaken by Asia. In terms of growth and sophistication, the European market is divided into EU-15, new EU-13 and East European markets, with the former presently having the highest value but the slowest growth. Coupled with the increasingly complex and unpredictable regulatory environment of the EU-28 countries, there are now very serious challenges to the number of new active ingredients that will come on to the market (Fig. 4). As can be seen in Fig. 4, Europe has

Table 5 Increase over time in the number of research candidates required in order to achieve a successful registration of an active ingredient in Europe

	1995	2000	2005–08	2010–14
Research candidates	52 500	139 429	140 000	159 574
Development targets	4	2	1.3	1.5
Active ingredient registration	1	1	1	1
Lead time (years)	8.3	9.1	9.8	11.3

Reproduced from Phillips McDougall (2016).

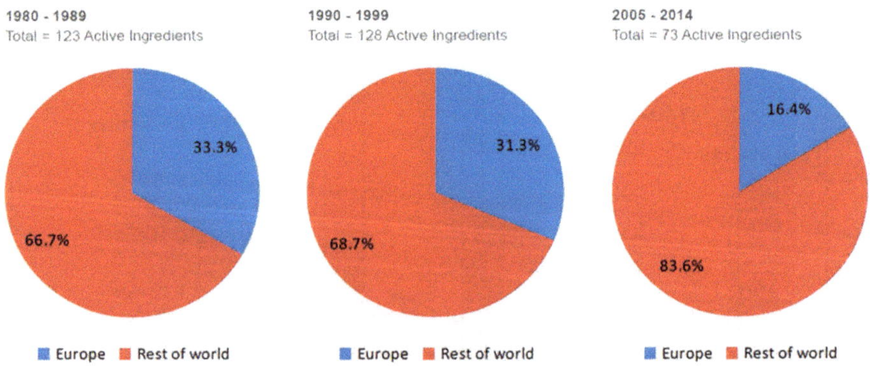

Figure 4 Split of active ingredient development between Europe and the rest of the world – 1980–2014 (Phillips McDougall, 2013).

seen a significant decrease (from 33.3% to 16.4%) in the number of active ingredients brought to market. In the period 2005–14, 14 of the active ingredients brought to market were fungicides; even if 25% of these were launched in Europe, at best that equates to only 3–4 new active ingredients in that time period.

6 Conclusion and future trends

The mantra of the need for food security and the world's ability to feed 5 billion people by 2050 is regularly debated, but the reality is that there is an imbalance between where the food is produced and where the food is actually needed. Europe, unlike many less developed regions of the world, has an efficient and effective agricultural industry supported by cutting-edge scientists and an agricultural industry willing and able to invest in the innovation pipeline. Regulation is of course vital and the first part of the objective of EU 1107/2009 is to 'to ensure a high level of protection of both human and animal health and the environment'. However, the 'extreme risk, adverse nature' of the regulations and the way that they are now being implemented across Europe, often driven as much by ideology as by science, runs the very real risk of reducing innovation and having a significant impact on the competitiveness of the agricultural industry across Europe. This poses a very real threat when the level of food production in the most productive area of the world could be severely impacted over the next few decades as these regulations start to take effect.

It would be a brave person to predict what the European agricultural industry will look like by 2030. The impact of the loss of CPPs is likely to be one of slow erosion and restriction rather than a very dramatic loss in most cases. At times, regulation can also encourage innovation, and new agronomic methods and technologies may be a positive way forward. However, as a part of an agricultural industry, we must wake up to what the changes may mean for the future and how we should deal with them. Innovation in CPPs may be slowing down but that also presents new incentives and opportunities to industry and scientists alike. It is essential that we take a realistic look at how effective some of these alternative approaches will be to control the levels of disease risk and the impact on yield that are likely to occur. Return on investment will come from different approaches and innovations, but this change of direction needs to start now if innovation is to help solve the agronomic challenges of the future. At present, many parts of the agricultural industry are working in isolation and often in competition with each other. It is only by working together that a truly integrated agronomic approach can be achieved and it is only with a clear, honest and realistic understanding of the challenges the industry faces that meaningful solutions will be found in the future.

7 References

Agrios, G. N. (2005), *Plant Pathology*, San Diego, CA: Elsevier- Academic Press.

Anon. (2014), European Commission: An overview of the establishment by the European Commission of a legislative-based strategy for endocrine disruptors, Brussels 2014.

Biffen, R. H. (1931), The cereal rusts and their control, *Transactions of the British Mycological Society* 16, 19–37.

Boorse, C. (1977), Health as a theoretical concept, *Philosophy of Science* 44, 542–73.

Brown, J. Chartrain, L., Lassere-Zuber, P. and Saintenac, C. (2015), Genetics of resistance to Zymoseptoria tritici and applications for wheat breeding, *Fungal Genetics and Biology* 79, 33–41.

Cook, R. J. and Veseth, R. J. (1992), *Wheat Health Management*, St Paul, MN: APS Press.

Copping, L. (2003), The evolution of crop protection companies, *Pesticide Outlook* 14, 276–8.

Doring, T. F., Pautasso, M., Finckh, M. R. and Wolfe, M. S. (2012), Concepts of plant health – reviewing and challenging the foundations of plant protection, *Plant Pathology* 61, 1–15.

Erlacher, E. and Wang, M. (2011), Regulation EC no. 1107/2009 and upcoming challenges for exposure assessment of plant protection products: Harmonisation or national modelling approaches, *Environmental Pollution* 159, 3357–63.

European Parliament and Council (2009), Regulation (EC) No 1107/2009 of the European Parliament and of the Council of 21 October 2009 concerning the placing of plant protection products on the market and repealing Council Directives 79/117/EEC and 91/414/EEC, *Official Journal of the European Union* L309, 1–50.

Feldman, M. (1995), Wheats, *in* Smartt, J. and Simmonds, N. W. (Eds), *Evolution of Crop Plants*, Harlow, UK: Longman Scientific and Technical, pp. 185–92.

Feldman, M. (2001), Origin of cultivated wheat, *in* Bonjean, A. P. and Angus, W. J. (Eds), *The World Wheat Book: A History of Wheat Breeding*, Paris, France: Lavoisier Publishing, pp. 3–56.

Fones, H. and Gurr, S. (2015), The impact of Septoria tritici Blotch disease on wheat: An EU perspective, *Fungal Genetics and Biology* 79, 3–7.

Food and Agricultural Organisation of the United Nations and World Health Organisation (2014), International Code of Conduct on Pesticide Management (20 September 2014), Available from http://www.fao.org

Grimmer, M., van den Bosch, F., Powers, S. and Paveley, N. (2015), Fungal resistance risk assessment based on traits associated with the rate of pathogen evolution, *Pest Management Science* 71(2), 207–15.

Handford, C. E., Elliot, C. T and Campbell, K. (2015), A review of the global pesticide legislation and the scale of challenge in reaching the global harmonisation of food safety standards, *Integrated Environmental Assessment and Management* 11(4), 525–36.

Horsfall, J. G. (1975), Fungi and fungicides: The story of a nonconformist, *Annual Review of Phytopathology* 13, 1–14.

Hughes, G., McRoberts, N. and Burnett, F. J. (1999), Decision-making and diagnosis in disease management, *Plant Pathology* 48, 147–53.

Ingram, J. (2011), A food systems approach to researching food security and its interactions with global environmental change, *Food Security* 3, 417–31.

Johnson, R. (1992), Reflections of a plant pathologist on breeding for resistance with emphasis on yellow rust and eyespot of wheat, *Plant Pathology* 41, 239–54.

Jorgensen, L. N., Jahn, M., Clark, B., Antichi, D., Goral, T., Schepers, H., Lucas, J., Rolland, B., Gouache, D. and Hornok, L. (2008), Best control practices of diseases in winter wheat. Endure deliverable D1.2. Available on-line www.eurowheat.org

Jorgensen, L. N., van den Bosch, F., Oliver, R. P., Heick, T. M. and Paveley, N. D. (2017), Targeting fungicide inputs according to need, *Annual Review of Phytopathology* 55, 181–203.

Kelman, A. and Peterson, P. D. (2002), Contributions of plant scientists to the development of the germ theory of disease, *Microbes and Infection* 4, 257–60.

King, R. (2014), The effect of loss of plant protection products on UK agriculture and horticulture and the wider economy. Report prepared for the AIC, NFU and CPA by Andersons, p. 67.

Kuck, K.-H. and Russell, P. E. (2006), FRAC: Combined resistance risk assessment, *Aspects of Applied Biology* 78, 3–10.

Leadbeater, A. J., West, S. J. E. and Pichon, E. (2000), Economics of cereal disease control – a European Perspective, *in Proceedings of the BCPC Conference – Pests and Diseases 2000*, Farnham, UK: BCPC, pp. 639–46.

Leroux, P., Albertini, C., Gautier, A., Gredt, M. and Walker, A.-S. (2007), Mutations in the CYP51 gene correlated with changes in sensitivity to sterol 14α-demethylation inhibitors in field isolates of Mycosphaerella graminicola, *Pest Management Science* 63(7), 688–98.

Lucas, J. A. (1998), *Plant Pathology and Plant Pathogens*, 3rd Edition, Oxford, UK: Blackwell Science, p. 274.

Lucas, J. and Fraaije, B. (2008), QoI resistance in Mycosphaerella graminicola: What have we learnt so far? *in* Dehne, H. W., Deising, H. B., Gisi, U., Kuck, K. H., Russell, P. E. and Lyr, H. (Eds), *Modern Fungicides and Antifungal Compounds*, Braunschweig, Germany: Deutsche Phytomedizinische Gesellschaft.

Lucas, J., Hawkins, N. and Fraaije, B. (2015), The evolution of fungicide resistance, *Advances in Applied Microbiology* 90, 29–92.

Manners, J. G. (1969), Presidential address, The rust diseases of wheat and their control, *Transactions of the British Mycological Society* 52, 177–86.

McCallan, S. E. A. (1967), History of fungicides, *in Fungicides, An Advanced Treatise*, volume I, New York, NY: Academic Press, pp. 1–37.

Oerke, E. C. (2006), Crop losses to pests, *Journal of Agricultural Science* 144, 31–43.

Orlob, G. B. (1973), Ancient and medieval plant pathology, *Pflanzenschutz-Nachrichten Bayer* 26, 265–94.

Phillips McDougall (2013), R&D Trends for chemical crop protection products and the position of the European Market: Report published by the European Crop Protection Association (ECPA), Phillips McDougall (part of the Agribusiness Intelligence trading division of Informa UK Limited), Available at: www.phillipsmcdougall.com

Phillips McDougall (2016), The cost of new agrochemical product discovery, development and registration in 1995, 2000, 2005-8 and 2010-2014, R&D expenditure in 2014 and expectations for 2019: A consultancy study for CropLife International, CropLife America and the European Crop Protection Association (ECPA), Phillips McDougall (part of the Agribusiness Intelligence trading division of Informa UK Limited), Available at: www.phillipsmcdougall.com

Prevost, B. (1807), *Memoire sur la Cause Immediate de la carie ou Charbon des bles, et de Plusieurs Autres Maladies des Plantes, et sur les Preservatifs la Carie*, Paris France: Bernard, Quai des Augustins. [English translation Keitt GW de (1939), *Memoir on the Immediate Cause of Bunt or Smut of Wheat and of Several other Diseases of Plants and Prevention of Bunt*, Menasha, WI: American Phytopathological Society, Phytopathology Classics No. 6].

Russell, P. E. (2001), Resistance management and the registration of pesticide products, *Europe Pesticide Outlook* 12, 56–9.

Russell, P. E. (2005), A century of fungicide evolution, *Journal of Agricultural Science* 143, 11–25.

Russell, P. E. (2006), The development of commercial disease control. BSPP Presidential Address, *Plant Pathology* 55, 585–94.

Saravy, S., Ficke, A., Aubertot, J.-N. and Hollier, C. (2012), Crop losses due to diseases and their implications for global food production losses and food security, *in Food Security: The Science, Sociology and Economics of Food Production and Access to Food*, Published by Springer (published online).

Schlösser, E. (1997), *Allgemeine Phytopathologie*, New York, NY: Georg Thieme Verlag.

Shewry, P. R. (2009), Wheat: Darwin review, *Journal of Experimental Botany* 60(6), 1537–53.

Smil, V. (2000), *Feeding the World: A Challenge for the 21st Century*, Cambridge, MA: The Massachusetts Institute of Technology Press.

Strange, R. and Scott, P. (2005), Plant disease: A threat to global food security, *Annual Review of Phytopathology* 43, 83–116.

Teng, P. S., Savary, S. and Revilla, I. (1993), Systems of plant protection, *in* Chadwick, D. J. and Marsh, J. (Eds), *Crop Protection and Sustainable Agriculture*, Chichester: Wiley, pp. 116–39.

Torriani, S. F. F., Melichar, J. P. E., Mills, C., Pain, N., Sierotzki, H. and Courbot, M. (2015), *Zymoseptoria tritici*: A major threat to wheat production, integrated approaches to control, *Fungal Genetics and Biology* 79, 8–12.

Traon, D. (2013), Ad-hoc study to support the initial establishment of the list of candidates for substitution as required in Article 80(7) of Regulation (EC) No 1107/2009: Final Report, European Commission Directorate General for Health and Consumers (DG SANCO), Brussels, Belgium, Available from the EC CIRCA platform: https://circabc.europa.eu, https://circabc.europa.eu

Williams, J. C. (2011), New EU pesticide legislation – the view of a manufacturer, *Aspects of Applied Biology* 106, 269–74.

Zadoks, J. C. (2008), *On the Political Economy of Plant Disease Epidemics. Capita Selecta in Historical Epidemiology*, Wageningen, the Netherlands: Wageningen Academic Publishers.

Chapter 12

Occurrence and avoidance of fungicide resistance in cereal diseases

Lise Nistrup Jørgensen, Aarhus University, Denmark; Richard Peter Oliver, Curtin University, Australia; and Thies Marten Heick, Aarhus University, Denmark

1 Introduction

Plant pathogenic fungi cause serious reductions in yield and quality of many cereal crops. Wheat (*Triticum aestivum*) and barley (*Hordeum vulgare*) are the most important grain cereals worldwide and play an essential role in food security (Anon., 2016). In 2014, wheat and barley were grown on 220 million and 49 million ha worldwide, yielding on average 3.3 and 2.9 tonnes/ha, respectively (Anon., 2016). In high-input agricultural systems found in Northwest Europe and in New Zealand, yields exceeding 10.0 tonnes/ha for wheat and 8.0 tonnes/ha for barley can be achieved in favourable years, indicating the present potential of both crops. A healthy crop stand throughout the entire growing season is a prerequisite for crops to achieve their full yield potential (Strange and Scott, 2005). However, plant diseases, weeds and insects can reduce yields significantly (Oerke, 2006) and are a real threat to global food security. The targeted use of fungicides can help to minimise those losses and thus is an invaluable part of the crop management toolbox to combat plant diseases efficiently and safely.

Several diseases are known to have major impacts in wheat and barley (Tables 1 and 2). In the maritime zone of Europe, the major wheat diseases are *Zymoseptoria tritici* causing Septoria tritici blotch (STB) and the rust diseases leaf (brown) rust (*Puccinia triticina*) and yellow (stripe) rust (*Puccinia striiformis*) (Jørgensen et al., 2014). In other parts of the world, the major diseases are tan spot (*Pyrenophora tritici-repentis*), stem rust (*Puccinia graminis*), head blight (*Fusarium* spp.), crown rot (*Fusarium pseudograminearum*) and spot blotch (mainly *Bipolaris sorokiniana*). The wheat form of powdery mildew (*Blumeria graminis*

http://dx.doi.org/10.19103/AS.2018.0039.13

Table 1 Major wheat diseases controlled by foliar fungicides

Disease preferred name	Abbreviation	Pathogen preferred name
Stem rust	SR	*Puccinia graminis*
Leaf rust	LR	*Puccinia triticina*
Yellow rust	YR	*Puccinia striiformis*
Powdery mildew	PM	*Blumeria graminis* f. sp. *tritici*
Spot blotch	SB	*Bipolaris sorokiniana*
Septoria tritici blotch	STB	*Zymoseptoria tritici*
Septoria nodorum blotch	SNB	*Parastagonospora nodorum*
Wheat blast	WB	*Magnaporthe oryzae tritici*
Tan spot	TS	*Pyrenophora tritici-repentis*
Crown rot	CR	*Fusarium pseudograminearum*
Take-all	TA	*Gaeumannomyces graminis*
Eye spot	ES	*Oculimacula yallundae*
Head blight	FHB	*Fusarium graminearum*, *F. culmorum* and other *Fusarium* species
Head blight	MHB	*Microdochium nivale* and *M. majus*

f. sp. *tritici*) is important in few places where modest temperatures and high humidity occur at the same time. Recently, wheat blast (*Magnaporthe oryzae tritici*) has emerged as a significant threat in South America and South Asia. In barley, the two net blotches (*Pyrenophora teres* f. *teres* and *Pyrenophora teres* f. *maculata*), scald (*Rhynchosporium commune*), leaf (brown) rust (*Puccinia hordei*), powdery mildew (*Blumeria graminis* f. sp. *hordei*) and ramularia leaf spot (RLS; *Ramularia collo-cygni*) are the most serious threats in most growing regions (Dewar et al., 2016). Furthermore, in diseases such as head blight, crown rot and spot blotch, the same pathogens cause disease in both wheat and barley.

Table 2 Major barley diseases controlled by foliar fungicides

Disease preferred name	Abbreviation	Pathogen preferred name
Leaf rust	LR	*Puccinia hordei*
Powdery mildew	PM	*Blumeria graminis* f. sp. *hordei*
Ramularia leaf spot	RLS	*Ramularia collo-cygni*
Net blotch (net form)	NFNB	*Pyrenophora teres* f. *teres*
Net blotch (spot form)	SFNB	*Pyrenophora teres* f. *maculata*
Rhynchosporium	Scald	*Rhynchosporium commune*
Spot blotch	SB	*Bipolaris sorokiniana* (but clearly a disease complex)
Crown rot	CR	*Fusarium pseudograminearum*
Head blight	FHB	*Fusarium graminearum*, *F. culmorum* and other *Fusarium* and *Microdochium* species

Fungicide use differs markedly around the world (Oliver and Hewitt, 2014). Fungicides were first used on cereals in Northwest Europe. In 1980, only about 10% of UK wheat crops received a foliar fungicide spray, but by 1990, this had risen to 95% and by 2005, it had reached nearly 100% (Lucas et al., 2015). The current use pattern in New Zealand is similar to that in Europe (Van Toor et al., 2013). In Australia, where yields are typically only about 2–4 tonnes/ha, about 50% of crops receive at least one foliar fungicide treatment and seed treatments are almost ubiquitous (Murray and Brennan, 2009, 2010). Information from other countries is sparse, but use is generally less intensive than in Northwest Europe.

The long-term use of fungicide classes based on a single mode of action has imposed consistent selection pressures on certain pathogens. Over time, this continuous selection pressure has resulted in many cases of fungicide resistance, leading to a reduction or the total loss of field performance of a range of active ingredients.

Resistance development is driven by multiple factors including the biology of pathogens, frequency of fungicide use, the mode of action of fungicides (single vs. multisite inhibitors) and usage patterns (used alone, in mixture and/or in alternations) (Brent et al., 1990; Russell, 1995; Grimmer et al., 2015). For many years now, the agro-chemical industry and regulators have advocated the adoption of so-called 'resistance management strategies' (Brent and Hollomon, 2007). These strategies take into account the risk factors and are enacted via specific recommendations on the product labels. In addition, local fungicide resistance expert groups [e.g. FRAG (Fungicide Resistance Action Group) – UK and Norbarag in the Nordic Baltic region] and advisors give recommendations on how to counteract resistance. However, carrying out effective resistance management has proven to be a difficult discipline and despite great efforts, resistance to fungicides has evolved in many pathogens.

In this chapter, the development of resistance to fungicides in cereal pathogens is briefly reviewed. The development of anti-resistance management strategies has largely evolved from studies of European cereal production and examples of how data from field trials have helped to illustrate impact from different control strategies are included as case studies.

2 Key fungicide classes for the control of cereal diseases

The most widely used fungicide classes for the control of diseases in cereals currently are 14α-demethylase inhibitors (DMIs; azoles; group G1), quinone-outside inhibitors (QoIs, strobilurins; group C3) and succinate dehydrogenase inhibitors (SDHIs, carboxamides; group C2). Resistance to methyl benzimidazole carbamate (MBC) (group B1) fungicides in eyespot and Septoria nodorum blotch ended their use on cereals in Europe in the 1980s (Brown et al., 1984; Horsten and Fehrmann, 1980). Since the start of that decade, DMIs have been the backbone of control of cereal diseases. QoIs were introduced in the late 1990s and despite widespread resistance in several pathogens, this group is still used for control of rusts in all cereals and leaf spot diseases in barley. SDHIs are the newest class of fungicides and their market shares have rapidly increased since 2010. All three fungicide classes have in common that they act on one specific target in the fungus, which leaves all active ingredients of those classes prone to resistance development.

According to the Fungicide Resistance Action Committee (FRAC), which coordinates international research on fungicide resistance, the risk of resistance development towards DMI fungicides was generally regarded to be 'medium' (FRAC, 2017; Brent et al., 1998). However, resistance to DMIs has now reached critical levels where major declines in sensitivity have occurred or are expected, which require regular resistance monitoring. Strobilurins were initially expected to have a 'moderate risk' of developing resistance, but as several diseases quickly developed high levels of resistance, they were reclassified to be at 'high risk' (Bartlett et al., 2002). The risk of resistance development for the SDHI class is estimated to be 'medium to high' (FRAC, 2016). Recent developments indicate that this assumption is justified, as SDHI resistance is now emerging in *Z. tritici*, *Pyrenophora teres* and *Ramularia collo-cygni* (Dooley et al., 2016a; Rehfus et al., 2016, 2018; Sierotzki and Scalliet, 2013).

Increasing problems with resistance towards QoIs, DMIs and lately also to SDHIs have put increasing focus on other compounds and especially the multisite inhibitors chlorothalonil (M5) and folpet (M4). These compounds are often used as a mixing partner with single-site fungicides.

An overview of different cereal diseases, groups of fungicides and their level of control and resistance situation is given in Table 3. The table reflects both the inherent sensitivity of the pathogen to the fungicide as well as the impact of evolved resistance. 'Red' indicates major problems with resistance and failure in field control. 'Green' indicates no shifting and no recognised problems with resistance. 'Orange' indicates confirmed cases of resistance, where field performance is still highly variable. As an example in barley, field control of net blotch by strobilurins is still very good despite development of specific mutations (F129L). The opposite is seen for control of SDHI-related mutation (C-H146R or C-H153R) on RLS, which in some countries/regions has led to total loss in efficacy (Hess, 2017), while other countries in the maritime zone still achieve good control of this disease using SDHIs and azoles.

New active ingredients are typically only released if they have a broad spectrum and control most, if not all, of the relevant diseases. This is certainly the case for the main

Table 3 Summary table of the efficacy profiles of the main fungicide classes currently in the maritime zone of Europe

	Disease	Strobilurins	Azoles	SDHIs	Multisite inhibitors
Wheat	Septoria tritici blotch	–	**(*)	****	**
	Yellow/Stripe rust	****	****	**(*)	*
	Brown rust	****	****	**(*)	*
	Tan spot	**(*)	**(*)	*	*
	Powdery mildew	–	***	*	*
Barley	Barley rust	****	****	**(*)	*
	Net blotch	***(*)	**(*)	****	*
	Rhynchosporium	****	****	****	*
	Ramularia leaf spot	–	**(*)	****	***
	Powdery mildew	***	***	*	*

'Red' indicates high degree of resistance, 'orange' indicates moderate or evolving resistance and 'green' indicates no indications of resistance.

Stars indicate differences in levels of efficacy from the most potent products belonging to the specific groups; – = efficacy not reliable due to resistance; * = low or uncertain efficacy based on intrinsic profile, **** = high efficacy. Stars in brackets indicate variable results.

fungicide classes (MBC, DMI, QoI, SDHI). However, the speed with which powdery mildews have developed resistance to these main classes has meant that fungicides targeting mainly these diseases (the compounds are then referred to as mildewcides) are in use in various parts of the world. Mildewcides are typically applied early in the season at localities with high risk of powdery mildew. The azanapthalene (E1) group of mildewcides now includes three compounds: quinoxyfen, proquinazid and quinofumelin. No case of field resistance has yet been reported, but laboratory mutants have been produced (Lee et al., 2008; Genet and Jaworska, 2009). The G2 fungicides inhibit sterol biosynthesis through action on both the Δ^{14}–reductase and Δ^{8}-Δ^{7}- isomerase steps (*erg2* and *erg24*). Although generally applied against powdery mildews, several of the mildewcides have a broader spectrum of control. The old morpholines and piperidine fungicides (e.g. fenpropimorph and fenpropidin) are out of patent and not widely available, but have been supplemented in recent years by the introduction of spiroxamine. Reduced field performances have been seen to morpholines, but only moderate shifting has been reported (Napier et al., 2000).

3 Cases of fungicide resistance

The maritime region of Europe is the epicentre of the cereal fungicide industry, typically receiving the earliest releases of new fungicides. The high disease pressure and high yield potentials mean that all active fungicides are intensively used. This is especially true for wheat and barley applications. Consequently, it is in Europe where fungicide resistance is most intensely studied and where new cases of fungicide resistance are most often discovered.

Life history parameters have been correlated with the speed with which resistance is detected after release (Grimmer et al., 2015). This has shown that resistance evolution is fastest in pathogens with the shortest latent periods. Hence, the first cereal pathogens to show resistance are normally the powdery mildews. The non-obligate leaf blight and blotch pathogens follow a few years later. Rusts are an interesting case that have shown very few cases of resistance despite widespread exposure to fungicides. Wild-type rusts are already resistant to MBC fungicides and the first generation of SDHIs such as carboxin. Furthermore, all rust species possess a variant of the QoI target gene, which means that the most likely resistant mutants are inviable (Grasso et al., 2006). The evolution of resistance to DMI fungicides has been slow but is now a significant problem in the control of Asian soybean rust. It remains to be seen whether resistance will develop at the predicted fast pace to the newer SDHIs (Oliver, 2014; Langenbach et al., 2016; Godoy et al., 2016; de Freitas et al., 2016).

4 Mechanisms of resistance

4.1 Target site mutations

Changes in target genes, and hence in amino acid sequences of the target protein/enzyme are the commonest observed cause of fungicide resistance. The altered protein sequence typically has a lower affinity for the fungicide and consequently more fungicide is needed to attain the same inhibitory effect. It is often surmised that the alteration has a deleterious effect on the enzyme's activity, which might be reflected in a fitness penalty acting on fungal reproduction in the absence of fungicide. This has rarely been demonstrated in the field and is discussed in Mikaberidze and McDonald (2015).

Each of the three current major fungicide classes (DMI, QoI and SDHI) targets a common, well-conserved protein in different fungal species: these are Cyp51B, CytB and three of the four subunit genes of succinate dehydrogenase (*Sdh-B*, *-C* and *-D*). Furthermore, orthologous amino acids are commonly observed to be selected by the same fungicide class in the different species. In the case of QoI, the *CytB* gene is exceptionally conserved and the major change seen is G143A in all species [see Oliver and Hewitt (2014) for an explanation of the nomenclature]. In the case of DMI and SDHI fungicides, the target genes show differences in the length of the genes so that orthologous amino acids are found in different positions. Mair and colleagues have published a set of protein sequence alignments to facilitate identification of orthologous mutations in different species to avoid confusion (Mair et al., 2016a).

4.2 QoI target site resistance

Resistance to QoI fungicides occurred within the first 2 years after their introduction in Europe (Sierotzki et al., 2000). The powdery mildews were the initially affected species, followed shortly after by STB in 2002 (Sierotzki, 2015). The affected strains were highly resistant and all carried the G143A mutation. By 2006, this mutation was found in more than 90% of STB strains in Northwest Europe.

The agro-chemical industry reacted promptly to this development and implemented regulations requiring that QoI fungicides were only to be sold in mixtures with fungicides of other classes. This regulation has likely slowed the development of resistance in other pathogens and in other parts of the world (Sierotzki, 2015). Nonetheless, all the relevant pathogens except the rusts have developed resistance in one or other region (Table 4). The failure of the rusts to develop resistance is linked to the presence of a fortuitous intron in the gene for *CytB* that splits codon 143 (Grasso et al., 2006). A similar intron is also found in net blotch (Sierotzki et al., 2007). Resistant mutants of net blotch show a much lower level of resistance and carry a different mutation: F129L (Semar et al., 2007). As a result, solo QoI fungicides are still able to exert useful control of this disease.

4.3 SDHI target site resistance

New fungicides belonging to the SDHIs have been released regularly since 2003 and today make up a substantial part of the fungicides used for the control of cereal diseases in Europe and elsewhere. The first cases of resistance were discovered much earlier, 5–7 years after the introduction of carboxin in the late 1960s, an SDHI used mainly as a seed dressing to control basidiomycete fungi. Resistance to the new generation of this fungicide group is caused by single mutations in the B, C and D subunits of the *Sdh* gene and give significant levels of resistance. So far, there are definite signs of fitness penalties to some of the mutations (Fraaije et al., 2012; Rehfus et al., 2018). The first cereal pathogen to display widespread resistance was net blotch. Strains carried single mutations in one of the subunit genes. The most common were B-H277Y, C-N75S, C-G79R, C-H134R, C-135R, D-D124N/E. D-H134R and D-D145G (Rehfus et al., 2016). More recently, resistance has emerged in *Z. tritici*, first in Ireland and in the UK (Dooley et al., 2016a), and now more widely in Europe (Rehfus et al., 2018).

Table 4 Fungicide resistance by class of fungicide

Pathogen	Fungicide group	Mode of resistance	Regions affected	Years to first detection of resistance	References
Blumeria graminis f. sp. *tritici*	QoI	Target site mutation	Europe, Australia	2	Fraaije et al. (2002), Robinson et al. (2002) and Lopez-Ruiz (pers. comm.)
	DMI	Target site mutation	China, Australia	7–10	Wyand and Brown (2005), Yan et al. (2009), Reis et al. (2013) and Lopez-Ruiz (pers. comm.)
Blumeria graminis f. sp. *graminis*	QoI	Target site mutation	Europe	2	Heaney et al. (2000)
	DMI	Target site mutation	Europe, Australia	3	Wyand and Brown (2005) and Tucker et al. (2013)
Zymoseptoria tritici	SDHI	Target site mutation	Europe (Ireland), UK	9	Dooley et al. (2016a,b)
	QoI	Target site mutation	N. Africa, Europe, N. America	4	Estep et al. (2013, 2015), Taher et al. (2014), Allioui et al. (2016), Cheval et al. (2017) and Fraaije et al. (2005)
	DMI	Target site mutation	Europe	>10, gradual	Cools and Fraaije (2008), Stammler and Semar (2011), Taher et al. (2014), Estep et al. (2015), Wieczorek et al. (2015) and Heick et al. (2017)
	All	Multidrug resistance	Europe	>20	Omrane et al. (2015, 2017)
Pyrenophora teres	SDHI	Target site mutation	Europe	7	Rehfus et al. (2016)
	DMI	Target site mutation and gene duplication	Australia	<10	Mair et al. (2016b)
	QoI	Target site mutation (F129L)	7	5	Semar et al. (2007)
Rhynchosporium commune	DMI	Target site mutation	Europe	10	Kendall et al. (1993) and Brunner et al. (2016)
	DMI	Over expression	Europe	>10	Hawkins et al. (2014)
	QoI	Target site mutation	Europe	10	Sierotzki (2015)

(*Continued*)

Table 4 (*Continued*)

Pathogen	Fungicide group	Mode of resistance	Regions affected	Years to first detection of resistance	References
Magnaporthe grisea	QoI	Target site mutation	Brazil	4	de Oliveira et al. (2015) and Castroagudín et al. (2015)
Microdochium nivale and *Microdochium majus*	QoI	Target site mutation	Europe, Japan	5	Walker et al. (2009), Kendall et al. (1994), Wheeler et al. (1995), Taggart et al. (1999) and Nielsen et al. (2013)
Parastagonospora nodorum	QoI	Target site mutation	Europe	8	Blixt et al. (2009)
Oculimacula yalundae	DMI	Target site mutation	Europe	5	Leroux and Gredt (1985, 1997), Leroux et al. (1988), Leroux and Marchegay (1991), Bateman et al. (1995), Dyer et al. (2000), Leroux et al. (2013), Walker et al. (2009)
Fusarium graminearum	DMI	?	USA	>5	Spolti et al. (2014)
Pyrenophora tritici-repentis	QoI	Target site mutation	Europe	5	Sierotzki (2015)
Ramularia collo-cygni	QoI	Target site mutation	Europe	4	Sierotzki (2015)
	DMI	Target site mutation	Europe	>20	FRAC (2016)
	SDHI	Target site mutation	Europe	>10	FRAC (2016) and Piotrowska et al. (2017)
Puccinia triticina and *P. striiformis*	DMI	Target site mutation	Europe	8	Stammler et al. (2009) and Bayles et al. (2000)
Puccinia triticina	DMI	Gene over-expression	Europe	8	Stammler et al. (2009)

4.4 DMI target site resistance

A more complex situation underlies resistance to DMIs, which has led to a slow but substantial decline in efficacy against STB and other diseases. DMI fungicides have only one target site, 14α-demethylase (*CYP51*) (Yang et al., 2008). *CYP51* is an essential enzyme for the synthesis of ergosterol, the predominant sterol of the fungal cell membrane. All species carry a Cyp51 paralog known as Cyp51B, but some species carry additional paralogs called A or C. A depletion of ergosterol leads to alterations of the fungal cell membrane and accumulation of toxic intermediates, which consequently inhibits cell growth (Balkis et al., 2002). Despite this presumed vulnerability, it was not until the mid-1990s that the first cases of fungicide resistance emerged in *Z. tritici*. Since then, several amino acid alterations have been identified, affecting one or more compounds of this fungicide class (Cools and Fraaije, 2013). Around 40 mutations have been described in Cyp51B (among them D134G, V136A/C, A379G, I381V and S524T), which occur in various combinations (Cools et al., 2013). Each combination of mutations has a different response to each of the several dozen DMI fungicides that have been used over the years (Cools et al., 2013). A model has been proposed in which the fungal population adapted first to the fungicides used in the 1970s such as triademinol and tebuconazole. When these fungicides were replaced with newer fungicides of this group, the mutant population adapted by evolving additional mutations. This process has continued as each new DMI active has predominated. This process fits the description of a 'rugged adaptive landscape' (Lucas et al., 2015; Hawkins and Fraaije, 2016). In most cases, mutants resistant to one fungicide are also significantly more resistant to others, but in some cases, mutants can be more sensitive to the fungicide than the wild type (Fraaije et al., 2007).

Other cereal pathogens have developed resistance to DMIs due to target site mutations including the pathogens causing powdery mildews, net blotch, scald and head blight in Australia, the USA and elsewhere. Mutations familiar from the *Z. tritici* gene also occur in these pathogens, such as Y137F and S524T (Tucker et al., 2015). Older actives such as propiconazole and tebuconazole are widely used outside Europe due to cost considerations. These fungicides have shown to provide the same selection pressure as was the case for *Z. tritici* in Europe decades earlier. These mutations have also been observed in cereal rusts albeit there is no significant field impact thus far (Stammler et al., 2009).

4.5 Enhanced gene expression for resistance to DMIs

An alternative mechanism conferring resistance to fungicides is the overexpression of the target site gene, resulting in a higher cellular concentration of the target protein/enzyme. Consequently, higher amounts of fungicide are needed to provide the same level of growth inhibition. Among cereal pathogens, so far only the DMIs have been affected by this resistance mechanism.

Strains of *Z. tritici* have been identified harbouring a 120 bp insert in the promoter region of the *CYP51B* gene, reducing the effect of DMIs (Cools et al., 2012; Chassot et al., 2008; Leroux and Walker, 2011). This results in modestly elevated resistance factors, however affecting all DMIs. A similar genotype was found in *Puccinia triticina* (Stammler et al., 2009). In *Rhynchosporium commune* (Hawkins et al., 2014), overexpression of *CYB51A* was linked to tebuconazole and propiconazole resistance.

4.6 Resistance due to efflux pump overexpression

Multidrug resistance (MDR) is the term used to describe decreased sensitivity to unrelated classes of fungicides. It is typically related to enhanced expression of MFS (major facilitator superfamily) efflux pumps, but the genotypic basis is often still unclear. In *Z. tritici*, a 519 bp insertion in the *MgMFS1* gene was correlated with *MgMFS1* overexpression and consequently to MDR (Omrane et al., 2015). The occurrence of MDR strains has not been widely investigated. Potentially, MDR can generate a major impact on the efficacy of several groups of modes of action and lead to a lower general control level (Omrane et al.,

2015). So far, little is known about how widespread MDR mechanisms are in pathogen populations of small-grain crops.

5 Elements of anti-resistance strategies

The aim of an anti-resistance strategy is to prolong the effective life of a fungicide, that is the time span from when a fungicide is used for the first time until it loses efficacy, even when used at the maximum permitted dose (Hobbelen et al., 2011). The selection for fungicide resistance can also be described as the increased prevalence of resistant strains to a fungicide in a given population. The equation $s = rR - rS$, where the selection coefficient (s) is defined as the difference between the average growth rate (r) of resistant and sensitive strains (R and S) (Milgroom, 1990), models the selection pressure.

The overall goal of resistance management is to minimise the selection coefficient 's'. Van den Bosch and colleagues (van den Bosch et al., 2014a) suggested the following strategies to reduce 's': (1) a reduction of both rR and rS or (2) a reduction of rR in relation to rS. The authors added the factor 'exposure time' (T) to Milgroom's equation ($sT = (rR - rS)T$), so that not only a reduction of rR will minimise s, but also a reduction of exposure time has a major impact. Among the principal tactics that reduced the build-up of fungicide resistance were the adjustment of doses, and mixing and replacing sprays with different modes of action, like combining DMIs, SDHIs and multisite inhibitors. It was concluded that dosage and mixing should be included when creating resistance management strategies. Hobbelen et al. (2013) tested different spraying strategies regarding the fitness cost of resistance, partial resistance to fungicides and differences in dose-response curves and decay rate when creating a mathematical model. The authors found that spraying strategies where fungicides were used as mixtures, or alternating the order of application should be preferred to consecutive use (Hobbelen et al., 2013). Those findings were confirmed by van den Bosch and colleagues reviewing experimental and modelling evidence (van den Bosch et al., 2014b). They concluded that adding a mixing partner to a fungicide at risk of resistance (full dose or with a reduced dose) minimises the rate of selection for fungicide resistance. Mixing two fungicides at risk might also help to slow down the process. In another article, van den Bosch and colleagues summarised results from papers dealing with different doses and their effects on resistance selection. It was shown that an increasing selection of fungicide resistance should be expected when fungicide doses are increased (van den Bosch et al., 2011). In terms of application timing, applying fungicides as soon as the STB infection risk is estimated to be high is recommended. If treatments are left until high disease pressure has developed, it is likely to reduce the effectiveness of treatment.

There are different approaches to optimise timing of fungicide applications. In Denmark, rainy days with a precipitation of a minimum of 1 mm are counted from the beginning of GS 32. If the threshold of four rainy days is reached, fungicide treatment should be carried out (Henriksen et al., 2000). Depending on the seasonal risk, the number of treatments might vary from one to three per season. In other countries, such as the UK and Ireland, both having generally higher disease pressure, spraying timings are more rigidly defined using leaf emergence as a treatment driver (Paveley et al., 2000). Fungicides are applied typically at the GS 31–32 (T1) and GS 39–45 (T2) to protect the crop during elongation and flag leaf emergence. In addition, an application at full heading or during flowering [around GS 59 (T3)] might be necessary to provide an adequate crop protection until harvest, including protection against mycotoxins. In the event of early and severe onset or prophylactically in a high-risk year, for example after a mild and wet winter, an early spraying around GS 29 (T0) might also be recommended to help reduce disease pressure from the early season start.

As shown in Table 3, certain diseases are under major pressure. Under the current disease pressure of *Z. tritici* and lately also of *Ramularia collo-cygni* and *Pyrenophora teres* and the dependence on a limited armoury of available active ingredients, it is crucial to promote well-balanced strategies that offer effective disease control as well as resistance management. In order to ensure the diversity of modes of action for the future, it is hoped that new effective groups will reach the market to help broaden the choice of control.

According to newest research described above, the following principles should be included when designing a spray programme:

1 Diversify the choice of effective fungicides in the control strategies.
2 Use fungicides always in mixtures or alternation with fungicides of a different mode of action, which do not have any positive cross-resistance patterns.
3 Limit the number of fungicide applications in the growing season and apply only when these treatments are required.
4 Apply fungicides at appropriate doses, adjusted to a given disease pressure.
5 Apply fungicides as protectants rather than as eradicants, as this provide better options for the use of reduced rates.
6 Incorporate integrated pest management (IPM) strategies (e.g. varietal resistance) in order to keep down the disease risk.

In order to be able to mix and alternate fungicides, a substantial number of different actives is needed. Several factors have conspired to reduce the number of actives that are available in a given region: stricter regulation in the European Union (EU), lack of registrations, local restrictions and price. With the present constraints of declining efficacy and complex cross-resistance patterns, more and more actives are at risk and it becomes more and more difficult to follow the recommendations mentioned above. This raises the question of how many actives are necessary to lower the risk of resistance to a minimum. An effective spray programme, which is set out to both provide effective control of a disease and reduce resistance, should in high-risk areas at a minimum be built up around four different modes of action, which need to be more or less equally efficient, as described by Rotteveel et al. (2011). In areas with less severe attack and need for lower intensity of spraying, the number may be slightly lower. In the current situation, this requirement is neither the case for control of, for example STB in wheat nor RLS in barley. These considerations emphasise the urgent need for new fungicides and even more for new modes of action. In the next few years, we can expect the release of a new azole (Revysol® from BASF) and a new mode of action (fenpicoxamid = Inatreq™ from DOW AgroSciences) (Owen et al., 2017). These products will help ease the burden; however, as with all single target actives, these two are also at risk for developing resistance. Hence, several additional actives would be desirable to ensure the sustainable protection against the major cereal diseases.

Specific recommendations to minimise resistance to fungicides are described broadly. In this paper, we have included a text box from the Agriculture and Horticulture Development Board (AHDB) webpage, which highlights all major elements.

Good resistance management is based on limiting the level of exposure of the target pathogen to the fungicide (AHDB, FRAG-UK).

- *Fungicide input is only one aspect of crop management and other integrated control measures should always be used, such as good hygiene through disposal of crop debris and control of volunteers, which may harbour disease.*
- *Always aim to select varieties exhibiting a high degree of resistance to diseases known to be prevalent in your area, in addition to the main agronomic factors you desire.*
- *Avoid growing large areas of any one variety, particularly in areas of high disease risk where the variety is known to be susceptible. Only use fungicides in situations where the risk or presence of disease warrants treatment.*
- *Use a dose that will give effective disease control and which is appropriate for the variety and disease pressure. Make full use of effective fungicides with different modes of action in mixtures or as alternate sprays.*
- *Ensure that mixing partners are used at doses that give similar efficacy and persistence.*
- *Monitor crops regularly for disease and treat before the infection becomes well established.*
- *Avoid repeated applications of the same product or mode of action and never exceed the maximum recommended number of applications. Make full use of effective fungicides with different modes of action in mixtures or as alternate sprays.*

5.1 Case study: field trials provide evidence on selection

STB has been the dominant leaf disease in wheat worldwide for more than three decades and a lot of research has been dedicated to elucidate the epidemiology of this fungus and to optimise its control (Orton et al., 2011). Several studies have demonstrated how important effective and well-coordinated disease control of STB is to avoid yield losses. However with the increasing problems with fungicide resistance, an additional factor needs to be addressed to also control major challenges that lie in managing fungicide resistance (Lucas et al., 2015).

Previous studies on the effect of fungicide application on fungicide sensitivity showed that strains of *Z. tritici* were more adapted to the fungicide post-treatment – that is selection had taken place (Leroux et al., 2006; Mavroeidi and Shaw, 2006; Dooley et al., 2016b). Mavroeidi and Shaw (2006) demonstrated that less-sensitive strains were selected by factors such as number of treatments and the use of higher doses. Hence, it is possible to manipulate the field population of a pathogen by modifying a spray plan. This approach was followed in field trials carried out in Denmark, Sweden and Lithuania in 2011–13. Leaves from different treatments were sampled at GS 75 and assessed for specific CYP51 mutations using pyrosequencing or quantitative polymerase chain reaction (qPCR). Fungicide efficacy and effect on yield parameters were also measured. The results illustrated that the application of different spray programmes resulted in selection of different *Z. tritici* variants (Wieczorek et al., 2015). Those trials were supported by four field trials carried out in Denmark in 2015 and 2016 that aimed at further testing of different spray tactics, typical for Denmark, using one to three applications (Heick et al., 2017). The fungicides included in the trials are given in Table 5. Leaf samples were analysed as mentioned above.

Selection of CYP51 mutations was significantly influenced by the repeated use of the same azole and number of active ingredients used in a spray strategy. As shown in Fig. 1 top, mixtures of SDHI + azoles selected less for specific mutations than solo actives used alone. A diverse choice of azoles also minimised the selection for new mutations. Similar findings were seen in four Danish field trials, from which it could also be seen that the most diverse choice of actives selected least for emerging mutations (Fig. 1 bottom). Azole mixtures with actives representing different cross-resistance groups reduced selection compared to repeated use of the same azole. Control strategies in which the number of sprays was reduced from three to two sprays per season were also very successful in minimising selection, as was replacement of the early azole treatments at GS 31–32 with the multisite inhibitor folpet (Heick et al., 2017).

Most tested treatments provided sufficient control and yields. There was a link between high disease control and selection for resistance, indicating that farmers increase their risk for developing resistance, if they apply too high input providing almost total control. The challenge is to define the right input providing optimal control, net yields and least selection.

Put together, these findings suggest that well thought-out spray programmes might delay the build-up of fungicide resistance. Data illustrated in this way are easy to disseminate to advisors and farmers, which should help to stimulate farmers to go for a diversified control strategy as part of an effective anti-resistance management. If resistance management is kept at very scientific and theoretical level, it is difficult to convince farmers that it matters to treat differently to their current practices.

The knowledge on elements influencing the selection must be incorporated together with the acquired knowledge on IPM approaches including the use of more resistant cultivars. In order to be acceptable solutions for the farmers, the strategies must provide both optimal disease and yield management as well as a stable anti-resistance strategy.

Table 5 Fungicides tested in field trials with winter wheat shown in Fig. 1

Fungicides	Active ingredient (a.i.)	Standard dose
Proline® EC250	Prothioconazole (250 g/L)	0.8 (200 g a.i./ha)
Bell®	Epoxiconazole + boscalid (67 + 233 g/L)	1.5 (100 + 350 g a.i./ha)
Prosaro® 250EC	Prothioconazole + tebuconazole (125 + 125 g/L)	1.0 (125 + 125 g a.i./ha)
Armure®	Difenoconazole + propiconazole (150 + 150 g/L)	0.8 (125 + 125 g a.i./ha)
Folpan® 500SC	Folpet (500 g/L)	1.5 (750 g a.i./ha)
Aviator® Xpro	Bixafen + prothioconazole (75 + 160 g/L)	1.25 (94 + 200 g a.i./ha)

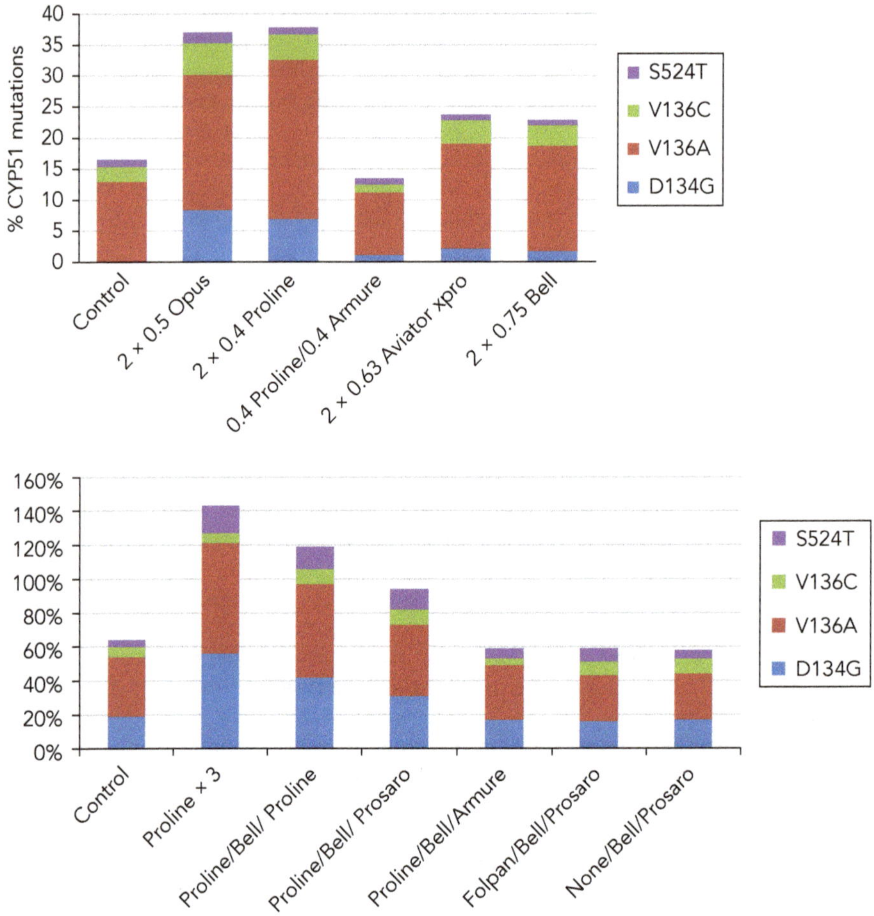

Figure 1 Selection of new CYP51 mutations in the *Z. tritici* population following the use of different DMI combinations. Top: average data from nine field trials in winter wheat carried out in Denmark, Sweden and Lithuania using two treatments. Bottom: average data from four winter wheat field trials with different control strategies using different degrees of diversification. Mutations are measured using pyrosequencing or qPCR (S524T).

5.2 The need for sufficient disease control agents

European Commission (EC) legislation has limited the availability of pesticide active substances used in plant protection products and further measures into the same direction are expected in the near future. The principle of risk assessment for approval of pesticide active substances was introduced in the Pesticide Authorisation Directive 91/414/EEC. This principle was modified by the introduction of Regulation (EC) 1107/2009, which applies 'hazard', the intrinsic toxicity of the active substance, rather than 'risk', the potential for hazard to occur, as the approval criterion. Potential impacts of EC pesticide legislation on agriculture in the EU have been estimated in various papers although major uncertainties still exist (Hillocks, 2012). The new regulation will without doubt have a significant impact on pesticide availability in the medium to long term. Furthermore, other regulations, such as the EU water directive (Water Framework Directive 2000/60/EC and Drinking Water Directive 1998/83/EC), have the potential to restrict pesticide use, as concerns regarding public health and economic costs associated with removing pesticides from water increase. It is expected that this rationale will further reduce the availability of effective pesticide active substances and modes of action, directly affecting crop protection and increasing pesticide resistance within pest and pathogen populations. Several papers have dealt with the specific impact from some of the reduction scenarios, which estimate that 20–50% of present actives will disappear following the new legislation (Hillocks, 2012; Jess et al., 2014).

Comparative assessment (CA) has been added to the approval system as a new harmonised requirement, aiming at further reducing the risks associated with pesticide use. According to Regulation EC 1107/2009, a CA is to be carried out when evaluating applications for plant protection products containing an active substance approved as a 'candidate for substitution'. A CA is required, as a minimum, at renewal or amendment of any authorisation of a product that contains a candidate for substitution. A list with candidates for substitution has been produced based on whether or not the active substance is an active substance that meets certain criteria, such as being significantly more toxic, meeting two PBT (persistent, bioaccumulative, toxic) criteria or classifies in a certain way. Subsequently, a decision on whether to authorise a new product or a product under renewal is made based on the comparison. Consequently, each EU member state is now obliged to conduct a risk assessment on the individual product, and in addition, a comparison of appropriate available risk assessments has to be carried out before deciding on whether to authorise the product in question. From an operational standpoint, it remains unclear precisely how this new requirement will influence fungicide availability, and thus agricultural production depending on new products or fungicides up for renewal. Risk of resistance development is seen as a factor which will play a major role when EU member states start carrying out CAs to decide whether a product is needed or not (Rotteveel et al., 2011).

The future challenge for agriculture in the EU is to sustain production and profitability using fewer actives and reduced pesticide inputs as guided within the framework of IPM, as proposed by the sustainable use of pesticides (Anon., 2009). The Directive requires the adoption of a National Action Plan for the sustainable use of pesticides. The plan should define a national strategy to achieve a sustainable use of pesticides and set down objectives, quantifiable measures and timeframes to reduce the risks associated with the use of pesticides. The approach used is variable depending on the specific country but uses typically a combination of statutory measures and voluntary industry-led initiatives.

5.3 Regulatory constraints on fungicides

The change in emphasis from risk- to hazard-based criteria for pesticide approvals is having a major impact upon the registration of fungicides needed for successful control of cereal diseases. Of particular significance would be the removal of active substances with potential endocrine-disrupting properties. This is likely to include several fungicides, and in particular, the loss of specific DMIs, including some of the more important azole fungicides, would significantly reduce the options for controlling the most economically destructive diseases of wheat and barley (STB, rhynchosporium in barley, rust diseases and Fusarium head blight). Current disease control programmes in cereals in the EU rely heavily upon the DMI fungicides, with most winter and spring crops receiving typically one to three applications per season (Jørgensen et al., 2017). The broad-spectrum activity of the DMIs coupled with their ability to provide protection and curative activity, makes them central for most applications.

The registration and commercialisation of the second generation of SDHI fungicides, including bixafen, fluxapyroxad, isopyrazam and penthiopyrad, have provided an alternative class of fungicides for the control of STB, RLS and barley net blotch (BNB). As QoIs are no longer effective and the efficacy of DMIs is declining for several disease like STB, it is likely that fungicide programmes will become heavily reliant on the SDHIs. Unfortunately, the development of resistance is already taking place in some populations of *Z. tritici*, *Pyrenophora teres* and *Ramularia collo-cygni*. Due to these changes, multisite inhibitors like chlorothalonil have become increasingly important for control of both STB and RLS. In some parts of Europe, most treatments per season are supplemented with a multisite inhibitor in order both to apply an anti-resistant strategy and to manage STB (Kildea, pers. comm.). However, in some countries like Denmark and Sweden, chlorothalonil was banned several years ago. In other countries, future regulations might likely restrict the use of chlorothalonil, constraining the creation of sound anti-resistance strategies even more. In future scenarios, it is very difficult to predict future options for control of STB, BNB and RLS, as much depends on how farmers adopt control measures as well as on the availability of genetic resistance and the availability of fungicides.

5.4 Alternative control measures

A major element of minimising fungicide inputs is to incorporate cultural factors in current control strategies in order to keep down disease pressure and the actual need for treatments. Crop rotation and the removal or incorporation of plant debris have not been shown to have any major impact on STB and RLS, whereas this is known to have impact on diseases like scald and net blotch (Turkington et al., 2012). Several of the leaf blotch diseases are mostly driven by humidity and rain parameters and less so by specific cultural methods. Varietal resistance is a major means for control of cereal diseases even though a fully resistant cultivar providing resistance to all major diseases does not exist. In most countries, evaluation of disease resistance is a requirement for new and already introduced cultivars to be adopted to or continued on the national recommended list. Full resistance towards STB does not exist and generally many cultivars having good resistance potential showed trade-offs such as relatively lower yields (Brown, 2015). This has thus been a discouragement for many farmers, who consequently decided against planting them. Recently, cultivars with both considerable resistance towards STB and acceptable yield have been introduced (Loyce et al., 2008). Cultivars with moderate to good resistance to STB are available although a sudden 'break down' in resistance might occur any time and has already been observed in several cases (Cowger et al., 2000). Breeding for disease resistance is also a major goal for barley breeders. However, no new cultivars have proven to provide full control of all relevant diseases (Anon., 2016). Even though growing of moderately resistant cultivars is not likely to diminish the need for chemical disease control completely, it helps planning fungicide treatments in a flexible manner and reducing the overall amounts of fungicides applied (Loyce et al., 2008; Jørgensen et al., 2008; Lynch et al., 2017).

Disease management by biological control agents (living organisms with the capacity to reproduce) and biopesticides (extracts of various species and pure compounds found in nature) is seen as an increasing market (Olsen, 2015). Several biological control agents have been tested for control of foliar diseases in wheat and barley, but so far only limited effects have been seen and control levels have generally been quite variable. The major constraints associated with biological control agents are to identify the right timing, the right dose and weather conditions (Reiss and Jørgensen, 2017). Most products that have been registered for the control of fungal pathogens are used under controlled conditions in the greenhouse (Kipngeno et al., 2015; Punja et al., 2016), where these problems are not as important. In the field inconsistent efficacy, limited control and high cost mean that these products are often used in combinations with traditional fungicides or cultural methods (Jacobsen et al., 2004). Several biological control agents have also been tested for control of foliar diseases in wheat and barley, but they are not used in practice in small-grain cereals yet.

6 Farmers' and advisors' perspectives

In order to understand how farmers spray in their cereal crops, reliable statistics are required. In many countries, such data are not available and often only chemical companies collect data to understand their market potential. These types of data are often not readily available in the public domain. In the sustainable use directive from EU, it is made obligatory that farmers keep records of their pesticide use [Regulation (EC) No 1185/2009]. In Denmark, it has furthermore been made obligatory that farmers upload information from their pesticide use per crop to a national database, providing a useful source for evaluating the use of pesticides at a regional basis (Kudsk et al., 2018). The collected data from Denmark shows that the intensity of fungicide use depends greatly on the region (Fig. 2). Some of the variations are linked to different cropping systems and

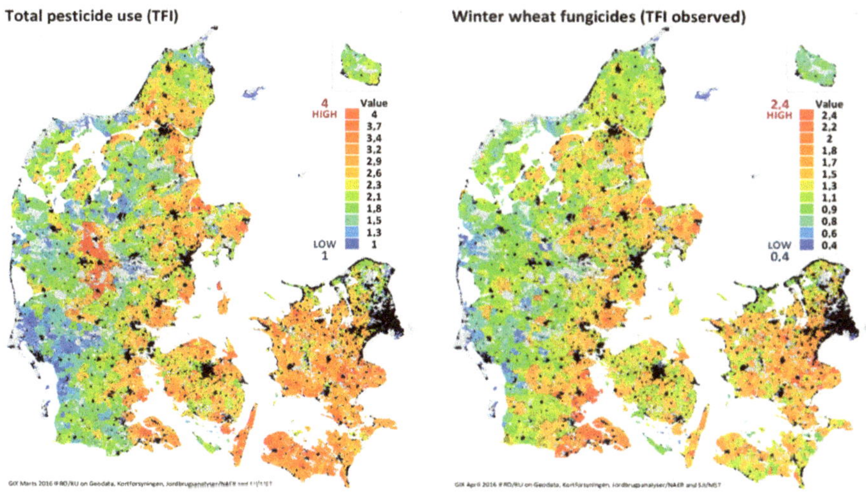

Figure 2 Variation in the total dose measured as treatments frequency index (TFI) applied for all pesticides in Denmark (left) and variation in fungicides applied in winter wheat (right). Data are based on farmers' spraying records from four consecutive seasons (2011–14) uploaded to a national database. The risk of STB varies among regions (Ørum, pers comm).

soil types, but others are linked to yield expectations or the risk for disease development as, for example STB (Jørgensen et al., 2017). Yields are generally better on the good soils where farmers also tend to apply higher inputs of fungicides.

The willingness of farmers to change their habits is generally higher, if the proposed change is well documented and changes do not have any negative implications on yields. In some cases, sudden changes in fungicide sensitivity have motivated farmers to respond promptly as in the way they choose their chemicals. As the first sensitivity changes were observed to QoI fungicides in the 1990s, sales data showed a major shift from using QoI products alone to using mixtures of azoles + QoIs (Jørgensen and Thygesen, 2006). Unfortunately, this was probably already too late as the *Z. tritici* population was already showing high levels of resistance and so counteractions are more difficult. This illustrates clearly that anti-resistance strategies should be in place before products are widely used.

Growers generally apply more fungicides than needed (Harwick et al., 2001; Wiik and Rosenqvist, 2010). This can be partly explained by the fact that it is quite difficult to know what the optimal input should be during a given season. In order to take account for a worst-case scenario, farmers will accept small levels of excess expenditure to avoid this risk. As a result, the risk-averse grower is likely to apply additional treatments or higher doses than needed in order to take account for high-risk seasons (te Beest et al., 2013). This partly explains the discrepancy between calculated long-term mean optimal dose and actual field-applied dose.

Agronomists term this 'insurance spraying' as a relatively small fungicide cost premium is paid each year to protect against occasional severe loss. The major drawback of this is the increased risk of resistance development, as both higher intensity of treatments and higher dose rates are seen to increase risk of resistance.

7 Where to look for further information

Several research groups across Europe engage in the investigation of fungicide resistance and of those factors, which play a major role in its development. Research is carried out by independent universities as well as major chemical companies, which have a specific scientific interest in monitoring the dynamics of fungicide resistance. This chapter describes the resistance situation for cereal pathogens seen in 2018. As cropping systems and disease management are very dynamic, it is recommended to always search for up-to-date information. Yearly updates are presented both by FRAC (www.FRAC 2017) and by the regional FRAG groups, which all have their webpages (e.g. AHDB, FRAG-UK; Norbarag). With the increasing pressure on how to reduce the dependency of fungicides and increasingly focus on how IPM can be better exploited, fungicide resistance management is seen as an important discipline in order to provide the most sustainable diseases management for future cereal production.

8 References

Allioui, N., Siah, A., Brinis, L., Reignault, P. and Halama, P. 2016. Identification of QoI fungicide-resistant genotypes of the wheat pathogen *Zymoseptoria tritici* in Algeria. *Phytopathologia Mediterranea*, 55, 89–97.

Anon. 2009. Directive 2009/128/EC of the European Parliament and of the Council of 21 October 2009 establishing a framework for Community action to achieve the sustainable use of pesticides. European Parliament and the Council of the European Union.

Anon. 2016. FAOSTAT. http://www.fao.org/faostat/en/#data/QC/metadata (accessed 10 January 2018).

Balkis, M. M., Leidich, S. D., Mukherjee, P. K. and Ghannoum, M. A. 2002. Mechanisms of fungal resistance. *Drugs*, 62, 1025–40.

Bartlett, D. W., Clough, J. M., Godwin, J. R., Hall, A. A., Hamer, M. and Parr-Dobrzanski, B. 2002. The strobilurin fungicides. *Pest Management Science*, 58, 649–62.

Bateman, G. L., Landau, S. and Welham, S. J. 1995. Sensitivity to prochloraz in populations of the eyespot fungus, *Pseudocercosporella herpotrichoides*, in relation to fungicide treatments and their efficacy in continuous winter wheat. *Annals of Applied Biology*, 126, 235–47.

Bayles, R. A., Stigwood P. L. and Clarkson, J. D. S. 2000. Shifts in sensitivity of *Puccinia striiformis* to DMI fungicides in the UK. *Acta Phytopathologica et Entomologica Hungarica*, 35, 381–2.

Blixt, E., Djurle, A., Yuen, J. and Olson, Å. 2009. Fungicide sensitivity in Swedish isolates of *Phaeosphaeria nodorum. Plant Pathology*, 58, 655–64.

Brent, K. J. and Hollomon, D. W. 1995. *Fungicide Resistance in Crop Pathogens: How Can it be Managed?* FRAC, p. 60. ISBN 90-72398-07-6.

Brent, K., Hollomon, D. and Shaw, M. 1990. Predicting the evolution of fungicide resistance. In: Green, M. B., LeBaron, H. M. and Moberg, W. K. (Eds), *Managing Resistance to Agrochemicals*. ACS Publications, Washington DC, pp. 303–19.

Brent, K. J., Hollomon, D. W. and Federation, G. C. P. 1998. *Fungicide Resistance: The Assessment of Risk*. Global Crop Protection Federation Brussels, Belgium.

Brown, J. K. M. 2015. Durable resistance of crops to disease: A Darwinian perspective. *Annual Review of Phytopathology*, 53, 513–39.

Brown, M. C., Taylor, G. S. and Epton, H. A. S. 1984. Carbendazim resistance in the eyespot pathogen *Pseudocercosporella herpotrichoides. Plant Pathology*, 33, 101–11.

Brunner, P. C., Stefansson, T. S., Fountaine, J., Richina, V. and McDonald, B. A. 2016. A global analysis of *CYP51* diversity and azole sensitivity in *Rhynchosporium commune*. *Phytopathology*, 106, 355–61.

Castroagudín, V. L., Ceresini, P. C., De Oliveira, S. C., Reges, J. T. A., Maciel, J. L. N., Bonato, A. L. V., Dorigan, A. F. and McDonald, B. A. 2015. Resistance to QoI fungicides is widespread in Brazilian populations of the wheat blast pathogen *Magnaporthe oryzae. Phytopathology*, 105, 284–94.

Chassot, C., Hugelshofer, U., Sierotzki, H. and Gisi, U. 2008. Sensitivity of *CYP51* genotypes to DMI fungicides in *mycosphaerella graminicola. Modern fungicides and antifungal compounds V: 15th International Reinhardsbrunn Symposium*, Friedrichroda, Germany, 6–10 May 2007. Deutsche Phytomedizinische Gesellschaft eV Verlag, pp. 129–36.

Cheval, P., Siah, A., Bomble, M., Popper, A. D., Reignault, P. and Halama, P. 2017. Evolution of QoI resistance of the wheat pathogen *Zymoseptoria tritici* in Northern France. *Crop Protection*, 92, 131–3.

Cools, H. J. and Fraaije, B. A. 2008. Are azole fungicides losing ground against Septoria wheat disease? Resistance mechanisms in *Mycosphaerella graminicola. Pest Management Science*, 64, 681–4.

Cools, H. J. and Fraaije, B. A. 2013. Update on mechanisms of azole resistance in *Mycosphaerella graminicola* and implications for future control. *Pest Management Science*, 69, 150–5.

Cools, H. J., Bayon, C., Atkins, S., Lucas, J. A. and Fraaije, B. A. 2012. Overexpression of the sterol 14α-demethylase gene (*MgCYP51*) in *Mycosphaerella graminicola* isolates confers a novel azole fungicide sensitivity phenotype. *Pest Management Science*, 68, 1034–40.

Cools, H. J., Hawkins, N. J. and Fraaije, B. A. 2013. Constraints on the evolution of azole resistance in plant pathogenic fungi. *Plant Pathology*, 62, 36–42.

Cowger, C., Hoffer, M. E., and Mundt, C. C. 2000. Specific adaptation by *Mycosphaerella graminicola* to a resistant wheat cultivar. *Plant Pathology*, 49, 445–51.

De Freitas, R. M. S., De Lima, L. E., Silva, R. S., Campos, H. D. and Perin, A. 2016. Fluxapyroxad in the Asian soybean rust control in the cerrado biome. *Revista Caatinga*, 29, 619–28.

De Oliveira, S. C., Castroagudín, V. L., Maciel, J. L. N., Pereira, D. A. S. and Ceresini, P. C. 2015. Cross-resistance to QoI fungicides azoxystrobin and pyraclostrobin in the wheat blast pathogen *Pyricularia oryzae* in Brazil. *Summa Phytopathologica*, 41, 298–304.

Dewar, A., Ferguson, A., Pell, J., Nicholls, C. and Watts, J. 2016. A review of pest management needs in cereals and oilseed rape in the UK. AHDB Research Review No. 86.

Dooley, H., Shaw, M. W., Mehenni-Ciz, J., Spink, J. and Kildea, S. 2016a. Detection of *Zymoseptoria tritici* SDHI-insensitive field isolates carrying the SdhC-H152R and SdhD-R47W substitutions. *Pest Management Science*, 72, 2203–7.

Dooley, H., Shaw, M. W., Spink, J. and Kildea, S. 2016b. The effect of succinate dehydrogenase inhibitor/azole mixtures on selection of *Zymoseptoria tritici* isolates with reduced sensitivity. *Pest Management Science*, 72, 1150–9.

Dyer, P. S., Hansen, J., Delaney, A. and Lucas, J. A. 2000. Genetic control of resistance to the sterol 14α-demethylase inhibitor fungicide prochloraz in the cereal eyespot pathogen *Tapesia yallundae*. *Applied and Environmental Microbiology*, 66, 4599–604.

Estep, L. K., Zala, M., Anderson, N. P., Sackett, K. E., Flowers, M., McDonald, B. A. and Mundt, C. C. 2013. First report of resistance to QoI fungicides in North American populations of *Zymoseptoria tritici*, causal agent of septoria tritici blotch of wheat. *Plant Disease*, 97, 1511.

Estep, L. K., Torriani, S. F. F., Zala, M., Anderson, N. P., Flowers, M. D., McDonald, B. A., Mundt, C. C. and Brunner, P. C. 2015. Emergence and early evolution of fungicide resistance in North American populations of *Zymoseptoria tritici*. *Plant Pathology*, 64, 961–71.

Fraaije, B. A., Butters, J. A., Coelho, J. M., Jones, D. R. and Hollomon, D. W. 2002. Following the dynamics of strobilurin resistance in *Blumeria graminis* f.sp. *tritici* using quantitative allele-specific real-time PCR measurements with the fluorescent dye SYBR Green I. *Plant Pathology*, 51, 45–54.

Fraaije, B. A., Cools, H. J., Fountaine, J., Lovell, D. J., Motteram, J., West, J. S. and Lucas, J. A. 2005. Role of ascospores in further spread of QoI-resistant cytochrome b alleles (G143A) in field populations of *Mycosphaerella graminicola*. *Phytopathology*, 95, 933–41.

Fraaije, B. A., Cools, H. J., Kim, S. H., Motteram, J., Clark, W. S. and Lucas, J. A. 2007. A novel substitution I381V in the sterol 14α-demethylase (*CYP51*) of *Mycosphaerella graminicola* is differentially selected by azole fungicides. *Molecular Plant Pathology*, 8, 245–54.

Fraaije, B. A., Bayon, C., Atkins, S., Cools, H. J., Lucas, J. A. and Fraaije, M. W. 2012. Risk assessment studies on succinate dehydrogenase inhibitors, the new weapons in the battle to control Septoria leaf blotch in wheat. *Molecular Plant Pathology*, 13, 263–75.

FRAC. 2016. Succinate Dehydrogenase Inhibitor (SDHI) Working Group 9th Meeting on December 2 2015. Accessed 10 January 2018.

FRAC. 2017. FRAC Code List 2016: Fungicides Sorted by Mode of Action (Including FRAC Code Numbering). http://www.frac.info/docs/default-source/publications/frac-code-list/frac-code-list-2017-final.pdf?sfvrsn = fab94a9a_2 (accessed 10 January 2018).

Genet, J. L. and Jaworska, G. 2009. Baseline sensitivity to proquinazid in *Blumeria graminis* f. sp. *tritici* and Erysiphe necator and cross-resistance with other fungicides. *Pest Management Science*, 65, 878–84.

Godoy, C. V., Seixas, C. D. S., Soares, R. M., Marcelino-Guimarães, F. C., Meyer, M. C. and Costamilan, L. M. 2016. Asian soybean rust in Brazil: Past, present, and future. *Pesquisa Agropecuaria Brasileira*, 51, 407–21.

Grasso, V., Palermo, S., Sierotzki, H., Garibaldi, A. and Gisi, U. 2006. Cytochrome *b* gene structure and consequences for resistance to Qo inhibitor fungicides in plant pathogens. *Pest Management Science*, 62, 465–72.

Grimmer, M. K., Van Den Bosch, F., Powers, S. J. and Paveley, N. D. 2015. Fungicide resistance risk assessment based on traits associated with the rate of pathogen evolution. *Pest Management Science*, 71, 207–15.

Hardwick, N. V., Jones, D. R. and Slough, J. E. 2001. Factors affecting diseases of winter wheat in England and Wales, 1989–98. *Plant Pathology*, 50, 453–62.

Hawkins, N. J. and Fraaije, B. 2016. Adaptive landscapes in fungicide resistance; fitness, epistasis, constraints and predictability. *18th International Reinhardsbrunn Symposium; Modern Fungicides and antifungal compounds*. Friedrichroda, Germany, 24–28 April 2016.

Hawkins, N. J., Cools, H. J., Sierotzki, H., Shaw, M. W., Knogge, W., Kelly, S. L., Kelly, D. E. and Fraaije, B. A. 2014. Paralog re-emergence: A novel, historically contingent mechanism in the evolution of antimicrobial resistance. *Molecular Biology and Evolution*, 31, 1793–802.

Heaney, S. P., Hall, A. A., Davies, S. A. and Olaya, G. 2000. Resistance to fungicides in the QoI-STAR cross-resistance group: Current perspectives. *The BCPC Conference: Pests and diseases, Volume 2 Proceedings of an international conference held at the Brighton Hilton Metropole Hotel*, Brighton, UK, 13–16 November 2000, pp. 755–62.

Heick, T. M., Justesen, A. F. and Jørgensen, L. N. 2017. Anti-resistance strategies for fungicides against wheat pathogen *Zymoseptoria tritici* with focus on DMI fungicides. *Crop Protection*, 99, 108–17.

Henriksen, K. E., Jørgensen, L. N. and Nielsen, G. 2000. PC-Plant Protection-a Danish tool to reduce fungicide input in cereals. *The BCPC Conference: Pests and diseases, Volume 3. Proceedings of an international conference held at the Brighton Hilton Metropole Hotel*, Brighton, UK, 13–16 November 2000. British Crop Protection Council, pp. 835–40.

Hess, M. 2017. Sensitivitätsverluste gegenüber den Ramularia-Blattflecken. *GetreideMagazin*, 2, 16–20.

Hillocks, R. 2012. Farming with fewer pesticides: EU pesticide review and resulting challenges for UK agriculture. *Crop Protection*, 31, 85–93.

Hobbelen, P. H. F., Paveley, N. D., Fraaije, B. A., Lucas, J. A. and Van Den Bosch, F. 2011. Derivation and testing of a model to predict selection for fungicide resistance. *Plant Pathology*, 60, 304–13.

Hobbelen, P. H. F., Paveley, N. D., Oliver, R. P. and Van Den Bosch, F. 2013. The usefulness of fungicide mixtures and alternation for delaying the selection for resistance in populations of *Mycosphaerella graminicola* on winter wheat: A modeling analysis. *Phytopathology*, 103, 690–707.

Horsten, J. and Fehrmann, H. 1980. Fungicide resistance of *Septoria nodorum* and *Pseudocercosporella herpotrichoides*: III. Survival ability of resistant strains in mixed populations/Fungizidresistenz von *Septoria nodorum* und *Pseudocercosporella herpotrichoides*: III. Überlebensfähigkeit resistenter Stämme in Mischpopulationen. *Zeitschrift für Pflanzenkrankheiten und Pflanzenschutz/Journal of Plant Diseases and Protection*, 87, 577–86.

Jacobsen, B. J., Zidack, N. K. and Larson, B. J. 2004. The role of Bacillus-based biological control agents in integrated pest management systems: Plant diseases. *Phytopathology*, 94, 1272–5.

Jess, S., Kildea, S., Moody, A., Rennick, G., Murchie, A. K. and Cooke, L. R. 2014. European Union policy on pesticides: Implications for agriculture in Ireland. *Pest Management Science*, 70, 1646–54.

Jørgensen, L. N. and Thygesen, K. 2006. Should strobilurins still be used in wheat? Danish and foreign experience. *Third Danish Plant Production Congress*, Denmark, 10–11 January, 2006. Danish Institute of Agricultural Sciences, pp. 485–6.

Jørgensen, L. N., Nielsen, G. C., Ørum, J. E., Jensen, J. E. and Pinnschmidt, H. O. 2008. Integrating disease control in winter wheat – Optimizing fungicide input. *Outlooks on Pest Management*, 19, 206–13.

Jørgensen, L. N., Hovmøller, M. S., Hansen, J. G., Lassen, P., Clark, B., Bayles, R., Rodemann, B., Flath, K., Jahn, M., Goral, T., Jerzy Czembor, J., Cheyron, P., Maumene, C., De Pope, C., Ban, R., Nielsen, G. C. and Berg, G. 2014. IPM strategies and their dilemmas including an introduction to www.eurowheat.org. *Journal of Integrative Agriculture*, 13, 265–81.

Jørgensen, L. N., Oliver, R. P., Van den Bosch, F. and Paveley, N. P. 2017. Targeting fungicides inputs according to need. *Annual Review of Phytopathology*, 55, 181–203.

Kendall, S. J., Hollomon, D. W., Cooke, L. R. and Jones, D. R. 1993. Changes in sensitivity to DMI fungicides in *Rhynchosporium secalis*. *Crop Protection*, 12, 357–62.

Kendall, S., Hollomon, D. W., Ishii, H. and Heaney, S. P. 1994. Characterisation of benzimidazole-resistant strains of *Rhynchosporium secalis*. *Pesticide Science*, 40, 175–81.

Kipngeno, P., Losenge, T., Maina, N., Kahangi, E. and Juma, P. 2015. Efficacy of *Bacillus subtilis* and *Trichoderma asperellum* against *Pythium aphanidermatum* in tomatoes. *Biological Control*, 90, 92–5.

Kudsk, P., Jørgensen, L. N. and Ørum, J. E. 2018. Pesticide Load – a new Danish pesticide risk indicator with multiple applications. *Land Use Policy*, 70, 384–93.

Langenbach, C., Campe, R., Beyer, S. F., Mueller, A. N. and Conrath, U. 2016. Fighting Asian soybean rust. *Frontiers in Plant Science*, 7, 797.

Lee, S., Gustafson, G., Skamnioti, P., Baloch, R. and Gurr, S. 2008. Host perception and signal transduction studies in wild-type *Blumeria graminis* f. sp. *hordei* and a quinoxyfen-resistant mutant implicate quinoxyfen in the inhibition of serine esterase activity. *Pest Management Science*, 64, 544–55.

Leroux, P. and Gredt, M. 1985. Resistance to sterol biosynthesis inhibitors in *Pseudocercosporella herpotrichoides*, causal agent of eyespot of cereals. *Caracteristiques Des Souches De Pseudocercosporella Herpotrichoides, Agent Du Piétin-Verse Des Cereales, Resistantes A Des Inhibiteurs De La Biosynthese Des Sterols*, 301, 785–8.

Leroux, P. and Gredt, M. 1997. Evolution of fungicide resistance in the cereal eyespot fungi *Tapesia yallundae* and *Tapesia acuformis* in France. *Pesticide Science*, 51, 321–7.

Leroux, P. and Marchegay, P. 1991. Resistance to prochloraz in strains of the cereal eyespot fungus *Pseudocercosporella herpotrichoides* isolated from winter wheat in France. *Caractérisation des souches de Pseudocercosporella herpotrichoides, agent du piétin-verse des céréales, résistantes au prochioraze, isolées en France sur blé tendre d'hiver*, 11, 767–76.

Leroux, P. and Walker, A. S. 2011. Multiple mechanisms account for resistance to sterol 14α-demethylation inhibitors in field isolates of *Mycosphaerella graminicola*. *Pest Management Science*, 67, 44–59.

Leroux, P., Gredt, M. and Boeda, P. 1988. Resistance to inhibitors of sterol biosynthesis in field isolates or laboratory strains of the eyespot pathogen *Pseudocercosporella herpotrichoides*. *Pesticide Science*, 23, 119–29.

Leroux, P., Walker, A., Albertini, C. and Gredt, M. 2006. Resistance to fungicides in French populations of *Septoria tritici*, the causal agent of wheat leaf blotch. *Aspects of Applied Biology*, 78, 153.

Leroux, P., Gredt, M., Remuson, F., Micoud, A. and Walker, A. S. 2013. Fungicide resistance status in French populations of the wheat eyespot fungi *Oculimacula acuformis* and *Oculimacula yallundae*. *Pest Management Science*, 69, 15–26.

Loyce, C., Meynard, G. C., Orum, J. E., Jensen, J. E. and Pinnschmidt, H. O. 2008. Interaction between cultivar and crop management effects on winter wheat diseases, lodging, and yield. *Crop Protection*, 27, 1131–42.

Lucas, J. A., Hawkins, N. J. and Fraaije, B. A. 2015. The evolution of fungicide resistance. *Advances in Applied Microbiology*, 90, 29–92.

Lynch, J. P., Glynn, E., Kildea, S. and Spink, J. 2017. Yield and optimum fungicide dose rates for winter wheat (*Triticum aestivum* L.) varieties with contrasting ratings for resistance to septoria tritici blotch. *Field Crops Research*, 204, 89–100.

Mair, W., Lopez-Ruiz, F., Stammler, G., Clark, W., Burnett, F., Hollomon, D., Ishii, H., Thind, T. S., Brown, J. K. M., Fraaije, B., Cools, H., Shaw, M., Fillinger, S., Walker, A. S., Mellado, E., Schnabel, G., Mehl, A. and Oliver, R. P. 2016a. Proposal for a unified nomenclature for target-site mutations associated with resistance to fungicides. *Pest Management Science*, 72, 1449–59.

Mair, W. J., Deng, W., Mullins, J. G., West, S., Wang, P., Besharat, N., Ellwood, S. R., Oliver, R. P. and Lopez-Ruiz, F. J. 2016b. Demethylase inhibitor fungicide resistance in *Pyrenophora teres* f. sp. *teres* associated with target site modification and inducible overexpression of *Cyp51*. *Frontiers in Microbiology*, 7, 1279.

Mavroeidi, V. I. and Shaw, M. W. 2006. Effects of fungicide dose and mixtures on selection for triazole resistance in *Mycosphaerella graminicola* under field conditions. *Plant Pathology*, 55, 715–25.

Mikaberidze, A. and McDonald, B. A. 2015. Fitness cost of resistance: Impact on management. In: Ishii, H. and Hollomon, D. W. (Eds), *Fungicide Resistance in Plant Pathogens: Principles and Guide to Practical Management*. Springer, Japan, pp. 77–89.

Milgroom, M. G. 1990. A stochastic model for the initial occurrence and development of fungicide resistance in plant pathogen populations. *Phytopathology*, 80, 410–16.

Murray, G. M. and Brennan, J. P. 2009. Estimating disease losses to the Australian wheat industry. *Australasian Plant Pathology*, 38, 558–70.

Murray, G. M. and Brennan, J. P. 2010. Estimating disease losses to the Australian barley industry. *Australasian Plant Pathology*, 39, 85–96.

Napier, B. A. S., Bayles, R. A. and Stigwood, P. L. 2000. Sensitivity of powdery mildew and yellow rust to DMI, morpholine and strobilurin fungicides in England and Scotland. *The BCPC Conference: Pests and diseases, Volume 1. Proceedings of an international conference held at the Brighton Hilton Metropole Hotel*, Brighton, UK, 13–16 November 2000, pp. 427–34.

Nielsen, L. K., Justesen, A. F., Jensen, J. D. and Jørgensen, L. N. 2013. Microdochium nivale and Microdochium majus in seed samples of Danish small grain cereals. *Crop Protection*, 43, 192–200.

Oerke, E.-C. 2006. Crop losses to pests. *The Journal of Agricultural Science*, 144, 31–43.

Oliver, R. P. 2014. A reassessment of the risk of rust fungi developing resistance to fungicides. *Pest Management Science*, 70, 1641–5.

Oliver, R. P. and Hewitt, H. G. 2014. *Fungicides in Crop Protection: Second Edition*. CABI, Wallingford UK.

Olsen, S. 2015. The biopesticide market – An analysis of the biopesticide market now and where it is going. *Outlooks on Pest Management*, 26, 203–7.

Omrane, S., Sghyer, H., Audéon, C., Lanen, C., Duplaix, C., Walker, A. S. and Fillinger, S. 2015. Fungicide efflux and the MgMFS1 transporter contribute to the multidrug resistance phenotype in *Zymoseptoria tritici* field isolates. *Environmental Microbiology*, 17, 2805–23.

Omrane, S., Audéon, C., Ignace, A., Duplaix, C., Aouini, L., Kema, G., Walker, A. S. and Fillinger, S. 2017. Plasticity of the MFS1 promoter leads to multidrug resistance in the wheat pathogen *Zymoseptoria tritici*. *Msphere*, 2. pii: e00393-17.

Orton, E. S., Deller, S. and Brown, J. K. M. 2011. *Mycosphaerella graminicola*: From genomics to disease control. *Molecular Plant Pathology*, 12, 413–24.

Owen, W. J., Yao, C., Myung, K., Kemmitt, G., Leader, A., Meyer, K. G., Bowling, A. J., Slanec, T. and Kramer, V. J. 2017. Biological characterization of fenpicoxamid, a new fungicide with utility in cereals and other crops. *Pest Management Science*, 73, 2005–16.

Paveley, N. D., Lockley, D., Vaughan, T. B., Thomas, J. and Schmidt, K. 2000. Predicting effective fungicide doses through observation of leaf emergence. *Plant Pathology*, 49, 748–66.

Piotrowska, M. J., Fountaine, J. M., Ennos, R. A., Kaczmarek, M. and Burnett, F. J. 2017. Characterisation of *Ramularia collo-cygni* laboratory mutants resistant to succinate dehydrogenase inhibitors. *Pest Management Science*, 73, 1187–96.

Punja, Z. K., Rodriguez, G. and Tirajoh, A. 2016. Effects of *Bacillus subtilis* strain QST 713 and storage temperatures on post-harvest disease development on greenhouse tomatoes. *Crop Protection*, 84, 98–104.

Rehfus, A., Miessner, S., Achenbach, J., Strobel, D., Bryson, R. and Stammler, G. 2016. Emergence of succinate dehydrogenase inhibitor resistance of *Pyrenophora teres* in Europe. *Pest Management Science*, 72, 1977–88.

Rehfus, A., Strobel, D., Bryson, R. and Stammler, G. 2018. Mutations in sdh genes in field isolates of *Zymoseptoria tritici* and impact on the sensitivity to various succinate dehydrogenase inhibitors. *Plant Pathology*, 67, 175–80.

Reis, E. M., Basso, D. F. and Zannata, M. 2013. Loss of sensitivity of *Blumeria graminis* f. sp. *tritici* to triadimenol applied as seed treatment. *Tropical Plant Pathology*, 38, 55–7.

Reiss, A. and Jørgensen, L. N. 2017. Biological control of yellow rust of wheat (*Puccinia striiformis*) with Serenade® ASO (*Bacillus subtilis* strain QST713). *Crop Protection*, 93, 1–8.

Robinson, H. L., Ridout, C. J., Sierotzki, H., Gisi, U. and Brown, J. K. M. 2002. Isogamous, hermaphroditic inheritance of mitochondrion-encoded resistance to Qo inhibitor fungicides in *Blumeria graminis* f. sp. *tritici*. *Fungal Genetics and Biology*, 36, 98–106.

Rotteveel, T., Jørgensen, L. N. and Heimbach, U. 2011. Resistance management in Europe: A preliminary proposal for the determination of a minimum number of active substances necessary to manage resistance. *EPPO Bulletin*, 41, 432–8.

Russell, P. 1995. Fungicide resistance: Occurrence and management. *The Journal of Agricultural Science*, 124, 317–23.

Semar, M., Strobel, D., Koch, A., Klappach, K. and Stammler, G. 2007. Field efficacy of pyraclostrobin against populations of *Pyrenophora teres* containing the F129L mutation in the cytochrome b gene. *Journal of Plant Diseases and Protection*, 114, 117–19.

Sierotzki, H. 2015. Respiration Inhibitors: Complex III. In: Ishii, H. and Hollomon, D. W. (Eds), *Fungicide Resistance in Plant Pathogens: Principles and Guide to Practical Management*. Springer, Japan, pp. 119–43.

Sierotzki, H. and Scalliet, G. 2013. A review of current knowledge of resistance aspects for the next-generation succinate dehydrogenase inhibitor fungicides. *Phytopathology*, 103, 880–7.

Sierotzki, H., Wullschleger, J. and Gisi, U. 2000. Point mutation in cytochrome b gene conferring resistance to strobilurin fungicides in *Erysiphe graminis* f. sp. *tritici* field isolates. *Pesticide Biochemistry and Physiology*, 68, 107–12.

Sierotzki, H., Frey, R., Wullschleger, J., Palermo, S., Karlin, S., Godwin, J. and Gisi, U. 2007. Cytochrome b gene sequence and structure of *Pyrenophora teres* and *P. tritici-repentis* and implications for QoI resistance. *Pest Management Science*, 63, 225–33.

Spolti, P., Del Ponte, E. M., Dong, Y., Cummings, J. A. and Bergstrom, G. C. 2014. Triazole sensitivity in a contemporary population of *Fusarium graminearum* from New York wheat and competitiveness of a tebuconazole-resistant isolate. *Plant Disease*, 98, 607–13.

Stammler, G. and Semar, M. 2011. Sensitivity of *Mycosphaerella graminicola* (anamorph: *Septoria tritici*) to DMI fungicides across Europe and impact on field performance. *EPPO Bulletin*, 41, 149–55.

Stammler, G., Cordero, J., Koch, A., Semar, M. and Schlehuber, S. 2009. Role of the Y134F mutation in cyp51 and overexpression of cyp51 in the sensitivity response of *Puccinia triticina* to epoxiconazole. *Crop Protection*, 28, 891–7.

Strange, R. N. and Scott, P. R. 2005. Plant disease: A threat to global food security. *Annual Review of Phytopathology*, 43, 83–116.

Taggart, P. J., Locke, T., Phillips, A. N., Pask, N., Hollomon, D. W., Kendall, S. J., Cooke, L. R. and Mercer, P. C. 1999. Benzimidazole resistance in *Rhynchosporium secalis* and its effect on barley leaf blotch control in the UK. *Crop Protection*, 18, 239–43.

Taher, K., Graf, S., Fakhfakh, M. M., Salah, H. B. H., Yahyaoui, A., Rezgui, S., Nasraoui, B. and Stammler, G. 2014. Sensitivity of *Zymoseptoria tritici* Isolates from Tunisia to pyraclostrobin, fluxapyroxad, epoxiconazole, metconazole, prochloraz and tebuconazole. *Journal of Phytopathology*, 162, 442–8.

Te Beest, D. E., Paverley, N. D., Shaw, M. W. and Van Den Bosch, F. 2013. Accounting for the economic risk caused by variation in disease severity in fungicide dose decisions, exemplified for *Mycosphaerella graminicola* on winter wheat. *Phytopathology*, 103, 666–72.

Tucker, M. A., Lopez, F. and Oliver, R. P. 2013. The impact of recently emerged mutations in Cyp51 in *Blumeria graminis* f. sp. *hordei* isolates in Western Australia. *Modern Fungicides and Antifungal Compounds VII. Proceedings of 17th International Reinhardsbrunn Symposium*, Friedrichroda Germany, 21–25 April 2013, p. 62.

Tucker, M. A., Lopez-Ruiz, F., Jayasena, K. and Oliver, R. P. 2015. Origin of fungicide-resistant barley powdery mildew in Western Australia: Lessons to be learned. In: Ishii, H. and Holloman, D. (Eds), *Fungicide Resistance in Plant Pathogens: Principles and Guide to Practical Management*. Springer, Japan, pp. 329–40.

Turkington, T., O'Donovan, J., Edney, M., Juskiw, P., McKenzie, R., Harker, K., Clayton, G., Xi, K., Lafond, G. and Irvine, R. 2012. Effect of crop residue, nitrogen rate and fungicide application on malting barley productivity, quality, and foliar disease severity. *Canadian Journal of Plant Science*, 92, 577–88.

Van Den Bosch, F., Paveley, N., Shaw, M., Hobbelen, P. and Oliver, R. 2011. The dose rate debate: Does the risk of fungicide resistance increase or decrease with dose? *Plant Pathology*, 60, 597–606.

Van Den Bosch, F., Oliver, R. P., Van Den Berg, F. and Paveley, N. 2014a. Governing principles can guide fungicide-resistance management tactics. *Annual Review Of Phytopathology*, 52, 175–95.
Van Den Bosch, F., Paveley, N., Van Den Berg, F., Hobbelen, P. and Oliver, R. 2014b. Mixtures as a fungicide resistance management tactic. *Phytopathology*, 104, 1264–73.
Van Toor, R. F., Viljanen-Rollinson, S. L. H., Rahman, A. and Teulon, D. A. J. 2013. Agrichemical use on wheat and barley crops in New Zealand in 2008–09. *New Zealand Journal of Crop and Horticultural Science*, 41, 9–22.
Walker, A. S., Auclair, C., Gredt, M. and Leroux, P. 2009. First occurrence of resistance to strobilurin fungicides in *Microdochium nivale* and *Microdochium majus* from French naturally infected wheat grains. *Pest Management Science*, 65, 906–15.
Wheeler, I., Kendall, S., Butters, J. and Hollomon, D. 1995. Detection of benzimidazole resistance in *Rhynchosporium secalis* using allele-specific oligonucleotide probes. *EPPO Bulletin*, 25, 113–16.
Wieczorek, T. M., Berg, G., Semaškienė, R., Mehl, A., Sierotzki, H., Stammler, G., Justesen, A. F. and Jørgensen, L. N. 2015. Impact of DMI and SDHI fungicides on disease control and CYP51 mutations in populations of *Zymoseptoria tritici* from Northern Europe. *European Journal of Plant Pathology*, 143, 861–71.
Wiik, L. and Rosenqvist, H. 2010. The economics of fungicide use in winter wheat in southern Sweden. *Crop Protection*. 29, 11–19.
Wyand, R. A. and Brown, J. K. M. 2005. Sequence variation in the CYP51 gene of *Blumeria graminis* associated with resistance to sterol demethylase inhibiting fungicides. *Fungal Genetics and Biology*, 42, 726–35.
Yan, L., Yang, Q., Zhou, Y., Duan, X. and Ma, Z. 2009. A real-time PCR assay for quantification of the Y136F allele in the CYP51 gene associated with *Blumeria graminis* f.sp. *tritici* resistance to sterol demethylase inhibitors. *Crop Protection*, 28, 376–80.
Yang, J., Liao, M. and Yang, S. 2008. Advances in sterol 14α-demethylase (CYP51). *Shengwu Gongcheng Xuebao/Chinese Journal of Biotechnology*, 24, 1681–8.

Chapter 13

Trends in exploring wheat and barley germplasm for novel disease resistance traits

Sambasivam Periyannan, Australian National University, Australia; and Lee Hickey, University of Queensland, Australia

1 Introduction

The emergence of highly virulent fungal pathogens presents a major challenge for the sustained production of wheat (*Triticum aestivum*) and barley (*Hordeum vulgare*) crops around the world. For instance, the Ug99 and TTTTF race of the stem rust pathogen *Puccinia graminis* f. sp. *tritici* identified in East Africa and Sicily, respectively, has become a serious threat to global food security due to its enhanced ability to infect modern wheat and barley cultivars (Mamo et al. 2015; Singh et al. 2015; Bhattacharya 2017). In addition, the emergence of blast disease in wheat has become a serious concern. New strains of blast recently detected in wheat fields of Brazil and Bangladesh have the potential to cause huge losses (Inoue et al. 2017). While the modern cultivars are highly prone to pathogen attack, historic germplasm collections of wheat and barley and their wild related species offer useful genetic variations that confer resistance against the prevailing fungal pathogen strains (Feuillet et al. 2008; Mamo et al. 2015; Das et al. 2016; Huang et al. 2016; Mondal et al. 2016). Such germplasm collections can be grouped into three categories, known as the primary, secondary and tertiary gene pools. The primary gene pool consists of landraces and progenitor species that have a genome structure resembling the cultivated crop. Hence, chromosomal segments carrying useful genetic variations can be readily introgressed into the targeted crop through direct hybridisation. In wheat, this group consists of the two broadly cultivated species, the hexaploid *T. aestivum* (AABBDD) and tetraploid *T. turgidum* (AABB) and their related diploid progenitors such as the diploid A (*T. monococcum*, *T. boeoticum* and *T. urartu*) and D (*Aegilops tauschii*) genome species.

http://dx.doi.org/10.19103/AS.2018.0039.15

For barley, this includes two species, namely *H. vulgare* (HH), which is cultivated worldwide, and *H. spontaneum* (HH), its progenitor species. Subsequently, the secondary gene pool consists of species that have at least one genome in common with the cultivated species, such as *T. timopheevii* (AAGG) and *Elymus* (HHStSt) for wheat and barley, respectively. Finally, the tertiary pool is represented by species carrying non-identical genomes like *Thinopyrum* species which carries the EE genome and has been exploited for improving traits in wheat (Feuillet et al. 2008). Although these valuable germplasm collections are preserved in gene banks around the world and offer useful genetic variations, only a handful of traits have been utilised for commercial breeding due to difficulties associated with (1) screening large numbers of accessions for disease resistance using diverse pathogen strains, (2) the lack of information on the genetic make-up and growth habits of accessions, (3) linkage drag where along with the targeted resistance loci linked undesirable traits are also transferred and (4) long breeding cycles due to biological constraints in crossing and lack of molecular markers linked with the trait (Feuillet et al. 2008; Dempewolf et al. 2017; Wang et al. 2017). However, with recent advancements in phenotyping, genotyping and other breeding technologies, more effective strategies have emerged to accelerate the identification and transfer of valuable disease resistance traits from relatively unexplored genetic resources into modern wheat and barley cultivars. Here, we provide an overview of these trends and highlight the future directions required to fully exploit historic germplasm collections in the pursuit for durable disease resistance required for sustainable crop production.

2 Advances in phenotyping germplasm for disease resistance

One of the initial steps in exploring germplasm for disease resistance is the initial purification and the evaluation for disease resistance through large-scale pathogen infection screening. As the individual lines of the germplasm are random collections originating from diverse geographical regions and sourced by various seed collectors and explorers, they are often impure and lack basic genetic information, such as growth habit. Often, the growth habit and morphological characteristics for each accession are only recorded during line purification through single seed descent. Such information is critical for effective screening of accessions using the pathogen. For instance, lines with winter growth habit require a period of cold to satisfy the vernalisation requirement in order to induce flowering, required for infecting the inflorescence, with disease such as Fusarium head blight. Traditionally, line purification and large-scale screening using the target pathogen was a tedious task requiring enormous resources such as large field or glasshouse space and labour. However, more recently, these practices have been simplified and fast-tracked through new technology and approaches like the Focused Identification of Germplasm Strategy (Bhullar et al. 2009), speed breeding technology (Watson et al. 2018) and high-throughput pathogen infection screening via imaging (Shakoor et al. 2017).

2.1 Focused identification of germplasm strategy

Focused identification of germplasm strategy (FIGS) offers a way to short-cut the screening process by avoiding the need to screen the entire germplasm collection. Instead, a subset of the germplasm is more efficiently targeted for trait screening, and can be applied to any trait, including disease resistance. Agro-ecological and trait information for the gene bank accessions is used to improve the likelihood of identifying accessions harbouring the desired trait (Bari et al. 2014). In wheat, FIGS has been successfully used for mining powdery mildew disease resistance, specifically, novel alleles for the *Pm3* locus. In this case, instead of screening 16000 wheat accessions, a subset composed of less than 10% of the total landraces was focused for the identification of novel alleles for *Pm3*. The subset was selected from places with climatic conditions in the geographical origin of these lines that favours the mildew pathogen (Bhullar et al. 2009). Subsequently, FIGS was also used for the identification of stem- and stripe-rust-resistant lines from wheat landrace collections maintained at the US Department of Agriculture (USDA) and International Centre for Agricultural Research in the Dry Areas organisations (Endresen et al. 2011; Bari et al. 2014). Similarly in barley, the technique was employed using climate data to effectively identify net blotch disease-resistant lines from a USDA collection (Endresen et al. 2011). Hence, with the increasing labour cost and urgent need to identify useful resistance for rapidly evolving pathogens, FIGS presents as a valuable tool to speed up identification of lines with novel disease for wheat and barley improvement.

2.2 Speed breeding

Speed breeding is a technique for rapid generation advance in wheat and barley performed in enclosed growth rooms or glasshouse facilities, whereby light and temperature conditions are optimised to accelerate plant development (Watson et al. 2018). The technique enables up to six generations of wheat or barley per year, compared to only one or two generations in the field. This tool facilitates rapid purification and phenotyping of germplasm collections for disease resistance traits. In wheat, the technique has been successfully employed for the identification and transfer of novel leaf rust and tan spot resistance traits from wheat accessions sourced from the Vavilov Institute of Plant Genetic Resources in St Petersburg, Russia (Fig. 1, Riaz et al. 2017; Dinglasan et al. 2018). Simultaneously, the controlled conditions provide an innovative phenotyping platform for disease infection screening in wheat. Here, the screening of seedlings and that of adult plants are effectively integrated within the same crop generation through manipulation of plant growth and sequential inoculations (Riaz et al. 2016; Dinglasan et al. 2016). The approach allows screening all-year-round and minimises errors due to the presence of other diseases or disease escape that are often challenges experienced in conventional field screening. Following the success in wheat, the technique was also applied in barley to rapidly transfer multiple disease resistance into a malting cultivar preferred by industry (Hickey et al. 2017). For targeted trait introgression, generation time can be further minimised when speed breeding is combined effectively with the fast generation cycling system where the embryos from premature seeds are cultured to reduce the growth cycle (Yan et al. 2017). This procedure can also help to overcome embryo abortion issues that occur in the hybrids between wild species and cultivated wheat.

2.3 High-throughput screening via imaging

One of the major expenses involved in exploring germplasm for disease resistance is due to the need for skilled labour and expertise to conduct the large-scale pathogen infection screening. An alternative approach to reduce this cost is to use automated and

Figure 1 Disease resistance screening of wheat landrace accessions from the Vavilov Institute of Plant Genetic Resources: (a) phenotyping resistance to stripe rust in a field screening nursery, (b) variation in stripe rust phenotypes on the flag leaf in the field nursery, (c) resistance to tan spot in a field screening nursery and (d) seedling assessment for tan spot resistance in the glasshouse.

high-throughput phenotyping platforms developed using advancements in imaging, sensing and computational techniques (Shakoor et al. 2017). For instance, multi- and hyperspectral imaging were recently applied for the assessment of Fusarium head blight and stripe rust disease incidence in wheat (Bauriegal et al. 2011; Devada et al. 2015). Similarly, fluorescence imaging was found useful for assessing powdery mildew and rust disease symptoms in wheat (Burling et al. 2011). For barley, imaging systems have been developed for foliar diseases such as powdery mildew, net blotch and brown rusts (Kuska et al. 2015; Wahabzada et al. 2015). More recently, a handheld crop senor that can detect green vegetation difference through sensing of normalised difference vegetation index was optimised for quantifying stripe rust disease resistance in wheat (Pretorius et al. 2017). Such a tool can more accurately quantify leaf area damage caused by disease and could be scaled up to screen large-field experiments using unmanned aerial vehicles.

3 Advances in genomics for discovery of disease resistance loci

Following line purification and screening germplasm for disease resistance, the second major step is to genetically position the targeted gene locus for identification of linked molecular markers for rapid trait introgression. This can be done effectively using linkage map and sequence information from the near-complete reference genomes of wheat, barley and its related species (Bevan et al. 2017). Apart from rapid transfer of the disease resistance traits into elite cultivars through marker-assisted selection, gene mapping further avoids repetition in the efforts to identify new sources, as germplasm collections often contain duplicated entries. Subsequently, with the advancements in sequencing, genome assembling and molecular marker techniques, linkage maps specific to germplasm collections and alien introgressed segments carrying novel traits are now available (Tiwari et al. 2014; Winfield et al. 2017).

3.1 Genome-wide association studies

With coordinated efforts, genomes of globally important crop plants and their wild related species were rapidly sequenced to further facilitate the identification of numerous single nucleotide polymorphic (SNP) markers. Subsequently, these SNP markers were employed for genotyping germplasm collections where marker-trait linkage analysis enables prompt identification of the genomic region underpinning trait expression. This process commonly known as Genome-wide association study (GWAS) has provided an effective tool for rapid characterisation and identification of novel disease resistance traits using wheat and barley landraces. For example, the 9K SNP array and a GWAS approach were employed in wheat to identify valuable leaf and stripe rust disease resistance loci in wheat landraces (Sthapit et al. 2014; Kertho et al. 2015; Maccaferri et al. 2015; Bulli et al. 2016). Alternative to SNP markers, the Diversity Arrays Technology genotyping-by-sequencing platform, where low copy fragments separated based on the restriction by a specific enzyme are captured and sequenced, also offers efficient genotyping of vast collections of accessions (Li et al. 2015). This system has been recently applied to effectively detect novel stripe rust and leaf rust resistance loci in Afghan and Russian wheat landraces (Manickavelu et al. 2016; Riaz et al. 2017). In barley, using SNP markers, GWAS was successfully used to identify spot blotch, stripe rust and leaf rust resistance loci in Latin American germplasm (Gutierrez et al. 2015). A similar approach was also used to detect Fusarium head blight resistance loci in barley landraces originating from Ethiopia and Eritrea (Mamo and Steffenson 2015).

3.2 Genomic selection

Genomic selection (GS) was originally developed and applied in animal breeding (Hayes and Goddard 2001), but is now considered one of the most promising tools for selecting desirable traits in plant breeding (Crossa et al. 2017). It serves to short-cut the breeding cycle by avoiding the need to phenotype before individuals are selected as parents for the next round of crossing. In the beginning, GS was used as a selection tool for predicting disease resistance in biparental populations (Rutkoski et al. 2011; Ornella et al. 2012). Later, it gained momentum for application to germplasm collections, such as diverse landraces from the Watkin's wheat collection, where the approach was used to predict rust resistance traits (Daetwyler et al. 2014). The prediction accuracy relies on the availability of a training population that has been genotyped and phenotyped for the trait of interest. The accuracy of prediction is affected by how closely related the training population is to the test material being predicted. This remains a major challenge for exploiting the prediction approach to mine germplasm collections, as by their very nature, accessions are extremely diverse. One approach could be phenotyping and genotyping a subset of the collection in order to predict the disease resistance for a larger number of accessions from the same geographical region, thereby reducing time and labour required to identify novel sources of disease resistance. In barley, GS is widely used for predicting disease resistance traits in breeding populations, while its application to predict disease resistance in germplasm collections is yet to be proven (Lorenz et al. 2012; Sallam et al. 2015).

3.3 Identification of markers and linkage maps specific to germplasm

As gene bank accessions largely comprise historic germplasm, gathered at various times and from different geographical locations, they tend to carry distinct genetic variations as compared to modern varieties. Characterisation and transfer of these novel traits into modern cultivars remains rapid where molecular markers linked with these traits were identified. However, the available SNP markers generated from the sequencing of a selected cultivar is less useful for mapping traits from historic germplasm due to huge genetic variations. Therefore, generation of germplasm-specific linkage mapping approaches such as the identification of SNP variations linked to various lines of the Watkins wheat landrace collections (Winfield et al. 2017) is highly useful. Subsequently, with the advancements in chromosome isolation and DNA sequencing, wheat chromosomes carrying alien segments were isolated and sequenced for identification of molecular markers specific to the alien segment. Using this approach, SNP markers were identified that are specific to rust resistance genes *Lr57* and *Yr40*, originating from the wild species *A. geniculata* (Tiwari et al. 2014). These tools serve to speed up the gene introgression process.

3.4 Identification of novel resistance through allele mining

Germplasm are valuable resources for mining novel variants of known disease resistance genes. As these lines have been less exposed to the prevailing pathogen, these sources remain effective and often provide broad-spectrum resistance. For instance, wheat germplasm collections originating from diverse wheat-growing regions were successfully mined for novel and effective variants of the mildew resistance gene *Pm3* (Bhullar et al. 2009, 2010). Subsequently, germplasm collections of the wild emmer wheat were exploited for mining novel variants of the wheat stripe rust resistance gene *Yr36* (Huang et al. 2016). Similarly in barley, the best known example is the identification of various alleles of the *Mla* locus resulting from screening barley germplasm maintained at IPK, Germany (Seeholzer et al. 2010).

4 Conclusion and future trends

Prior to the assembly of the near-complete reference genomes, characterisation and mapping of novel disease resistance traits in wheat and barley were based on low-resolution genetic linkage maps. But in recent years, more and more SNP-based markers have been identified and mapped from these large and complex genomes as the near-complete reference was made available through the use of modern sequencing and computational techniques (Bevan et al. 2017; Yuan et al. 2017). Subsequently, a number of cultivars and germplasm related to the two crops have been sequenced to identify a high degree of allelic variations for gene sequences and DNA markers. This rapid genomic characterisation facilitates faster transfer of novel traits from germplasm as SNP markers tightly linked to the targeted trait can be readily identified for their use in marker-assisted breeding (Golicz et al. 2016; Hurgobin and Edwards 2017). Integrating these new genomic tools with rapid generation advance systems, such as speed breeding, could lead to unparalleled rapid gene discovery and gene deployment (Li et al. 2018).

Even with the identification of closely linked molecular markers, transfer of novel disease resistance traits from germplasm requires repeated cycles of crossing of the targeted line with elite cultivars. An alternative for this tedious task is to isolate and transfer the gene directly into the targeted cultivars using transgenic approaches. Isolation of the gene sequence responsible for the disease resistance trait also provides gene-specific markers for the perfect selection of resistant lines carrying the targeted traits. With the advancements in chromosome isolation, DNA sequencing and targeted gene capture, simplified resistance gene cloning approaches such as targeted chromosome-based cloning via long-range assembly, mutational mapping, mutant chromosome sequencing and resistance gene enrichment and sequencing have been developed (Periyannan 2017). An interesting feature of these novel strategies is the reduced dependency on chromosome recombination and genome reference sequence, making them highly suitable for isolating disease resistance particularly from the wild species germplasm (Wulff and Moscou 2014; Periyannan 2017). Thus, along with improved genotyping and phenotyping methods, rapid gene cloning could open the door for gene editing approaches to engineer more robust genetic resistance to combat aggressive crop pathogens in the near future (Andolfo et al. 2016).

5 Where to look for further information

Germplasm resources at USA: https://www.ars-grin.gov/

Wheat germplasm resources network: http://archive.gramene.org/species/triticum/wheat_germplasm.html

Barley germplasm resources network: http://archive.gramene.org/species/hordeum/barley_germplasm.html

Wheat germplasm at International Maize and Wheat Improvement Center (CIMMYT): https://www.cimmyt.org/germplasm-bank/

Focused Identification of Germplasm Strategy (FIGS): http://www.icarda.org/figs/teaser

Diversity Arrays Technology (DArT): http://www.diversityarrays.com/dart-application-dartseq

6 References

Andolfo, G., Iovieno, P., Frusciante, L. and Ercolano, M. R. (2016). Genome-editing technologies for enhancing plant disease resistance. *Front. Plant Sci.* 7, 1813.

Bari, A., Amri, A., Street, K., Mackay, M., De Pauw, E., Sanders, R., Nazari, K., Humeid, B., Konopka, J. and Alo, A. (2014). Predicting resistance to stripe (yellow) rust (*Puccinia striiformis*) in wheat genetic resources using focused identification of germplasm strategy. *J. Agr. Sci.* 152, 906–16.

Bauriegel, E., Glebel, A., Geyer, M., Schmidt, U. and Herppich, W. (2011). Early detection of fusarium infection in wheat using hyper-spectral imaging. *Comput. Electron. Agric.* 75, 304–12.

Bevan, M. W., Uauy, C., Wulff, B. B. H., Zhou, J., Krasileva, K. and Clark, M. D. (2017). Genomic innovation for crop improvement. *Nature* 543, 346–54.

Bhattacharya, S. (2017). Wheat rust back in Europe. *Nature* 542, 145–6.

Bhullar, N. K., Street, K., Mackay, M., Yahiaoui, N. and Keller, B. (2009). Unlocking wheat genetic resources for the molecular identification of previously undescribed functional alleles at the *Pm3* resistance locus. *Proc. Natl. Acad. Sci. U.S.A.* 106, 9519–24.

Bhullar, N. K., Zhang, Z., Wicker, T. and Keller, B. (2010). Wheat gene bank accessions as a source of new alleles of the powdery mildew resistance gene *Pm3*: A large scale allele mining project. *BMC Plant Biol.* 10, 88.

Bulli, P., Zhang, J., Chao, S., Chen, X. and Pumphrey, M. (2016). Genetic architecture of resistance to stripe rust in a global winter wheat germplasm collection. *G3* 6, 2237–53.

Burling, K., Hunsche, M. and Noga, G. (2011). Use of blue-green and chlorophyll fluorescence measurements for differentiation between nitrogen deficiency and pathogen infection in winter wheat. *J. Plant Physiol.* 168, 1641–8.

Crossa, J., Perez-Rodriguez, P., Cuevas, J., Montesinos-Lopez, O., Jarquin, D., Campos, G., Burgueno, J., Gonzalez-Camacho, J. M., Perez-Elizalde, S., Beyene, Y., Dreisigacker, S., Singh, R., Zhang, X., Gowda, M., Roorkiwal, M., Rutkoski, J. and Varshney, R. K. (2017). Genomic selection in plant breeding: Methods, models, and perspectives. *Trends Plant Sci.* 22, 961–75.

Daetwyler, H. D., Bansal, U. K., Bariana, H. S., Hayden, M. J. and Hayes, B. J. (2014). Genomic prediction for rust resistance in diverse wheat landraces. *Theor. Appl. Genet.* 127, 1795–803.

Das, M. K., Bai, G., Mujeeb-Kazi, A. and Rajaram, S. (2016). Genetic diversity among synthetic hexaploidy wheat accessions (*Triticum aestivum*) with resistance to several fungal diseases. *Genet. Resour. Crop Evol.* 63, 1285–96.

Dempewolf, H., Baute, G., Anderson, J., Kilian, B., Smith, C. and Guarino, L. (2017). Past and future use of wild relatives in crop breeding. *Crop Sci.* 57, 1070–82.

Devadas, R., Lamb, D., Backhouse, D. and Simpfendorfer, S. (2015). Sequential application of hyperspectral indices for delineation of stripe rust infection and nitrogen deficiency in wheat. *Precis. Agric.* 16, 477–91.

Dinglasan, E. G., Godwin, I. D., Mortlock, M. Y. and Hickey, L. T. (2016). Resistance to yellow spot in wheat grown under accelerated growth conditions. *Euphytica* 209, 693–707.

Dinglasan, E. G., Godwin, I. D., Phan, H. T. T., Tan, K. C., Platz, G. J. and Hickey, L. T. (2018). Vavilov wheat accessions provide useful sources of resistance to tan spot (syn. yellow spot) of wheat. *Plant Pathol.* doi:10.1111/ppa.12822.

Endresen, D. T. F., Street, K., Mackay, M., Bari, A. and De Pauw, E. (2011). Predictive association between biotic stress traits and ecogeographic data for wheat and barley landraces. *Crop Sci.* 51, 2036–55.

Feuillet, C., Langridge, P. and Waugh, R. (2008). Cereal breeding takes a walk on the wild side. *Trends Genet.* 24, 24–32.

Golicz, A. A., Batley, J. and Edwards, D. (2016) Towards plant pangenomics. *Plant Biotech. J.* 14, 1099–105.

Gutierrez, L., German, S., Pereyra, S., Hayes, P. M., Perez, C. Z., Capettini, F., Locatelli, A., Berberian, N. M., Falconi, E. E., Estrada, R., Fros, D., Gonza, V., Altamirano, H., Huerta-Espino, J., Neyra, E., Orjeda, G., Sandoval-Islas, S., Singh, R., Turkington, K. and Castro, A. J. (2015). Multi-environment multi-QTL association mapping identifies disease resistance QTL in barley germplasm from Latin America. *Theor. Appl. Genet.* 128, 501–16.

Hayes, B. J. and Goddard, M. E. (2001). Prediction of total genetic value using genome-wide dense marker maps. *Genetics* 157, 1819–29.

Hickey, L. T., Germán, S., Pereyra, S., Diaz, J., Ziems, L. A., Fowler, R., Platz, G. J., Franckowiak, J. D. and Dieters, M. (2017). Speed breeding for multiple disease resistance in barley. *Euphytica* 213, 64.

Huang, L., Raat, D., Sela. H., Klymiuk. V., Lidzbarsky, G., Feng, L., Krugman, T. and Fahima, T. (2016). Evolution and adaptation of wild emmer wheat populations to biotic and abiotic stresses. *Ann. Rev. Phytopathol.* 54, 279–301.

Hurgobin, B. and Edwards, D. (2017). SNP discovery using a pangenome: Has the single reference approach become obsolete? *Biology* 6, 21.

Inoue, Y., Vy, T. T. P., Yoshida, K., Asano, H., Mitsuoka, C., Asuke, S., Anh, V. L., Cumagun, C. J. R., Chuma, I., Terauchi, R., Kato, K., Mitchell, T., Valent, B., Farman, M. and Tosa, Y. (2017). Evolution of the wheat blast fungus through functional losses in a host specificity determinant. *Science* 357, 80–3.

Kertho, A., Mamidi, S., Bonman, J. M., McClean, P. E. and Acevedo, M. (2015). Genome-wide association mapping for resistance to leaf and stripe rust in winter-habit hexaploidy wheat landraces. *PLoS ONE* 10, 129580.

Kuska, M., Wahabzada, M., Leucker, M., Dehne, H. W., Kersting, K., Oerke, E. C., Steiner, U. and Mahlein, A. K. (2015). Hyperspectral phenotyping on the microscopic scale: Towards automated characterization of plant-pathogen interactions. *Plant Methods* 11, 28.

Li, H., Vikram, P., Singh, R. P., Kilian, A., Carling, Y., Song, J., Burgueno-Ferreira, J. A., Bhavani, S., Huerta-Espino, J., Payne, T., Sehgal, D., Wenzl, P. and Singh, S. (2015). A high density GBS map of bread wheat and its application for dissecting complex disease resistance traits. *BMC Genomics* 16, 216.

Li, H., Rasheed, A., Hickey, L. T. and He, Z. (2018). Fast-forwarding genetic gain. *Trends Plant Sci.* (in press).

Lorenz, A. J., Smith, K. P. and Jannink, J. L. (2012). Potential and optimization of genomic selection for fusarium head blight resistance in six-row barley. *Crop Sci.* 52, 1609–21.

Maccaferri, M., Zhang, J., Bulli, P., Abate, Z., Chao, S., Cantu, D., Bossolini, E., Chen, X., Pumphrey, M. and Dubcovsky, J. (2015). A genome-wide association study of resistance to stripe rust (*Puccinia striiformis* f. sp. *tritici*) in a worldwide collection of hexaploidy spring wheat (*Triticum aestivum* L.). *G3* 5, 449–65.

Mamo, B. E. and Steffenson, B. J. (2015). Genome-wide association mapping of fusarium head blight resistance and agromorphological traits in barley landraces from Ethiopia and Eritrea. *Crop Sci.* 55, 1494–512.

Mamo, B. E., Smith, K. P., Brueggeman, R. S. and Steffenson, B. J. (2015). Genetic characterization of resistance to wheat stem rust race TTKSK in landrace and wild barley accessions identifies the *rpg4/Rpg5* locus. *Phytopathology* 105, 99–109.

Manickavelu, A., Joukhadar, R., Jighly, A., Lan, C., Huerta-Espino, J., Stanikzai, A. S., Kilian, A., Singh, R. P. and Ban, T. (2016). Genome wide association mapping of stripe rust resistance in Afghan wheat landraces. *Plant Sci.* 252, 222–9.

Mondal, S., Rutkoski, J. E., Velu, G., Singh, P. K., Crespo-Herrera, L. A., Guzman, C., Bhavani, S., Lan, C., He, X. and Singh, R. P. (2016). Harnessing diversity in wheat to enhance grain yield, climate resilience, disease and insect pest resistance and nutrition through conventional and modern breeding approaches. *Front. Plant Sci.* 7, 991.

Ornella, L., Singh, S., Perez, P., Burgueno, J., Singh, R., Tapia, E., Bhavani, S., Dreisigacker, S., Braun, H., Mathews, K. and Crossa, J. (2012). Genomic prediction of genetic values for resistance to wheat rusts. *Plant Genome* 5, 136–48.

Periyannan. S. (2017). Sustaining global agriculture through rapid detection and deployment of genetic resistance to deadly crop diseases. *New Phytol.* doi:10.1111/nph.14928.

Pretorius, Z. A., Lan, C. X., Prins, R., Knight, V., McLaren, N. W., Singh, R. P., Bender, C. M. and Kloppers, F. J. (2017). Application of remote sensing to identify adult plant resistance loci to stripe rust in two bread wheat mapping populations. *Precis. Agric.* 18, 411–28.

Riaz, A., Periyannan, S., Aitken, E. and Hickey, L. (2016). A rapid phenotyping method for adult plant resistance to leaf rust in wheat. *Plant Methods* 12, 17.

Riaz, A., Athiyannan, N., Periyannan, S., Afanasenko, O., Mitrofanova, O., Aitken, E. A. B., Lagudah, E. and Hickey, L. T. (2017). Mining Vavilov's treasure chest of wheat diversity for adult plant resistance to *Puccinia triticina*. *Plant Dis.* 101, 317–23.

Rutkoski, J. E., Heffner E. L. and Sorrells, M. E. (2011). Genomic selection for durable stem rust resistance in wheat. *Euphytica* 179, 161–73.

Sallam, A. H., Endelman, J. B., Jannink, J. L. and Smith, K. P. (2015). Assessing genomic selection prediction accuracy in a dynamic barley breeding population. *Plant Genome* doi:10.3835/plantgenome2014.05.0020.

Seeholzer, S., Tsuchimatsu, T., Jordan, T., Bieri, S., Yang, W., Jahoor, A., Shimizu, k. K., Keller, B. and Schulze-Lefert, P. (2010). Diversity at the *Mla* powdery mildew resistance locus from cultivated barley reveals sites of positive selection. *Mol. Plant Microbe Interact.* 23, 497–509.

Shakoor, N., Lee, S. and Mockler, T. C. (2017). High throughput phenotyping to accelerate crop breeding and monitoring of diseases in the field. *Curr. Opin. Plant Biol.* 38, 184–92.

Singh, R. P., Hodson, D. P., Jin, Y., Lagudah, E. S., Ayliffe, M. A., Bhavani, S., Rouse, M. N., Pretorius, Z. A., Szabo, L. J., Huerta-Espino, J. and Basnet, B. R. (2015). Emergence and spread of new races of wheat stem rust fungus: Continued threat to food security and prospects of genetic control. *Phytopathology* 105, 872–84.

Sthapit, J., Newcomb, M., Bonman, J. M., Chen, X. and Se, D. R. (2014). Genetic diversity for stripe rust resistance in wheat landraces and identification of accessions with resistance to stem rust and stripe rust. *Crop Sci.* 54, 2131–9.

Tiwari, V. K., Wang, S., Sehgal, S., Vrana, J., Friebe, B., Kubalakova, M., Chhuneja, P., Dolezel, J., Akhunov, E., Kalia, B., Sabir, J. and Gill, B. S. (2014). SNP discovery for mapping alien introgressions in wheat. *BMC Genomics* 15, 273.

Wahabzada, M., Mahlein, A. K., Bauckhage, C., Steiner, U., Oerke, E. C. and Kersting, K. (2015). Metro maps of plant disease dynamics-automated mining of differences using hyperspectral images. *PLoS ONE* 10, 116902.

Wang, C., Hu, S., Gardner, C. and Lubberstedt, T. (2017). Emerging avenues for utilization of exotic germplasm. *Trends Plant Sci.* 22, 624–37.

Watson, A., Ghosh, S., Williams, M., Cuddy, W. S., Simmonds, J., Rey, M. D., Hatta, M. A. M., Hinchliffe, A., Steed, A., Reynolds, D., Adamski, N., Breakspear, A., Korolev, A., Rayner, T., Dixon, L. E., Riaz, A., Martin, W., Ryan, M., Edwards, D., Batley, J., Raman, H., Carter, J., Rogers, C., Domoney, C., Moore, G., Harwood, W., Nicholson, P., Dieters, M. J., DeLacy, I. H., Zhou, J., Uauy, C., Boden, S. A., Park, R. F., Wulff, B. B. H. and Hickey, L. T. (2018). Speed breeding is a powerful tool to accelerate crop research and breeding. *Nat. Plants* 4, 23–9.

Winfield, M. O., Allen, A. M., Wilkinson, P. A., Burridge, A. J., Barker, G. L. A., Coghill, J., Waterfall, C., Wingen, L. U., Griffiths, S. and Edwards, K. J. (2017). High-density genotyping of the A. E. Watkins collection of hexaploidy landraces identifies a large molecular diversity compared to elite bread wheat. *Plant Biotech. J.* 16, 165–75.

Wulff, B. B. H. and Moscou, M. J. (2014). Strategies for transferring resistance into wheat: From wide crosses to GM cassettes. *Front. Plant Sci.* 5, 692.

Yan, G., Liu, H., Wang, H., Lu, Z., Wang, Y., Mullan, D., Hamblin, J. and Liu, C. (2017). Accelerated generation of selfed pure line plants for gene identification and crop breeding. *Front. Plant Sci.* 8, 1786.

Yuan, Y., Bayer, P. E., Batley, J. and Edwards, D. (2017). Improvements in genomic technologies: application to crop genomics. *Trends Biotech.* 35, 547–58.

Chapter 14

Developments in diagnostic techniques for cereal pathogens

Sadia Iqbal and Michael G. K. Jones, Western Australian State Agricultural Biotechnology Centre - Murdoch University, Australia

1 Introduction

Accurate detection and identification of cereal diseases is important because this information is vital for undertaking appropriate control measures. Diagnostic methods have advanced rapidly in many areas of biology, and this is also the case in agriculture. Methods used to identify crop diseases have progressed from examining the morphology and symptoms of a disease with a magnifying glass. Although this is often still the basis for rapid field assessment, molecular diagnostic techniques are now used routinely to screen plant materials and soils for the presence of disease-causing organisms. Some benefits of molecular diagnostics include:

- Identification of the presence of a disease-causing organism before symptoms are apparent on plants.
- Replacing testing methods which involve infection of indicator plants or plating on selective media, thus reducing incubation times required to identify the presence of a pathogen or to confirm observed symptoms.

http://dx.doi.org/10.19103/AS.2018.0039.18

- Correct identification of the causal organism both at the level of species and often the race of a pathogen.
- Molecular diagnostics can now often be done rapidly with a higher throughput of samples compared to manual sampling and/or culture techniques.
- Diagnostic tests can be multiplexed to check for the presence of one or several organisms at the same time, including the presence of fungal pathogens, nematodes, viruses, viroids, bacteria or phytoplasmas.
- The presence of previously unknown pests or diseases can be uncovered using new nucleic acid sequencing technologies.

Although pre-sowing seed treatments, such as seed coating, are used widely for cereals to avoid or reduce effects of soil-borne and seed-borne pathogens, routine inspections should nevertheless be carried out to detect any symptoms at early seedling stages. Accurate identification of a disease or pest is essential for the implementation of appropriate, effective control measures as early as possible, especially as part of an integrated pest management strategy.

2 Prevention strategies and pre-symptom identification

In many countries, disease-forecasting systems are in place, based on both weather conditions and the detection of disease-causing agents such as spores and insects, using a network of traps. These systems are used to inform growers in advance of potential pest and disease risks for their crops. For example, humid weather promotes the spread of some fungal and bacterial diseases, and increasing numbers of aphids presage aphid infestations and outbreaks of viral diseases they may transmit. Such information can be used by growers to take preventive measures to reduce the risk of disease spread through their fields by planning field operations accordingly.

Collection and identification of airborne biological agents ('bioaerosols') can be undertaken to monitor spores or other disease agents present in the air long before disease symptoms appear on a crop. Spore samplers are used to trap such samples either from air using adhesive tapes or agar plates with constant air flow to collect samples for analysis. The collected samples are then examined to identify pests and pathogens present, and to assess risks by quantifying active inocula. Aphid numbers can also be monitored in the same way. Samples collected can also be subjected to nucleic-acid-based detection methods such as polymerase chain reaction (PCR) and qRT-PCR to detect specific pathogens (Walsh et al. 2005; Falacy et al. 2007).

The use of disease-free seed is often the key to ensuring a healthy start for a crop. A wide array of tests including visual, chemical or physical tests can be used to identify seed affected by pathogens: for example, seed floating in sodium chloride, optical sorters, seed staining for fungal pathogens, sorting of seed to exclude disease-affected grain of cereals such as wheat and barley. Seed dressing treatments are used as a preventive measure for most fungal pathogens to control soil-borne or seed-borne diseases. Fungicide application at the seedling stage is also recommended for wheat to prevent development of powdery mildew, leaf rust and loose smut (https://www.agric.wa.gov.au/barley/seed-dressing-and-furrow-fungicides-cereals-western-australia). The potential disease risk in a field can be assessed by testing the soil and the seed before sowing.

2.1 Visual identification

After disease symptoms have appeared in a crop, the most common method used to identify the cause is visual observation and diagnosis by an expert. This requires a field visit by a plant pathologist or trained agronomist, visual observation and a survey of the field to assess the severity of the disease or pest infestation. This approach is still widely practised in most developed counties.

Although visual symptoms and their recognition by a trained person is usually reliable, it is nevertheless time-consuming. Commercially based sample analysis in some cases has replaced or is used to confirm the visual diagnosis: plant samples with disease symptoms and/or insect/pest specimens, or soil samples, are sent to the local Department of Agriculture or to a commercial testing service. The relevant research and extension department conducts tests on the samples received and provides a report to the grower on the identification, relative risk of crop losses and management and control of identified diseases. Based on the test results, agriculture departments may also issue area risk predictions and warnings so that growers can look out for specific symptoms to aid early detection. An example is the 'Rust Bust' programme, an initiative of the Australian Cereal Rust Control Program, which is supported by the Australian Grains Research and Development Corporation, to increase awareness and encourage growers to use available tools for rust management.

In Australia, various disease-testing services also offer pre-season seed and soil testing as well as disease identification based on sample testing. These include AgWest Plant Laboratories, Department of Primary Industries and Regional Development, PreDicta B from Primary Industries and Regions South Australia for diseases of cereals such as tan spot and white grain disorder, crown rot and the nematode *Pratylenchus* spp., provided by the South Australian Research and Development Institute, TASAG ELISA Testing Services, Department of Primary Industries, Parks, Water and Environment (Tasmania), New South Wales Department of Primary Industries and Plant Pest Diagnostic Service at the Department of Agriculture and Fisheries, Queensland.

2.2 Morphological identification

Morphological identification of diseases is not limited to symptoms on plants. For detection of fungal pathogens which can be cultured in the lab, structures such as conidia and hyphae are studied microscopically to look for markers specific to the fungal species set as a standard by taxonomists. The disadvantage of culture-based methods is that they are time-consuming, whereas fungal spread in the field can be rapid under favourable conditions. Technology such as computerized image analysis can fast-track the identification of disease with fewer resources: for example, leaf blight, leaf spot, powdery mildew, leaf rust and smut in cereals have been identified with an accuracy of up to 91% by image processing and pattern recognition technology (Pujari et al. 2015).

For insects and other pests, the specimens can be collected and examined for morphological features for species identification, while identification of viral diseases is often based on typical symptoms evident on infected plants.

Using morphological symptoms for disease identification is limited by the available information, and there is potential for misidentification of causal organisms if symptoms of a new disease mimic an existing one. Environmental factors also affect the development of disease symptoms, and pathogens present in a complex may be missed or misidentified.

3 Protein-based identification and serological methods

Because plant viruses cannot be cultured *in vitro,* and are composed of nucleic acid genomes protected by capsid proteins, techniques based on detection of capsid proteins have long been used for their detection. Serological or immuno-diagnostic assays include enzyme-linked immunosorbent assays (ELISA), immunofluorescence, immunofluorescence colony staining, immuno-strip tests and techniques such as immune-capture electron microscopy and mass spectroscopy.

3.1 Disease diagnostics based on identifying characteristic proteins

Various forms of ELISA have been used widely, especially to detect plant virus diseases. Other methods of detection include immunofluorescence and flow cytometry. Matrix-assisted laser desorption/ionization-time of flight (MALDI-TOF) mass spectrometry can also be used to identify pests via their protein profiles. However, since the advent of molecular biology, molecular techniques are being applied increasingly for diagnosing diseases in both the animal and plant kingdoms.

3.1.1 Enzyme-linked immunosorbent assay

There are many variants to the ELISA technique, but in its simplest 'direct' form, an antigen or epitope is attached to the base of 96 multi-well plates and binds a specific antibody, which has been conjugated to a 'tag' enzyme such as horseradish peroxidase or alkaline phosphatase. After washing away unbound antibodies, a substrate of the bound enzyme is added which is converted to a coloured product, and which can be quantified by spectrometry in a plate reader. The detection can be made more sensitive using an indirect ELISA, in which a specific but un-conjugated antibody binds the antigen. A secondary conjugated antibody, designed to bind the host species of the first antibody, is added and binds the first antibody. Since more than one secondary antibody can bind the primary antibody, this amplifies the colour reaction. In sandwich ELISA, the well is coated with the capture antibody. The antigen or test sample extract is added and is bound by the immobilized antibody. A second conjugated antibody is added that binds to the captured antigen in the sample, followed by addition of the substrate to the conjugated enzyme to generate a colour. The main advantage of sandwich ELISA is that it is two to five times more sensitive, and more specific, than direct or indirect ELISAs. Samples are usually regarded as positive if the optical density of the coloured product is twice that of control wells.

ELISA has been used to detect plant viruses since the 1970s, but requires the availability of an appropriate antiserum or antibody. Its use has expanded with the availability of more than 800 antisera available to detect both plant viruses and other plant diseases. For example, ELISA can now be used to detect fungal diseases in cereals like common bunt of wheat (Eibel et al. 2005). The methodology used in ELISA assays has been streamlined for more rapid detection by the addition of magnetic beads or particles which reduce washing and centrifugation steps. There are also variations that enable ELISA to be used for in-field detection, such as Tissue Blot Immunoassays, which can be used to detect cereal viruses, for example, Barley yellow dwarf virus (Makkouk and Comeau 1994). Monoclonal antibodies, which are more specific than polyclonal antibodies, can discriminate between different virus serotypes, such as Cereal yellow dwarf virus-RPV and Barley yellow dwarf virus-MAV (Kumari et al. 2006). The use of lateral flow devices or 'dipsticks' enables rapid detection in the field. For dipsticks, a membrane containing a line of primary antibodies, and a positive control line, is dipped in an extract of plant material, rinsed and transferred to the substrate solution. Colour development at the antibody line can be seen in a few minutes or quantified using a handheld reflectometer.

Immunofluorescence has also been developed to detect the rust causing fungus *Puccinia triticina* in wheat (Gao et al. 2013). Infected samples are prepared as fixed slides and fluorescent dyes are applied which conjugate to the specific antibody, and target molecules can be identified by microscopy (Kennedy et al. 1999).

3.1.2 Serological test kits

Serological testing principles have been developed into test kits available for growers to do preliminary on-site tests. Diagnostic companies like AgDia, Adgen, BIOREBA, Loewe, Pocket Diagnostic and Sigma provide test kits for detection of viruses, bacteria and fungal pathogens in plants. The same principle combined with chromatography is employed in immunochromatographic strips which can detect mycotoxins like alternariol from plant samples (Kong et al. 2017). Commercial kits (RIDASCREEN® and RIDA®QUICK) based on enzyme immunoassays and immunochromatography are available for quantitative analysis of fumonisin which is a mycotoxin of *Fusarium* (R-Biopharm). Such kits can be used to detect contamination in cereal seeds and feed derived from cereals.

3.1.3 Mass spectroscopy

MALDI-TOF mass spectroscopy can also be used to identify plant pests and diseases. There are two main applications – protein profiling and protein mass fingerprinting. In protein profiling, proteins are extracted simply from a whole organism, such as a nematode or an aphid, or a bacterial culture, and the extract is mixed with a matrix and spotted onto a MALDI-TOF plate, which is then loaded into the mass spectrometer. A protein profile is generated which is characteristic of that organism, and in some cases can be used to identify different races of a pest, for example, oat and lucerne races of the stem nematode *Ditylenchus dipsaci* (Perera et al. 2009). It can also differentiate between different strains of bacteria associated with plant roots such as wheat (Stets et al. 2013). This technology is useful to identify disease complexes and mixtures of disease-causing pests, for example, wheat seed-gall nematode *Anguina tritici* even when present as a mixture with other nematodes and different species of aphids (Perera et al. 2005a,b).

Protein mass fingerprinting has been used to identify plant viruses. A simple one-dimensional gel of a symptomatic leaf can be used to identify the capsid protein, which can be eluted, digested with trypsin and a peptide mass fingerprint generated by MALDI-TOF MS that is diagnostic for a particular virus (Luo et al. 2010).

4 Nucleic-acid-based techniques

It is evident that nucleic-acid-based diagnostics, especially those methods based on the PCR, have revolutionized the field of diagnostics. Such PCR-based techniques rely on detection of pathogen-specific DNA or RNA sequences, with standard PCR, quantitative PCR and more recently 'digital droplet' PCR being used widely for identification of diseases.

Detection of plant disease using PCR techniques requires knowledge of pathogen gene sequences, most commonly of the 16S/18S ribosomal RNA genes or internal transcribed spacer (ITS) regions, which can be amplified using primers which bind flanking conserved regions.

The application of such techniques requires extraction of nucleic acids and amplification of target sequences. For lab-based detection methods, a wide range of kits and methods are available. For field-based detection of nucleic acids, extraction methods such as lateral flow devices, nitrocellulose membrane-based DNA extraction, Solid Phase Reversible Immobilization (SPRI), use of common filter pipette tips in conjunction with SPRI technology and magnetic SPRI bead-based extraction have been tested successfully (DeAngelis et al. 1995; Tomlinson et al. 2010a; Wee et al. 2015). In addition to being economically feasible and portable, sample processing time has been reduced to as little as 30 sec using a cellulose-based dipstick method (Zou et al. 2017). In some cases where PCR inhibitors are not limiting, a DNA extraction step can be avoided using assays known as 'BIO PCR' in which a small sample is directly added to the PCR reaction where cell lysis occurs during initial denaturing, making DNA available for the following PCR cycles. Various nucleic-acid-based studies used to detect cereal pathogens are summarized in Table 1.

4.1 PCR-based assays

Nucleic-acid-based assays have been developed based on intergenic spacer and ITS regions (of rDNA units), mitochondrial DNA, sequence-characterized amplified region (SCAR) markers or single copy genes involved in the biosynthesis of mycotoxins (Edwards et al. 2002; González-Jaén et al. 2004). Such assays can be designed to identify particular species, for example, *Tilletia horrida* in rice causing kernel smut (ITS region of DNA), *T. controversa* Kuhn causing dwarf bunt of wheat (SCAR-PCR), cereal and legume root lesion nematode *Pratylenchus thornei* (SCAR-PCR, Carrasco-Ballesteros et al. 2007; Liu et al. 2009; Chen et al. 2016). Conventional PCR has also been used to differentiate between fungal subspecies, for example, *Pyrenophora teres f. teres and P. teres f. maculate* causing barley net blotch (Lu et al. 2010). The ITS region has multiples copies, is highly conserved between species and has been widely used for disease detection and identification in cereals (White et al. 1990; Kochanova et al. 2004). A divergence of 3% or more in ITS sequences of fungal pathogens is considered to differentiate different species (O'Brien et al. 2005; Walker et al. 2008; Hughes et al. 2009). PCR assays based on the TEF1-α (Translational Elongation Factor 1α) in *Fusarium* species have successfully differentiated toxigenic *Fusarium* species (Sampietro et al. 2010). Reverse transcription PCR (RT-PCR) has been applied successfully to detect many plant viruses such as Barley yellow dwarf virus (Lacroix et al. 2016). Multiplex detection of more than one pathogen in a single PCR reaction has also been developed and can save time and costs. For example, multiplex PCR can be used to detect stem rust (*Puccinia graminis f. sp. Tritici*), leaf rust (*Puccinia triticina*) and powdery mildew (*Blumeria graminis f. sp. Tritici*, Chen et al. 2015). Other pathogens detectable in pre-symptomatic leaves by multiplex PCR assays are *Septoria tritici*, *Parastagonospora nodorum* (Berk), *Puccinia striiformis* West and *Puccinia recondita* Rob ex Desm (Fraaije et al. 2001).

The amplified product from a PCR reaction can be visualized after electrophoresis on agarose gels combined with an intercalating dye. DNA detection test strips™ can also detect the amplified product without the need for equipment and reagents required for gel electrophoresis (Knoll et al. 2002).

Table 1 Summary of some nucleic-acid-based pathogen detection studies in cereals

Nucleic-acid-based assay	Pathogen	Crop/Plant	Plant part/ Stage	References
BIO-PCR	*X. oryzae* pv. oryzae	Rice	Seed	Sakthivel et al. (2001)
Multiplex PCR-rDNA,TR16,FUM5	*Fusarium* spp.	Maize	Cornmeal	Bluhm et al. (2002)
BIO-PCR	*Acidovorax avenae ssp.* avenae	Rice	Seed	Song et al. (2004)
Multiplex PCR	*F. culmorum*, *F. graminearum* and *F. sporotrichioides*	Wheat	Seed	Demeke et al. (2005)
Conventional PCR	*Fusarium* spp.	Wheat, corn, barley, oat, rye	Seed	
Real-time PCR	*Mycosphaerella graminicola*	Wheat	Pre-symptomatic leaves	Guo et al. (2006)
Real-time PCR	*Pyrenophora teres*	Barley	Leaves	Leisova et al. (2006)
Conventional PCR	*Fusarium spp.*	Maize	Seed	Jurado et al. (2006)
Conventional PCR	*Septoria tritici*	Wheat	Seed	Consolo et al. (2009)
Multiplex PCR	Barley and cereal yellow dwarf viruses, Wheat spindle streak virus, Wheat streak mosaic virus and Soil-borne wheat mosaic virus	Oat	Leaf	Deb and Anderson (2008)
Taqman real-time PCR	*Tilletia controversa*	Wheat	Various plant growth stages	Yuan et al. (2009)
Multiplex real-time PCR	Wheat streak mosaic virus (WSMV) and Triticum mosaic virus (TriMV)	Wheat	Seedlings	Price et al. (2010)
Real-time Bio-PCR	*Burkholderia glumae*	Rice	Seed	Kim et al. (2011)
Real-time PCR	*X. oryzae* pv. oryzae, *X. oryzae* pv. oryzicola, *B. glumae*	Rice	Seed	Lu et al. (2014)
Multiplex PCR	*X. oryzae* pv. Oryzae + *X. oryzae* pv. Oryzicola + *B. glumae*			
Multiplex PCR	*P. graminis f. sp. tritici*, *P. triticina* and *B. graminis f. sp. tritici*	Wheat	Seedling leaves	Chen et al. (2015)
Multiplex real-time PCR	*Fusarium spp.*	Maize	Kernel	Preiser et al. (2015)
Multiplex PCR	*X. oryzae* pv. *Oryzae* + *B. glumae* + *A. avenae* subsp. *avenae*	Rice	Seed	Kang et al. (2016)
Multiplex PCR	Barley and Cereal yellow dwarf viruses, Wheat spindle streak virus, Wheat streak mosaic virus and Soil-borne wheat mosaic virus	Oat	Leaf	Deb and Anderson (2008)

4.2 Fluorescence-based PCR assays

Other nucleic-acid-based methods include the 'PicoGreen' assay, which uses a fluorometric microtitre plate with a cyanine dye that specifically binds to double-stranded DNA, and DNA levels as low as 10 pg can be detected. Real-time fluorescence-based PCR uses 'Taqman' probes to target a specific region within the target amplicon generated by forward and reverse primers as in a PCR (Fraaije et al. 1999). Fluorescence released from the probe bound to the amplicon during the reaction can be measured and quantified. Multiplexing can be done by using probes labelled with different fluorescent dyes (FAM, ROX, TET, TAM), for example a five-plex, real-time PCR assay successfully detected and distinguished *T. indica* and other *Tilletia* species *(T. walkeri, T. ehrhartae, T. horrida and a group comprising T. caries, T. laevis, T. controversa, T. bromi* and *T. fusca*) in wheat grains, and four different viruses simultaneously from potato (Mortimer-Jones et al. 2009; Tan et al. 2009).

Real-time PCR is very sensitive and specific, and diagnostic assays including TaqMan assays have been developed to detect diseases such as seed-borne rusts and soil-borne mosaic viruses in cereals, with sensitivities down to 0.65 pg/µL for stem rust of wheat (Ratti et al. 2004; McNeil et al. 2004; Liu et al. 2015). The use of digital droplet PCR has made this technique even more sensitive and quantifiable, and is the preferred method of detection due to high throughput, increasing data on pathogen genome sequences and fast processing of a large number of samples.

A stand-alone real-time microchip PCR system consisting of a PCR reaction chamber microchip with integrated thin-film heater, a compact fluorescence detector to detect amplified DNA, a microcontroller to control the entire thermocycling operation with data acquisition capability and a battery has also been developed to detect *Pseudomonas* and *Fusarium* species, demonstrating the portability of this technology (Koo et al. 2013). Nucleic-acid-based methods can not only identify primary plant pests, but also help to identify parasitoids which live inside the pest and help in the biological control of these pests, for example in cereal aphids (Ye et al. 2017a,b).

4.3 *In situ* hybridization

In situ hybridization (ISH) is a molecular detection technique which uses probes labelled with digoxigenin or biotin-conjugated nucleotides that bind to a known target sequence. Fluorescence labelled probes enhance the sensitivity of this technique (Fluorescence *In Situ* Hybridization, FISH) by sensitive detection of hybridization signal and better resolution of images. 'FISH' is used more for chromosome mapping and localization of gene expression, and has been employed to detect microorganisms. Rust causing pathogens in plant tissues can be detected by biotin-streptavidin ISH (Ellison et al. 2016). FISH can be used to detect virus presence and localization in plants such as Tomato yellow leaf curl virus (Kliot et al. 2014). It has also successfully detected brown rot causing bacteria from potato tissue by targeting 23 rRNA (Wullings et al. 1998).

5 Point-of-care diagnostics

To reduce the cost and equipment required for assays described above, point-of-care diagnostic assays are now being used which do not require sophisticated equipment and can be done rapidly and cheaply in the field (Yager et al. 2006): for example isothermal amplification of DNA in combination with lateral flow strips and portable fluorometers. This approach has been used successfully to detect pathogen DNA, although most research is on animal pathogens (Chow et al. 2008; Lutz et al. 2010; Mahalanabis et al. 2010; Rohrman and Richards-Kortum 2012).

5.1 Loop-mediated isothermal amplification

Loop-mediated isothermal amplification (LAMP) eliminates the need for a thermal cycler because it can be carried out at a constant temperature, and is sensitive enough to detect low amounts of input DNA. This makes it ideal for detection of plant pathogens in the field (Fukuta et al. 2014; Tomlinson et al. 2010b; Bhat et al. 2013). In wheat, this assay could detect wheat stripe rust pathogen *Puccinia striiformis* f.sp. *tritici* as early as 24 h after infection with as low as 2 pg/μL of template DNA (Huang et al. 2011). LAMP uses two long outer primers and two short inner primers that recognize six specific sequences in the target DNA. However, this is also a limitation, since suitable expertise is required both for primer design and assay optimization. It has been used to detect potato leafroll virus and *Fusarium oxysporum* using dyes for visual detection of DNA (Ahmadi et al. 2013; Almasi et al. 2013; Zhang et al. 2013a). When combined with ELISA, LAMP makes it possible to process hundreds of samples simultaneously within a few hours with the detection sensitivity as low as a single copy of target DNA (Lee et al. 2014; Ravan and Yazdanparast 2012). Reverse transcription LAMP can be used to detect viruses, for example, BYDV from oats and wheat (Zhao et al. 2010). This technology has been successfully tested in the field at the Centre for AgriBioscience (AgriBio) in Victoria, Australia. A LAMP-Biosensor technology variation integrates electrochemical sensors and LAMP technology which rapidly detects target DNA with the sensitivity of as low as ten copies of pathogen DNA (Hsieh et al. 2012).

5.2 Helicase-dependent amplification

Helicase-dependent amplification (HDA) assays are less complicated to design than LAMP assays since only two primers are required as for PCR, but it uses a DNA helicase to separate the two strands of DNA instead of temperature cycling. Primer annealing and extension is facilitated by the isothermal conditions, while single-stranded binding protein and MutL endonuclease are added to prevent rehybridization of complementary ssDNA. Amplicons can be detected within 60 min by this method (Vincent et al. 2004). A limitation of this assay is that optimization of enzyme activity (helicase and DNA polymerase) is harder for longer (>120 bp) targets (Guatelli et al. 1990). Although it has not yet been widely used for plant pathogen diagnostics, it has been used to identify *Phytophthora* and has been developed as a microfluidic chip which combines DNA extraction and HDA (Mahalanabis et al. 2010; Schwenkbier et al. 2015).

Isothermal amplification of DNA approaches have led to the development of other assays like rolling circle amplification (RCA) also known as the 'padlock probe' assay and Recombinase Polymerase Amplification (RPA). RCA generates long DNA fragments by strand displacement activity of DNA polymerase. It has high specificity and when combined with dot-blot hybridization can detect a pathogen in as low as 1 pg of plant DNA, for example *Xanthomonas oryzae* pv. oryzae and *X. oryzae* pv. Oryzicola in rice seed (Tian et al. 2014). RPA is probably considered the most sensitive isothermal technique because it can be combined with reverse transcriptase (RT-RPA), ELISA (RPA-ELISA) and gold nanoparticle electrochemical sensors. This makes it 10 000 times more sensitive than conventional PCR when used to detect the plant pathogen *Pseudomonas syringae* before disease symptoms appear (Lau et al. 2017). Detection of plant pathogens involving isothermal amplification of RNA based on transcription has also been developed, and is known as nucleic-acid-sequence-based amplification (Compton 1991). This method is useful for detection of single-stranded RNA viruses.

6 Next-generation sequencing for plant diagnostics

The nucleic-acid-based methods for plant disease detection discussed so far all rely on the availability of existing information on DNA or RNA sequences of pathogens. Sequence diversity of pathogens can limit such methods, and if there is no sequence information available, such methods cannot be used. However, rapid advances in sequencing technology and the development of next-generation sequencing (NGS) can solve these issues. NGS enables detection of previously unknown plant pathogens, or unknown variations in known pathogens (Rott et al. 2017). Sequencing a plant genome or transcriptome (RNA-Seq) also provides information on biochemical pathways and mechanisms of pathogen interactions (Luan et al. 2011; Zhang et al. 2013b). For example, RNA-seq was used to detect about 40 viruses in garlic cloves imported to Australia – some new to science and some not previously recorded in Australia (Wylie et al. 2014; Jo et al. 2017). Different sample preparation methods, sequencing technologies and analysis approaches have been developed for specific requirements (reviewed by Boonham et al. 2014). A comparison of workflows for different forms of virus diagnostics is provided in Fig. 1.

7 Non-invasive imaging techniques

With the development of novel imaging techniques, non-invasive methods are being developed to identify diseases using various forms of spectral or hyperspectral analysis. Reflection, transmission and absorption of light occur when light interacts with plants to generate characteristic profiles, and these profiles are modified by the presence of plant pathogens and host resistance responses (Thomas et al. 2017). Depending on specific equipment available, parameters such as leaf chlorophyll content, transpiration rate and leaf surface temperature can be measured. Hyperspectral analysis of plant stresses is well developed, and applications now include the identification of plant diseases. Such imaging can help determine disease spread across regions or fields and the severity of disease spread. These data can also be combined with other disease predicting models for disease forecasting.

Visible and near-infrared wavelengths can discriminate leaf pigments, and diseases which develop symptoms on a leaf can be detected early based on differences in chlorophyll content (Moshou et al. 2005). Powdery mildew in barley can be detected by hyperspectral reflectance and tomato mosaic tobamovirus symptoms in *Nicotiana debneyi* can be detected as early as 10 days after inoculation when symptom development is not visible (Polischuk et al. 1997). Experiments on sugar beet suggest that spectral indices can be used to differentiate between diseases such as leaf spot, rust and powdery mildew (Mahlein et al. 2013). In wheat, various studies have exploited spectral reflectance to detect head blight, leaf rust and stem rust (Bravo et al. 2003; Bauriegel et al. 2010; Ashourloo et al. 2014).

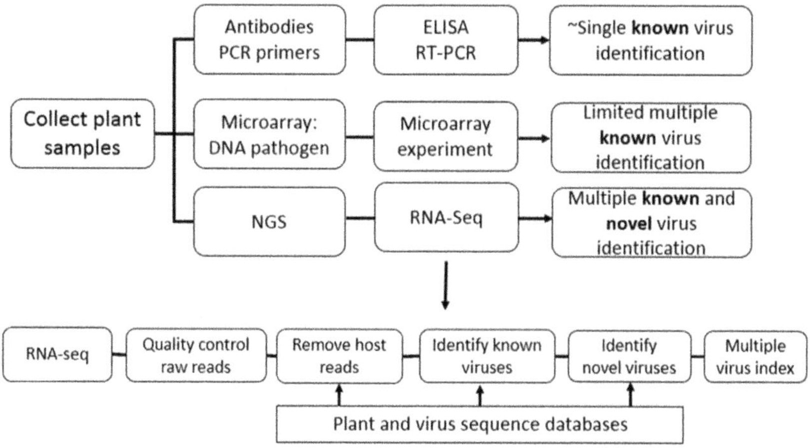

Figure 1 A comparison of flow diagrams for procedures followed to detect plant viruses using ELISA, RT-PCR, microarrays and next-generation sequencing. Source: Modified from Boonham et al. (2014).

Thermal imaging, based on differences in leaf surface temperature, is also a tool used for disease detection. Stomatal conductance is affected by diseases such as downy mildew and can be quantified by thermal imaging (Oerke et al. 2006). More advanced hyperspectral imaging cameras facilitate data collection in three dimensions, with X- and Y- axes for spatial and Z- for spectral data and can be used for plant phenotyping, plant health and disease identification such as rice panicle blast in large-scale agriculture (Kobayashi et al. 2001; Ustin et al. 2009; Bock et al. 2010). Spectral imaging using drones can be effective to detect and map diseases over large cropping areas, where scouting is not feasible and pathogen-infected patches of the field may remain unidentified until severe symptoms have developed.

7.1 Plantix smart crop assistant

In this age of androids and artificial intelligence, mobile phone technology is also being applied to detect plant diseases, for example in the form of a crop advisory App called 'Plantix'. This uses image recognition technology, and by comparing the image uploaded by a grower with images in its information library, it can suggest solutions to the problem depending on the type of pathogen or nutrient deficiency detected. This type of App can also connect growers and extension workers to communicate, and to be aware of disease warnings relevant to their crops (https://plantix.net). The use of smart phones and mobile apps for disease detection, presence or expected presence is set to expand rapidly.

8 Conclusions

Increased trade, intensive air travel and a changing climate pose increasing biosecurity risks for introduction of new pests and diseases for cereals. The recent introduction of the Russian wheat aphid into Australia is one such example. This calls for accurate and rapid diagnostics

Table 2 A comparison of some methods used for pathogen diagnostics

Method	Advantage	Disadvantage
ELISA (enzyme-linked immunosorbent assay)	Sensitive and specific	Requires antibodies specific to antigens present in each pathogen
Serological test kits	Easy-to-use compact testing equipment	Expensive for large-scale use. Not available for many pathogens
Mass Spectrometry (protein profiling)	Highly sensitive and rapid. Can differentiate between races of a pest	A reference profile is required for identification
PCR (polymerase chain reaction)	Highly sensitive and specific, multiplexing can be done	Nucleic acid extraction required in most cases. Prior knowledge of diagnostic pathogen sequences needed to design primers. Sequencing required to identify new species. Equipment required for thermal cycling
Real-time PCR	More sensitive (0.65 pg/µL) than conventional PCR, quantitative data can be generated; can be multiplexed	Nucleic acid extraction required. Assay optimization needed with prior knowledge of pathogen sequences. More expensive reagents and equipment may be needed
LAMP (loop-mediated isothermal amplification)	Isothermal method (no expensive equipment required). Highly specific and sensitive (single copy of target DNA can be detected)	Complicated primer design requires more experienced personnel. Assay specific to a pathogen
HAD (helicase-dependent amplification)	Isothermal method, simple primer design	Enzymes (helicase and endonuclease) required, optimization difficult for longer (>120 bp) targets
RCA (rolling circle amplification)	Very sensitive (1 pg DNA). Can be combined with other techniques such as dot-blot hybridization, ELISA and electrochemical sensors	An additional denaturation step is required before isothermal amplification
NGS (next-generation sequencing)	No prior knowledge about the pathogen is required, new pathogens can be discovered	Expensive sample processing, data analysis can be lengthy

so that the best measures to control, monitor and manage new incursions can be deployed rapidly. A comparison of various methods of disease diagnostics is presented in Table 2.

9 Where to look for further information

There are some recent reviews describing the latest trends in molecular diagnostics for plant pathogens including:

Ray, M., Ray, A., Dash, S., Mishra, A., Achary, K. G., Nayak, S. and Singh, S. 2017. Fungal disease detection in plants: Traditional assays, novel diagnostic techniques and biosensors. *Biosensors and Bioelectronics*, 87, 708–23. doi:10.1016/j.bios.2016.09.032.

Donoso, A. and Valenzuela, S. 2018. In-field molecular diagnosis of plant pathogens: recent trends and future perspectives. *Plant Pathology*. doi:10.1111/ppa.12859.

Various agriculture departments in Australia and around the world provide testing services and issue warnings for pathogens actively damaging crops in an area. For example, the Department of Economic Development, Jobs, Transport and Resources in Victoria, Australia and Diagnostic Laboratory Services in Western Australia, the Plant and Pest Diagnostic Clinic at the University of Nebraska Lincoln and UMass Extension Plant Diagnostic Lab in the USA.

http://agriculture.vic.gov.au/agriculture/pests-diseases-and-weeds/diagnostic-services
https://www.agric.wa.gov.au/bacteria/ddls-plant-pathology-services
https://cropwatch.unl.edu/plantdisease/unl-diagnostic-clinic-lincoln
https://ag.umass.edu/services/plant-diagnostics-laboratory

These departments also issue disease guides every year to report on occurrence of disease and pests. More information about disease testing services in Australia can be found at the Grains Research and Development Corporation (Field Crop Diseases) website:

https://extensionaus.com.au/field-crop-diseases/
https://extensionaus.com.au/field-crop-diseases/disease-testing-services-australia/

10 References

Ahmadi, S., Almasi, M. A., Fatehi, F., Struik, P. C. and Moradi, A. 2013. Visual detection of *Potato leafroll virus* by one-step reverse transcription loop-mediated isothermal amplification of DNA with hydroxynaphthol blue dye. *Journal of Phytopathology*, 161, 120–4.

Almasi, M. A., Erfan Manesh, M., Jafary, H. and Dehabadi, S. M. H. 2013. Visual detection of *Potato leafroll virus* by loop-mediated isothermal amplification of DNA with the GeneFinder™ dye. *Journal of Virological Methods*, 192, 51–4.

Ashourloo, D., Mobasheri, M. and Huete, A. 2014. Developing two spectral disease indices for detection of wheat leaf rust (*Puccinia triticina*). *Remote Sensing*, 6, 4723.

Bauriegel, E., Giebel, A. and Herppich, W. 2010. Rapid *Fusarium* head blight detection on winter wheat ears using chlorophyll fluorescence imaging. *Journal of Applied Botany and Food Quality*, 83(2), 196–203.

Bhat, A. I., Siljo, A. and Deeshma, K. P. 2013. Rapid detection of *Piper yellow mottle virus* and *Cucumber mosaic virus* infecting black pepper (*Piper nigrum*) by loop-mediated isothermal amplification (LAMP). *Journal of Virological Methods*, 193, 190–6.

Bluhm, B. H., Cousin, M. A. and Woloshuk, C. P. 2004. Multiplex real-time PCR detection of fumonisin-producing and trichothecene-producing groups of *Fusarium* species. *Journal of Food Protection*, 67, 536–43.

Bock, C. H., Poole, G. H., Parker, P. E. and Gottwald, T. R. 2010. Plant disease severity estimated visually, by digital photography and image analysis, and by hyperspectral imaging. *Critical Reviews in Plant Sciences*, 29, 59–107.

Boonham, N., Kreuze, J., Winter, S., Van Der Vlugt, R., Bergervoet, J., Tomlinson, J. and Mumford, R. 2014. Methods in virus diagnostics: From ELISA to next generation sequencing. *Virus Research*, 186, 20–31.

Bravo, C., Moshou, D., West, J., McCartney, A. and Ramon, H. 2003. Early disease detection in wheat fields using spectral reflectance. *Biosystems Engineering*, 84, 137–45.

Carrasco-Ballesteros, S., Castillo, P., Adams, B. J. and Pérez-Artés, E. 2007. Identification of *Pratylenchus thornei*, the cereal and legume root-lesion nematode, based on SCAR-PCR and satellite DNA. *European Journal of Plant Pathology*, 118, 115–25.

Chen, S., Cao, Y.-Y., Li, T.-Y. and Wu, X.-X. 2015. Simultaneous detection of three wheat pathogenic fungal species by multiplex PCR. *Phytoparasitica*, 43, 449–60.

Chen, Y., Yang, X., Yao, J., Kyaw, E. P., Zhang, A.-F., Li, Y.-F., Gu, C.-Y., Zang, H.-Y. and Gao, T.-C. 2016. Simple and rapid detection of *Tilletia horrida* causing rice kernel smut in rice seeds. *Scientific Reports*, 6, 33258.

Chow, W. H. A., McCloskey, C., Tong, Y., Hu, L., You, Q., Kelly, C. P., Kong, H., Tang, Y.-W. and Tang, W. 2008. Application of isothermal helicase-dependent amplification with a disposable detection device in a simple sensitive stool test for toxigenic *Clostridium difficile*. *The Journal of Molecular Diagnostics: JMD*, 10, 452–8.

Compton, J. 1991. Nucleic acid sequence-based amplification. *Nature*, 350, 91.

Consolo, V. F., Albani, C. M., Berón, C. M., Salerno, G. L. and Cordo, C. A. 2009. A conventional PCR technique to detect *Septoria tritici* in wheat seeds. *Australasian Plant Pathology*, 38, 222–7.

DeAngelis, M. M., Wang, D. G. and Hawkins, T. L. 1995. Solid-phase reversible immobilization for the isolation of PCR products. *Nucleic Acids Research*, 23, 4742–3.

Deb, M. and Anderson, J. M. 2008. Development of a multiplexed PCR detection method for Barley and *Cereal yellow dwarf viruses*, *Wheat spindle streak virus*, *Wheat streak mosaic virus* and *Soil-borne wheat mosaic virus*. *Journal of Virological Methods*, 148, 17–24.

Demeke, T., Clear, R. M., Patrick, S. K. and Gaba, D. 2005. Species-specific PCR-based assays for the detection of *Fusarium* species and a comparison with the whole seed agar plate method and trichothecene analysis. *International Journal of Food Microbiology*, 103, 271–84.

Edwards, S. G., O'Callaghan, J. and Dobson, A. D. W. 2002. PCR-based detection and quantification of mycotoxigenic fungi. *Mycological Research*, 106, 1005–25.

Eibel, P., Wolf, G. A. and Koch, E. 2005. Detection of *Tilletia caries*, causal agent of common bunt of wheat, by ELISA and PCR. *Journal of Phytopathology*, 153, 297–306.

Ellison, M. A., McMahon, M. B., Bonde, M. R., Palmer, C. L. and Luster, D. G. 2016. *In situ* hybridization for the detection of rust fungi in paraffin embedded plant tissue sections. *Plant Methods*, 12, 37.

Falacy, J. S., Grove, G. G., Mahaffee, W. F., Galloway, H., Glawe, D. A., Larsen, R. C. and Vandemark, G. J. 2007. Detection of *Erysiphe necator* in air samples using the polymerase chain reaction and species-specific primers. *Phytopathology*, 97, 1290–7.

Fraaije, B. A., Lovell, D. J., Rohel, E. A. and Hollomon, D. W. 1999. Rapid detection and diagnosis of *Septoria tritici* epidemics in wheat using a polymerase chain reaction/PicoGreen assay. *Journal of Applied Microbiology*, 86, 701–8.

Fraaije, B. A., Lovell, D. J., Coelho, J. M., Baldwin, S. and Hollomon, D. W. 2001. PCR-based assays to assess wheat varietal resistance to blotch (*Septoria tritici* and *Stagonospora nodorum*) and rust (*Puccinia striiformis* and *Puccinia recondita*) diseases. *European Journal of Plant Pathology*, 107, 905–17.

Fukuta, S., Takahashi, R., Kuroyanagi, S., Ishiguro, Y., Miyake, N., Nagai, H., Suzuki, H., Tsuji, T., Hashizume, F., Watanabe, H. and Kageyama, K. 2014. Development of loop-mediated isothermal amplification assay for the detection of *Pythium myriotylum*. *Letters in Applied Microbiology*, 59, 49–57.

Gao, L., Chen, W., Liu, T. and Liu, B. 2013. An immunofluorescence assay for the detection of wheat rust species using monoclonal antibody against urediniospores of *Puccinia triticina*. *Journal of Applied Microbiology*, 115, 1023–8.

González-Jaén, M. T., Mirete, S., Patiño, B., López-Errasquín, E. and Vázquez, C. 2004. Genetic markers for the analysis of variability and for production of specific diagnostic sequences in fumonisin-producing strains of *Fusarium verticillioides*. *European Journal of Plant Pathology*, 110, 525–32.

Guatelli, J. C., Whitfield, K. M., Kwoh, D. Y., Barringer, K. J., Richman, D. D. and Gingeras, T. R. 1990. Isothermal, *in vitro* amplification of nucleic acids by a multienzyme reaction modeled after retroviral replication. *Proceedings of the National Academy of Sciences of the United States of America*, 87, 1874–8.

Guo, J.-R., Schnieder, F. and Verreet, J.-A. 2006. Presymptomatic and quantitative detection of *Mycosphaerella graminicola* development in wheat using a real-time PCR assay. *FEMS Microbiology Letters*, 262, 223–9.

Hsieh, K., Patterson, A. S., Ferguson, B. S., Plaxco, K. W. and Soh, H. T. 2012. Rapid, sensitive, and quantitative detection of pathogenic DNA at the point of care through microfluidic electrochemical quantitative loop-mediated isothermal amplification. *Angewandte Chemie*, 124, 4980–4.

Huang, C., Sun, Z., Yan, J., Luo, Y., Wang, H. and Ma, Z. 2011. Rapid and precise detection of latent infections of wheat stripe rust in wheat leaves using loop-mediated isothermal amplification. *Journal of Phytopathology*, 159, 582–4.

Hughes, K. W., Petersen, R. H. and Lickey, E. B. 2009. Using heterozygosity to estimate a percentage DNA sequence similarity for environmental species' delimitation across basidiomycete fungi. *New Phytologist*, 182, 795–8.

Jo, Y., Choi, H., Kim, S.-M., Kim, S.-L., Lee, B. C. and Cho, W. K. 2017. The pepper virome: Natural co-infection of diverse viruses and their quasispecies. *BMC Genomics*, 18, 453.

Jones, S., Baizan-Edge, A., MacFarlane, S. and Torrance, L. (2017). Viral diagnostics in plants using next generation sequencing: Computational analysis in practice. *Frontiers in Plant Science*, 8, 1770. doi:10.3389/fpls.2017.01770.

Jurado, M., Vázquez, C., Marín, S., Sanchis, V. and Teresa González-Jaén, M. 2006. PCR-based strategy to detect contamination with mycotoxigenic *Fusarium* species in maize. *Systematic and Applied Microbiology*, 29, 681–9.

Kang, I. J., Kang, M.-H., Noh, T.-H., Shim, H. K., Shin, D. B. and Heu, S. 2016. Simultaneous detection of three bacterial seed-borne diseases in rice using multiplex polymerase chain reaction. *The Plant Pathology Journal*, 32, 575–9.

Kennedy, Wakeham and Cullington 1999. Production and immunodetection of ascospores of *Mycosphaerella brassicicola*: Ringspot of vegetable crucifers. *Plant Pathology*, 48, 297–307.

Kim, B. K., Cho, M. S., Kim, M. H., Choi, H. J., Kang, M. J., Shim, H. S., Ahn, T.-Y., Kim, J. and Park, D. S. 2011. Rapid and specific detection of *Burkholderia glumae* in rice seed by real-time Bio-PCR using species-specific primers based on an *rhs* family gene. *Plant Disease*, 96, 577–80.

Kliot, A., Kontsedalov, S., Lebedev, G., Brumin, M., Cathrin, P. B., Marubayashi, J. M., Skaljac, M., Belausov, E., Czosnek, H. and Ghanim, M. 2014. Fluorescence *in situ* Hybridizations (FISH) for the localization of viruses and endosymbiotic bacteria in plant and insect tissues. *Journal of Visualized Experiments: JoVE* (84), 51030.

Knoll, S., Vogel, R. F. and Niessen, L. 2002. Identification of *Fusarium graminearum* in cereal samples by DNA detection Test Strips™. *Letters in Applied Microbiology*, 34, 144–8.

Kobayashi, T., Kanda, E., Kitada, K., Ishiguro, K. and Torigoe, Y. 2001. Detection of rice panicle blast with multispectral radiometer and the potential of using airborne multispectral scanners. *Phytopathology*, 91, 316–23.

Kochanova, M., Zouhar, M., Prokinova, E. and Rysanek, P. 2004. Detection of *Tilletia controversa* and *Tilletia caries* in wheat by PCR method. *Plant Soil Environment*, 50, 75–7.

Kong, D., Xie, Z., Liu, L., Song, S., Zheng, Q. and Kuang, H. 2017. Development of an immunochromatographic assay for the detection of alternariol in cereal and fruit juice samples. *Food and Agricultural Immunology*, 28, 1082–93.

Koo, C., Malapi-Wight, M., Kim, H. S., Cifci, O. S., Vaughn-Diaz, V. L., Ma, B., Kim, S., Abdel-Raziq, H., Ong, K., Jo, Y.-K., Gross, D. C., Shim, W.-B. and Han, A. 2013. Development of a real-time microchip PCR system for portable plant disease diagnosis. *PLoS ONE*, 8, e82704.

Kumari, S. G., Muharram, I., Makkouk, K. M., Al-Ansi, A., El-Pasha, R., Al-Motwkel, W. A. and Kassem, A. H. 2006. Identification of viral diseases affecting barley and bread wheat crops in Yemen. *Australasian Plant Pathology*, 35, 563–8.

Lacroix, C., Renner, K., Cole, E., Seabloom, E. W., Borer, E. T. and Malmstrom, C. M. 2016. Methodological guidelines for accurate detection of viruses in wild plant species. *Applied and Environmental Microbiology*, 82(6), 1966–75.

Lau, H. Y., Wu, H., Wee, E. J. H., Trau, M., Wang, Y. and Botella, J. R. 2017. Specific and sensitive isothermal electrochemical biosensor for plant pathogen DNA detection with colloidal gold nanoparticles as probes. *Scientific Reports*, 7, 38896.

Lee, M.-F., Chen, J.-Y. and Peng, C.-F. 2014. Study on the LAMP-PCR-hybridization–thermal melt–ELISA method for molecular detection of multidrug resistance in *Mycobacterium tuberculosis* isolates. *Biomarkers and Genomic Medicine*, 6, 126–32.

Leisova, L., Minarikova, V., Kucera, L. and Ovesna, J. 2006. Quantification of *Pyrenophora teres* in infected barley leaves using real-time PCR. *Journal of Microbiological Methods*, 67, 446–55.

Liu, J. H., Gao, L., Liu, T. G. and Chen, W. Q. 2009. Development of a sequence-characterized amplified region marker for diagnosis of dwarf bunt of wheat and detection of *Tilletia controversa* Kühn. *Letters in Applied Microbiology*, 49, 235–40.

Liu, M., McCabe, E., Chapados, J. T., Carey, J., Wilson, S. K., Tropiano, R., Redhead, S. A., Lévesque, C. A. and Hambleton, S. 2015. Detection and identification of selected cereal rust pathogens by TaqMan® real-time PCR. *Canadian Journal of Plant Pathology*, 37, 92–105.

Lu, S., Platz, G. J., Edwards, M. C. and Friesen, T. L. 2010. Mating type locus-specific polymerase chain reaction markers for differentiation of *Pyrenophora teres* f. *teres* and *P. teres* f. *maculata*, the causal agents of barley net blotch. *Phytopathology*, 100, 1298–306.

Lu, W., Pan, L., Zhao, H., Jia, Y., Wang, Y., Yu, X. and Wang, X. 2014. Molecular detection of *Xanthomonas oryzae* pv. *oryzae*, *Xanthomonas oryzae* pv. *oryzicola*, and *Burkholderia glumae* in infected rice seeds and leaves. *The Crop Journal*, 2, 398–406.

Luan, J.-B., Li, J.-M., Varela, N., Wang, Y.-L., Li, F.-F., Bao, Y.-Y., Zhang, C.-X., Liu, S.-S. and Wang, X.-W. 2011. Global analysis of the transcriptional response of whitefly to tomato yellow leaf curl china virus reveals the relationship of coevolved adaptations. *Journal of Virology*, 85, 3330–40.

Luo, H., Wylie, S. J. and Jones, M. G. K. 2010. Identification of plant viruses using one-dimensional gel electrophoresis and peptide mass fingerprints. *Journal of Virological Methods*, 165, 297–301.

Lutz, S., Weber, P., Focke, M., Faltin, B., Hoffmann, J., Muller, C., Mark, D., Roth, G., Munday, P., Armes, N., Piepenburg, O., Zengerle, R. and Von Stetten, F. 2010. Microfluidic lab-on-a-foil for nucleic acid analysis based on isothermal recombinase polymerase amplification (RPA). *Lab on a Chip*, 10, 887–93.

Mahalanabis, M., Do, J., Almuayad, H., Zhang, J. Y. and Klapperich, C. M. 2010. An integrated disposable device for DNA extraction and helicase dependent amplification. *Biomedical Microdevices*, 12, 353–9.

Mahlein, A. K., Rumpf, T., Welke, P., Dehne, H. W., Plümer, L., Steiner, U. and Oerke, E. C. 2013. Development of spectral indices for detecting and identifying plant diseases. *Remote Sensing of Environment*, 128, 21–30.

Makkouk, K. M. and Comeau, A. 1994. Evaluation of various methods for the detection of *Barley yellow dwarf virus* by the tissue-blot immunoassay and its use for virus detection in cereals inoculated at different growth stages. *European Journal of Plant Pathology*, 100, 71–80.

McNeil, M., Roberts, A. M. I., Cockerell, V. and Mulholland, V. 2004. Real-time PCR assay for quantification of *Tilletia caries* contamination of UK wheat seed. *Plant Pathology*, 53, 741–50.

Mortimer-Jones, S. M., Jones, M. G. K., Jones, R. A. C., Thomson, G. and Dwyer, G. I. 2009. A single tube, quantitative real-time RT-PCR assay that detects four potato viruses simultaneously. *Journal of Virological Methods*, 161, 289–96.

Moshou, D., Bravo, C., Oberti, R., West, J., Bodria, L., McCartney, A. and Ramon, H. 2005. Plant disease detection based on data fusion of hyper-spectral and multi-spectral fluorescence imaging using Kohonen maps. *Real-Time Imaging*, 11, 75–83.

O'Brien, H. E., Parrent, J. L., Jackson, J. A., Moncalvo, J.-M. and Vilgalys, R. 2005. Fungal community analysis by large-scale sequencing of environmental samples. *Applied and Environmental Microbiology*, 71, 5544–50.

Oerke, E. C., Steiner, U., Dehne, H. W. and Lindenthal, M. 2006. Thermal imaging of cucumber leaves affected by downy mildew and environmental conditions. *Journal of Experimental Botany*, 57, 2121–32.

Perera, M. R., Vanstone, V. A. and Jones, M. G. K. 2005a. A novel approach to identify plant parasitic nematodes using matrix-assisted laser desorption/ionization time-of-flight mass spectrometry. *Rapid Communications in Mass Spectrometry*, 19, 1454–60.

Perera, M. R., Vargas, R. D. F. and Jones, M. G. K. 2005b. Identification of aphid species using protein profiling and matrix-assisted laser desorption/ionization time-of-flight mass spectrometry. *Entomologia Experimentalis et Applicata*, 117, 243–7.

Perera, M., Taylor, S., Vanstone, V. and Jones, M. 2009. Protein biomarkers to distinguish oat and lucerne races of the stem nematode, *Ditylenchus dipsaci*, with quarantine significance for Western Australia. *Nematology*, 11, 555–63.

Polischuk, V. P., Shadchina, T. M., Kompanetz, T. I., Budzanivskaya, I. G., Boyko, A. L. and Sozinov, A. A. 1997. Changes in reflectance spectrum characteristic of Nicotiana debneyi plant under the influence of viral infection. *Archives of Phytopathology and Plant Protection*, 31, 115–19.

Preiser, V., Goetsch, D., Sulyok, M., Krska, R., Mach, R. L., Farnleitner, A. and Brunner, K. 2015. The development of a multiplex real-time PCR to quantify *Fusarium* DNA of trichothecene and fumonisin producing strains in maize. *Analytical Methods*, 7, 1358–65.

Price, J. A., Smith, J., Simmons, A., Fellers, J. and Rush, C. M. 2010. Multiplex real-time RT-PCR for detection of *Wheat streak mosaic virus* and *Triticum mosaic virus*. *Journal of Virological Methods*, 165, 198–201.

Pujari, J. D., Yakkundimath, R. and Byadgi, A. S. 2015. Image processing based detection of fungal diseases in plants. *Procedia Computer Science*, 46, 1802–8.

Ratti, C., Budge, G., Ward, L., Clover, G., Rubies-Autonell, C. and Henry, C. 2004. Detection and relative quantitation of *Soil-borne cereal mosaic virus* (SBCMV) and *Polymyxa graminis* in winter wheat using real-time PCR (TaqMan®). *Journal of Virological Methods*, 122, 95–103.

Ravan, H. and Yazdanparast, R. 2012. Development and evaluation of a loop-mediated isothermal amplification method in conjunction with an enzyme-linked immunosorbent assay for specific detection of *Salmonella* serogroup D. *Analytica Chimica Acta*, 733, 64–70.

Rohrman, B. A. and Richards-Kortum, R. R. 2012. A paper and plastic device for performing recombinase polymerase amplification of HIV DNA. *Lab on a chip*, 12, 3082–8.

Rott, M., Xiang, Y., Boyes, I., Belton, M., Saeed, H., Kesanakurti, P., Hayes, S., Lawrence, T., Birch, C., Bhagwat, B. and Rast, H. 2017. Application of next generation sequencing for diagnostic testing of tree fruit viruses and viroids. *Plant Disease*, 101, 1489–99.

Sakthivel, N., Mortensen, C. and Mathur, S. 2001. Detection of *Xanthomonas oryzae* pv. *oryzae* in artificially inoculated and naturally infected rice seeds and plants by molecular techniques. *Applied Microbiology and Biotechnology*, 56, 435–41.

Sampietro, D. A., Marín, P., Iglesias, J., Presello, D. A., Vattuone, M. A., Catalan, C. A. N. and Gonzalez Jaen, M. T. 2010. A molecular based strategy for rapid diagnosis of toxigenic *Fusarium* species associated to cereal grains from Argentina. *Fungal Biology*, 114, 74–81.

Schwenkbier, L., Pollok, S., Konig, S., Urban, M., Werres, S., Cialla-May, D., Weber, K. and Popp, J. 2015. Towards on-site testing of *Phytophthora* species. *Analytical Methods*, 7, 211–17.

Song, W. Y., Kim, H. M., Hwang, C. Y. and Schaad, N. W. 2004. Detection of *Acidovorax avenae* ssp. *avenae* in rice seeds using BIO-PCR. *Journal of Phytopathology*, 152, 667–76.

Stets, M. I., Pinto, A. S., Huergo, L. F., De Souza, E. M., Guimarães, V. F., Alves, A. C., Steffens, M. B. R., Monteiro, R. A., Pedrosa, F. D. O. and Cruz, L. M. 2013. Rapid identification of bacterial isolates from wheat roots by high resolution whole cell MALDI-TOF MS analysis. *Journal of Biotechnology*, 165, 167–74.

Tan, M.-K., Ghalayini, A., Sharma, I., Yi, J., Shivas, R., Priest, M. and Wright, D. 2009. A one-tube fluorescent assay for the quarantine detection and identification of *Tilletia indica* and other grass bunts in wheat. *Australasian Plant Pathology*, 38, 101–9.

Thomas, S., Wahabzada, M., Kuska, M. T., Rascher, U. and Mahlein, A.-K. 2017. Observation of plant–pathogen interaction by simultaneous hyperspectral imaging reflection and transmission measurements. *Functional Plant Biology*, 44, 23–34.

Tian, Y., Zhao, Y., Xu, R., Liu, F., Hu, B. and Walcott, R. R. 2014. Simultaneous detection of *Xanthomonas oryzae* pv. *oryzae* and *X. oryzae* pv. *oryzicola* in rice seed using a padlock probe-based assay. *Phytopathology*, 104, 1130–7.

Tomlinson, J. A., Boonham, N. and Dickinson, M. 2010a. Development and evaluation of a one-hour DNA extraction and loop-mediated isothermal amplification assay for rapid detection of phytoplasmas. *Plant Pathology*, 59, 465–71.

Tomlinson, J. A., Dickinson, M. J. and Boonham, N. 2010b. Detection of *Botrytis cinerea* by loop-mediated isothermal amplification. *Letters in Applied Microbiology*, 51, 650–7.

Ustin, S. L., Gitelson, A. A., Jacquemoud, S., Schaepman, M., Asner, G. P., Gamon, J. A. and Zarco-Tejada, P. 2009. Retrieval of foliar information about plant pigment systems from high resolution spectroscopy. *Remote Sensing of Environment*, 113, S67–S77.

Vincent, M., Xu, Y. and Kong, H. 2004. Helicase-dependent isothermal DNA amplification. *EMBO Reports*, 5, 795–800.

Walker, J. F., Miller J. R., O. K. and Horton, J. L. 2008. Seasonal dynamics of ectomycorrhizal fungus assemblages on oak seedlings in the southeastern Appalachian Mountains. *Mycorrhiza*, 18, 123–32.

Walsh, K., Korimbocus, J., Boonham, N., Jennings, P. and Hims, M. 2005. Using real-time PCR to discriminate and quantify the closely related wheat pathogens *Oculimacula yallundae* and *Oculimacula acuformis*. *Journal of Phytopathology*, 153, 715–21.

Wee, E. J. H., Lau, H. Y., Botella, J. R. and Trau, M. 2015. Re-purposing bridging flocculation for on-site, rapid, qualitative DNA detection in resource-poor settings. *Chemical Communications*, 51, 5828–31.

West, J. S., Atkins, S. D., Emberlin, J. and Fitt, B. D. L. 2008. PCR to predict risk of airborne disease. *Trends in Microbiology*, 16, 380–7.

White, Bruns, T., Lee, S. and Taylor, J. 1990. Amplification and direct sequencing of fungal ribosomal RNA Genes for phylogenetics. In: Innis, M. A., D. H. Gelfand, J. J. Sninsky and T. J. White (Eds), *PCR Protocols: A Guide to Methods and Applications*. Academic Press, Inc, New York, NY, pp. 315–22.

Wullings, B. A., Van Beuningen, A. R., Janse, J. D. and Akkermans, A. D. L. 1998. Detection of *Ralstonia solanacearum*, which causes brown rot of potato, by fluorescent *in situ* hybridization with 23S rRNA-targeted probes. *Applied and Environmental Microbiology*, 64, 4546–54.

Wylie, S. J., Li, H., Saqib, M. and Jones, M. G. K. 2014. The global trade in fresh produce and the vagility of plant viruses: A case study in garlic. *PLoS ONE*, 9, e105044.

Yager, P., Edwards, T., Fu, E., Helton, K., Nelson, K., Tam, M. R. and Weigl, B. H. 2006. Microfluidic diagnostic technologies for global public health. *Nature*, 442, 412.

Ye, Z., Vollhardt, I. M. G., Girtler, S., Wallinger, C., Tomanovic, Z. and Traugott, M. 2017a. An effective molecular approach for assessing cereal aphid-parasitoid-endosymbiont networks. *Scientific Reports*, 7, 3138.

Ye, Z., Vollhardt, I. M. G., Tomanovic, Z. and Traugott, M. 2017b. Evaluation of three molecular markers for identification of European primary parasitoids of cereal aphids and their hyperparasitoids. *PLoS ONE*, 12, e0177376.

Yuan, Q., Nian, S., Yin, Y., Li, M., Cai, J. and Wang, Z. 2009. Development of a PCR-based diagnostic tool specific to wheat dwarf bunt, caused by *Tilletia* controversa. *European Journal of Plant Pathology*, 124, 585–94.

Zhang, X., Zhang, H., Pu, J., Qi, Y., Yu, Q., Xie, Y. and Peng, J. 2013a. Development of a real-time fluorescence loop-mediated isothermal amplification assay for rapid and quantitative detection of *Fusarium oxysporum* f. sp. *cubense* tropical race 4 in soil. *PLoS ONE*, 8, e82841.

Zhang, Z., Zhang, P., Li, W., Zhang, J., Huang, F., Yang, J., Bei, Y. and Lu, Y. 2013b. De novo transcriptome sequencing in *Frankliniella occidentalis* to identify genes involved in plant virus transmission and insecticide resistance. *Genomics*, 101, 296–305.

Zhao, K., Liu, Y. and Wang, X. 2010. Reverse transcription loop-mediated isothermal amplification of DNA for detection of Barley yellow dwarf viruses in China. *Journal of Virological Methods*, 169, 211–14.

Zou, Y., Mason, M. G., Wang, Y., Wee, E., Turni, C., Blackall, P. J., Trau, M. and Botella, J. R. 2017. Nucleic acid purification from plants, animals and microbes in under 30 seconds. *PLoS Biology*, 15, e2003916.

Chapter 15

Natural antifungal compounds for the control of diseases in wheat and other cereals

A. Schouten, Wageningen University, The Netherlands

1 Introduction

As terrestrial plants are sessile, they reside in the same habitat during their entire life span. As a consequence, they have to challenge emerging adverse biotic and abiotic conditions. In coping with biotic stress situations, like herbivores, pathogens or other competing plants, they rely heavily on an array of secondary metabolites, that is, metabolites that are not considered essential for basic growth (Fraenkel, 1959). Over the years, such defence compounds have been defined as phytoalexins, phytoanticipins (VanEtten et al., 1994), allelochemicals or even semiochemicals (Dewhirst and Pickett, 2010), depending on the focal point of the described research. Such annotations may, however, be prone to confusion, as the definition may alter over time or be biased because the individual compounds have, yet undiscovered, broader activities, thus providing additional biological and ecological roles.

The first issue is particularly true for the term 'allelopathy', which was initially defined by Molisch in 1937 as the biochemical queues released by one plant that prevents the development of neighbouring plants (Molisch, 1937). Over the years, the definition altered and involved negative chemical interactions among organisms (Whittaker and Feeny, 1971), which was subsequently further extended by including the positive interactions among organisms (Rice, 1984). In 1996, the International Allelopathy Society suggested

http://dx.doi.org/10.19103/AS.2018.0039.22

to define allelopathy as 'any process involving the secondary metabolites produced by plants, microorganisms, viruses, and fungi that influence the growth and development of agricultural and biological systems (excluding animals), including positive and negative effects' (Torres et al., 1996; Willis, 2007). Nowadays, scientists gradually move away from this rather general definition back to the classical definition of allelopathy as postulated by Molisch (Willis, 2007). In this review, allelopathy is used in the classical sense.

The second issue primarily concerns the definitions of phytoanticipins and phytoalexins, as they suggest that particular defence metabolites are constitutively present inside the plants or *de novo* synthesized upon pathogen attack, respectively (VanEtten et al., 1994). Yet, such differences in timing of accumulation may depend entirely on the plant species, which may even be further influenced by the interaction with particular non-pathogenic microorganisms. For example, sakuranetin is considered a phytoalexin in rice and, as it is constitutively present, a phytoanticipin in blackcurrants, and maize roots accumulate 2,4-dihydroxy-7-methoxy-1,4-benzoxazin-3-one (DIMBOA) by interacting with the arbuscular mycorrhizal symbiont *Glomus mosseae* (Song et al., 2011). In this background, the distinction made between phytoalexin and phytoanticipin would therefore be rather trivial (Meyer et al., 2016).

The latter example also illustrates that plants do not form a discrete operational entity when dealing with adverse biotic situations, but can be affected by the various (beneficial) microbes present in their environment, leading to disease resistance and a better overall performance when attacked. And it is becoming increasingly evident that the non-pathogenic microbial population residing in the phyllosphere, the bulk soil and, particularly, the rhizosphere is a crucial element in biotic and abiotic stress management by plants.

This chapter therefore not only focuses on the antifungal defence compounds that can be synthesized by wheat and other cereals themselves, but also emphasizes the influence of the microbial community in the rhizosphere and phyllosphere on the accumulation and conversion of plant-derived defence secondary metabolites, the additional antimicrobial compounds produced by the associated microorganisms and the counter defences mounted by plant pathogenic fungi towards several of these natural antifungal compounds.

2 Plant defence compounds

In their defence towards fungi, cereals may employ various constitutively expressed or *de-novo*-synthesized secondary metabolites (Meyer et al., 2016): alkaloids, phenolic compounds, like hydroxybenzoic and hydroxycinnamic acids, tannins, proanthocyanidins, phenolic lipids, (prenylated) flavonoids and coumarins (Cushnie and Lamb, 2005, 2011). Due to the various derivatives that can be formed, each of which may even have a different set of biological activities; it is, however, often not clear what the contribution of the individual compound is in plant defence. The best studied cereal on this matter, barley, can produce a myriad of secondary metabolites, some of which have shown antifungal activity like gramine, against powdery mildew (*Blumeria graminis* f. sp. *hordei*), hordatine A and B against powdery mildew and foot rot (*Helminthosporium sativus*), putresceine against powdery mildew, spermidine against powdery mildew and brown rust (*Puccinia hordei*), spermine against powdery mildew and p-coumaroylhydroxyagmatine against powdery mildew (Kremer and Ben-Hammouda, 2009). For other cereals, including wheat, the characterization of antifungal compounds is rather limited (Ahuja et al., 2012). Like wild barley species, wheat can synthesize benzoxazinoids, which individually can have varying biological activities, including antifungal activity. More recently, serotonins with antifungal activity were discovered in wheat. Wheat also accumulates relatively high amounts of ferulic acid, a strong antioxidant, with antifungal activity.

2.1 Benzoxazinoids

The best studied group of bioactive compounds in Poaceae is the rather unique family of benzoxazinoids, which were reported first in 1955 (Virtanen and Hietala, 1955). Over the past decades, a significant number of papers, already calculated at exceeding 550 in 2009 (Niemeyer, 2009), have been published on the biosynthesis, the complex chemistry, the accumulation and the biological role and activity of the benzoxazinoids, which have subsequently been carefully reviewed (Niemeyer, 1988, 2009; Makowska et al., 2015; Wouters et al., 2016; Schulz et al., 2013; Du Fall and Solomon, 2011). The biosynthesis pathway has been found in maize (*Zea mays*), wheat (*Triticum aestivum*), rye (*Secale cereale*) and the wild barley, *Hordeum lechleri*, although not in oat, rice and sorghum, and, remarkably, cultivated barley (Niemeyer, 2009; Kremer and Ben-Hammouda, 2009). Benzoxazinoids evolve from indole-3-glycerol phosphate, which is initially converted into indole in the chloroplast by the enzyme BX1 (Fig. 1). After translocation into the endoplasmic reticulum, the indole is further processed in three successive steps into 2-hydroxy-2-1,4-benzoxazin-3-one (HBOA) through the sequential activity of different monooxygenases, BX2, BX3, BX4 and BX5. In the cytoplasm, DIBOA is glucosylated (DIBOA-Glc) by means of a glucosyltransferase, BX8/9, after which it can be further derivatized, for example, by the dioxygenases, BX6 and the O-methyltransferase, BX7, into DIMBOA-Glc (Niemeyer, 2009; Makowska et al., 2015; Wouters et al., 2016). The active derivatives can be divided into three classes: hydroxamic acids, lactams and methyl derivatives (Niemeyer, 2009). When glucosylated, they are considered biologically inactive and stored in the vacuole. Upon tissue damage, specific glucosidases, located in the plastids, convert the benzoxazinoid glucosides into the bioactive form (Niemeyer, 2009; Wouters et al., 2016).

In the plant, benzoxazinoid accumulation can vary with the cultivar, the developmental stage of the plant and the location within the plant (leaves, roots), and can be further influenced by temperature, water and nitrogen stress, photoperiod, light intensity, UV radiation and the interaction with (beneficial) microbes (Niemeyer, 2009; Song et al., 2011; Neal et al., 2012). The plant hormones, jasmonic acid and methyl jasmonate, increased the accumulation of HDMBOA-glucoside in the leaves of maize (Oikawa et al., 2001). In wheat, both jasmonate and methyl jasmonate application also increased the DIMBOA concentration in seedlings (Slesak et al., 2001) and roots (Zbeng et al., 2008), and *cis*-jasmone application increased the DIMBOA concentration in the aerial parts (Bassioli Moraes et al., 2008).

With respect to biological role, they are best studied for their allelochemical activity, preventing other, competing, plants (weeds) from developing (Bertin et al., 2003; Schulz et al., 2013; Barnes and Putnam, 1987; Pérez and Ormeño-Núñez, 1993; Niemeyer, 1988, 2009). However, decades ago, they were also found to be associated with the detoxification of herbicides (Shimabukuro, 1967; Ioannou et al., 1980) and the protection against insect herbivores (Klun and Brindley, 1966; Klun and Robinson, 1969; Guthrie et al., 1986; Thackray et al., 1990; Long et al., 1978; Niemeyer, 1988; Gianoli and Niemeyer, 1998), plant pathogenic bacteria (Hartman et al., 1975; Corcuera et al., 1978; Niemeyer,

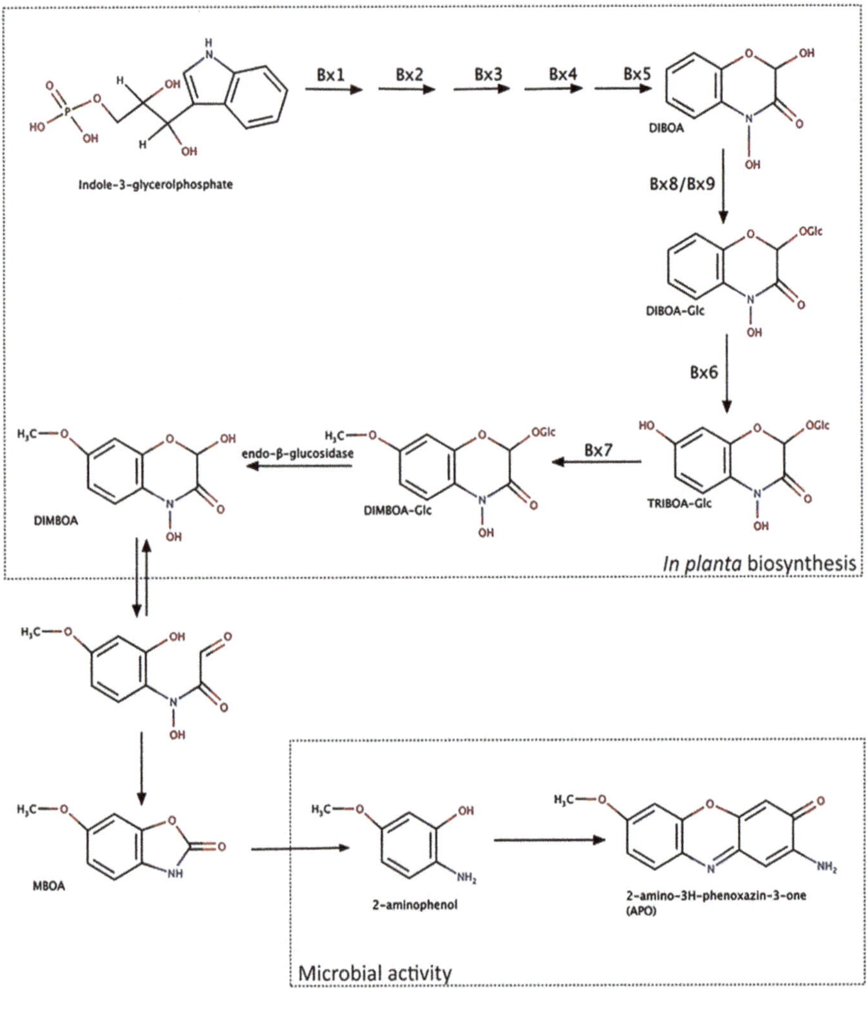

Figure 1 General pathway for the biosynthesis of the antifungal metabolites DIBOA and DIMBOA in the plant, MBOA and, through microbial activity, APO (modified from Niemeyer, 2009; Wouters et al., 2016).

1988) and fungi (Virtanen and Hietala, 1955; Wahlroos and Virtanen, 1959; Couture et al., 1971; Baker and Smith, 1977; Søltoft et al., 2008; Long et al., 1978). Despite all the research, insight into the role of benzoxazinoid hydroxamic acids is still rather fragmented. This is caused by: (1) the various benzoxazinoids that can be synthesized by the various species of Poaceae and their different activities, together with the bioactive derivatives that can evolve from each of them; (2) the differential *in planta* accumulation of benzoxazinoid hydroxamic acids in place and time; (3) the induction of additional defence mechanisms (ISR and/or SAR) or metabolites; (4) the role of (beneficial) microorganisms in benzoxazinoid metabolization and (5) the ability of pathogenic fungi or fungal strains to deal with benzoxazinoid hydroxamic acids, for example, by further converting them.

2.1.1 The antifungal activities of benzoxazinoids and their derivatives in wheat, maize and rye

Generally, DIMBOA and DIBOA are the most common benzoxazinoids encountered in cereals. And, although the ratios and concentrations may vary, the most abundant are considered DIMBOA in wheat and maize extracts, followed by DIBOA for wheat and DIBOA and DIM_2BOA for maize. In rye extracts, next to DIBOA, DIMBOA may be present (Niemeyer, 2009). However, Søltoft and colleagues (Søltoft et al., 2008) also reported the biosynthesis of the methyl derivatives, HMBOA, HDMBOA, and their respective glucosides, in several Danish wheat cultivars, with the concentration of HMBOA-glc generally being higher than that of DIMBOA-glc.

Direct antifungal activity of the benzoxazinoids was reported towards *Helminthosporium turcicum* (Couture et al., 1971), *Parastagonospora nodorum* (Baker and Smith, 1977; Du Fall and Solomon, 2013), *Microdochium nivale* (currently *Fusarium nivale*) (Virtanen and Hietala, 1955; Wahlroos and Virtanen, 1959) and *Rhizoctonia solani* (Song et al., 2011). When testing individual benzoxazinoids *in vitro*, DIBOA proved to be the most potent growth inhibitor of *Gaeumannomyces graminis* var. *tritici* (*Ggt*) strains, followed by DIMBOA, whereas their respective conversion products, BOA and MBOA, were far less potent (Wilkes et al., 1999). The exact method by which the benzoxazinoids affect fungi is still rather unresolved. Studies with *Candida albicans* showed a higher antifungal activity of hydroxamic acids when compared to lactams (Bravo and Lazo, 1996). The more lipophilic the benzoxazinoid is, the higher its antifungal activity, suggesting that membrane transfer plays an important role in the antifungal activity (Bravo et al., 1997; Wouters et al., 2016). Nevertheless, the *in planta* fungal resistance has in several cases been associated with the accumulation of particular benzoxazinoids (Niemeyer, 1988; Søltoft et al., 2008). An in-depth principal component analysis (PCA) revealed positive correlations between the resistance to the causal agents of *Fusarium* head blight (FHB) disease, by applying a mixture of *Fusarium avenaceum*, *Fusarium culmorum*, *Fusarium graminearum* (teleomorph *Gibberella zeae*) and *Microdochium nivale*, and the concentrations of the benzoxazinoids DIMBOA-glc, HMBOA-glc, HMBOA, DIBOA-glc, HBOA-glc and DIM2BOA-glc (Søltoft et al., 2008). Next to direct antifungal activity, benzoxazinoids can additionally provide indirect resistance towards fungi, as their accumulation also can trigger other defence mechanisms or, in a more latent way, prime the plant for defence (Ahmad et al., 2011; Song et al., 2011).

2.1.2 The relevance of benzoxazinoids in plant resistance: fungal defences and microbial antagonism

Both BOA and MBOA can be converted by microbial activity in the soil into the more stable aminophenoxazines, APO (Fig. 1) and AMPO, respectively (Niemeyer, 2009). APO was shown to be toxic to *Fusarium verticillioides*, although this fungus can convert BOA, via the transition into the 2-aminophenol intermediate, into, for the fungus, less toxic N-(2-hydroxyphenyl) malonamic acid (HPMA) and 2-acetamidophenol (BOA-X), thus preventing APO from being synthesized (Fig. 2) (Glenn et al., 2003). Remarkably, in turn, the

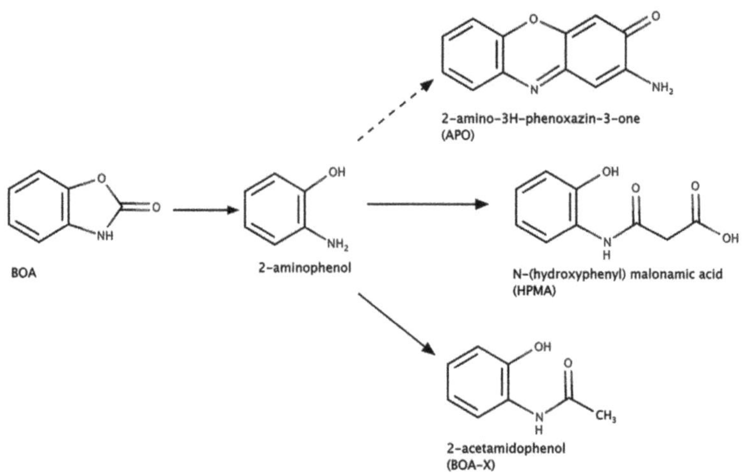

Figure 2 *Fusarium verticillioides* can convert BOA, via the transition into the 2-aminophenol intermediate, into HPMA and BOA-X. This derivatization prevents APO, which is toxic for *F. verticillioides*, from being synthesized (modified from Glenn et al., 2003).

bacterium *Bacillus mojavensis*, which is not sensitive to APO, can interfere with this fungal detoxification process by channelling BOA, through the 2-aminophenol intermediate, into the production of APO, thus controlling *Fusarium verticillioides* proliferation in maize and, subsequently, the accumulation of mycotoxins (Bacon et al., 2007).

This example illustrates the extreme complexity of determining the relevance of benzoxazinoids or their derivatives as effective antifungal compounds. And it does not stand on its own. *Fusarium moniliforme* and *F. subglutinans* isolates from maize, too, could detoxify BOA and MBOA into HPMA and N-(2-hydroxy-4-methoxyphenyl) malonamic acid (HMPMA), respectively (Richardson and Bacon, 1995; Yue et al., 1998; Vilich et al., 1999). *G. graminis* var. *graminis* (*Ggg*) isolates could convert both BOA and MBOA in the same way. *Ggt* could only convert BOA, although less efficiently than *Ggg*, whereas *G. graminis* var. *avenae* (*Gga*) could convert neither BOA nor MBOA. The ability to detoxify BOA and MBOA corroborated with the level of disease symptoms on various wheat cultivars, but not in oats (Friebe et al., 1998), which does not produce benzoxazinoids. This suggests that the ability to detoxify benzoxazinoids determines host range, like what was found for *Gga*, which can detoxify the saponin, avenacin A1, thus being able to infect oats (Bowyer et al., 1995). However, *F. culmorum* could only convert BOA, and this did not seem to correlate with the severity of disease symptoms on the roots (Friebe et al., 1998). And, although all tested *F. moniliforme* isolates from maize could convert BOA and MBOA, an *F. moniliforme* isolate from rice could not, but still cause disease (Richardson and Bacon, 1995).

2.2 Serotonins

Relatively recent, additional compounds, also synthesized from indole-3-glycerol phosphate, with antimicrobial activity were described for wheat (Du Fall and Solomon, 2013). The accumulation of serotonin (also known as 5-hydroxytryptamine or 5-HT) and its derivatives N-feruloyltryptamine and N-feruloylserotonin (Fig. 3) was triggered by the toxin and effector, ToxA, which can be produced by *Phaeosphaeria nodorum* and *Pyrenophora tritici-repentis*. Serotonin was shown to interfere with sporulation of *Phaeosphaeria nodorum* (Du Fall and Solomon, 2013) and *Bipolaris sorokiniana* (Ishihara et al., 2017). However, for *Phaeosphaeria nodorum*, serotonin exposure seemed to lead to an increase in accumulation of the mycotoxin altenariol (Du Fall and Solomon, 2013). Through the action of other effectors, for example, SnTox3, the SnToxA-induced serotonin defence pathway in wheat may be suppressed again (Du Fall and Solomon, 2013), thus illustrating the ability of a pathogen to subsequently outmanoeuvre plant defence strategies.

2.3 Ferulic acid

Ferulic acid (4-hydroxy-3-methoxycinnamic acid) (Fig. 4) can be synthetized, like all hydroxycinnamic acid derivatives, from the amino acid phenylalanine. It is relatively abundantly present in plants, particularly in cereals, although it is rarely found in a free state and predominantly conjugated with mono-, oligo- and polysaccharides, polyamines and lipids. For the plant, ferulic acid is considered important for its antioxidative and

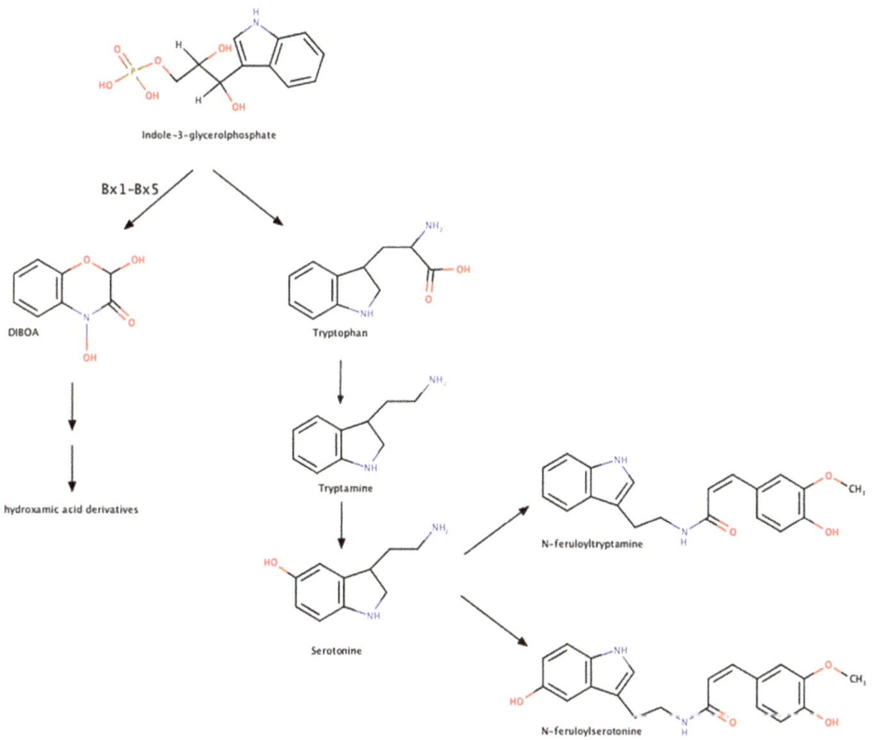

Figure 3 The biosynthesis of serotonin and its derivatives in wheat. Like the biosynthesis of benzoxazinoids, indole-3-glycerolphosphate is the key starting molecule, but this is now converted into tryptophan in the initial step [modified from Du Fall and Solomon, 2013, and molecule structures according to the PubChem database (https://pubchem.ncbi.nlm.nih.gov)].

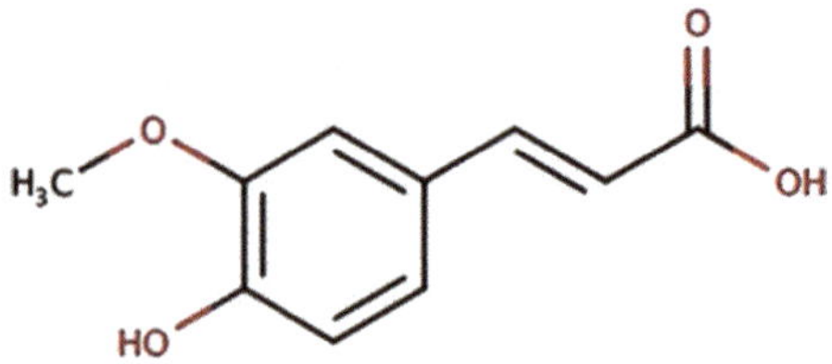

Figure 4 The chemical structure of ferulic acid.

antimicrobial activity (Ou and Kwok, 2004). Next to antibacterial activity, *in vitro* antifungal activity was demonstrated against *Sclerotinia sclerotiorum*, *Fusarium oxysporum*, *Alternaria* sp., *Botrytis cinerea* and *Penicillium digitatum* (Lattanzio et al., 1994; Ou and Kwok, 2004). Seed treatment of two spring wheat cultivars with chitosan promoted germination and the synthesis of phenolic acids, especially ferulic acid, and lignin in the primary leaf. The treatment moderated seed-borne *F. graminearum* infection and also repressed spreading of the fungus to the primary roots of the germinating seedlings (Reddy et al., 1999). Based on *in vitro* studies with several wheat cultivars, the accumulation of ferulic acid synthesis in the period from anthesis until approximately 20 days after anthesis seemed to correlate with resistance to *Fusarium* (McKeehen et al., 1999). This suggestion was recently more or less confirmed, as wheat genotypes with the highest ferulic acid contents in wheat spikes were the most resistant towards FHB disease in a field situation (Martin et al., 2017). *In vitro*, ferulic acid was shown to be an efficient inhibitor, in a dose-dependent manner, of the mycotoxin type B trichothecene biosynthesis, which include deoxynivalenol (DON) and its acetylated forms 3-acetyl-4-deoxynivalenol and 15-acetyl-4-deoxynivalenol (3- and 15-ADON), and nivalenol (NIV) and its acetylated form 4-acetylnivalenol or fusarenone X (FX), in *F. culmorum* and *F. graminearum* strains, by reducing *Tri*-gene transcription (Boutigny et al., 2009).

3 Outsourcing defences: recruiting soil microbes

Other than the above-mentioned compounds, it seems that wheat has lost, through domestication, or has always lacked the ability to produce antimicrobial agents, like avenacin A1, avenacoside B, gramine, hordenine, scopoletin, sorgoleone, momilactone B or geranic acid, that can be found in other cereal crops (Belz, 2007; Du Fall and Solomon, 2011), to defend itself against fungal diseases. However, particularly in the rhizosphere, wheat seems to have mastered to manipulate the microbial community for organizing the protection. Such outsourcing of defences against pathogens by means of recruiting antibiotic-producing microorganisms and stimulating the production of relevant levels of antibiotics may be a very effective and cost-efficient strategy for the protection against fungal plant pathogens.

Outsourcing of defences by wheat is a classical and well-studied example of biological disease control. The knowledge on this was initiated in the 1960s by the observation that in some important growing areas, like the Inland Pacific Northwest of the United States, the Netherlands and the United Kingdom, wheat and, also, barley could be successfully cultivated as a monoculture for many decades without showing the detrimental effects of the ascomycete root pathogen *G. graminis* var. *tritici*, the causal agent of take-all disease (Gerlagh, 1968; Shipton, 1972; Baker and Cook, 1974). Typically, take-all could only inflict serious biological and economic damage in the first three to five years after initiating a wheat or barley monoculture in a field, and, additionally, the take-all disease could reappear when this monoculture growing strategy was interrupted, for example, by fallow or crop rotation (Baker and Cook, 1974; Cook et al., 1995; Cook, 2007). Already early on, it was recognized that the soil played an important role in this take-all decline (TAD) effect, changing from disease conducive to disease suppressive over the growing seasons. Disease-suppressive soils could be transferred to another, disease-conducive, field, thus providing an accelerated development of suppressiveness in the conducive field (Baker and Cook, 1974; Cook et al., 1995; Cook, 2007). Soils lost their disease suppressive abilities after pasteurization, indicating a microbial factor in the TAD event (Shipton, 1975; Cook and Rovira, 1976; Raaijmakers and Weller, 1998). Indeed, the uninterrupted wheat cultivation caused the accumulation of antibiotic-producing pseudomonads in the rhizosphere. A similar observation was made for the reduction of Rhizoctonia bare patch disease, caused by the basidiomycete *R. solani*, anastomosis group 8 (AG-8), in wheat: A continuous culture of wheat caused, after an initial increase, a decline in disease incidence to undetectable levels (Roget, 1995). This suppressiveness could be transferred to non-suppressive soils and removed by radioactive irradiation or pasteurization (Wiseman et al., 1996). The microorganism(s) and mechanism(s) involved in the Rhizoctonia bare patch decline were not identified, but the involvement of *Trichoderma* spp. was suggested. However, bacteria also may play a role. When applied as a seed inoculation in wheat, a *Bacillus* sp., strain L324-92, could suppress root diseases caused by, *G. graminis* var. *tritici*, *R. solani* AG-8 and several *Pythium* species (Kim et al., 1997). The mode of action by which the L324-92 repressed wheat root diseases has thus far not been resolved. However, recently, cyclic lipopetides, produced by *Bacillus amyloliquefaciens* S76-3 isolated from wheat spikes, showed antagonistic activity towards *F. graminearum* (Gong et al., 2015). In another study, phenazine-producing recombinant derivatives from *Pseudomonas fluorescens* strain Q8r1-96 were significantly more effective against *R. solani* AG-8 root rot than the parental strain (Huang et al., 2004).

Not only in the rhizosphere but also in the phyllosphere, bacteria may be important allies in providing resistance. In a greenhouse study, inoculation of spikes with *Pseudomonas chlororaphis* subsp. *aurantiaca* Pa40 showed wheat sharp eyespot disease (*Rhizoctonia cerealis*) suppression, which was similar to a chemical treatment with validamycin A (Jiao et al., 2013). Molecular studies indicated that Pa40 contained genes coding for enzymes involved in the biosynthesis of pyrrolnitrin and 2-hydroxylated phenazine.

In all, there is strong evidence that wheat has developed a remedy against fungal diseases through facilitating the accumulation of particular antibiotic-producing rhizosphere bacteria that antagonize fungal root pathogens. The extensive research on the phenomena in wheat has led to the insight that, depending on the growing area, individual soil-borne *Pseudomonas* species or their isolates can mediate TAD by producing different compounds with antifungal activity, of which the unrelated phloroglucinol, phenazines, pyoluteorin, pyrrolnitrin and cyclic lipopeptides (CLPs) (Fig. 5) are currently the best described. And, in its wake, the research also showed that several of these compounds, like phenazine-1-carboxylic acid, pyrrolnitrin and CLPs, have antifungal activities against other pathogens, both in the rhizosphere and phyllosphere. These compounds, their biosynthesis and antifungal activity will now be discussed in more detail.

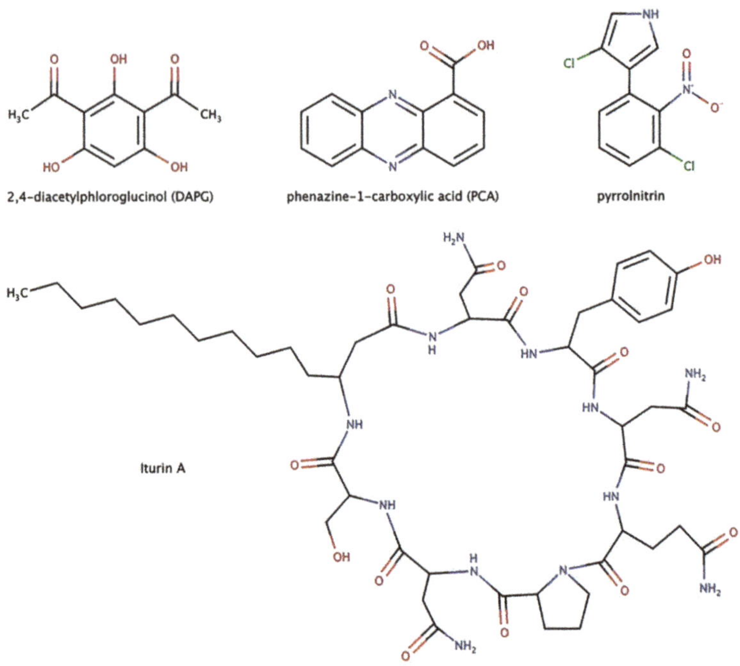

Figure 5 The chemical structures of the bacterial antibiotics 2,4-diacetylphloroglucinol (DAPG), phenazine-1-carboxylic acid (PCA), pyrrolnitrin and the cyclic lipopeptide, iturin A, which are involved in antifungal defence in wheat. The structure of iturin A is modified from Gong et al., 2015.

3.1 Phenazine-1-carboxylic acid (PCA)

Phenazine-1-carboxylic acid (PCA) is one of the naturally occurring phenazines, currently comprising a family of about 150 heterocyclic-nitrogen-containing compounds, which are synthesized by several gram-positive and gram-negative bacteria, of which *Pseudomonas* and *Streptomyces* are the best known (Blankenfeldt and Parsons, 2014; McDonald et al., 2001; Mavrodi et al., 2006; Chin-A-Woeng et al., 2003; Turner and Messenger, 1986). Phenazines are synthesized from chorismate, provided by the shikimate pathway, through the synthesis of various precursors by several enzymes, consecutively being annotated as PhzE, PhzD, PhzF, PhzB and PhzG, together with several non-catalysed steps (Blankenfeldt and Parsons, 2014). In this process, through condensation, PHzB merges two 6-amino-5-oxocyclohex-2-ene-1-carboxylic acid (AOCHC) molecules head-to-head into hexahydrophenazine-1,6-dicarboxylic acid (HHPDC) (Blankenfeldt and Parsons, 2014). Another enzyme, PhzA, is approximately 70% identical to PhzB, but it was not shown to be involved in the AOCHC condensation assay (Ahuja et al., 2008). However, although currently not clarified, PhzA seems to play a role in phenazine biosynthesis by pseudomonads, as the absence of PhzA reduced the PCA production to 25% (McDonald et al., 2001).

Most phenazines have antimicrobial activity towards fungi and both gram-positive and gram-negative bacteria (Toohey et al., 1965) due to their redox activity (Laursen and Nielsen, 2004; Price-Whelan et al., 2006). Phenazine accumulation is considered an important means to suppress take-all by *Pseudomonas fluorescens* 2-79, which produces PCA, and *Pseudomonas aureofaciens* 30-84, which produces PCA, 2-hydroxyphenazine-1-carboxylic acid, and 2-hydroxyphenazine (Thomashow and Weller, 1988; Pierson and Thomashow, 1992). Several wheat-colonizing *Pseudomonas fluorescens* and *Pseudomonas chlororaphis* strains produce PCA, and it was shown that PCA-deficient transposon mutants were unable to inhibit mycelial growth of *Ggt* and to suppress take-all disease, thus implicating PCA as the active compound in TAD. Next to *Ggt*, *R. solani*, *Cochliobolus sativus* and several species of the oomycete *Pythium* (*Pythium aristosporum*, *Pythium heterothallicum* and *Pythium volutum*) proved to be highly sensitive to PCA, with complete mycelial growth inhibition at 1 µg/ml *in vitro* (Gurusiddaiah et al., 1986).

3.2 2,4-Diacetylphloroglucinol (DAPG)

Extensive studies in Washington State and the Netherlands showed that TAD evolved because of the accumulation of *Pseudomonas fluorescens* strains, like CHA0, Q8r1-96 (currently designated as *Pseudomonas brassicacearum*), SSB17 and Q2-87, all producing 2,4-diacetylphloroglucinol (DAPG) instead of PCA. In the producing *Pseudomonas* species, the DAPG biosynthesis locus contains six genes, of which *phlA*, *phlB*, *phlC* and *phlD* code for the biosynthetic enzymes, *phlE* codes for a transporter protein and *phlF* codes for a repressor protein. The three enzymes, encoded by *phlA*, *phlC* and *phlB*, function collectively and appear to have a dual function in DAPG biosynthesis: (1) the condensation reaction of, probably, acetoacetyl-CoA into and 8-carbon polyketide and (2) the acetylation of the monoacetylphloroglucinol into DAPG. The enzyme, encoded by *phlD*, converts the 8-carbon polyketide into MAPG (Bangera and Thomashow, 1999). All six genes are globally conserved among the DAPG-producing pseudomonads encountered in soils (Weller et al., 2007).

Microorganisms show a variation in their sensitivity to DAPG (Keel et al., 1992; Bangera and Thomashow, 1999; de Souza et al., 2003; Isnansetyo et al., 2003; Schouten et al., 2004; Kamei and Isnansetyo, 2003). *Ggt* is considered highly sensitive to DAPG, with an EC_{90} averaging 5 µg/ml among different isolates (Kwak et al., 2009), whereas *Rhizoctonia solani* was rather insensitive to DAPG (Keel et al., 1992). *Pythium* species varied significantly in their sensitivity to DAPG, with minimal inhibitory concentration (MIC) values ranging from 8 µg/ml for *Pythium volutum*, isolated from wheat, to 72 µg/ml for *Pythium ultimum*, isolated from lily and tulip (de Souza et al., 2003). *Pythium deliense*, isolated from cucumber, was virtually insensitive to DAPG with a MIC value greater than 300 µg/ml.

3.3 Cyclic lipopeptides

CLPs are complex molecules comprised of a fatty acid chain attached to a cyclic peptide chain. The peptide chains are synthesized by several large multifunctional enzymes, known as nonribosomal peptide synthetases (NRPS), each of which can catalyse one to five sequential steps in the biosynthesis of a particular CLP, including fusing the lipid moiety to the first peptide, peptide elongation and peptide cyclization (Gewolb, 2002; Sieber and Marahiel, 2005; Raaijmakers et al., 2010). The structural variety of the CLP that can be synthesized is virtually infinite, as they can be composed of a varying number and arrangement of amino acids, in D-or L-configuration, which sometimes even are chemically modified, a fatty acid moiety that can vary in length and composition, and varying cyclization arrangements (Raaijmakers et al., 2010). The resulting structural variations suggest different functions and for the best studied *Pseudomonas* and *Bacillus* species synthesizing them, CLPs are associated with biofilm formation (attachment), motility and competition with coexisting plants (virulence), nematodes and microbes, including bacteria, fungi, oomycetes and protozoa (Andersen et al., 2003; Raaijmakers et al., 2010).

A viscosin-like CLP isolated from *Pseudomonas fluorescens* strain HC1-07, isolated from wheat-growing areas in Hebei province, China, showed antifungal activity against *Ggt* and *R. solani AG-8*. Mutagenesis studies showed that the CLP *viscB* biosynthesis gene mutant not only lost the ability to swarm, swim and form a biofilm, but also exhibited a reduction in efficacy to control take-all and Rhizoctonia root rot. Subsequent *viscB* gene complementation in this mutant fully restored the biological control activity again (Yang et al., 2014).

More recently, a *Bacillus amyloliquefaciens* strain, S76-3, was isolated from *F. graminearum*-infected wheat spikes of the FHB susceptible wheat cultivar Annong 8455. Three different CLPs, iturin A (Fig. 5), plipastatin A and surfactin, were resolved from this isolate, each belonging to a different class of CLP. In *in vitro* assays, iturin A and plipastatin A killed *F. graminearum* conidia at concentrations of 50 and 100 μg/ml, respectively, and caused disruption of cell walls and membrane structures of conidia and hyphae, resulting in distortions of the hyphal structures and branching (Gong et al., 2015).

3.4 Pyrrolnitrin

Pyrrolnitrin was found to be synthesized by several *Pseudomonas* species, *Burkholderia cepacia*, *Myxococcus fulvus*, *Corallococcus exiguus*, *Cystobacter ferrugineus* and *Enterobacter agglomerans* (reviewed by van Pée and Ligon, 2000) and, more recently, *Acinetobacter haemolyticus* (Mujumdar et al., 2014). For *Pseudomonas fluorescens* isolate BL914, a cluster of four genes were identified, coding for PrnA, PrnB, PrnC and PrnD. These enzymes convert the tryptophan precursor into pyrrolnitrin (Hammer et al., 1997; van Pée and Ligon, 2000). This antibiotic showed antifungal activity towards several fungal plant pathogens, such as the soil-borne *R. solani* and *Fusarium sambucinum* and the foliar fungi *F. graminearum*, *F. culmorum* and *Pyrenophora tritici-repentis*. The actual role of pyrrolnitrin in the biology of the producing organism is often not that clear, as the pyrrolnitrin-producing isolates can also produce other antibiotics, like PCA and HCN (Pfender et al., 1993; Mujumdar et al., 2014). Both pyrrolnitrin and the producing *Pseudomonas fluorescens* strain Pf-5 could repress ascocarp formation of *Pyrenophora tritici-repentis* under laboratory circumstances, but neither the antibiotic nor the bacterium could repress pathogen development on wheat straw when a field situation was simulated (Pfender et al., 1993). This may be because of its light sensitivity, which also thwarted the application of pyrrolnitrin as a fungicide, although stable synthetic derivatives, like fenpiclonil and fludioxonil, were developed from it (van Pée and Ligon, 2000).

4 Facilitating the outsourcing of defences

To control pathogens in the roots, leaves or spikes, it would suffice to have wheat plants efficiently facilitate the proliferation of the relevant antibiotic-producing isolates of *Pseudomonas* or *Bacillus* species and the production of their antibiotics in the rhizosphere or phyllosphere. However, how wheat, barley or any other plant for that matter are facilitating this is still poorly understood. Plants, pathogens and antagonistic microorganisms all seem to govern the involved processes and subsequent success of the microbial antagonism, which can be further modulated by environmental conditions (Duffy et al., 1999; De La Fuente et al., 2006; Hartmann et al., 2009; Bakker et al., 2013). Nevertheless, the production of antibiotics seems to play an important role. When compared to phenazine-lacking mutants, phenazine-producing *Pseudomonas aureofaciens* and *Pseudomonas fluorescens* strains showed better wheat-root-colonizing abilities and also better persisted in the rhizosphere (Mazzola et al., 1992). Roots release a range of components, including root cap, border and other cells, insoluble mucilage, soluble exudates and volatile organic carbons (Bais et al., 2006; Hartmann et al., 2009; Jones et al., 2009; Walker et al., 2003), which all affect the microbiome composition in the rhizosphere. When grown in the same soil under identical conditions, wheat, maize, rape and barrel clover (*Medicago truncatula*), all developed different microbial communities in their rhizospheres (Haichar et al., 2008). In the rhizosphere of wheat, it was observed that irrigation enhanced the accumulation of DAPG-producing pseudomonads, whereas dry conditions enhanced the accumulation of phenazine-producing pseudomonads (Mavrodi et al., 2012a,b,c). This would make sense for the plant, as *R. solani* is more sensitive to phenazines and the most threatening root pathogen under dry conditions (Mavrodi et al., 2012b). When wheat roots were infected by *Ggt*, the root-associated antagonistic *Pseudomonas fluorescens* isolate Pf29Arp changed its own gene expression, probably due to the development of necrotic roots (Barret et al., 2009). At the same time, Pf29Arp triggered host defences, causing the wheat to better resist *Ggt* (Daval et al., 2011).

In a split root experiment with barley, the infection by *Pythium ultimum* at the inducer side caused an increase in DAPG biosynthesis of *Pseudomonas fluorescens* CHA0 at the responder side and root exudates positively affected *PhlA* expression. Combined, this indicates that the increase of the antifungal activity of the bacterium required a systemic plant signal and a direct interaction between antagonist and pathogen was not relevant (Jousset et al., 2011).

To achieve suppression of take-all, the minimal density was determined at 10^5 colony forming units (cfu) per gram roots for isolate Q2-87 (Raaijmakers and Weller, 1998). Experiments by seed inoculation of 27 wheat cultivars with two *Pseudomonas fluorescens* DAPG-producing strains, Q8r1-96 (an aggressive root colonizer) and Q2-87 (a less aggressive root colonizer), showed differences in root colonization levels among the bacterial strains and cultivars (Okubara et al., 2004). The biosynthesis of antibiotics, too, seems to be governed by both bacterial density and plant genotype. Further studies revealed that the *in vitro* DAPG production between the two strains varied significantly, being 1850 μg/ml for Q8r1-96 and 19.4 μg/ml by Q2-87. However, DAPG deposits recovered from roots were 1946, 1650 and 2767 ng/g for Q8r1-96 and 1468, 366 and 80 ng/g for Q2-87 on the cultivars Buchanan, Finley and Tara, respectively. Thus, similar amounts were deposited on Buchanan by both strains, whereas strain Q8r1-96 deposited significantly more DAPG on Tara and Finley roots than Q2-87 (Okubara and Bonsall, 2008). Maurhofer and colleagues established that, in the wheat rhizosphere, the unrelated DAPG-producing *Pseudomonas* strains, Q2-87 and CHA0, stimulated each other to synthesize DAPG (Maurhofer et al., 2004).

4.1 Adaptation of beneficial antagonists to cereals

At the isolate level, too, rhizosphere antagonists may have adapted to their host plant. This is illustrated by studies on diversity among the DAPG-producing *Pseudomonas fluorescens* population, and through the phylogenetic analysis of the *phlD* gene, which encodes one of the crucial enzymes in the DAPG biosynthesis. Based on this, 22 genotypes, designated A to T, PfY and PfZ, are currently recognized (Landa et al., 2006; Weller et al., 2007). The isolates Q8r1-96 and Q2-87 were classified as genotypes D and B, respectively (Landa et al., 2003; Weller et al., 2007). Genotype D was the most dominant genotype in the TAD fields of Washington State (Raaijmakers and Weller, 2001). In a long-term wheat monoculture field plots suppressive to take-all, next to D, the genotypes A, O, P and Q in Mount Vernon, WA (Landa et al., 2002), and the genotypes I and J in North Dakota (McSpadden Gardener et al., 2000; Landa et al., 2006) were encountered as well. In both the soybean and maize rhizosphere, next to D, genotypes A and S were predominant, in which genotype S resembles genotype D (McSpadden Gardener et al., 2005). By contrast, in the pea rhizosphere, genotypes D and P had been enriched (Landa et al., 2002).

For maize, it was determined that DIMBOA, which can be released by the roots into the rhizosphere, particularly attracted the DIMBOA-tolerant *Pseudomonas putida* KT2440 (Neal et al., 2012), which is a strain that can, like some other fluorescent pseudomonads, produce pyoverdine siderophores (Baune et al., 2017). The strong iron chelating activity of pyoverdines can starve competing organisms of iron, and, in this way, interfere with the proliferation of plant pathogenic fungi (Ma et al., 2017). Remarkably, maize roots can accumulate DIMBOA through a symbiotic relationship with the arbuscular mycorrhizal fungus, *Glomus mosseae*, making them more resistant to the pathogen *Rhizoctonia solani* (Song et al., 2011). Next to induced defence responses, direct effects may also play a role, as *R. solani* is susceptible to DIMBOA (Song et al., 2011).

5 Resistance to microbial antagonism

As indicated previously, several *Fusarium* species can convert the precursor of a toxic metabolite, APO, into a harmless metabolite (Glenn et al., 2003). And, identical to the established mechanisms in the fields of medical microbiology and bioremediation, plant pathogens can also enzymatically degrade or inactivate antibiotic compounds, alter the target sites inside their cells and expel antibiotic compounds by employing active efflux (Alexander et al., 1999; De Waard, 1997; Fleissner et al., 2002; Morrissey and Osbourn, 1999; Schoonbeek et al., 2002; Steffens et al., 1996; VanEtten et al., 1989, 2001). This leads to variations in susceptibility to antifungal metabolites produced by antagonistic bacteria (Duffy et al., 2003). Isolates of *Ggt* differed considerably in their susceptibility towards DAPG and phenazines (Mazzola et al., 1995) and, as already mentioned, *Pythium* species and their isolates also varied considerably regarding DAPG sensitivity (de Souza et al., 2003). Several pathogenic and non-pathogenic *Fusarium oxysporum* isolates could fully detoxify DAPG by deacetylation and subsequent aromatic ring-breaking (Schouten et al., 2004). In *Botrytis cinerea*, the efflux pump BcAtrB was involved in the non-degradative resistance against both phenazines (Schoonbeek et al., 2002) and DAPG (Schouten et al., 2008). At the same time, *B. cinerea* is capable of degrading DAPG by the combined action of the laccase BcLCC2 and one of its substrates, tannic acid, most likely through the generation of radicals (Schouten et al., 2008).

It was also demonstrated that a pyridine-carboxylic acid, fusaric acid, that can be produced by various pathogenic and non-pathogenic *Fusarium oxysporum* isolates that were collected from wheat and other crop plants and soil (Notz et al., 2002; Schouten et al., 2004), could specifically interfere with DAPG biosynthesis in the antagonistic strain

Pseudomonas fluorescens CHA0 (Duffy and Défago, 1997a) by repressing the expression of *phlA* (Notz et al., 2002; Schnider-Keel et al., 2000). By obstructing fusaric acid production through zinc amendment, *phlA* repression could subsequently be mitigated, improving the biocontrol capabilities of CHA0 again (Duffy and Défago, 1997b). Remarkably, the DAPG biosynthesis was relatively unaffected by fusaric acid in other DAPG-producing *Pseudomonas* strains (Duffy et al., 2003, 2004), indicating that fusaric acid can be rendered obsolete in the defence of plant pathogens against antibiosis.

Although variations in sensitivity can be observed, there are nevertheless no indications that *Ggt* populations, as whole, become less sensitive to 2,4-DAPG during a prolonged wheat monoculture (Kwak et al., 2009). It has been suggested that most antagonistic microbes suppress the proliferation of plant pathogens through an array of mechanisms, which may be direct or indirect (Handelsman and Stabb, 1996), thus making the development of resistance unachievable for the plant pathogen. A mutant of *Pseudomonas aeruginosa* 7NSK2, deficient in the production of the phenazine derivative pyocyanin, lost the ability to generate induced systemic resistance (ISR) against *Botrytis cinerea* in tomato and bean. The subsequent ISR, triggered by 7NSK2, was attributed to the plant cell damage, caused by the combined action of the siderophore pyochelin and pyocyanin, leading to the production of active oxygen species (Audenaert et al., 2002). DAPG-deficient CHA0 mutants, too, were less capable of triggering ISR responses in *Arabidopsis*, indicating that the antibiotic itself also triggers plants (Iavicoli et al., 2003). The antagonistic activity of *Pseudomonas chlororaphis* subsp. *aurantiaca* Pa40 toward *R. cerealis* was suggested a concerted action of antibiotics, HCN, proteases and siderophores, as well as the triggering of plant defence responses (Jiao et al., 2013).

The selection pressure on plant pathogens is also less intense because only small amounts of the compounds are produced at microsites by the biocontrol agent (Séveno et al., 2002) and growers change wheat varieties regularly for economic reasons or pest control (Kwak et al., 2009). As both wheat and barley species and their varieties can differentially modulate the accumulation of phlD-populations and their genotypes (Bergsma-Vlami et al., 2005; De La Fuente et al., 2006; Mazzola et al., 2004; Picard et al., 2004; Landa et al., 2006) and subsequently affect the DAPG accumulation in the rhizosphere (Okubara and Bonsall, 2008), the selection pressure on the total pathogen population will even be further diminished as compared to the inundative application of pesticides (Séveno et al., 2002).

6 Conclusions and future trends

Wheat and other cereals can defend themselves against plant pathogens both by producing antimicrobial secondary metabolites and by recruiting antibiotic-producing microorganisms. For a long time, in wheat, only benzoxazinoids were considered to be the prime compounds involved in resisting fungal pathogens, although the relevance of the individual benzoxazinoid derivatives in defence against fungal pathogens has not really been resolved. As already indicated, there are several reasons for this. There are different benzoxazinoid derivatives that can be present in wheat and other cereals, each of which can have different biological activities, their accumulation in the plant can be transient and localized and microbial activity can convert benzoxazinoids into constituents that can be either less or more toxic to particular plant pathogenic fungi. The antifungal activity of ferulic acid is being recognized once again, providing new options for developing fungal-disease-resistant cultivars through breeding programmes. The additional value of such cultivars is the notion that ferulic acid has general beneficial health effects for humans (Ou and Kwok, 2004) and can also prevent mycotoxins from accumulating, even post-harvest (Boutigny et al., 2009). Only recently, the antifungal activity of serotonin (Du Fall and Solomon, 2013; Ishihara et al., 2017) and related compounds (Ishihara et al., 2017) were described, suggesting that wheat, barley and possibly other cereals may synthesize additional compounds with yet undiscovered antimicrobial activities.

Next to arbuscular mycorrhizae (AMs) and rhizobia (Denison and Kiers, 2011), it is now well established that plants can initiate and maintain intricate relationships with microorganisms, particularly bacteria and fungi, that can benefit the plant with respect to biotic and abiotic stress situations. Particularly, the example of TAD shows that the exploitation of microbial antagonism can be a very effective strategy, too, for the protection of wheat against plant pathogenic fungi and oomycetes through the synthesis of specific antibiotics. There is ample evidence that wheat facilitates their establishment, although the exact mechanism by which this is achieved is still unclear. Such lack of knowledge frustrates the development of disease-resistant cultivars through breeding programmes. The interaction with particular endophytes, which are defined as microorganisms that colonize plants without provoking disease symptoms (Wilson, 1995; Stone et al., 2004; Schulz and Boyle, 2005), can improve in disease and pest defences, plant vigour, growth and crop yield (Sikora, 1992; Rodriguez et al., 2008; Aly et al., 2011; Schouten, 2016), similar to what has been described for the rhizosphere antagonists (Bakker et al., 2013; Pieterse and Dicke, 2007; Pineda et al., 2010). These endophytes, too, can produce metabolites that have direct as well as indirect effects on pathogens, as was shown for an endophytic *F. oxysporum* isolate, which produced various nematotoxic metabolites (Bogner et al., 2017). Two of these compounds, indole acetic acid (IAA) and 4-hydroxybenzoic acid (4-HBA), are also known plant hormones that can increase plant vigour against biotic and abiotic stress. Similar to other auxins, the effect of IAA in plants varies and depends on the concentration. It generally stimulates growth, fruit development and the development of lateral roots (Ivanchenko et al., 2008). Hormonal cross-talk in the primary root was observed between auxin and the jasmonate/ethylene (JA/ET) defence pathways (Ortega-Martinez et al., 2007; Pieterse et al., 2012) and auxins positively modified stress tolerance of the plant (Kerchev et al., 2015). Other than plants, many soil-borne bacteria and fungi are capable of producing IAA (Duca et al., 2014) and there is proof that such microorganisms can modify root architecture (Zamioudis et al., 2013). When 4-HBA was exogenously added, drought tolerance of winter wheat and the freezing tolerance of spring wheat were improved, whereas its structural analogue, salicylic acid, diminished the freezing tolerance of winter wheat and the drought tolerance of spring wheat (Horváth et al., 2007). Currently, there is little information on the role of endophytes and their metabolites against fungal pathogens in wheat. In greenhouse experiments, the basidiomycete *Piriformospora indica*, a root-associated endophyte with a large host range (Qiang et al., 2012), significantly reduced disease symptom severity of a typical leaf (*Blumeria graminis* f. sp. *tritici*), stem base (*Pseudocercosporella herpotrichoides*) and root (*F. culmorum*) pathogen. Field experiments showed that, in the presence of the endophyte, the leaf pathogen was not at all impaired, and the severity of the disease caused by *Pseudocercosporella herpotrichoides* was significantly reduced and an increase in numbers of sheath layers and hydrogen peroxide concentration after *B. graminis* infection were identified. The latter suggested that the endophyte either triggered systemic defence responses or primed the host plant (Serfling et al., 2007). Studies with *Piriformospora indica* in barley imply that roots are partly protected from necrotrophic pathogen *F. culmorum* by mobilizing the plant's antioxidant capabilities (Harrach et al., 2013). However, *Piriformospora indica*, which originates from a sub-tropical arid region (Verma et al., 1998), did not seem to be well adapted to the ecological conditions that can be found in European countries with temperate climatic conditions (Serfling et al., 2007; Rabiey et al., 2017).

The section on benzoxazinoids showed that, by combining the biosynthesis pathways of microbe and plant, other compounds, sometimes with antifungal properties, can be synthesized, which cannot be produced by the plant alone. The same applies for endophytes and only recently, this role of microorganism in chemical conversions of plant metabolites is becoming appreciated, not in the least by the realization that such modifications can lead to novel constituents with yet undiscovered pharmacological qualities (Ludwig-Müller, 2015; Rodriguez et al., 2017). And, as illustrated previously, tripartite interactions at the level of biosynthesis of secondary metabolites with antifungal activity alone can be rather complex and surprising. Such metabolomic research is still in its infancy and is definitely challenging. Nevertheless, biological, molecular, biochemical and microscopical analysis methods have continuously been refined and integrated. For example, it is now possible to localize and quantify the *in vivo* accumulation of particular secondary metabolites, as was exhibited for the occurrence and distribution of the phytoalexin anigorufone in banana by using the matrix-free laser desorption/ionization Fourier transform ion cyclotron resonance mass spectrometric imaging (LDI-FT-ICR-MSI) technique (Hölscher et al., 2014). In all, multidisciplinary research will be crucial in the future for identifying the role of constituents in the protection of wheat against plant pathogenic fungi, regardless of their origin.

7 Acknowledgements

I apologize in advance to all investigators whose research was not cited because of space limitations.

8 Where to look for further information

Next to cereals there are many other plants capable of producing allelochemicals and the number of compounds with allelopathic activity is enormous. For a broader view on allelopathy and allelochemicals, please, refer to Latif et al. (2017). The International Allelopathy Society (http://allelopathy-society.osupytheas.fr) has a very broad focus, covering allelopathy and allelochemical interactions among plants, microorganisms and other macrofauna in both terrestrial and aquatic ecosystems. This society has been in existence for more than 20 years, and currently has over 1000 members from more than 50 countries. The society supports the *Journal of Allelochemical Interactions*, launched in 2014, organizes a congress every three years and has now initiated an EU COST Action programme, named ALLFORAGR, to further intensify collaborations with regard to the identification, release mechanisms, selectivity and persistence, mode-of-action and the expression of allelochemicals.

The research on and application of microorganisms for controlling pests and diseases in an integrated pest management (IPM) approach is expanding, as it is considered a potentially accessible and cost-effective strategy, particularly for poor farmers in third-world countries. This expansion is further accelerated by the fact that conventional chemical pesticides are becoming ineffective due to resistance development and their application is heavily debated and sometimes even banned due to environmental and human health concerns (for the EU, consult http://ec.europa.eu/food/plant/pesticides/eu-pesticides-database/public). For many years, the International Organisation for Biological Control (IOBC) has been promoting environmentally safe methods for pest and disease control (http://www.iobc-global.org). This is a global voluntary organization, bringing together scientists studying methods for biological control from various angles, which are communicated through workshops, newsletters and its official journal, *BioControl*.

The CGIAR (formerly the Consultative Group for International Agricultural Research) comprises a consortium of 15 research centres, known as the CGIAR Consortium of International Agricultural Research Centres, and is funded by national governments, multilateral funding and development agencies and private foundations (http://www.cgiar.org). The International Maize and Wheat Improvement Centre, better known by its Spanish acronym CIMMYT (for Centro Internacional de Mejoramiento de Maíz y Trigo) (http://www.cimmyt.org), is part of this consortium. Recently, several CGIAR centres and their partners collaborated further by initiating the Systemwide Program on Integrated Pest Management (SP-IPM) to achieve more productive and healthy agroecosystems by accelerating the development of innovative and adapted technologies for the improvement of pest and disease management (http://www.spipm.cgiar.org and http://www.spipm.cgiar.org/members/centers). A book providing an interesting read on the diverse topics related to biocontrol has been edited by Vincent et al. (2007), and it includes a historical perspective on the TAD research by Cook (2007). Meghvansi and Varma (2015) published a broad overview of soil suppressiveness and its role in IPM.

9 References

Ahmad, S., Veyrat, N., Gordon-Weeks, R., et al. (2011). Benzoxazinoid metabolites regulate innate immunity against aphids and fungi in maize. *Plant Physiology* 157: 317–27.

Ahuja, E. G., Janning, P., Mentel, M., et al. (2008). PhzA/B catalyzes the formation of the tricycle in phenazine biosynthesis. *Journal of the American Chemical Society* 130: 17053–61.

Ahuja, I., Kissen, R. and Bones, A. M. (2012). Phytoalexins in defense against pathogens. *Trends in Plant Science* 17: 73–90.

Alexander, N. J., McCormick, S. P. and Hohn, T. M. (1999). TRI12, a trichothecene efflux pump from *Fusarium sporotrichioides*: gene isolation and expression in yeast. *Molecular and General Genetics* 261: 977–84.

Aly, A. H., Debbab, A. and Proksch, P. (2011). Fungal endophytes: unique plant inhabitants with great promises. *Applied Microbiology and Biotechnology* 90: 1829–45.

Andersen, J. B., Koch, B., Nielsen, T. H., et al. (2003). Surface motility in *Pseudomonas* sp. DSS73 is required for efficient biological containment of the root-pathogenic microfungi *Rhizoctonia solani* and *Pythium ultimum*. *Microbiology-Sgm* 149: 37–46.

Audenaert, K., Pattery, T., Cornelis, P., et al. (2002). Induction of systemic resistance to *Botrytis cinerea* in tomato by *Pseudomonas aeruginosa* 7NSK2: role of salicylic acid, pyochelin, and pyocyanin. *Molecular Plant-Microbe Interactions* 15: 1147–56.

Bacon, C. W., Hinton, D. M., Glenn, A. E., et al. (2007). Interactions of *Bacillus mojavensis* and *Fusarium verticillioides* with a benzoxazolinone (BOA) and its transformation product, APO. *Journal of Chemical Ecology* 33: 1885–97.

Bais, H. P., Weir, T. L., Perry, L. G., et al. (2006). The role of root exudates in rhizosphere interactions with plants and other organisms. *Annual Review of Plant Biology* 57: 233–66.

Baker, K. F. and Cook, R. J. (1974). *Biological Control of Plant Pathogens*. San Francisco, CA: W.H. Freeman and Company.

Baker, E. A. and Smith, I. M. (1977). Antifungal compounds in winter-wheat resistant and susceptible to *Septoria nodorum*. *Annals of Applied Biology* 87: 67–73.

Bakker, P. A. H. M., Berendsen, R. L., Doornbos, R. F., et al. (2013). The rhizosphere revisited: root microbiomics. *Frontiers in Plant Science* 4: 165.

Bangera, M. G. and Thomashow, L. S. (1999). Identification and characterization of a gene cluster for synthesis of the polyketide antibiotic 2,4-diacetylphloroglucinol from *Pseudomonas fluorescens* Q2-87. *Journal of Bacteriology* 181: 3155–63.

Barnes, J. P. and Putnam, A. R. (1987). Role of benzoxazinones in allelopathy by rye (*Secale cereale* L.). *Journal of Chemical Ecology* 13: 889–906.

Barret, M., Frey-Klett, P., Guillerm-Erckelboudt, A. Y., et al. (2009). Effect of wheat roots infected with the pathogenic fungus *Gaeumannomyces graminis* var. *tritici* on gene expression of the biocontrol bacterium *Pseudomonas fluorescens* Pf29Arp. *Molecular Plant-Microbe Interactions* 22: 1611–23.

Bassioli Moraes, M. C., Birkett, M. A., Gordon-Weeks, R, et al. (2008). *cis*-Jasmone induces accumulation of defence compounds in wheat, *Triticum aestivum*. *Phytochemistry* 69: 9–17.

Baune, M., Qi, Y. L., Scholz, K., et al. (2017). Structural characterization of pyoverdines produced by *Pseudomonas putida* KT2440 and *Pseudomonas taiwanensis* VLB120. *Biometals* 30: 589–97.

Belz, R. G. (2007). Allelopathy in crop/weed interactions – an update. *Pest Management Science* 63: 308–26.

Bergsma-Vlami, M., Prins, M. E. and Raaijmakers, J. M. (2005). Influence of plant species on population dynamics, genotypic diversity and antibiotic production in the rhizosphere by indigenous *Pseudomonas* spp. *FEMS Microbiology Ecology* 52: 59–69.

Bertin, C., Yang, X. and Weston, L. A. (2003). The role of root exudates and allelochemicals in the rhizosphere. *Plant and Soil* 256: 67–83.

Blankenfeldt, W. and Parsons, J. F. (2014). The structural biology of phenazine biosynthesis. *Current Opinion in Structural Biology* 29: 26–33.

Bogner, C. W., Kamdem, R. S. T., Sichtermann, G., et al. (2017). Bioactive secondary metabolites with multiple activities from a fungal endophyte. *Microbial Biotechnology* 10: 175–88.

Boutigny, A. L., Barreau, C., Atanasova-Penichon, V., et al. (2009). Ferulic acid, an efficient inhibitor of type B trichothecene biosynthesis and *Tri* gene expression in *Fusarium* liquid cultures. *Mycological Research* 113: 746–53.

Bowyer, P., Clarke, B. R., Lunness, P., et al. (1995). Host-range of a plant-pathogenic fungus determined by a saponin detoxifying enzyme. *Science* 267: 371–4.

Bravo, H. R. and Lazo, W. (1996). Antialgal and antifungal activity of natural hydroxamic acids and related compounds. *Journal of Agricultural and Food Chemistry* 44: 1569–71.

Bravo, H. R., Copaja, S. V. and Lazo, W. (1997). Antimicrobial activity of natural 2-benzoxazolinones and related derivatives. *Journal of Agricultural and Food Chemistry* 45: 3255–7.

Chin-A-Woeng, T. F. C., Bloemberg, G. V. and Lugtenberg, B. J. J. (2003). Phenazines and their role in biocontrol by *Pseudomonas* bacteria. *New Phytologist* 157: 503–23.

Cook, R. J. (2007). Take-all decline: model system in the science of biological control and clue to the success of intensive cropping. In: Vincent, C., Goettel, M. and Lazarovits, G. (Eds), *Biological Control: A Global Perspective*. Wallingford, UK: CABI, pp. 399–414.

Cook, R. J. and Rovira, A. D. (1976). Role of bacteria in biological-control of *Gaeumannomyces graminis* by suppressive soils. *Soil Biology & Biochemistry* 8: 269–73.

Cook, R. J., Thomashow, L. S., Weller, D. M., et al. (1995). Molecular mechanisms of defense by rhizobacteria against root disease. *Proceedings of the National Academy of Sciences of the United States of America* 92: 4197–201.

Corcuera, L. J., Woodward, M. D., Helgeson, J. P., et al. (1978). 2,4-dihydroxy-7-methoxy-2h-1,4-benzoxazin-3(4h)-one, and inhibitor from *Zea mays* with differential activity against soft rotting *Erwinia* species. *Plant Physiology* 61: 791–5.

Couture, R. M., Routley, D. G. and Dunn, G. M. (1971). Role of cyclic hydroxamic acids in monogenic resistance of maize to *Helminthosporium turcicum*. *Physiological Plant Pathology* 1: 515–21.

Cushnie, T. P. T. and Lamb, A. J. (2005). Antimicrobial activity of flavonoids. *International Journal of Antimicrobial Agents* 26: 343–56.

Cushnie, T. P. T. and Lamb, A. J. (2011). Recent advances in understanding the antibacterial properties of flavonoids. *International Journal of Antimicrobial Agents* 38: 99–107.

Daval, S., Lebreton, L., Gazengel, K., et al. (2011). The biocontrol bacterium *Pseudomonas fluorescens* Pf29Arp strain affects the pathogenesis-related gene expression of the take-all fungus *Gaeumannomyces graminis* var. *tritici* on wheat roots. *Molecular Plant Pathology* 12: 839–54.

De La Fuente, L., Landa, B. B. and Weller, D. M. (2006). Host crop affects rhizosphere colonization and competitiveness of 2,4-diacetylphloroglucinol-producing *Pseudomonas fluorescens*. *Phytopathology* 96: 751–62.

de Souza, J. T., Arnould, C., Deulvot, C., et al. (2003). Effect of 2,4-diacetylphloroglucinol on *Pythium*: cellular responses and variation in sensitivity among propagules and species. *Phytopathology* 93: 966–75.

De Waard, M. A. (1997). Significance of ABC transporters in fungicide sensitivity and resistance. *Pesticide Science* 51: 271–5.

Denison, R. F. and Kiers, E. T. (2011). Life histories of symbiotic rhizobia and mycorrhizal fungi. *Current Biology* 21: R775–85.

Dewhirst, S. Y. and Pickett, J. A. (2010). Production of semiochemical and allelobiotic agents as a consequence of aphid feeding. *Chemoecology* 20: 89–96.

Du Fall, L. A. and Solomon, P. S. (2011). Role of cereal secondary metabolites involved in mediating the outcome of plant-pathogen interactions. *Metabolites* 1: 64–78.

Du Fall, L. A. and Solomon, P. S. (2013). The necrotrophic effector SnToxA induces the synthesis of a novel phytoalexin in wheat. *New Phytologist* 200: 185–200.

Duca, D., Lorv, J., Patten, C. L., et al. (2014). Indole-3-acetic acid in plant-microbe interactions. *Antonie Van Leeuwenhoek International Journal of General and Molecular Microbiology* 106: 85–125.

Duffy, B. K. and Défago, G. (1997a). A *Fusarium* pathogenicity factor blocks antibiotic biosynthesis by *Pseudomonas fluorescens* biocontrol strains. *Phytopathology* 87: S26.

Duffy, B. K. and Défago, G. (1997b). Zinc improves biocontrol of *Fusarium crown* and root rot of tomato by *Pseudomonas fluorescens* and represses the production of pathogen metabolites inhibitory to bacterial antibiotic biosynthesis. *Phytopathology* 87: 1250–7.

Duffy, B. K., Défago, G and Forrer, H. R. (1999). Environmental factors modulating antibiotic and siderophore biosynthesis by *Pseudomonas fluorescens* biocontrol strains. *Applied and Environmental Microbiology* 65: 2429–38.

Duffy, B., Schouten, A. and Raaijmakers, J. M. (2003). Pathogen self-defense: mechanisms to counteract microbial antagonism. *Annual Review of Phytopathology* 41: 501–38.

Duffy, B., Keel, C. and Defago, G. (2004). Potential role of pathogen signaling in multitrophic plant-microbe interactions involved in disease protection. *Applied and Environmental Microbiology* 70: 1836–42.

Fleissner, A., Sopalla, C. and Weltring, K. M. (2002). An ATP-binding cassette multidrug-resistance transporter is necessary for tolerance of *Gibberella pulicaris* to phytoalexins and virulence on potato tubers. *Molecular Plant-Microbe Interactions* 15: 102–8.

Fraenkel, G. S. (1959). Raison d'être of secondary plant substances. *Science* 129: 1466–70.

Friebe, A., Vilich, V., Henning, L., et al. (1998). Detoxification of benzoxazolinone allelochemicals from wheat by *Gaeumannonmyces graminis* var. *tritici*, *G. graminis* var. *graminis*, *G. graminis* var. *avenae*, and *Fusarium culmorum*. *Applied and Environmental Microbiology* 64: 2386–91.

Gerlagh, M. (1968). Introduction of *Ophiobolus graminis* into new polders and its decline *Netherlands Journal of Plant Pathology* 74 Suppl. 2: 97.

Gewolb, J. (2002). Working outside the protein-synthesis rules. *Science* 295: 2205–7.
Gianoli, E. and Niemeyer, H. M. (1998). DIBOA in wild Poaceae: sources of resistance to the Russian wheat aphid (*Diuraphis noxia*) and the greenbug (*Schizaphis graminum*). *Euphytica* 102: 317–21.
Glenn, A. E., Meredith, F. I., Morrison, W. H., et al. (2003). Identification of intermediate and branch metabolites resulting from biotransformation of 2-benzoxazolinone by *Fusarium verticillioides*. *Applied and Environmental Microbiology* 69: 3165–9.
Gong, A.-D., Li, H.-P., Yuan, Q.-S., et al. (2015). Antagonistic mechanism of Iturin A and Plipastatin A from *Bacillus amyloliquefaciens* S76-3 from wheat spikes against *Fusarium graminearum*. *Plos One* 10: e0116871.
Gurusiddaiah, S., Weller, D. M., Sarkar, A., et al. (1986). Characterization of an antibiotic produced by a strain of *Pseudomonas fluorescens* inhibitory to *Gaeumannomyces graminis* var. *tritici* and *Pythium* spp. *Antimicrobial Agents and Chemotherapy* 29: 488–95.
Guthrie, W. D., Wilson, R. L., Coats, J. R., et al. (1986). European corn-borer (Lepidoptera, Pyralidae) leaf-feeding resistance and DIMBOA content in inbred lines of dent maize grown under field versus greenhouse conditions. *Journal of Economic Entomology* 79: 1492–6.
Haichar, F. E., Marol, C., Berge, O., et al. (2008). Plant host habitat and root exudates shape soil bacterial community structure. *ISME Journal* 2: 1221–30.
Hammer, P. E., Hill, D. S., Lam, S. T., et al. (1997). Four genes from *Pseudomonas fluorescens* that encode the biosynthesis of pyrrolnitrin. *Applied and Environmental Microbiology* 63: 2147–54.
Handelsman, J. and Stabb, E. V. (1996). Biocontrol of soilborne plant pathogens. *Plant Cell* 8: 1855–69.
Harrach, B. D., Baltruschat, H., Barna, B., et al. (2013). The mutualistic fungus *Piriformospora indica* protects barley roots from a loss of antioxidant capacity caused by the necrotrophic pathogen *Fusarium culmorum*. *Molecular Plant-Microbe Interactions* 26: 599–605.
Hartman, J. R., Kelman, A. and Upper, C. D. (1975). Differential inhibitory activity of a corn extract to Erwinia spp. causing soft rot. *Phytopathology* 65: 1082–8.
Hartmann, A., Schmid, M., van Tuinen, D., et al. (2009). Plant-driven selection of microbes. *Plant and Soil* 321: 235–57.
Hölscher, D., Dhakshinamoorthy, S., Alexandrov, T., et al. (2014). Phenalenone-type phytoalexins mediate resistance of banana plants (*Musa* spp.) to the burrowing nematode *Radopholus similis*. *Proceedings of the National Academy of Sciences of the United States of America* 111: 105–10.
Horváth, E., Pál, M., Szalai, G., et al. (2007). Exogenous 4-hydroxybenzoic acid and salicylic acid modulate the effect of short-term drought and freezing stress on wheat plants. *Biologia Plantarum* 51: 480–7.
Huang, Z. Y., Bonsall, R. F., Mavrodi, D. V., et al. (2004). Transformation of *Pseudomonas fluorescens* with genes for biosynthesis of phenazine-1-carboxylic acid improves biocontrol of rhizoctonia root rot and *in situ* antibiotic production. *FEMS Microbiology Ecology* 49: 243–51.
Iavicoli, A., Boutet, E., Buchala, A., et al. (2003). Induced systemic resistance in *Arabidopsis thaliana* in response to root inoculation with *Pseudomonas fluorescens* CHA0. *Molecular Plant-Microbe Interactions* 16: 851–8.
Ioannou, Y. M., Dauterman, W. C. and Tucker, W. P. (1980). Degradation of diazinon by 2,4-dihydroxy-7-methoxy-2h-1,4-benzoxazin-3(4h)-one in maize. *Phytochemistry* 19: 1607–11.
Ishihara, A., Kumeda, R., Hayashi, N., et al. (2017). Induced accumulation of tyramine, serotonin, and related amines in response to *Bipolaris sorokiniana* infection in barley. *Bioscience Biotechnology and Biochemistry* 81: 1090–8.
Isnansetyo, A., Cui, L. Z., Hiramatsu, K., et al. (2003). Antibacterial activity of 2,4-diacetylphloroglucinol produced by *Pseudomonas* sp. AMSN isolated from a marine alga, against vancomycin-resistant *Staphylococcus aureus*. *International Journal of Antimicrobial Agents* 22: 545–7.
Ivanchenko, M. G., Muday, G. K. and Dubrovsky, J. G. (2008). Ethylene-auxin interactions regulate lateral root initiation and emergence in *Arabidopsis thaliana*. *Plant Journal* 55: 335–47.
Jiao, Z., Wu, N., Hale, L., et al. (2013). Characterisation of *Pseudomonas chlororaphis* subsp. *aurantiaca* strain Pa40 with the ability to control wheat sharp eyespot disease. *Annals of Applied Biology* 163: 444–53.

Jones, D. L., Nguyen, C. and Finlay, R. D. (2009). Carbon flow in the rhizosphere: carbon trading at the soil-root interface. *Plant and Soil* 321: 5–33.

Jousset, A., Rochat, L., Lanoue, A., et al. (2011). Plants respond to pathogen infection by enhancing the antifungal gene expression of root-associated bacteria. *Molecular Plant-Microbe Interactions* 24: 352–8.

Kamei, Y. and Isnansetyo, A. (2003). Lysis of methicillin-resistant *Staphylococcus aureus* by 2,4-diacetylphloroglucinol produced by *Pseudomonas* sp. AMSN isolated from a marine alga. *International Journal of Antimicrobial Agents* 21: 71–4.

Keel, C., Schnider, U., Maurhofer, M., et al. (1992). Suppression of root diseases by *Pseudomonas fluorescens* CHA0 – Importance of the bacterial secondary metabolite 2,4-diacetylphloroglucinol. *Molecular Plant-Microbe Interactions* 5: 4–13.

Kerchev, P., De Smet, B., Waszczak, C., et al. (2015). Redox strategies for crop improvement. *Antioxidants & Redox Signaling* 23: 1186–205.

Kim, D. S., Cook, R. J. and Weller, D. M. (1997). *Bacillus* sp. L324-92 for biological control of three root diseases of wheat grown with reduced tillage. *Phytopathology* 87: 551–8.

Klun, J. A. and Brindley, T. A. (1966). Role of 6-methoxybenzoxazolinone in inbred resistance of host plant (maize) to first-brood larvae of european corn borer. *Journal of Economic Entomology* 59: 711–18.

Klun, J. A. and Robinson, J. F. (1969). Concentration of 2 1,4-benzoxazinones in dent corn at various stages of development of plant and its relation to resistance of host plant to European corn borer. *Journal of Economic Entomology* 62: 214–20.

Kremer, R. J. and Ben-Hammouda, M. (2009). Allelopathic plants. 19. Barley (*Hordeum vulgare* L). *Allelopathy Journal* 24: 225–42.

Kwak, Y. S., Bakker, P. A. H. M., Glandorf, D. C. M., et al. (2009). Diversity, virulence, and 2,4-diacetylphloroglucinol sensitivity of *Gaeumannomyces graminis* var. *tritici* isolates from Washington state. *Phytopathology* 99: 472–9.

Landa, B. B., Mavrodi, O. V., Raaijmakers, J. M., et al. (2002). Differential ability of genotypes of 2,4-diacetylphloroglucinol-producing *Pseudomonas fluorescens* strains to colonize the roots of pea plants. *Applied and Environmental Microbiology* 68: 3226–37.

Landa, B. B., Mavrodi, D. M., Thomashow, L. S., et al. (2003). Interactions between strains of 2,4-diacetylphloroglucinol-producing *Pseudomonas fluorescens* in the rhizosphere of wheat. *Phytopathology* 93: 982–94.

Landa, B. B., Mavrodi, O. V., Schroeder, K. L., et al. (2006). Enrichment and genotypic diversity of phlD-containing fluorescent *Pseudomonas* spp. in two soils after a century of wheat and flax monoculture. *FEMS Microbiology Ecology* 55: 351–68.

Latif, S., Chiapusiox, G. and Weston, L. A. (2017). Allelopathy and the role of allelochemicals in plant defence. *Advances in Botanical Research* 82: 19–54.

Lattanzio, V., De Cicco, V., Di Venere, D., et al. (1994). Antifungal activity of phenolics against fungi commonly encountered during storage. *Italian Journal of Food Science* 1: 23–30.

Laursen, J. B. and Nielsen, J. (2004). Phenazine natural products: biosynthesis, synthetic analogues, and biological activity. *Chemical Reviews* 104: 1663–85.

Long, B. J., Dunn, G. M. and Routley, D. G. (1978). Relationship of hydroxamate concentration in maize and field reaction to *Helminthosporium turcicum*. *Crop Science* 18: 573–5.

Ludwig-Müller, J. (2015). Plants and endophytes: equal partners in secondary metabolite production? *Biotechnology Letters* 37: 1325–34.

Ma, R. Q., Cao, Y, Cheng, Z. Q., et al. (2017). Identification and genomic analysis of antifungal property of a tomato root endophyte *Pseudomonas* sp. p21. *Antonie Van Leeuwenhoek International Journal of General and Molecular Microbiology* 110: 387–97.

Makowska, B., Bakera, B. and Rakoczy-Trojanowska, M. (2015). The genetic background of benzoxazinoid biosynthesis in cereals. *Acta Physiologiae Plantarum* 37: 1–12.

Martin, C., Schnee, S., Lintz, J., et al. (2017). Role of ferulic acid in Fusarium head blight resistance of wheat spikes. *Future IPM 3.0 Towards a Sustainable Agriculture*. Riva del Garda, Italy: IOBC-WPRS.

Maurhofer, M., Baehler, E., Notz, R., et al. (2004). Cross talk between 2,4-diacetylphloroglucinol-producing biocontrol pseudomonads on wheat roots. *Applied and Environmental Microbiology* 70: 1990–8.

Mavrodi, D. V., Blankenfeldt, W. and Thomashow, L. S. (2006). Phenazine compounds in fluorescent *Pseudomonas* spp. biosynthesis and regulation. *Annual Review of Phytopathology* 44: 417–45.

Mavrodi, D. V., Mavrodi, O. V., Parejko, J. A., et al. (2012a). Accumulation of the antibiotic phenazine-1-carboxylic acid in the rhizosphere of dryland cereals. *Applied and Environmental Microbiology* 78: 804–12.

Mavrodi, O. V., Mavrodi, D. V., Parejko, J. A., et al. (2012b). Irrigation differentially impacts populations of indigenous antibiotic-producing *Pseudomonas* spp. In the rhizosphere of wheat. *Applied and Environmental Microbiology* 78: 3214–20.

Mavrodi, O. V., Walter, N., Elateek, S., et al. (2012c). Suppression of *Rhizoctonia* and *Pythium* root rot of wheat by new strains of *Pseudomonas*. *Biological Control* 62: 93–102.

Mazzola, M., Cook, R. J., Thomashow, L. S., et al. (1992). Contribution of phenazine antibiotic biosynthesis to the ecological competence of fluorescent pseudomonads in soil habitats. *Applied and Environmental Microbiology* 58: 2616–24.

Mazzola, M., Fujimoto, D. K., Thomashow, L. S., et al. (1995). Variation in sensitivity of *Gaeumannomyces graminis* to antibiotics produced by fluorescent *Pseudomonas* spp. and effect on biological control of take-all of wheat. *Applied and Environmental Microbiology* 61: 2554–9.

Mazzola, M., Funnell, D. L. and Raaijmakers, J. M. (2004). Wheat cultivar-specific selection of 2,4-diacetylphloroglucinol-producing fluorescent *Pseudomonas* species from resident soil populations. *Microbial Ecology* 48: 338–48.

McDonald, M., Mavrodi, D. V., Thomashow, L. S., et al. (2001). Phenazine biosynthesis in *Pseudomonas fluorescens*: branchpoint from the primary shikimate biosynthetic pathway and role of phenazine-1,6-dicarboxylic acid. *Journal of the American Chemical Society* 123: 9459–60.

McKeehen, J. D., Busch, R. H. and Fulcher, R. G. (1999). Evaluation of wheat (*Triticum aestivum* L.) phenolic acids during grain development and their contribution to *Fusarium* resistance. *Journal of Agricultural and Food Chemistry* 47: 1476–82.

McSpadden Gardener, B. B., Schroeder, K. L., Kalloger, S. E., et al. (2000). Genotypic and phenotypic diversity of *phlD*-containing *Pseudomonas* strains isolated from the rhizosphere of wheat. *Applied and Environmental Microbiology* 66: 1939–46.

McSpadden Gardener, B. B., Gutierrez, L. J., Joshi, R., et al. (2005). Distribution and biocontrol potential of phlD(+) pseudomonads in corn and soybean fields. *Phytopathology* 95: 715–24.

Meghvansi, M. K. and Varma, A. (2015). Organic amendments and soil suppressiveness in plant disease management. *Soil Biology*, Vol. 46, Springer, Heidelberg. doi: 10.1007/978-3-319-23075-7.

Meyer, J., Murray, S. L. and Berger, D. K. (2016). Signals that stop the rot: regulation of secondary metabolite defences in cereals. *Physiological and Molecular Plant Pathology* 94: 156–66.

Molisch, H. (1937). *Der Einfluß einer Pflanze auf die andere, Allelopathie*. Jena, Germany: Gustav Fischer.

Morrissey, J. P. and Osbourn, A. E. (1999). Fungal resistance to plant antibiotics as a mechanism of pathogenesis. *Microbiology and Molecular Biology Reviews* 63: 708–24.

Mujumdar, S. S., Bashetti, S. P. and Chopade, B. A. (2014). Plasmid pUPI126-encoded pyrrolnitrin production by *Acinetobacter haemolyticus* A19 isolated from the rhizosphere of wheat. *World Journal of Microbiology & Biotechnology* 30: 495–505.

Neal, A. L., Ahmad, S., Gordon-Weeks, R., et al. (2012). Benzoxazinoids in root exudates of maize attract Pseudomonas putida to the rhizosphere. *Plos One* 7: e35498.

Niemeyer, H. M. (1988). Hydroxamic acids (4-hydroxy-1,4-benzoxazin-3-ones), defense chemicals in the Gramineae. *Phytochemistry* 27: 3349–58.

Niemeyer, H. M. (2009). Hydroxamic acids derived from 2-Hydroxy-2 H-1,4-Benzoxazin-3(4 H)-one: key defense chemicals of cereals. *Journal of Agricultural and Food Chemistry* 57: 1677–96.

Notz, R., Maurhofer, M., Dubach, H., et al. (2002). Fusaric acid-producing strains of *Fusarium oxysporum* alter 2,4-diacetylphloroglucinol biosynthetic gene expression in *Pseudomonas fluorescens*

CHA0 in vitro and in the rhizosphere of wheat. *Applied and Environmental Microbiology* 68: 2229–35.

Oikawa, A., Ishihara, A., Hasegawa, M., et al. (2001). Induced accumulation of 2-hydroxy-4,7-dimethoxy-1,4-benzoxazin-3-one glucoside (HDMBOA-Glc) in maize leaves. *Phytochemistry* 56: 669–75.

Okubara, P. A. and Bonsall, R. F. (2008). Accumulation of *Pseudomonas*-derived 2,4-diacetylphloroglucinol on wheat seedling roots is influenced by host cultivar. *Biological Control* 46: 322–31.

Okubara, P. A., Kornoely, J. P. and Landa, B. B. (2004). Rhizosphere colonization of hexaploid wheat by *Pseudomonas fluorescens* strains Q8r1-96 and Q2-87 is cultivar-variable and associated with changes in gross root morphology. *Biological Control* 30: 392–403.

Ortega-Martinez, O., Pernas, M., Carol, R. J., et al. (2007). Ethylene modulates stem cell division in the *Arabidopsis thaliana* root. *Science* 317: 507–10.

Ou, S. Y. and Kwok, K. C. (2004). Ferulic acid: pharmaceutical functions, preparation and applications in foods. *Journal of the Science of Food and Agriculture* 84: 1261–9.

Pérez, F. J. and Ormeño-Núñez, J. (1993). Weed growth interference from temperate cereals: the effect of a hydroxamic-acids-exuding rye (*Secale cereale* L.) cultivar. *Weed Research* 33: 115–19.

Pfender, W. F., Kraus, J. and Loper, J. E. (1993). A genomic region from *Pseudomonas fluorescens* pf-5 required for pyrrolnitrin production and inhibition of *Pyrenophora tritici-repentis* in wheat-straw. *Phytopathology* 83: 1223–8.

Picard, C., Frascaroli, E. and Bosco, M. (2004). Frequency and biodiversity of 2,4-diacetylphloroglucinol-producing rhizobacteria are differentially affected by the genotype of two maize inbred lines and their hybrid. *FEMS Microbiology Ecology* 49: 207–15.

Pierson, L. S. and Thomashow, L. S. (1992). Cloning and heterologous expression of the phenazine biosynthetic locus from *Pseudomonas aureofaciens* 30–84. *Molecular Plant-Microbe Interactions* 5: 330–9.

Pieterse, C. M. J. and Dicke, M. (2007). Plant interactions with microbes and insects: from molecular mechanisms to ecology. *Trends in Plant Science* 12: 564–9.

Pieterse, C. M. J., Van der Does, D., Zamioudis, C., et al. (2012). Hormonal modulation of plant immunity. *Annual Review of Cell and Developmental Biology* 28: 489–521.

Pineda, A., Zheng, S. J., van Loon, J. J. A., et al. (2010). Helping plants to deal with insects: the role of beneficial soil-borne microbes. *Trends in Plant Science* 15: 507–14.

Price-Whelan, A., Dietrich, L. E. P. and Newman, D. K. (2006). Rethinking 'secondary' metabolism: physiological roles for phenazine antibiotics. *Nature Chemical Biology* 2: 71–8.

Qiang, X., Weiss, M., Kogel, K. H., et al. (2012). *Piriformospora indica* – a mutualistic basidiomycete with an exceptionally large plant host range. *Molecular Plant Pathology* 13: 508–18.

Raaijmakers, J. M. and Weller, D. M. (1998). Natural plant protection by 2,4-diacetylphloroglucinol-producing *Pseudomonas* spp. in take-all decline soils. *Molecular Plant-Microbe Interactions* 11: 144–52.

Raaijmakers, J. M. and Weller, D. M. (2001). Exploiting genotypic diversity of 2,4-diacetylphloroglucinol-producing *Pseudomonas* spp.: characterization of superior root-colonizing *P. fluorescens* strain Q8r1–96. *Applied and Environmental Microbiology* 67: 2545–54.

Raaijmakers, J. M., de Bruijn, I., Nybroe, O., et al. (2010). Natural functions of lipopeptides from *Bacillus* and *Pseudomonas*: more than surfactants and antibiotics. *FEMS Microbiology Reviews* 34: 1037–62.

Rabiey, M., Ullah, I., Shaw, L. J., et al. (2017). Potential ecological effects of *Piriformospora indica*, a possible biocontrol agent, in UK agricultural systems. *Biological Control* 104: 1–9.

Reddy, M. V. B., Arul, J., Angers, P., et al. (1999). Chitosan treatment of wheat seeds induces resistance to *Fusarium graminearum* and improves seed quality. *Journal of Agricultural and Food Chemistry* 47: 1208–16.

Rice, E. L. (1984). *Allelopathy*. New York, NY: Academic Press.

Richardson, M. D. and Bacon, C. W. (1995). Catabolism of 6-methoxy-benzoxazolinone and 2-belazoxazolinone by *Fusarium moniliforme*. *Mycologia* 87: 510–17.

Rodriguez, R. J., Henson, J., Van Volkenburgh, E., et al. (2008). Stress tolerance in plants via habitat-adapted symbiosis. *ISME Journal* 2: 404–16.

Rodriguez, P., Gonzalez, D. and Rodríguez Giordano, S. (2017). Endophytic microorganisms: a source of potentially useful biocatalysts. *Journal of Molecular Catalysis B: Enzymatic* 133: S569–81. https://doi.org/10.1016/j.molcatb.2017.02.013.

Roget, D. K. (1995). Decline in root rot (*Rhizoctonia solani* AG-8) in wheat in a tillage and rotation experiment at Avon, South Australia. *Australian Journal of Experimental Agriculture* 35: 1009–13.

Schnider-Keel, U., Seematter, A., Maurhofer, M., et al. (2000). Autoinduction of 2,4-diacetylphloroglucinol biosynthesis in the biocontrol agent *Pseudomonas fluorescens* CHA0 and repression by the bacterial metabolites salicylate and pyoluteorin. *Journal of Bacteriology* 182: 1215–25.

Schoonbeek, H. J., Raaijmakers, J. M. and De Waard, M. A. (2002). Fungal ABC transporters and microbial interactions in natural environments. *Molecular Plant-Microbe Interactions* 15: 1165–72.

Schouten, A. (2016). Mechanisms involved in nematode control by endophytic fungi. *Annual Review of Phytopathology* 54: 121–42.

Schouten, A., van den Berg, G., Edel-Hermann, V., et al. (2004). Defense responses of *Fusarium oxysporum* to 2,4-diacetylphloroglucinol, a broad-spectrum antibiotic produced by *Pseudomonas fluorescens. Molecular Plant-Microbe Interactions* 17: 1201–11.

Schouten, A., Maksimova, O., Cuesta-Arenas, Y., et al. (2008). Involvement of the ABC transporter BcAtrB and the laccase BcLCC2 in defence of *Botrytis cinerea* against the broad-spectrum antibiotic 2,4-diacetylphloroglucinol. *Environmental Microbiology* 10: 1145–57.

Schulz, B. and Boyle, C. (2005). The endophytic continuum. *Mycological Research* 109: 661–86.

Schulz, M., Marocco, A., Tabaglio, V., et al. (2013). Benzoxazinoids in rye allelopathy – From discovery to application in sustainable weed control and organic farming. *Journal of Chemical Ecology* 39: 154–74.

Serfling, A., Wirsel, S. G. R., Lind, V., et al. (2007). Performance of the biocontrol fungus *Piriformospora indica* on wheat under greenhouse and field conditions. *Phytopathology* 97: 523–31.

Séveno, N. A., Kallifidas, D., Smalla, K., et al. (2002). Occurrence and reservoirs of antibiotic resistance genes in the environment. *Reviews in Medical Microbiology* 13: 15–27.

Shimabukuro, R. H. (1967). Atrazine metabolism and herbicidal selectivity. *Plant Physiology* 42: 1269–76.

Shipton, P. J. (1972). Take-all in spring-sown cereals under continuous cultivation: disease progress and decline in relation to crop succession and nitrogen. *Annals of Applied Biology* 71: 33–46.

Shipton, P. J. (1975). Take-all decline during cereal monoculture. In: Bruehl, G. W. (Ed.), *Biology and Control of Soil-borne Plant Pathogens*. St. Paul, MN: The American Phytopathological Society, pp. 137–44.

Sieber, S. A. and Marahiel, M. A. (2005). Molecular mechanisms underlying nonribosomal peptide synthesis: approaches to new antibiotics. *Chemical Reviews* 105: 715–38.

Sikora, R. A. (1992). Management of the antagonistic potential in agricultural ecosystems for the biological control of plant parasitic nematodes. *Annual Review of Phytopathology* 30: 245–70.

Slesak, E., Slesak, M. and Gabrys, B. (2001). Effect of methyl jasmonate on hydroxamic acid content, protease activity, and bird cherry-oat aphid *Rhopalosiphum padi* (L.) probing behavior. *Journal of Chemical Ecology* 27: 2529–43.

Søltoft, M., Jorgensen, L. N., Svensmark, B., et al. (2008). Benzoxazinoid concentrations show correlation with Fusarium Head Blight resistance in Danish wheat varieties. *Biochemical Systematics and Ecology* 36: 245–59.

Song, Y. Y., Cao, M., Xie, L. J., et al. (2011). Induction of DIMBOA accumulation and systemic defense responses as a mechanism of enhanced resistance of mycorrhizal corn (*Zea mays* L.) to sheath blight. *Mycorrhiza* 21: 721–31.

Steffens, J. J., Pell, E. J. and Tien, M. (1996). Mechanisms of fungicide resistance in phytopathogenic fungi. *Current Opinion in Biotechnology* 7: 348–55.

Stone, J. K., Polihook, J. D. and White, J. R. J. F. (2004). Endophytic fungi. In: Mueller, G., Bills, G. F. and Foster, M. S. (Eds), *Biodiversity of Fungi: Inventory and Monitoring Methods*. Burlington, Canada: Elsevier Inc., pp. 241–70.

Thackray, D. J., Wratten, S. D., Edwards, P. J., et al. (1990). Resistance to the aphids *Sitobion avenae* and *Rhopalosiphum padi* in Gramineae in relation to hydroxamic acid levels. *Annals of Applied Biology* 116: 573–82.

Thomashow, L. S. and Weller, D. M. (1988). Role of a phenazine antibiotic from pseudomonas-fluorescens in biological-control of *Gaeumannomyces graminis* var *tritici*. *Journal of Bacteriology* 170: 3499–508.

Toohey, J. I., Nelson, C. D. and Krotkov, G. (1965). Toxicity of phenazine carboxylic acids to some bacteria algae higher plants and animals. *Canadian Journal of Botany* 43: 1151–5.

Torres, A., Olivia, R. M., Castellano, D., et al. (1996). A science of the future. *1st World Congress on Allelopathy*. Cadiz, Spain: University of Cadiz, p. 278.

Turner, J. M. and Messenger, A. J. (1986). Occurrence, biochemistry and physiology of phenazine pigment production. *Advances in Microbial Physiology* 27: 211–75.

van Pée, K. H. and Ligon, J. M. (2000). Biosynthesis of pyrrolnitrin and other phenylpyrrole derivatives by bacteria. *Natural Product Reports* 17: 157–64.

VanEtten, H. D., Matthews, D. E. and Matthews, P. S. (1989). Phytoalexin detoxification – Importance for pathogenicity and practical implications. *Annual Review of Phytopathology* 27: 143–64.

VanEtten, H. D., Mansfield, J. W., Bailey, J. A., et al. (1994). 2 classes of plant antibiotics: phytoalexins versus phytoanticipins. *Plant Cell* 6: 1191–2.

VanEtten, H., Temporini, E. and Wasmann, C. (2001). Phytoalexin (and phytoanticipin) tolerance as a virulence trait: why is it not required by all pathogens? *Physiological and Molecular Plant Pathology* 59: 83–93.

Verma, S., Varma, A., Rexer, K. H., et al. (1998). Piriformospora indica, gen. et sp. nov., a new root-colonizing fungus. *Mycologia* 90: 896–903.

Vilich, V., Lohndorf, B., Sikora, R. A., et al. (1999). Metabolism of benzoxazolinone allelochemicals of *Zea mays* by *Fusarium subglutinans*. *Mycological Research* 103: 1529–32.

Vincent, C., Goettel, M. and Lazarovits, G., Eds. (2007). *Biological Control: A Global Perspective*. CABI, Wallingford, UK. doi: 10.1079/9781845932657.0000.

Virtanen, A. I. and Hietala, P. K. (1955). 2(3)-Benzoxazolinone, an anti-fusarium factor in rye seedlings. *Acta Chemica Scandinavica* 9: 1543–4.

Wahlroos, O. and Virtanen, A. I. (1959). The precursors of 6-methoxy-benzoxazolinone in maize and wheat plants, their isolation and some of their properties. *Acta Chemica Scandinavica* 13: 1906–8.

Walker, T. S., Bais, H. P., Grotewold, E., et al. (2003). Root exudation and rhizosphere biology. *Plant Physiology* 132: 44–51.

Weller, D. M., Landa, B. B., Mavrodi, O. V., et al. (2007). Role of 2,4-diacetylphloroglucinol-producing fluorescent *Pseudomonas* spp. in the defense of plant roots. *Plant Biology* 9: 4–20.

Whittaker, R. H. and Feeny, P. P. (1971). Allelochemics: chemical interactions between species. *Science* 171: 757–70.

Wilkes, M. A., Marshall, D. R. and Copeland, L. (1999). Hydroxamic acids in cereal roots inhibit the growth of take-all. *Soil Biology and Biochemistry* 31: 1831–6.

Willis, R. J. (2007). *The History of Allelopathy*. Dordrecht, the Netherlands: Springer.

Wilson, D. (1995). Endophyte – the evolution of a term, and clarification of its use and definition. *Oikos* 73: 274–6.

Wiseman, B. M., Neate, S. M., Keller, K. O., et al. (1996). Suppression of *Rhizoctonia solani* anastomosis group 8 in Australia and its biological nature. *Soil Biology & Biochemistry* 28: 727–32.

Wouters, F. C., Gershenzon, J. and Vassão, D. G. (2016). Benzoxazinoids: reactivity and modes of action of a versatile class of plant chemical defenses. *Journal of the Brazilian Chemical Society* 27: 1379–97.

Yang, M. M., Wen, S. S., Mavrodi, D. V., et al. (2014) Biological control of wheat root diseases by the CLP-producing strain *Pseudomonas fluorescens* HC1-07. *Phytopathology* 104: 248–56.
Yue, Q., Bacon, C. W. and Richardson, M. D. (1998). Biotransformation of 2-benzoxazolinone and 6-methoxy-benzoxazolinone by *Fusarium moniliforme*. *Phytochemistry* 48: 451–4.
Zamioudis, C., Mastranesti, P., Dhonukshe, P., et al. (2013). Unraveling root developmental programs initiated by beneficial *Pseudomonas* spp. bacteria. *Plant Physiology* 162: 304–18.
Zbeng, Y. Q., Zhao, Y., Liu, X. G., et al. (2008). Chemical inducement of 2,4-dihydroxy-7-methoxy-1,4-benzoxazin-3-one (DIMBOA) in wheat seedlings. *Allelopathy Journal* 21: 263–71.

Chapter 16

The role of crop rotation, intercropping and tillage practices for foliar disease management of wheat and barley

T. K. Turkington, Agriculture and Agri-Food Canada, Canada; K. Xi, Alberta Agriculture and Forestry, Canada; and H. R. Kutcher, University of Saskatchewan, Canada

1 Introduction

There have been a range of studies assessing the factors which predispose modern agriculture to biotic stresses such as plant diseases (Apple, 1977; Barnes, 1964; Cowling, 1978; Harlan, 1972; Paddock, 1967; Zadoks and Schein, 1979; Zhan et al., 2015). Boudreau and Mundt (1997) have characterized modern crop production as highly disrupted and simplified agro-ecosystems in which the stabilizing mechanisms that function in natural ecosystems have been significantly weakened. This can be seen, for example, in the use of uniform cultivars with reduced genetic diversity (Cowling, 1978; Harlan, 1972; Zadoks and Schein, 1979; Zhan et al., 2015). The narrow genetic basis for disease resistance in modern cultivars increases the potential for pathogens to overcome this resistance (Harlan, 1972). Modern agriculture results in greater plant uniformity, higher planting density and reduced rotations (which increased spatial and temporal uniformity) (Apple, 1977; Harlan, 1972; Wolfe, 1978; Zadoks and Schein, 1979; Zhan et al., 2015). This results in large blocks of genetically and morphologically uniform plants grown close together, resulting in dense crop canopies with stable microenvironments, which can allow pathogens to adapt quickly and attack a large number of plants at a similar state of development (Zadoks and Schein, 1979).

To counteract the predisposition of modern agro-ecosystems to plant diseases it has been suggested that cropping systems try to emulate some of the stabilizing

http://dx.doi.org/10.19103/AS.2018.0039.21
Published by Burleigh Dodds Science Publishing Limited, 2019.

characteristics of natural ecosystems (Apple, 1977; Zadoks and Schein, 1979). Zadoks and Schein (1979) have described a natural ecosystem as one in which 'disease is always present, but the host's genetic makeup, the population's genetic diversity and the dispersion of genotypes combine to prevent extreme disease growth rates'. Diversity is an integral component of a natural ecosystem that can be used to provide greater stability in modern agro-ecosystems with regard to disease management. In this chapter, diversity in time and space will be discussed in relation to managing cereal leaf diseases, primarily in wheat and barley.

2 Increasing temporal diversity: crop rotation

Karlen et al. (1994) have described the shift away from the use of legumes and other crops in rotations after the Second World War, and the increasing reliance on synthetic fertilizers and pesticides to lessen the need for management via 'extended rotations'. Farmers increasingly concentrated on growing a few crop species in a 'crop monoculture', sometimes referred to as a fixed cropping system or rotation (Finckh and Wolfe, 1997; Black et al., 1974; Tanaka et al., 2002). Because they are highly simplified, with a small number of crop components, these systems are more vulnerable to abiotic (weather, soil conditions etc.) and biotic (diseases, insects etc.) stresses (Finckh and Wolfe, 1997; Tanaka et al., 2002; Wolfe, 2002).

Crop rotation can be one of the most effective and sustainable methods of managing a number of plant diseases. However, it must be stressed that important factors such as relative commodity prices, net returns, ease of production/marketing, availability of storage facilities and on-farm feed/forage requirements are also very important considerations for farmers when planning crop rotations of sufficient length for disease management (Bailey et al., 2000; Cook, 2000; Cook and Veseth, 1991; Harker et al., 2015; Johnston et al., 2005; Karlen et al., 1994; Kutcher et al., 2011, 2013; Smith et al., 2013, 2015; Tanaka et al., 2002).

Crop rotation can be thought of as a biological method of disease management (Cook and Veseth, 1991). Rotation to a non-host(s) for a sufficient period allows enough time for decomposition of infected crop residues, and/or a reduction in the viability of pathogen survival structures and the pathogen's ability to produce inoculum, thus eliminating a potential source of disease. As Cook and Veseth (1991) indicated in their book *Wheat Health Management* '. . . rotation allows time for natural enemies to destroy the pathogens of one crop while one or preferably two unrelated crops are grown'. They also indicated that rotation acts like a natural type of 'soil fumigation' where the collective activity of 'antibiotic, predatory, and competitive organisms' help to eliminate plant pathogens from soil and infected crop residues.

The success of rotations as a form of biological disease control is dependent on a number of factors including time, the environment, the nature of the pathogen being managed and the characteristics of the host crop. For rotation to be an effective disease management strategy, farmers need to keep in mind that certain requirements or conditions must be met. Firstly, the majority of the disease inoculum must be from the field itself with limited movement of plant pathogen inoculum from adjacent fields. Consequently, crop rotation will tend to be most effective for those diseases that are soil- or residue-borne. Crop rotation will not be effective for diseases where seed-borne pathogens are an important source of disease, such as the cereal smuts. Furthermore,

Published by Burleigh Dodds Science Publishing Limited, 2019.

rotation will not be effective for cereal rusts as their causal agents are mobile and readily move from one field to another or over long distances. For example, cereal rusts, especially for wheat and barley, will overwinter on cereals and grasses in the southern United States and northern Mexico, while stripe rust can also overwinter in the Pacific Northwest (PNW) and California (Mathre, 1997; Menzies and Gilbert, 2003; Tekauz, 2003; Wiese, 1987). Urediniospores are blown northward by wind currents, affecting successive northerly winter and spring cereal crops (Aylor, 1990, 2003; Brown and Hovmøller, 2002; Chen, 2005; Eversmeyer et al., 1984; Eversmeyer and Kramer, 2000; Menzies and Gilbert, 2003; Nagarajan and Singh, 1990). However, viable overwintering inoculum in winter wheat is considered to play a role in causing local epidemics of wheat stripe rust (*Puccinia striiformis* Westend.) in central Alberta (Kumar et al., 2013). Thus, rotation away from wheat and barley may help manage stripe rust in these types of regions. Similarly, in winter-grown short season crops, as in Australia, the elimination of the 'Green Bridge', susceptible hosts growing during the summer, can substantially reduce disease levels.

A second major consideration regarding the use of a diverse crop rotation to manage disease is that the host range of the pathogen must be narrow. Crop rotation amongst different small grain cereals such as wheat, barley and oat as well as corn is an effective strategy for management of leaf spot diseases including tan spot, the septorias, scald and the net blotches. However, rotations amongst these crops is not an effective strategy for a disease like *Fusarium* head blight, caused by *Fusarium graminearum* Schwabe *sensu lato*, as this pathogen has an extensive host range that includes all these small grain cereals and corn (Mathre, 1997; Wiese, 1987; Sutton, 1982; Parry et al., 1995; Stack, 2000; Farr et al., 1989; Ginns, 1986). The final condition for rotation to be effective is that pathogens are unable to survive for long periods in the absence of a suitable host. Thus, there should be a rapid reduction in pathogen viability and its capacity to produce infective structures, that is spores, within a relatively short period of time. Rotation will not be as effective for those pathogens that have the capacity to produce long-lived resting structures or where they can survive in infected plant tissues that resist decay. For example, survival of *F. graminearum* in crop residue is highest in plant tissues that are resistant to decay, especially the node tissues of small grain cereals (Sutton, 1982; Burgess and Griffin, 1968).

2.1 Limiting cereal leaf diseases using crop rotation

Crop rotation can be an effective tool to lower the risk of cereal leaf diseases in wheat and barley (Bailey et al., 2000; Johnston et al., 2005; Krupinsky et al., 2006; Kutcher et al., 2011; Menzies and Gilbert, 2003; Tekauz, 2003; Turkington et al., 2006, 2012; Wiese, 1987). Some studies have shown that rotation is not always effective. As an example, Bailey et al. (2000) and Johnston et al. (2005) found that planting barley on non-host residues did not necessarily decrease the risk of net blotch of barley. In contrast, Turkington et al. (2006) found that in commercial barley fields in Alberta, from 1995 to 1997, lower levels of scald (*Rhynchosporium secalis* (Oudem.) J. J. Davis) and net blotch (*Pyrenophora teres* Drechs.) tended to occur when barley was planted in fields previously cropped to a non-host.

Krupinsky et al. (2004) found lower barley leaf spot levels, primarily net blotch, when barley was grown on non-host residue, especially wheat, canola and field peas. In a large crop sequencing study, Krupinsky et al. (2006) found that barley productivity was reduced when grown on barley residues versus non-host residues. More recently, Kutcher et al. (2011) found increased barley leaf spot severity, primarily net blotch, when barley was grown on barley residue versus non-host residues.

Turkington et al. (2012) have also found the largest impact on the barley leaf spot complex resulted from crop rotation versus fungicide application or nitrogen fertilizer rate. The largest reductions in disease and increases in grain yield and kernel size resulted from planting barley on either field pea or canola residue versus barley on barley residue. In general, for scald a rotation of 1–2 years away from a susceptible host may be sufficient to allow for the destruction of most infected plant tissue, since the scald pathogen primarily infects leaves and leaf sheaths, which decompose fairly rapidly (Mathre, 1997; Mayfield and Clare, 1984a,b). However, a longer rotation will be needed for net blotch, since the fungal pathogen that causes this disease also infects barley stems, which are more resistant to decay and will persist, especially the node tissue, for a longer period of time (Duczek et al., 1999; Mathre, 1997).

Crop rotation is also important in the management of wheat leaf spot diseases. Bailey et al. (1992) found that leaf spot severity was higher in wheat following spring or winter wheat compared with field peas or a summer fallow period. Bockus and Claassen (1992) found reduced tan spot (*Pyrenophora tritici-repentis* (Died.) Drechsler) levels with wheat grown on sorghum versus wheat residues, while wheat yields increased with rotation to sorghum, but only when adequate moisture levels occurred following the sorghum crop. In the Saskatchewan Parkland, Pedersen and Hughes (1992) found that when environmental conditions were favourable for disease development *Septoria nodorum* blotch (*Parastagonospora nodorum* (Berk.) Quaedvlieg, Verkley, and Crous) severity was reduced in wheat by a rotation with 2 years between wheat crops, but not for rotations with only 1 year between wheat crops. However, when environmental conditions were less conducive to disease, a rotation with 1 year out of wheat reduced *Septoria nodorum* blotch severity to the same extent as a rotation with 2 years between wheat crops.

In research trials and in commercial wheat fields in southwestern Saskatchewan, Fernandez et al. (1998) reported that leaf spot severity was not always lower in wheat planted after a single non-host compared with continuous wheat. They concluded that at least 2 years out of wheat were required for decomposition of infected residues to reduce wheat leaf spot levels. In a subsequent survey of commercial wheat fields, Fernandez et al. (2009) found a limited effect of rotation on wheat leaf spot severity, while in another study Fernandez et al. (2016) found reduced leaf spot severity when wheat followed a non-host or fallow.

Bailey et al. (2001) reported reduced leaf disease when wheat was grown on summer fallow or field pea residue versus wheat residue, and that wheat yields were increased when wheat was not grown on wheat residues. Krupinsky et al. (2004) found lower wheat leaf spot levels (tan spot and *Septoria nodorum* blotch), when wheat was grown on non-host residue, especially canola, barley and flax, but yields tended to be similar to wheat on wheat residue. In a large crop sequencing study, Krupinsky et al. (2006) also found reduced wheat productivity when grown on wheat residues versus non-host residues. Johnston et al. (2005) found that increased leaf spot severity and reduced yields were observed when wheat was planted on wheat residues versus a non-host.

In a review of the impact of break crops on wheat, Angus et al. (2015) reported improved wheat yields following non-host break crops versus wheat on wheat based on a combined analysis of data from Australia, Europe and North America. They focused on

Published by Burleigh Dodds Science Publishing Limited, 2019.

the potential impact of break crops primarily on root diseases as well as the supply of nitrogen and moisture. In Latvia, Bankina et al. (2015) found that rotation of wheat with non-host crops reduced tan spot, but was less effective for *Septoria tritici* blotch, except when there was more than 1 year between wheat crops. In subsequent research, Bankina et al. (2018) also found increased tan spot levels under continuous wheat versus rotations with non-hosts, especially under reduced tillage with a disc implement versus ploughing. Andert et al. (2016) investigated the impact of crop rotation on the intensity of fungicide use in Germany for a range of crops and found that continuous production of the same crop type for 3 years increased use intensity in winter wheat, winter barley and fall rye versus more diverse rotations. In Uruguay, Mazzilli et al. (2016) analysed the impact of the previous crop on wheat yields and found yield losses associated with planting wheat on wheat residue. However, yields increased when a disease-resistant cultivar was grown and the authors attributed the negative impact of growing wheat on wheat to the increased impact of leaf diseases.

For some cereal leaf spot diseases, the potential for seed-borne inoculum and seed-to-seedling transmission may negate the benefit of extended rotation intervals. For example, Turkington et al. (2006) found substantial levels of both scald and net blotch developed in fields not planted to barley or forage grasses for the previous 4 years. Although they did not measure seed-borne inoculum levels, in another study of seed-borne barley pathogens Turkington et al. (2002) found *Pyrenophora teres* (cause of net blotch) infected seed as much as 89%, with average annual levels of infection that ranged from 15% to 37%. In Alberta, Skoropad (1959), Lee et al. (1999) and Xi and Burnett (1997) found that *Rhynchosporium secalis*, the causal agent of scald, infected barley awns, lemmas, paleas and pericarps. The movement of scald into scald-free areas via infected seed has been suggested by a number of researchers (Habgood, 1971; Reed, 1957; Skoropad, 1959). Research in the 1950s and 1970s demonstrated transmission of *R. secalis* from infected seeds to seedlings mainly under greenhouse conditions (Habgood, 1971; Jackson and Webster, 1976; Kay and Owen, 1973; Reed, 1957; Skoropad, 1959). Reed (1957) also observed that infected seed that remained on the soil surface after seeding could act as a source of inoculum. In some countries, including India, New Zealand and England seed-borne inoculum of net blotch is considered to be an important source of disease (Hampton, 1980; Jordan, 1981; Sheridan et al., 1983; Shipton et al. 1973; Singh and Chand 1985).

Research from the United States has indicated that seed infection of wheat may be a factor to consider when using rotation as a means of cereal leaf spot management. In Florida, Luke et al. (1983) found decreased development of *Septoria nodorum* blotch, caused by *P. nodorum*, with either a 1 or 2 year rotation using uninfected seed, but not when *P. nodorum* infected seed was used. Similar results were also found by Milus and Chalkley (1997) in Arkansas where low (7%) and high (34–40%) levels of seed infection with *P. nodorum* led to significant leaf infection in an experiment planted in an area not cropped to wheat for at least the previous 4 years. They concluded that crop rotation should be combined with appropriate seed treatment to help prevent outbreaks of *Septoria nodorum* blotch. In New York State, Shah et al. (1995) found significant relationships between *P. nodorum* seed infection levels of up to 40% and subsequent development of *Septoria nodorum* blotch in winter wheat. Seed to seedling transmission of *P. tritici-repentis*, the causal agent of tan spot of wheat, has been demonstrated under greenhouse conditions and under laboratory conditions and in artificial potting media under outside conditions (Luz et al., 1998; Schilder and Bergstrom, 1995).

3 Increasing spatial diversity: intercropping

Most research on intercropping has focused on agronomic factors and crop productivity (Boudreau and Mundt, 1997). However, intercropping of different crop species to increase the amount of diversity within an individual field can also improve disease management (Boudreau and Mundt, 1997; Finckh and Wolfe, 1998; Kantor, 1999; Sullivan, 2001; Wolfe, 2002). Four main types of intercropping systems have been identified:

- Mixed intercropping, where two or more crops are grown in the same field with no distinct planting pattern.
- Row intercropping, where two or more crops are grown in the same field with at least one of the components grown in distinct rows.
- Strip intercropping, where two or more crops are grown in the same field, but in strips that are wide enough to facilitate the use of farm equipment.
- Relay intercropping, where an additional crop is planted into an existing plant stand.

Burdon (1978) and Boudreau and Mundt (1997) have reviewed the ways intercropping can limit disease development. Because the proportion of susceptible host tissue decreases within the crop, intercropping reduces the production, amount and effectiveness of inoculum available for further spread and development within the crop. Secondly, increasing the space between susceptible hosts within crops also results in pathogen inoculum having to travel greater distances, impeding disease development. A third mechanism involves interception or filtering of pathogen propagules by the non-host component of the intercrop. As an example, Bannon and Cooke (1998) found that the clover in an intercrop with wheat reduced the dispersal of *Zymoseptoria tritici* (Desm.) Quaedvlieg & Crous pycnidiospores horizontally from an inoculum source or vertically from lower infected leaves. A fourth mechanism, cross-protection, may be similar to the induced resistance mechanism that has been suggested to occur for mixtures of the same crop species where non-virulent pathogen races trigger defence responses in the host, which then limit the impact of virulent races; however, cross-protection would involve non-host crops and non-pathogenic organisms (Boudreau and Mundt, 1997; Brown, 1975; Burdon, 1978). One other potential mechanism may occur via an influence on the microenvironment. Variability within the intercrop as a result of the presence of morphologically different crop components or an influence via an individual component of the intercrop canopy may produce less favourable microenvironmental conditions, leading to a reduction in disease development (Boudreau and Mundt, 1997; Burdon, 1978).

Hummel et al. (2009) studied the impact of intercropping wheat and canola on crop productivity and disease management. There was a negative relationship between wheat leaf spot severity (mainly the *Septoria* leaf spot complex) and increasing proportions of canola in the intercropping treatments. However, at one site, levels of disease were higher for the intercropped treatment versus wheat alone, while at the other site disease levels were similar. Overall, yield differences between intercropped and monoculture wheat were limited and Hummel et al. (2009) concluded there was insufficient benefit to recommend adoption of intercropping. Pridham and Entz (2008) found that intercropping of wheat with flax or field pea reduced leaf disease levels in wheat, but yields were variable and not consistently better for the intercropping treatments. Bulson et al. (1997) compared wheat only with varying proportions of wheat and field bean and found that intercropping increased the level of powdery mildew (*Blumeria graminis* (DC) Speer f. sp. *Tritici* emend.

É. J. Marchal). Although yields were highest for the wheat alone sown at the recommended density, land equivalent ratios (LER) were higher with the intercrop, especially when wheat was sown at the recommended density of 75%. Bedoussac et al. (2008) found that durum wheat diseases were not reduced by growing an intercrop of durum and winter pea. More recently, Chapagain (2014) found that intercropping of wheat and common bean or faba bean did not influence the level of disease compared with wheat monoculture. In contrast, Zhu et al. (2017) found intercropping of wheat and faba bean reduced powdery mildew levels and improved yield.

In barley, Kinane and Lyngkaer (2002) detected significant reductions in net blotch severity only when barley was intercropped with field pea, while none of the intercropping treatments reduced leaf rust severity. Chapagain (2014) reported that intercropping of barley and field pea did not influence the level of disease compared with barley monoculture. Vilich-Meller (1992) found that intercropping of spring barley or oat reduced powdery mildew (*B. graminis* f.sp. *hordei*) in barley and leaf blotch (*Drechslera avenacea* (M.A. Curtis ex Cooke) Shoemaker) in oat. The impact of intercropping barley with a living mulch of kura clover was studied by Kosinski et al. (2011) in 2006 and 2007. In both years they found that spot-form net blotch severity was significantly reduced in rotations containing kura clover intercropped with barley versus barley alone. However, silage yields in 2006 were compromised by competition from the Kura clover in rotation compared with the barley only. In contrast, slight improvements in overall silage yield occurred with the intercropping treatment in 2007, although Kura clover made up the majority of the silage yield. Recently, a 9 year trial based on a 3 year rotational cycle was completed at two sites in Alberta, Canada, that assessed the impact of intercropping of barley variety mixtures, and variety rotation on silage production (Turkington et al., 2015a,b). For both locations leaf disease, primarily net-form net blotch, was generally highest for continuous production of the same barley variety, and lowest for mixtures or intercrops with barley, oat and triticale, especially where the variety components changed each year. Silage yields tended to be lowest for the continuous barley with the same variety, highest for the intercropping treatments with the same or different varieties, and intermediate for barley mixtures, especially when disease levels were increased. Overall results indicated that the addition of diversity in terms of different crop types and barley genetics can help to reduce the level of leaf disease and improve silage productivity.

4 Increasing genetic diversity: gene deployment

As noted earlier, there has been a reduction in genetic diversity in modern crop production in favour of uniform cultivars that meet a variety of agronomic, yield and quality requirements (Cowling, 1978). The development of a narrow range of modern high-yielding cultivars has been at the expense of genetic variability, presumably reducing resistance to plant pests (Harlan, 1972; Cowling, 1978; Wolfe, 2002). To address the problem of limited genetic diversity within our modern cropping systems, a number of authors have recommended the use of multilines, variety mixtures, gene deployment as well as the use of intercropping (Apple, 1977; Boudreau and Mundt, 1997; Browning and Frey, 1969; Cowling, 1978; Cox et al., 2004; Finckh et al., 2000; Finckh and Wolfe, 1998; Frey et al., 1977; Fry, 1982; Garrett and Mundt, 1999; Harlan, 1972; McDonald and Linde, 2002a,b; Newton et al., 2009; Priestly, 1981; Wolfe, 1985, 2002; Zadoks and Schein, 1979).

Gene deployment has been suggested as a way to add diversity to our modern cropping systems (Adugna, 2004; Burdon et al., 2014; Finckh and Wolfe, 1998; Frey et al., 1977; Fry, 1982; Priestly, 1981; McDonald and Linde, 2002a,b; REX Consortium, 2016; Zhan et al., 2015). Gene deployment can be achieved via spatial deployment locally or regionally or gene rotation over time and may involve near-isogenic lines or the use of diverse crop varieties with different resistance genes. For example, resistance genes deployed in one area may be effective against pathogen races originating from another region where crop varieties with different resistance genes are being grown. Gene deployment has been suggested as a strategy to increase the durability of resistance genes and reduce the spread and development of cereal rusts in North America (Frey et al., 1973; Fry, 1982). On a smaller spatial scale, Priestley (1981) suggested that disease spread can be restricted if neighbouring fields are planted to cultivars with different resistance genes. Priestley (1981) and Priestly and Wolfe (1977) developed diversification schemes to assist farmers with on-farm deployment of barley and winter wheat cultivars among farms within a single year and in successive years to reduce the risk of powdery mildew and stripe rust, respectively.

Temporal deployment of cultivars with different resistance genes (gene rotation) may also be a potential method of adding a small level of diversity to a monoculture system (Finckh and Wolfe, 1998), where crop rotation may not be a viable option. For example, diversity in time with regard to different crop species for an individual field may be difficult if not impossible to implement for livestock operations, given on-farm feed requirements and market factors. Continuous barley production leads to a build-up of disease in these fields and a general reduction in yield potential over the long-term (Mathre, 1997; Tekauz, 2003). One option that farmers may use to help reduce leaf diseases, while maintaining productivity, is to introduce some level of diversity by deploying different varieties in subsequent growing seasons. By changing varieties, the farmer has the potential to change the disease resistance genes in use each year, thereby reducing the impact of disease. For example, Turkington et al. (2005) demonstrated that growing the same barley variety for 3 years in a row resulted in significantly higher scald and net blotch severity and lower yield and grain size compared to barley variety rotation. However, reductions in disease and improvements in yield and grain size were highest when barley was grown on triticale versus barley on barley residue.

5 The role of conservation tillage

The use of tillage as a form of sanitation has been advocated as a traditional method of cereal leaf disease control (Maloy, 2005; Mathre, 1997; Menzies and Gilbert, 2003; Palti, 1981; Tekauz, 2003; Turkington, 2003; Wiese, 1987). Tillage may be extensive, for example mouldboard ploughing or the use of a disker followed by several cultivator passes and harrowing to prepare the soil for seeding as well as weed management (Daigh and DeJong-Hughes, 2017; DeJong-Hughes and Daigh, 2017a). Alternatively, tillage may be less aggressive and accomplished using chisel ploughs/shovel-equipped cultivators with several passes over the field followed by harrowing and seeding. The combination of breaking up crop residues into smaller pieces coupled with mixing and burial of crop residues in the soil can help to accelerate decomposition of pathogen-infected crop residues, thus removing a potential source of inoculum. Tillage also places infected crop residues beneath a layer of soil where pathogen fruiting bodies and/or spores are unable to be dispersed and infect above-ground plant parts (Palti, 1981). Unfortunately, tillage can have significant negative effects on soil health and sustainability and promotes water evaporation (Daigh and DeJong-Hughes, 2017; DeJong-Hughes and Daigh, 2017b; Derpsch and Freidrich, 2009; Derpsch, 2015; Friedrich et al., 2012; Soane et al., 2012).

As a result there has been growing interest in and implementation of conservation tillage for crop production (Awada et al., 2014; Derpsch, 2015; Derpsch and Friedrich, 2009; Kassam et al., 2015). Conservation tillage is seen as a method of ensuring environmentally sustainable agricultural production that enhances soil quality, prevents erosion, promotes soil biological activity and diversity, and reduces labour and fuel costs (Awada et al., 2014; Busari et al., 2015; Daigh and DeJong-Hughes, 2017; DeJong-Hughes and Daigh, 2017b; Derpsch, 2015; FAO, 2015; Friedrich et al., 2012; Soane et al., 2012). In contrast, conventional tillage represents a significant disruption to the soil and soil surface environments where the stabilizing characteristics and mechanisms that function in natural ecosystems are greatly impaired. Unfortunately, the direct impact of conservation tillage is the retention of crop residues and pathogen propagules at or just below the soil surface, which may then act as an inoculum source, thus increasing the risk of disease (Awada et al., 2014; Coutts and Smith, 1991; Evans and Fleury, 1993; Leduc, 1994; Watkins and Boosalis, 1994). An overview of the potential impact of conservation versus conventional tillage on cereal leaf diseases will be presented.

Under conservation tillage, moisture levels are increased at the soil surface (Steiner, 1994), the rate of evaporation is reduced (Sadler and Turner, 1994; Steiner, 1994) and there is a general reduction in temperature and dampening of daily fluctuations (Clegg and Francis, 1994; Horton et al., 1994). Increased moisture levels and cooler temperatures close to the soil surface may increase disease risk (Rothrock, 1992). Alternatively, there may be increased population levels, activity and diversity of various organisms in the soil surface environment and this may help to reduce the risk of disease (Cochran et al., 1994; Cook and Baker, 1983; Cook and Veseth, 1991; Lupwayi et al., 1998). Stubble retention on the soil surface may decrease the amount of photosynthetically active radiation at the soil surface as a result of shading, but residue will also offer physical protection of the surface helping to prevent wind and water erosion (Alberts and Neibling, 1994; Fryrear and Bilbro, 1994; Wilkins et al., 1988). As Watkins and Boosalis (1994) have suggested, retention of stubble on the soil surface may influence overall plant health and this may either enhance or restrict the plant's ability to withstand pathogen attack.

In general, the influence of conservation tillage on above-ground plant diseases may be fairly obvious; the more infected crop residue on the soil surface the greater the disease risk (Watkins and Boosalis, 1994). Residue-borne barley and wheat leaf disease pathogens sporulate on infected plant residues and these spores then serve as primary sources of inoculum (Mathre, 1997; Menzies and Gilbert, 2003; Tekauz, 2003; Wiese, 1987). Because there is an abundant reservoir of infected residue on the soil surface under conservation tillage, disease may appear earlier and subsequent development may be accelerated producing greater disease levels at crop maturity. Pittelkow et al. (2015) conducted a meta-analysis using a global dataset and while not directly comparing no-till and conventional tillage in terms of plant disease levels, they found an overall 5.7% reduction in yield under no-till versus conventional tillage. However, when they focused on dry land crop production in semi-arid regions, no-till resulted in a 7.3% increase in yield when it was combined with residue retention and crop rotation and they attributed this to 'improved water infiltration and greater soil moisture conservation'. Research investigating the impact of conservation tillage on cereal leaf diseases has been conducted since the mid- to late 1980s in the prairies of western Canada, where there has been extensive adoption of conservation tillage starting in the 1990s. Traditionally in western Canada the most common form of conventional tillage is the use of a heavy duty cultivator (chisel plough); mouldboard ploughing is uncommon (Green, 1984). In western Canada, typical conventional tillage practices leave between 50% and 80% residue cover after each tillage operation, depending on the tillage implement used (Jensen and Timmermans, 1991; MAFRI, 2008), but these residues still act as a significant source of pathogen inoculum. Although ploughing leaves behind a smaller amount of residue, subsequent tillage operations related to seedbed preparation likely bring infected crop residues back to the soil surface.

From 1994 to 2001, Bailey et al. (2000) and Kutcher et al. (2011) studied the impact of various crop and pest management practices on managing leaf diseases in cereals. Tillage treatments included conventional tillage with tillage operations in the previous fall and the following spring before seeding; minimum tillage with only spring tillage prior to seeding; and zero tillage with no tillage operations before seeding. Levels of net blotch in barley and a combination of tan spot and the *Septoria* leaf spot complex in wheat were similar for all tillage regimes, while barley and wheat yields were similar amongst tillage systems in most years except barley in 1994 and 2001 where yields were significantly higher for zero tillage versus conventional tillage. The latter year was very dry with low severity of leaf spot diseases and the yield benefit of zero tillage was attributed to the increase in soil moisture under zero tillage. In the latter study, Kutcher et al. (2011) detected similar levels of the leaf spot complex in wheat among tillage systems.

The lack of a tillage effect on cereal leaf diseases was also previously reported by Bailey et al. (1992), where the tillage system did not significantly affect tan spot severity in 3 of 4 years of the trial. In 1992 and 1993, Bailey (1996) reported slightly higher overall leaf spot severity under conventional tillage versus zero tillage, while isolation of the causal agents of speckled leaf blotch (*S. tritici*) and tan spot (*P. tritici-repentis*) was slightly higher under zero tillage. In contrast no differences in isolation of the causal agent of leaf/glume blotch (*P. nodorum*) occurred, while isolation of the spot blotch fungus (*Bipolaris sorokiniana* (Sacc.) Shoemaker) was lower overall, and was slightly lower under zero tillage. From 1991 to 1998, Bailey et al. (2001) found wheat leaf spot levels and grain yields were similar under zero, minimum and conventional tillage. In a survey of commercial wheat fields in Saskatchewan, Fernandez et al. (2009) reported that there was limited impact of tillage system on overall leaf spot severity, but there were effects on the percentage isolation of causal agents. Isolation of the causal agent of tan spot increased under conservation tillage, while isolation of the spot blotch pathogen was increased under conventional tillage, with isolation of the causal agents of the *Septoria* leaf spot complex reduced under zero tillage. Similar results were found in Manitoba by Gilbert and Woods (2001) where under conservation tillage there was a decrease in isolation of the causal agents of the *Septoria* leaf spot complex, but an increase in isolation of the tan spot fungus compared with conventional tillage. In a survey of commercial barley fields in 1995–7 in Alberta, Turkington et al. (2006) reported that the probability of higher levels of scald or net blotch in fields under conventional, minimum or zero tillage was similar regardless of the tillage system used.

In earlier work by Jordan and Allen (1984) in the United Kingdom, higher net blotch inoculum loads, disease levels and earlier disease development were reported when crop residues remained on the soil surface versus ploughing of crop residues into the soil. Bockus and Claassen (1992) found that the ploughing under of crop residues in a continuous wheat rotation significantly reduced tan spot levels compared with no till, with intermediate disease levels for the chisel plough treatment. Wheat yield was significantly higher in 2 of 4 years for the plough treatment compared with no till, while yields for the chisel plough treatment were significantly higher than no till in 1 of 4 years. In contrast, under a wheat–sorghum rotation differences in tan spot levels among tillage treatments were less consistent, while yields were similar among tillage regimes. In a study not directly comparing conventional and conservation tillage, Adee and Pfender (1989) suggested that the level of primary inoculum, that is tan spot infected crop residues, was related to disease development, and management of infected residues was recommended to reduce tan spot development and impact. Rees et al. (1982) also did not directly compare conservation tillage and conventional tillage regimes in Australia, but did demonstrate that tan spot levels and percentage yield loss increased as the amount of infected residue applied to plot areas increased. Schuh (1990) compared tan spot development in four commercial winter wheat fields, two under conventional tillage (i.e. mouldboard ploughing followed by tillage to prepare the seedbed) and two under conservation tillage (disking followed by chisel ploughing). In general, disease levels were lower for the conventional versus conservation tillage treatments. Jørgensen and Olsen (2007) also found that reduced tillage with only harrowing resulted in significantly higher levels of tan spot in winter wheat planted into winter wheat residues, while *Septoria* leaf spot and powdery mildew were lower compared with the ploughed treatment. In both years of their study in Denmark, winter wheat yields were higher for the ploughed treatment versus reduced tillage.

In the Northern Great Plains of the United States, Krupinsky et al. (2007) evaluated the effect of tillage and nitrogen regimes on wheat leaf spot severity (tan spot and the *Septoria* complex) from 1986 to 1996. Overall, no till/direct seeding had limited to no impact on leaf spot severity, especially during dry years, 1988–92. However, no till/direct seeding did result in higher leaf spot severity in wet years, but only in 15 of 47 evaluations, 9 of 61 evaluations and 7 of 50 evaluations, for the spring wheat annual, spring wheat fallow and the winter wheat sunflower rotations, respectively. Furthermore, there were interactions between tillage regime and nitrogen fertilizer rate, whereby no till/direct seeding had higher leaf spot levels compared with conventional and minimum tillage regimes when the nitrogen rate was low, but at higher nitrogen rates disease levels were similar. In Poland, Sawinska et al. (2006) found that levels of powdery mildew, tan spot and *Septoria* leaf blight in winter wheat were increased under conventional tillage versus direct seeding, especially with adequate moisture. Andert et al. (2016) investigated the impact of crop rotation and tillage on the intensity of fungicide use in Germany for a range of crops and found that ploughing, compared to not ploughing, reduced fungicide intensity in winter types of wheat, barley and rye. Carignano et al. (2008) investigated the impact of tillage/residue management (no till, reduced till and burning of residue), fungicide application, nitrogen fertility and cultivar on leaf spot severity and productivity in a winter wheat on winter wheat rotation over 2 years at a total of five site years in Kansas. They found a significant increase in leaf spot severity under no till, while the difference in yield between no till and reduced till, although significant, was relatively small. Significant interactions were detected between tillage regime, and cultivar or fungicide. In general, differences among tillage systems were greatest when no fungicide was applied and a susceptible cultivar was grown, while disease levels and productivity were similar among tillage systems when either a fungicide or resistant cultivar was used.

6 Conclusions and future trends

As has been noted, modern agro-ecosystems are often composed of either monocultures or rotations with a small number of crop components and uniform cultivars, resulting in reduced genetic diversity and a narrow genetic base which compromises disease resistance (Cowling, 1978; Harlan, 1972; Wolfe, 2002; Zadoks and Schein, 1979). These simplified agro-ecosystems are vulnerable to abiotic and biotic stresses, and often allow pest species which can exploit these systems to proliferate rapidly (Finckh and Wolfe, 1997; Tanaka et al., 2002; Wolfe, 2002). The use of techniques such as variety mixtures, intercropping, rotation and gene deployment can help to increase genetic diversity, particularly if combined (Chaube and Singh, 1991).

On-farm implementation of techniques to enhance diversity need not be expensive or cumbersome. For example, the addition of diversity for plant disease management within a forage production system can be achieved through the use of mixtures or intercropping. Agronomically adapted and superior crop cultivars can be combined to capitalize on the benefits of adding diversity to this type of cropping system. Unfortunately, the potential use of diversity via mixtures or intercropping may be more problematic, especially in the short term in agro-ecosystems focused on food production where end users are more concerned about uniformity and specific commodity characteristics (Finckh et al., 2000; Mundt, 2002; Newton et al., 1998; Newton and Swanston, 1999; Wolfe, 2000, 2002). However, close collaboration among researchers, farmers and commodity end users, coupled with the use of crop components that are agronomically compatible and that have complementary quality characteristics, may help to illustrate the potential of genetic diversity in the form of intercropping.

Crop rotation still has a significant role to play in adding diversity to current and future cropping systems. However, crop rotation will always need to be considered within the context of the myriad other factors that influence a farmer's crop choice and land use decisions (Cook and Veseth, 1991; Karlen et al., 1994; Tanaka et al., 2002). More than 80 years ago, Howard and Stakman (1921a) indicated that rotations need to be based on crops that are profitable for the farmer, otherwise the benefit in terms of disease management is negated. Nevertheless farmers can make more informed decisions about the use of rotations when trying to balance competing goals in terms of production, economic and sustainability requirements. Parker (1915) in his book *Field Management and Crop Rotation* outlined the important role of crop rotation and this quote still applies today:

> Crop rotation in itself is not the cure-all for unproductive land or the absolute key to profits from high priced agricultural land. But crop rotation is the chief factor in a combination of good farming practices that will maintain the productivity of the soil, and around which intensive systems of farming may be developed that will yield the maximum crop value per acre at the minimum of expense. Crop rotation is to general field agriculture what the foundation is to the house, the solid base on which we may successfully rear a permanent superstructure designed in a hundred different ways according to our individual requirements and desires.

For farmers to truly capitalize on the potential benefits of using diversity for disease management, they need to have a good understanding of the life cycle of the pathogen(s) they are trying to manage. The importance of knowledge of the plant

Published by Burleigh Dodds Science Publishing Limited, 2019.

disease issues that challenge farmers is not a new concept. In the book *Farm Economy: A Cyclopedia of Agriculture for the Practical Farmer and His Family*, Howard and Stakman (1921b) stressed the importance of knowledge regarding the pathogen(s) of concern in the following quote:

> It therefore becomes necessary to know what kind of germ is causing a particular disease, and further to know the habits of this particular germ. This is all the more true, since the different kinds of germs may have very different habits, and therefore must be controlled by different methods. It is absolutely necessary therefore for a farmer, before he attempts to control a disease, to know what germ causes the disease and how it acts in order that he may apply his preventative measures at the proper time and in the proper place.

Knowledge regarding how a pathogen survives and develops will illustrate where strategies in addition to genetic diversity and rotation need to be used. Moreover, farmers are typically faced with more than one disease, thus an integrated approach will be needed; one that is based on a diversity of disease and crop management strategies.

The retention of crop residues at or just below the soil surface may have both a direct and an indirect influence on the development of cereal leaf diseases (Watkins and Boosalis, 1994). Overall, crop production factors such as environmental variation among individual fields, regions and years, crop rotation, seed-borne inoculum and choice of cultivar (level of disease resistance) will likely have a larger impact on the risk of disease development compared with the type of tillage system that is used (Bailey, 1996; Bailey and Duczek, 1996; Krupinsky et al., 2002). Overall, there is potential for residue-borne cereal leaf diseases to be effectively managed under conservation tillage by the use of sound rotations and agronomic practices, and careful choice of variety.

7 Where to look for further information

The References section contains key publications and websites that provide introductions and more in-depth overviews of subject areas related to crop rotation, conservation tillage and intercropping as well as their impact in relation to cereal leaf spot diseases.

Additional useful sources of information include:

- http://asi.ucdavis.edu/programs/sarep/about/copy_of_what-is-sustainable-agriculture/practices/conservation-tillage
- http://www.fao.org/docrep/012/al298e/al298e.pdf
- http://www.fao.org/conservation-agriculture/overview/principles-of-ca/en/
- http://publications.gov.sk.ca/documents/20/85517-principle%20practices%20crop%20rotation%202017.pdf
- http://apsnet.org/edcenter/Pages/default.aspx
- http://www.agr.gc.ca/eng/science-and-innovation/agricultural-practices/soil-and-land/soil-management/issues-management-problems-and-solutions-for-maintaining-a-zero-tillage-system-and-other-beneficial-soil-management-practices-1of10/?id=1238006852354
- https://grdc.com.au/resources-and-publications

8 References

Adee, E. A. and Pfender, W. F. (1989), The effect of primary inoculum level of *Pyrenophora tritici-repentis* on tan spot epidemic development in wheat, *Phytopathology*, 79, 873–7.

Adugna, A. (2004), Alternative approaches in deploying genes for disease resistance in crops, *Asian J. Plant Sci.*, 3, 618–23.

Alberts, E. E. and Neibling, W. H. (1994), Influence of crop residues on water erosion, *in* P. W. Unger (Ed.), *Managing Agricultural Residues*, CRC Press, Inc., Boca Raton, FL, pp. 19–39.

Andert, S., Bürger, J., Stein, S. and Gerowitt, B. (2016), The influence of crop sequence on fungicide and herbicide use intensities in north German arable farming, *Eur. J. Agron.*, 77, 81–9.

Angus, J. F., Kirkegaard, J. A., Hunt, J. R., Ryan, M. H., Ohlander, C. L. and Peoples, M. B. (2015), Break crops and rotations for wheat, *Crop Pasture Sci.*, 66, 523–52.

Apple, J. L. (1977), The theory of disease management, *in* J. G. Horsfall and E. B. Cowling (Eds), *Plant Disease an Advance Treatise: How Disease is Managed*, Volume 1, Academic Press Inc., New York, NY, pp. 79–101.

Awada, L., Lindwall, C. W. and Sonntag, B. (2014), Conservation tillage impacts on soil, crop and the environment, *Int. Soil Water Conserv. Res.*, 2, 47–65.

Aylor, D. E. (1990), The role of intermittent wind in the dispersal of fungal pathogens, *Annu. Rev. Phytopathol.*, 28, 73–92.

Aylor, D. E. (2003), Spread of plant disease on a continental scale: Role of aerial dispersal of pathogens, *Ecology*, 84, 1989–97.

Bailey, K. L. (1996), Diseases under conservation tillage systems, *Can. J. Plant Sci.*, 76, 635–9.

Bailey, K. L. and Duczek, L. J. (1996), Managing cereal diseases under reduced tillage, *Can. J. Plant Pathol.*, 18, 159–67.

Bailey, K. L., Mortensen, K. and Lafond, G. P. (1992), Effects of tillage systems and crop rotations on root and foliar diseases of wheat, flax, and peas in Saskatchewan, *Can. J. Plant Sci.*, 72, 583–91.

Bailey, K. L., Johnston, A. M., Kutcher, H. R., Gossen, B. D. and Morrall, R. A. A. (2000), Managing crop losses from foliar diseases with fungicides, rotation, and tillage in the Saskatchewan Parkland, *Can. J. Plant Sci.*, 80, 169–75.

Bailey, K. L., Gossen, B. D., Lafond, G. P., Watson, P. R. and Derksen, D. A. (2001), Effect of tillage and crop rotation on root and foliar diseases of wheat and pea in Saskatchewan from 1991 to 1998: Univariate and multivariate analyses, *Can. J. Plant Sci.*, 81, 789–803.

Bankina, B., Bimšteine, G., Ruža, A., Kreita, D., Katamadze, M. and Paura, L. (2015), Crop rotation – The main factor influencing the development of wheat leaf blotch, in *Proceedings of the 25th NJF Congress*, Riga, Latvia, 16–18 June 2015, pp. 65–9.

Bankina, B., Bimšteine, G., Arhipova I., Kaņeps, J. and Stanka, T. (2018), Importance of agronomic practice on the control of wheat leaf diseases, *Agriculture*, 8, 56. https://doi.org/10.3390/agriculture8040056 (accessed 10 June 2018).

Bannon, F. J. and Cooke, B. M. (1998), Studies on dispersal of *Septoria tritici* pycnidiospores in wheat–clover intercrops, *Plant Pathol.*, 47, 49–56.

Barnes, E. H. (1964), Changing plant disease losses in a changing agriculture, *Phytopathology*, 54, 1314–19.

Bedoussac, L., Matura, M. and Hemptinne, J.-L. and Justes, E. (2008), Is durum wheat-winter pea intercropping efficient to reduce pests and diseases? *in 10th Congress of European Society for Agronomy*, 15–19 September 2008, Bologna, Italy.

Black, A. L., Siddoway, F. H. and Brown, P. L. (1974), Summer fallow in the northern Great Plains (winter wheat), *in Summer Fallow in the Western United States*, USDA-ARS Conserv. Res. Rep. no. 17, USDA, Washington DC, pp. 36–50.

Bockus, W. W. and Claassen, M. M. (1992), Effects of crop rotation and residue management practices on severity of tan spot of winter wheat, *Plant Dis.*, 76, 633–6.

Boudreau, M. A. and Mundt, C. C. (1997), Ecological approaches to disease control, *in* N. A. Rechcigl and J. E. Rechcigl (Eds), *Environmentally Safe Approaches to Crop Disease Control*, CRC Press LLC, Boca Raton, FL, pp. 33–62.

Published by Burleigh Dodds Science Publishing Limited, 2019.

Brown, J. F. (1975), Factors affecting the relative ability of strains of fungal pathogens to survive in populations, *J. Aust. Inst. Agric. Sci.*, 41, 3–11.

Brown, J. K. M. and Hovmøller, M. S. (2002), Aerial dispersal of pathogens on the global and continental scales and its impact on plant disease, *Science*, 297, 537–41.

Browning, J. A. and Frey, K. J. (1969), Multiline cultivars as a means of disease control, *Annu. Rev. Phytopathol.*, 7, 355–82.

Bulson, H. A. J., Snaydon, R. W. and Stopes, C. E. (1997), Effects of plant density on intercropped wheat and field beans in an organic farming system, *J. Agric. Sci.*, 128, 59–71.

Burdon, J. J. (1978), Mechanisms of disease control in heterogeneous plant populations – an ecologist's view, *in* P. R. Scott and A. Bainbridge (Eds) *Plant Disease Epidemiology*, Blackwell Science Publishing, Oxford, UK, pp. 193–200.

Burdon, J. J., Barrett, L. G., Rebetzke, G. and Thrall, P. H. (2014), Guiding deployment of resistance in cereals using evolutionary principles, *Evol. Appl.*, 7, 609–24.

Burgess, L. W. and Griffin, D. M. (1968), The recovery of *Gibberella zeae* from wheat straws, *Aust. J. Exp. Agric. Anim. Husb.*, 8, 364–70.

Busari, M. A., Kukal, S. S., Kaur, A., Bhatt, R. and Dulazi, A. A. (2015), Conservation tillage impacts on soil, crop and the environment, *Int. Soil Water Conserv. Res.*, 3, 119–29.

Carignano, M., Staggenborg, S. A. and Shroyer, J. P. (2008), Management practices to minimize tan spot in a continuous wheat rotation, *Agron. J.*, 100, 145–53.

Chapagain, T. (2014), Intercropping wheat and barley with nitrogen fixing legume species in low input organic systems, Ph.D. Thesis, Faculty of Graduate and Postdoctoral Studies, University of British Columbia, 192pp.

Chaube, H. S. and Singh, U. S. (1991), *Plant Disease Management: Principles and Practices*, CRC Press, Boca Raton, FL.

Chen, X. M. (2005), Epidemiology and control of stripe rust [*Puccinia striiformis* f. sp. *tritici*] on wheat, *Can. J. Plant Pathol.*, 27, 314–37.

Clegg, M. D. and Francis, C. A. (1994), Crop management, *in* J. L. Hatfield and D. L. Karlen (Eds), *Sustainable Agriculture Systems*, CRC Press, Inc., Boca Raton, FL, pp. 135–56.

Cochran, V. L., Sparrow, S. D. and Sparrow, E. B. (1994), Residue effects on soil micro- and macro-organisms, *in* P. W. Unger (Ed.), *Managing Agricultural Residues*, CRC Press, Inc., Boca Raton, FL, pp. 163–84.

Cook, R. J. (2000), Advances in plant health management in the twentieth century, *Annu. Rev. Phytopathol.*, 38, 95–116.

Cook, R. J. and Baker, K. F. (1983), *The Nature and Practice of Biological Control of Plant Pathogens*, APS Press, St Paul, MN, 539pp.

Cook, R. J. and Veseth, R. J. (1991), *Wheat Health Management. Plant Health Management Series*, APS Press, St Paul, MN, 152pp.

Coutts, G. R. and Smith, R. K. (1991), *Zero Tillage Production Manual*, Manitoba North Dakota Zero Tillage Farmers Association, Brandon, Manitoba, 42pp.

Cowling, E. B. (1978), Agricultural and forest practices that favor epidemics, *in* J. G. Horsfall and E. B. Cowling (Eds), *Plant Disease an Advance Treatise: How Disease Develops in Populations*, Volume 2, Academic Press Inc., New York, NY, pp. 361–81.

Cox, C. M., Garrett, K. A., Bowden, R. L., Fritz, A. K., Dendy, S. P. and Heer, W. F. (2004), Cultivar mixtures for the simultaneous management of multiple diseases: Tan spot and leaf rust of wheat, *Phytopathology*, 94, 961–9.

Daigh, A. and DeJong-Hughes, J. (2017), A brief history of soil tillage and tillage research, *Upper Midwest Tillage Guide: Tillage Guide Part 1*, University of Minnesota Extension, https://www.extension.umn.edu/agriculture/soils/tillage/tillage-guide-history/docs/upper-midwest-tillage-guide-history.pdf (accessed 9 March 2018).

DeJong-Hughes, J. and Daigh, A. (2017a), Tillage implements, purpose and ideal use, *Upper Midwest Tillage Guide: Tillage Guide Part 2*, University of Minnesota Extension, https://www.extension.umn.edu/agriculture/soils/tillage/tillage-guide-implements/docs/upper-midwest-tillage-guide-implements.pdf (accessed 9 March 2018).

DeJong-Hughes, J. and Daigh, A. (2017b), Reducing tillage intensity, *Upper Midwest Tillage Guide: Tillage Guide Part 3*, University of Minnesota Extension, https://www.extension.umn.edu/agriculture/soils/tillage/tillage-guide-intensity/docs/upper-midwest-tillage-guide-intensity.pdf (accessed 9 March 2018).

Derpsch, R. (2015), Frontiers in conservation tillage and advances in conservation tillage, Food and Agriculture Organization of the United Nations, Agriculture and Consumer Protection Department, Conservation Agriculture, http://www.fao.org/ag/ca/6b.html (accessed 9 March 2018).

Derpsch, R. and Friedrich, T. (2009), Global overview of conservation agriculture adoption, *IV World Congress on Conservation Agriculture*, New Delhi, India, February 2009, Food and Agriculture Organization of the United Nations, Agriculture and Consumer Protection Department, Conservation Agriculture, CA adoption worldwide, http://www.fao.org/ag/ca/6b.html and http://www.fao.org/ag/ca/doc/Derpsch-Friedrich-Global-overview-CA-adoption3.pdf (accessed 9 March 2018).

Duczek, L. J., Sutherland, K. A., Reed, S. L., Bailey, K. L. and Lafond, G. P. (1999), Survival of leaf spot pathogens on crop residues of wheat and barley in Saskatchewan, *Can. J. Plant Pathol.*, 21, 165–73.

Evans, R. and Fleury, D. (1993), *Conservation Farming Guide*, Alberta Conservation Tillage Society, October 1993, 48pp.

Eversmeyer, M. G. and Kramer, C. L. (2000), Epidemiology of wheat leaf rust and stem rust in the central Great Plains of the USA, *Annu. Rev. Phytopathol.*, 38, 491–513.

Eversmeyer, M. G., Kramer C. L. and Browder, L. E. (1984), Presence, viability, and movement of *Puccinia recondita* and *P. graminis* inoculum in the Great Plains, *Plant Dis.*, 68, 392–95.

FAO (2015), Conservation agriculture, Food and Agriculture Organization of the United Nations. Agriculture and Consumer Protection Department, http://www.fao.org/ag/ca/index.html (accessed 9 March 2018).

Farr, D. F., Bills, G. F., Chamuris, G. P. and Rossman, A. Y. (1989), *Fungi on Plants and Plant Products in the United States*, American Phytopathology Society Press, St Paul, MN, 1252pp.

Fernandez, M. R., Zentner, R. P., McConkey, B. G. and Campbell, C. A. (1998), Effects of crop rotations and fertilizer management on leaf spotting diseases of spring wheat in southwestern Saskatchewan, *Can. J. Plant Sci.*, 78, 489–96.

Fernandez, M. R., Pearse, P. G., Holzgang, G., Basnyat, P. and Zentner, R. P. (2009), Impacts of agronomic practices on the leaf spotting complex of common wheat in eastern Saskatchewan, *Can. J. Plant Sci.*, 89, 717–30.

Fernandez, M. R., Wang, H., Cutforth, H. and Lemke, R. (2016), Climatic and agronomic effects on leaf spots of spring wheat in the western Canadian prairies, *Can. J. Plant Sci.*, 96, 895–907.

Finckh, M. R. and Wolfe, M. S. (1997), The use of biodiversity to restrict plant diseases and some consequences for farmers and society, *in* L. E. Jackson (Ed.), *Ecology in Agriculture*, Academic Press, San Diego, CA, pp. 199–233.

Finckh, M. R. and Wolfe, M. S. (1998), Diversification strategies, *in* D. G. Jones (Ed.), *The Epidemiology of Plant Diseases*, Chapman and Hall, London, UK, pp. 231–59.

Finckh, M. R., Gacek, E. S., Goyeau, H., Lannou, C., Merz, U., Mundt, C. C., Munk, L., Nadziak, J., Newton, A. C., de Vallavieille-Pope, C. and Wolfe, M. S. (2000), Cereal variety and species mixtures in practice, with emphasis on disease resistance, *Agronomie*, 20, 813–37.

Frey, K. J., Browning, J. A. and Simons, M. D. (1973), Management of host resistance genes to control disease loss, *Z. Pflanzenkr. Pflanzenschutz*, 80, 160–80.

Frey, K. J., Browning, J. A. and Simons, M. D. (1977), Management systems for host genes to control disease loss, *Ann. N.Y. Acad. Sci.*, 287, 255–74.

Friedrich, T., Derpsch, R. and Kassam, A. (2012), Overview of the Global Spread of Conservation Agriculture, *Field Actions Science Reports* [Online], Special Issue 6 | 2012, Online since 06 November 2012, connection on 18 June 2018. URL: http://journals.openedition.org/factsreports/1941 (accessed 18 June 2018).

Fry, W. E. (1982), *Principles of Plant Disease Management*, Academic Press, New York, NY.

Published by Burleigh Dodds Science Publishing Limited, 2019.

Fryrear, D. W. and Bilbro, J. D. (1994), Wind erosion control with residues and related practices, *in* P. W. Unger (Ed.), *Managing Agricultural Residues*, CRC Press, Inc., Boca Raton, FL, pp. 7–17.

Garrett, K. A. and Mundt, C. C. (1999), Epidemiology in mixed host populations, *Phytopathology*, 89, 984–90.

Gilbert, J. and Woods, S. M. (2001), Leaf spot diseases of spring wheat in southern Manitoba farm fields under conventional and conservation tillage, *Can. J. Plant Sci.*, 81, 551–9.

Ginns, J. H. (1986), Compendium of plant disease and decay fungi in Canada, 1960–1980, Publication 1813, Research Branch, Agriculture Canada.

Green, F. M. (1984), Machinery for tillage and planting, Agdex 740-2, May 1984, Alberta Agriculture, Edmonton, Alberta.

Habgood, R. M. (1971), The transmission of *Rhynchosporium secalis* by infected barley seed, *Plant Pathol.*, 20, 80–1.

Hampton, J. G. (1980), The role of seed-borne inoculum in the epidemiology of net blotch of barley in New Zealand, *N. Z. J. Exp. Agric.*, 8, 297–9.

Harker, K. N., O'Donovan, J. T., Turkington, T. K., Blackshaw, R. E., Lupwayi, N. Z, Smith, E. G., Johnson, E. N., Gan, Y., Kutcher, H. R., Dosdall, L. M. and Peng, G. (2015), Canola rotation frequency impacts canola yield and associated pests, *Can. J. Plant Sci.*, 95, 9–20.

Harlan, J. R. (1972), Genetics of disaster, *J. Environ. Quality*, 1, 212–15.

Horton, R., Kluitenberg, G. J. and Bristow, K. L. (1994), Surface crop residue effects on the soil surface energy balance, *in* P. W. Unger (Ed.), *Managing Agricultural Residues*, CRC Press, Inc., Boca Raton, FL, pp. 143–62.

Howard, C. W. and Stakman, E. C. (1921a), Rotation as a means of blight-control, *in Farm Economy: A Cyclopedia of Agriculture for the Practical Farmer and his Family*, Better Farming Association, Minneapolis, MN, pp. 41–4.

Howard, C. W. and Stakman, E. C. (1921b), Diseases of cereal crops, *in Farm Economy: A Cyclopedia of Agriculture for the Practical Farmer and his Family*, Better Farming Association, Minneapolis, MN, pp. 762–76.

Hummel, J. D., Dosdall, L. M., Clayton, G. W., Turkington, T. K., Lupwayi, N. Z., Harker, K. N. and O'Donovan, J. T. (2009), Canola–wheat intercrops for improved agronomic performance and integrated pest management, *Agron. J.*, 101, 1190–7.

Jackson, L. F. and Webster, R. K. (1976), Seed and grasses as possible sources of *Rhynchosporium secalis* for barley in California, *Plant Dis. Reptr.*, 60, 233–6.

Jensen, T. and Timmermans, J. (1991), Conservation tillage, Agdex 516-3, September 1991, Agri-fax, Alberta Agriculture, Edmonton, Alberta.

Johnston, A. M., Kutcher, H. R. and Bailey, K. L. (2005), Impact of crop sequence decisions in the Saskatchewan Parkland, *Can. J. Plant Sci.*, 85, 95–102.

Jordan, V. W. L. (1981), Aetiology of barley net blotch caused by *Pyrenophora teres* and some effects on yield, *Plant Pathol.*, 30, 77–87.

Jordan, V. W. L. and Allen, E. C. (1984), Barley net blotch: Influence of straw disposal and cultivation methods on inoculum potential, and on incidence and severity of autumn disease, *Plant Pathol.*, 33, 547–59.

Jørgensen, L. N. and Olsen, L. V. (2007). Control of tan spot (*Drechslera tritici-repentis*) using cultivar resistance, tillage methods and fungicides, *Crop Prot.*, 26, 1606–16.

Kantor, S. (1999), Intercropping, agriculture and natural resources, Fact Sheet #531, Cooperative Extension, Washington State University, Renton, WA.

Karlen, D. L., Varvel, G. E., Bullock, D. G. and Cruse, R. M. (1994), Crop rotations for the 21st century, *Advances in Agronomy*, 53, 1–45.

Kassam, A., Friedrich, T., Derpsch, R. and Kienzle, J. (2015), Overview of the worldwide spread of conservation agriculture, *Field Actions Sci. Rep.: J. Field Actions*, 8, 1–10.

Kay, J. G. and Owen., H. (1973), Transmission of *Rhynchosporium secalis* on barley grain, *Trans. Br. Mycol. Soc.*, 60, 405–11.

Kinane, J. and Lyngkaer, M. F. (2002), Effect of barley-legume intercrop on disease frequency in an organic farming system, *Proc. 6th Conf. EFPP 2002, Prague. Plant Prot. Sci.*, 38, 227–31.

Kosinski, S. M., King, J. R., Harker, K. N., Turkington, T. K. and Spaner, D. (2011), Barley and triticale underseeded with a kura clover living mulch: Effects on weed pressure, disease incidence, silage yield, and forage quality, *Can. J. Plant Sci.*, 91, 677–87.

Krupinsky, J. M., Bailey, K. L., McMullen, M. P., Gossen, B. D., Turkington, T. K. (2002), Managing plant disease risk in diversified cropping systems, *Agron. J.*, 94, 198–209.

Krupinsky, J. M., Tanaka, D. L., Lares, M. T. and Merrill, S. D. (2004), Leaf spot diseases of barley and spring wheat as influenced by preceding crops, *Agron. J.*, 96, 259–66.

Krupinsky, J. M., Tanaka, D. L., Merrill, S. D., Liebig, M. A. and Hanson, J. D. (2006), Crop sequence effects of 10 crops in the northern Great Plains, *Agric. Syst.*, 88, 227–54.

Krupinsky, J. M., Halvorson, A. D., Tanaka, D. L. and Merrill, S. D. (2007), Nitrogen and tillage effects on wheat leaf spot diseases in the Northern Great Plains, *Agron. J.*, 99, 562–9.

Kumar, K., Holtz, M. D., Xi, K. and Turkington, T. K. (2013), Overwintering potential of the stripe rust pathogen (*Puccinia striiformis*) in central Alberta, *Can. J. Plant Pathol.*, 35, 304–14.

Kutcher, H. R., Johnston, A. M., Bailey, K. L. and Malhi, S. S. (2011), Managing crop losses from plant diseases with foliar fungicides, rotation, and tillage on a Black Chernozem in Saskatchewan, Canada, *Field Crops Res.*, 124, 205–12.

Kutcher, H. R., Brandt, S. A., Smith, E. G., Ulrich, D., Malhi, S. S. and Johnston, A. M. (2013), Blackleg disease of canola mitigated by resistant cultivars and four-year crop rotations in western Canada, *Can. J. Plant Pathol.*, 35, 209–21.

Leduc, P. (Ed.). (1994), *Direct Seeding Manual*, Prairie Agricultural Machinery Institute and Saskatchewan Soil Conservation Association, Humboldt, Sask.

Lee, H. K., Tewari, J. P. and Turkington, T. K. (1999), Histopathology and isolation of *Rhynchosporium secalis* from infected barley seed, *Seed Sci. Tech.*, 27, 477–82.

Luke, H. H., Pfahler, P. L. and Barnett, R. D. (1983), Control of *Septoria nodorum* on wheat with crop rotation and seed treatment, *Plant Dis.*, 67, 949–51.

Lupwayi, N. Z., Rice, W. A. and Clayton, G. W. (1998), Soil microbial diversity and community structure under wheat as influenced by tillage and crop rotation, *Soil Biol. Biochem.*, 30, 1733–41.

Luz, W. C. da, Bergstrom, G. C. and Stockwell, C. A. (1998), Seed-applied bioprotectants for control of seedborne *Pyrenophora tritici-repentis* and agronomic enhancement of wheat, *Can. J. Plant Pathol.*, 20, 384–6.

Maloy, O. C. (2005), Plant disease management, *The Plant Health Instructor,* doi:10.1094/PHI-I-2005-0202-01, https://www.apsnet.org/edcenter/intropp/topics/Documents/PlantDiseaseManagement.aspx, (accessed 18 June 2018).

Manitoba Agriculture, Food and Rural Initiatives (MAFRI). (2008), Soil management guide, https://www.gov.mb.ca/agriculture/environment/soil-management/soil-management-guide/pubs/soil-management-guide.pdf (accessed 11 June 2018).

Mathre, D. E. (1997), *Compendium of Barley Diseases*, 2nd Edition, APS Press, St Paul, MN, 90pp.

Mayfield, A. H. and B. G. Clare. (1984a), Survival over summer of *Rhynchosporium secalis* in host debris in the field, *Aust. J. Agric. Res.*, 35, 789–97.

Mayfield, A. H. and B. G. Clare. (1984b), Effects of common stubble treatments and sowing sequences on scald disease (*Rhynchosporium secalis*) in barley crops, *Aust. J. Agric. Res.*, 35, 799–805.

Mazzilli, S. R., Ernst, O. R., de Mello, V. P. and Perez, C. A. (2016), Yield losses on wheat cops associated to the previous winter crop: Impact of agronomic practices based on on-farm analysis, *Eur. J. Agron.*, 75, 99–104.

McDonald, B. A. and Linde, C. (2002a), Pathogen population genetics, evolutionary potential, and durable resistance, *Annu. Rev. Phytopathol.*, 40, 349–79.

McDonald, B. A and Linde, C. (2002b), The population genetics of plant pathogens and breeding strategies for durable resistance, *Euphytica*, 124, 163–80.

Menzies, J. and Gilbert, J. (2003), Diseases of wheat, *in* Bailey, K. L., Gossen, B. D., Gugel, R. K. and Morrall, R. A. A. (Eds), *Diseases of Field Crops in Canada*, 3rd Edition, Canadian Phytopathological Society, Saskatoon, Saskatchewan, pp. 94–128.

Milus, E. A. and Chalkley, D. B. (1997), Effect of previous crop, seedborne inoculum, and fungicides on development of Stagonospora blotch, *Plant Dis.*, 81, 1279–83.

Mundt, C. C. (2002), Use of multiline cultivars and cultivar mixtures for disease management, *Annu. Rev. Phytopathol.*, 40, 381–410.

Nagarajan, S. and Singh, D. V. (1990), Long-distance dispersion of rust pathogens, *Annu. Rev. Phytopathol.* 28, 139–53.

Newton A. C. and Swanston, J. S. (1999), Cereal variety mixtures reducing inputs and improving yield and quality – why isn't everybody growing them? Scottish Crop Research Institute Annual Report for 1998.

Newton, A. C., Swanston, J. S., Guy, D. C. and Ellis, R. P. (1998), The effect of cultivar mixtures on malting quality in winter barley, *J. Inst. Brew.*, 104, 41–5.

Newton, A. C., Begg, G. S. and Swanston, J. S. (2009), Deployment of diversity for enhanced crop function, *Ann. Appl. Biol.*, 154, 309–22.

Paddock, W. C. (1967), Phytopathology in a hungry world, *Ann. Rev. Phytopathol.*, 5, 375–90.

Palti, J. (1981), *Cultural Practices and Infectious Crop Diseases*, Springer, Berlin, 243pp.

Parker, E. C. (1915), *Field Management and Crop Rotation*, Webb Publishing Co., St Paul, MN.

Parry, D. W., Jenkinson, P. and McLeod, L. (1995), *Fusarium* ear blight (scab) in small grain cereals – a review, *Plant Pathology*, 44, 207–38.

Pedersen, E. A. and Hughes, G. R. (1992), The effect of crop rotation on development of the septoria disease complex on spring wheat in Saskatchewan, *Can. J. Plant Pathol.*, 14, 152–8.

Pittelkow, C. M., Liang, X., Linquist, B. A., Van Groenigen, K. J., Lee, J., Lundy, M. E., Van Gestel, N., Six, J., Venterea, R. T. and van Kessel, C. (2015), Productivity limits and potentials of the principles of conservation agriculture, *Nature*, 517, 365–8.

Pridham, J. C. and Entz, M. H. (2008), Intercropping spring wheat with cereal grains, legumes and oilseeds fails to improve productivity under organic management, *Agron. J.*, 100, 1436–42.

Priestly, R. H. (1981), Choice and deployment of resistant cultivars for cereal disease control, *in* J. F. Jenkyn and R. T. Plumb (Ed.), *Strategies for the Control of Cereal Disease*, Blackwell Science Publishing, Oxford, UK.

Priestly, R. H. and Wolfe, M. S. (1977), Crop protection by cultivar diversification, *in Proceedings 1977 British Crop Protection Conference – Pests and Diseases*, Volume 1, pp. 135–40.

Reed, H. E. (1957), Studies on barley scald, Bull. Univ. Tenn. Agric. Exp. Stn., No. 268.

Rees, R. G., Platz, G. J. and Mayer, R. J. (1982), Yield losses in wheat from yellow spot: Comparison of estimates derived from single tillers and plots, *Aust. J. Agric. Res.*, 33, 899–908.

REX Consortium (2016), Combining selective pressures to enhance the durability of disease resistance genes, *Front. Plant Sci.*, 7, 1916.

Rothrock, C. S. (1992), Tillage systems and plant disease, *Soil Sci.*, 154, 308–15.

Sadler, E. J. and Turner, N. C. (1994), Water relationships in a sustainable agriculture system, *in* J. L. Hatfield and D. L. Karlen (Eds), *Sustainable Agriculture Systems*, CRC Press, Inc., Boca Raton, FL, pp. 21–46.

Sawinska, Z., Malecka, I. and Blecharczyk, A. (2006), Impact of previous crops and tillage systems on health status of winter wheat, *Electron. J. Pol. Agric. Univ.*, 9, #51, http://www.ejpau.media.pl/volume9/issue4/art-51.html (accessed 11 June 2018).

Schilder, A. M. C. and Bergstrom, G. C. (1995), Seed transmission of *Pyrenophora tritici-repentis*, causal fungus of tan spot of wheat, *Eur. J. Plant Pathol.*, 101, 81–91.

Schuh, W. (1990), The influence of tillage systems on incidence and spatial pattern of tan spot of wheat, *Phytopathology*, 80, 804–7.

Shah, D., Bergstrom, G. C. and Ueng, P. P. (1995), Initiation of Septoria nodorum blotch epidemics in winter wheat by seedborne *Stagonospora nodorum*, *Phytopathology*, 85, 452–7.

Sheridan, J. E., Grbavac, N. and Ballard, L. (1983), Strategies for controlling net blotch of barley, *Proc. 36th N. Z. Weed and Pest Control Conf.*, pp. 242–6.

Shipton, W. A., Khan, T. N. and Boyd, W. J. R. (1973), Net blotch of barley, *Rev. Plant Pathol.*, 52, 269–90.

Singh, M. and Chand, J. N. (1985), Studies on the survival of *Helminthosporium teres* incitant of net blotch of barley, *Indian Phytopath.*, 38, 659–61.

Skoropad, W. P. (1959), Seed and seedling infection of barley by *Rhynchosporium secalis*, *Phytopathology*, 49, 623–6.

Smith, E. G., Kutcher, H. R., Brandt, S. A., Ulrich, D., Malhi, S. S. and Johnston, A. M. (2013), The profitability of short-duration canola and pea rotations in western Canada, *Can. J. Plant Sci.*, 93, 933–40.

Smith, E. G., Turkington, T. K., O'Donovan, J. T., Edney, J. T., Juskiw, P. E., McKenzie, R. H., Clayton, G. W., Xi, K., May, W. E., Irvine, R. B., Brandt, S., Johnson, E. N. and Perkovic, S. (2015), Influence of production systems on return and risk from malting barley production in Western Canada, *Can. J. Plant Sci.*, 96, 339–46.

Soane, B. D., Ball, B. C., Arvidsson, J., Basch, G., Moreno, F. and Roger-Estrade, J. (2012), No-till in northern, western and south-western Europe: A review of problems and opportunities for crop production and the environment, *Soil Till. Res.*, 118, 66–87.

Stack, R. W. (2000), Return of an old problem: Fusarium head blight of small grains, *APSnet Plant Health Reviews*, http://www.apsnet.org/publications/apsnetfeatures/Pages/headblight.aspx, (accessed 15 August 2018).

Steiner, J. L. (1994), Crop residue effects on water conservation, *in* P. W. Unger (Ed.), *Managing Agricultural Residues*, CRC Press, Inc., Boca Raton, FL, pp. 41–76.

Sullivan, P. (2001), Intercropping principles and practices: Agronomy systems guide, Appropriate technology transfer for rural areas, National Center for Appropriate Technology, Fayetteville, AR, http://www.attra.org/attra-pub/intercrop.html (accessed 18 June 2018).

Sutton, J. C. (1982), Epidemiology of wheat head blight and maize ear rot caused by *Fusarium graminearum*, *Can. J. Plant Pathol.*, 4, 195–209.

Tanaka, D. L., Krupinsky, J. M., Liebig, M. A., Merrill, S. D., Ries, R. E., Hendrickson, J. R., Johnson, H. A. and Hanson, J. D. (2002), Dynamic cropping systems, *Agron. J.*, 94, 957–61.

Tekauz, A. (2003), Diseases of barley, *in* K. L. Bailey, B. D. Gossen, R. K. Gugel and R. A. A. Morrall (Ed.), *Diseases of Field Crops in Canada*, 3rd Edition, Canadian Phytopathological Society, Saskatoon, Saskatchewan, pp. 30–53.

Turkington, T. K. (2003), Diseases of field crops in Canada, *in* K. L. Bailey, B. D. Gossen, R. K. Gugel, and R. A. A. Morrall (Ed.), *Diseases of Field Crops in Canada*, 3rd Edition, Canadian Phytopathological Society, Saskatoon, Saskatchewan, pp. 9–14.

Turkington, T. K., Clear, R. M., Burnett, P. A., Patrick, S. K., Orr, D. D. and Xi, K. (2002), Fungal plant pathogens infecting barley and wheat seed from Alberta, 1995–1997, *Can. J. Plant Pathol.*, 24, 302–8.

Turkington, T. K., Xi, K., Tewari, J. P., Lee, H. K., Clayton, G. W. and Harker, K. N. (2005), Cultivar rotation as a strategy to reduce leaf diseases under barley monoculture, *Can. J. Plant Pathol.*, 27, 283–90.

Turkington, T. K., Xi, K., Clayton, G. W., Burnett, P. A., Klein-Gebbinck, H., Lupwayi, N. Z., Harker, K. N. and O'Donovan, J. T. (2006), Impact of crop management on leaf diseases in Alberta barley fields, 1995–1997, *Can. J. Plant Pathol.*, 28, 441–9.

Turkington, T. K., O'Donovan, J. T., Edney, M. J., Juskiw, P. E., McKenzie, R. H., Harker, K. N., Clayton, G. W., Xi, K., Lafond, G. P., Irvine, R. B., Brandt, S., Johnson, E. N., May, W. E. and Smith, E. (2012), Effect of crop residue, nitrogen rate and fungicide application on malting barley productivity, quality, and foliar disease severity, *Can. J. Plant Sci.*, 92, 577–88.

Turkington, T. K., Xi, K., Harker, K. N., O'Donovan, J. T., Blackshaw, R., McAllister, T. and Lupwayi, N. Z. (2015a), The impact of barley variety rotation, mixtures, and intercropping on leaf disease and silage production, *Can. J. Plant Pathol.*, 38, 131–2 (Abstr.).

Turkington, T. K., Xi, K., Harker, K. N., O'Donovan, J. T., Blackshaw, R., McAllister, T. and Lupwayi, N. Z. (2015b), The impact of barley variety rotation, mixtures, and intercropping on leaf disease and silage production, *Proceedings of the XVIII International Plant Protection Congress (IPPC) 2015*, 24–27 August 2015, Free University Berlin, Henry Ford Building, Berlin, Germany, ID: O NOC I-1 (oral presentation).

Vilich-Meller, V. (1992), Mixed cropping of cereals to suppress plant diseases and omit pesticide applications, *Bio. Agri. Hort.*, 8, 299–308.

Watkins, J. E. and Boosalis, M. G. (1994), Plant disease incidence as influenced by conservation tillage systems, *in* P. W. Unger (Ed.), *Managing Agricultural Residues*, CRC Press, Inc., Boca Raton, FL, pp. 261–83.

Wiese, M. V. (1987), *Compendium of Wheat Diseases*, 2nd Edition, APS Press, St Paul, MN, 112pp.

Wilkins, D. E., Klepper, B. L. and Rasmussen, P. E. (1988), Management of grain stubble for conservation-tillage systems, *Soil Till. Res.*, 12, 25–35.

Wolfe, M. S. (1978), Some practical implications of the use of cereal variety mixtures, *in* P. R. Scott and A. Bainbridge (Eds), *Plant Disease Epidemiology*, Blackwell Science Publishing, Oxford, UK, pp. 201–7.

Wolfe, M. S. (1985), The current status and prospects of multiline cultivars and variety mixtures for disease resistance, *Annu. Rev. Phytopathol.*, 23, 251–73.

Wolfe, M. S. (2000), Crop strength through diversity, *Nature*, 406, 681–2.

Wolfe, M. S. (2002), The role of functional biodiversity in managing pest and diseases in organic production systems, *in the BCPC Conference, Pest and Diseases 2002*, Volume 1, pp. 531–616, Proceedings of the British Crop Protection Council, Brighton, UK, 18–21 November 2002, The British Crop Protection Council, Surrey, UK.

Xi, K. and Burnett, P. A. (1997), Staining paraffin-embedded sections of scald of barley before paraffin removal, *Biotech. Histochem.*, 72, 173–7.

Zadoks, J. C. and Schein, R. D. (1979), *Epidemiology and Plant Disease Management*, Oxford University Press, New York.

Zhan, J., Thrall, P. H., Papaïx, J., Xie, L. and Burdon, J. J. (2015), Playing on a pathogen's weakness: Using evolution to guide sustainable plant disease control strategies, *Annu. Rev. Phytopathol.*, 53, 19–43.

Zhu, J-H., Dong, Y., Xiao, J-X., Zheng, Y. and Tang, L. (2017), Effects of N application on wheat powdery mildew occurrence, nitrogen accumulation and allocation in intercropping system, *Chinese J. Applied Ecol.*, 28, 3985–93. In Chinese with English abstract.

Index

Lightning Source UK Ltd.
Milton Keynes UK
UKHW02n1501191018
330829UK00001B/4/P